# Metabolic Syndrome

Akhlaq A. Farooqui

# Metabolic Syndrome

## An Important Risk Factor for Stroke, Alzheimer Disease, and Depression

Springer

Akhlaq A. Farooqui
Ohio State University
Columbus, OH, USA

ISBN 978-1-4614-7317-6 ISBN 978-1-4614-7318-3 (eBook)
DOI 10.1007/978-1-4614-7318-3
Springer New York Heidelberg Dordrecht London

Library of Congress Control Number: 2013937291



Printed on acid-free paper

Springer is part of Springer Science+Business Media (www.springer.com)

*This monograph is dedicated to Tahira, my wife, companion, colleague, and friend.*

# Preface

Brain function depends on the complex architecture of neuronal networks and this complexity arises from the morphological intricacy that neurons acquire during the course of differentiation, a process, which is regulated by growth factors, cytokines, transcription factors and soluble as well as membrane-bound receptors. The metabolic syndrome is a highly prevalent condition that affects a considerable number of adult Americans and has become increasingly relevant to many disciplines in clinical medicine. At the molecular level, metabolic syndrome is accompanied not only by dysregulation in the expression of adipocytokines, cytokines, and chemokines, but also by alterations in levels of lipid mediators. These changes modulate immune response and inflammation that lead to alterations in the hypothalamic "bodyweight/appetite/satiety set point" resulting in the initiation and development of the metabolic syndrome. Metabolic syndrome is a risk factor for neurological disorders such as stroke, depression, and Alzheimer disease (AD). Having the metabolic syndrome approximately doubles the risk of having a stroke, AD, and depression compared with age-matched healthy humans.

Recent studies have also indicated that patients with metabolic syndrome and AD show insulin resistance. Insulin not only modulates levels of β-amyloid protein aggregation, the primary constituent of senile plaques and neuropathological hallmarks of AD, but also inhibits the activity of glycogen synthase kinase-3, preventing the phosphorylation of tau protein and thereby inhibiting the formation of neurofibrillary tangles, the other neuropathological hallmark of AD. Psychosocial factors activate the HPA axis, producing hypersecretion of corticotrophin-releasing hormone, adrenocorticotropic hormone, and cortisol. This dysregulation of the HPA axis promotes deposition of visceral adipose tissue which secretes adipocytokines and inflammatory cytokines. These adipokines and cytokines have been implicated in insulin resistance, an important component of the metabolic syndrome. The molecular mechanism underlying the mirror relationship between metabolic syndrome and neurological disorders is not fully understood. However, all cellular and biochemical alterations observed in metabolic syndrome such as impairment of endothelial cell function, abnormality in essential fatty acid metabolism, and alterations in lipid mediators along with abnormal insulin/leptin signaling, and insulin resistance may represent a pathological bridge between metabolic syndrome and neurological disorders such as stroke, AD, and depression.

Information on the link among metabolic syndrome, stroke, AD, and depression is scattered throughout the literature mainly in the form of original papers and some reviews. Many books are available on the biochemical aspects of metabolic syndrome, but at present there are no books on the relationship between metabolic syndrome and neurological disorders. As baby boomers grow older, the enormous impact of metabolic syndrome and the above mentioned neurological disorders will be felt by the American society. At present more than 800,000 people suffer from stroke every year (approximately 1 person every 45 s), approximately 5.1 million people over the age of 65 in the USA suffer from AD, and about 38 million US adults have experienced depression at some point during their lifetime. The projected cost to Medicare for treating these neurological diseases is estimated to be about 5 trillion dollars by 2050. This number does not include other visceral and neurological diseases. Such an amount will not only burst National Institute of Health budget but will seriously affect the US economy. Available drugs may not reverse the stroke, AD, and depression. However, healthy diet, regular exercise, and retardation of metabolic syndrome may produce beneficial effects not only on motor and cognitive functions but also on memory deficits that occur to some extent during normal aging and to a large extent in stroke, AD, and depression. This monograph provides readers with a comprehensive and cutting-edge description of links among metabolic syndrome, stroke, AD, and depression in a manner that is useful not only to students and teachers but also to researchers, dietitians, nutritionists, and physicians.

This monograph has ten chapters. The first chapter describes the effects of lifestyle on metabolic syndrome and neurological disorders. Chapter 2 describes glucose and fructose-induced toxicity in the liver and brain. Chapter 3 deals with the effects of essential fatty acid metabolism on metabolic syndrome and neurological disorders. Chapters 4 and 5 focus on cutting edge information on the contribution of lipid mediators in the pathogenesis of metabolic syndrome and neurological disorders and molecular aspects of obesity and insulin resistance in metabolic syndrome and neurological disorders. Chapter 6 describes the effect of dietary phytochemicals on metabolic syndrome and neurological disorders. Chapters 7 and 8 are devoted to molecular mechanisms associated with the metabolic link among metabolic syndrome, stroke, and AD. Chapter 9 describes the cutting-edge information on signal transduction processes and the metabolic link between metabolic syndrome and depression. Chapter 10 provides readers with a perspective that will be important for future research work on the relationship between metabolic syndrome and neurological disorders. My presentation and demonstrated ability to present complicated information on signal transduction processes in metabolic syndrome and neurological disorders will make this book particularly accessible to neuroscience graduate students, teachers, and fellow researchers. It can be used as a supplemental text for a range of neuroscience and biochemistry courses. Clinicians, neuroscientists, neurologists, and pharmacologists will find this book useful for understanding the molecular aspects of metabolic syndrome and its impact on the brain and its vulnerability to neurological disorders. To the best of my knowledge, this monograph will be the first to provide a comprehensive description of signal transduction processes associated with the relationship between metabolic syndrome and neurological disorders.

The choices of topics presented in this monograph are personal. They are based not only on my interest in the biochemistry of metabolic syndrome, stroke, AD, and depression but also on areas where major progress has been made. I have tried to ensure uniformity and mode of presentation as well as a logical progression of subjects from one topic to another and have provided the extensive bibliography. For the sake of simplicity and uniformity a large number of figures with chemical structures of drugs used for the treatment of metabolic syndrome and neurological disorders along with line diagrams of colored signal transduction pathways are also included. I hope that my attempt to integrate and consolidate the knowledge on metabolic links among metabolic syndrome, stroke, AD, and depression will initiate more studies on molecular mechanisms that link metabolic syndrome with neurological disorders. This knowledge may be useful in developing novel therapeutic interventions for metabolic syndrome, stroke, Alzheimer's disease, and depression.

Columbus, OH, USA Akhlaq A. Farooqui

# Acknowledgments

I thank my wife, Tahira, for critical reading of this monograph, offering valuable advice, useful discussion, and evaluation of subject matter. Without her help and participation, this monograph neither could nor would have been completed. I would also like to express my gratitude to Melissa Higgs and Ann H. Avouris of Springer, New York, for their quick responses to my queries and professional manuscript handing. It has been a pleasure working with them for many years.

# Contents

# List of Abbreviations

| | |
|---|---|
| PtdCho | Phosphatidylcholine |
| PtdEtn | Phosphatidylethanolamine |
| PlsCho | Choline plasmalogen |
| PlsEtn | Ethanolamine plasmalogen |
| PtdIns | Phosphatidylinositol |
| PtdIns4*P* | Phoshatidylinositol 4-phosphate |
| PtdIns(4,5)$P_2$ | Phosphatidylinositol 4,5-bisphosphate |
| PtdH | Phosphatidic acid |
| PtdSer | Phosphatidylserine |
| AD | Alzheimer disease |
| AGE | Advanced glycation endproducts |
| ALA | α-Linolenic acid |
| APP | Amyloid precursor protein |
| ARA | Arachidonic acid |
| Cer | Ceramide |
| COX | Cyclooxygenase |
| DHA | Docosahexaenoic acid |
| EPA | Eicosapentaenoic acid |
| EPOX | Epoxygenase |
| HDL | High density lipoprotein |
| IGF | Insulin growth factor |
| IL | Interleukin |
| Ins-1,4,5-$P_3$ | Inositol-1,4,5-trisphosphate |
| IR | Insulin receptor |
| LA | Linoleic acid |
| LDL | Low density lipoprotein |
| LOX | Lipoxygenase |
| LTP | Long-term potentiation |
| MetS | Metabolic syndrome |
| PKA | Protein kinase A |

| | |
|---|---|
| PKC | Protein kinase C |
| $PLA_2$ | Phospholipase $A_2$ |
| PLC | Phospholipase C |
| PLD | Phospholipase D |
| RAGE | Receptor for advanced glycation endproducts |
| Sph | Sphingosine |
| TLR | Toll-like receptor |
| TNF-α | Tumor necrosis factor-alpha |

# Chapter 1
# Lifestyle as a Risk Factor for Metabolic Syndrome and Neurological Disorders

## 1.1 Introduction

Health-related lifestyle is defined as the way in which humans live and maintain their health for longevity. Diet plays an important role in maintaining good health and longevity. Diet components not only provide energy and building material to the body but also have ability to prevent and protect against acute and chronic diseases. Good nutrition, daily exercise, and adequate sleep are the foundations for maintaining optimal health (Fig. 1.1). Managing stress in positive ways instead of activities, such as snacking, smoking and drinking alcohol. These activities increase stress in our cells and body tissues. At the molecular level stress may promote chronic diseases, such as diabetes, metabolic syndrome (MetS), cancer, and neurological disorders, such as stroke, Alzheimer disease (AD), Parkinson disease (PD), and depression. These diseases are multifactorial conditions, which are influenced by aging, diet, genetic factors, and exposure to toxic environment. Thus, risk factors for neurodegenerative diseases include old age, positive family history, unhealthy lifestyle (consumption of energy-dense food and sedentary lifestyle), and exposure to toxic environment (Farooqui 2010). Risk factors for neurodegenerative diseases not only cause an increase in inflammatory signaling in the brain and visceral organs but also produce immune system dysfunction throughout the body. Chronic inflammation is characterized by long-standing chronic activation of microglia in the brain and macrophages in the visceral organs and sustained activation of NF-κB and generation of inflammatory mediators. The sustained expression and release of inflammatory mediators cause an imbalance in the inflammatory cycle homeostasis by activating additional microglia in the brain and macrophages in the visceral tissues, promoting further release of inflammatory cytokines, such as tumor necrosis factor-α (TNF-α), interleukin-1β (IL-1β), and interleukin-6 (IL-6) (Farooqui 2010).

It is estimated that in Paleolithic period human derived their energy from diet that contained protein (37 %), carbohydrates (41 %), and fat (22 %). Humans have evolved consuming diet with approximately equal amounts of dietary DHA and ARA. In contrast, current Western diet (American diet) contains about half of total

A.A. Farooqui, *Metabolic Syndrome: An Important Risk Factor for Stroke, Alzheimer Disease, and Depression*, DOI 10.1007/978-1-4614-7318-3_1,

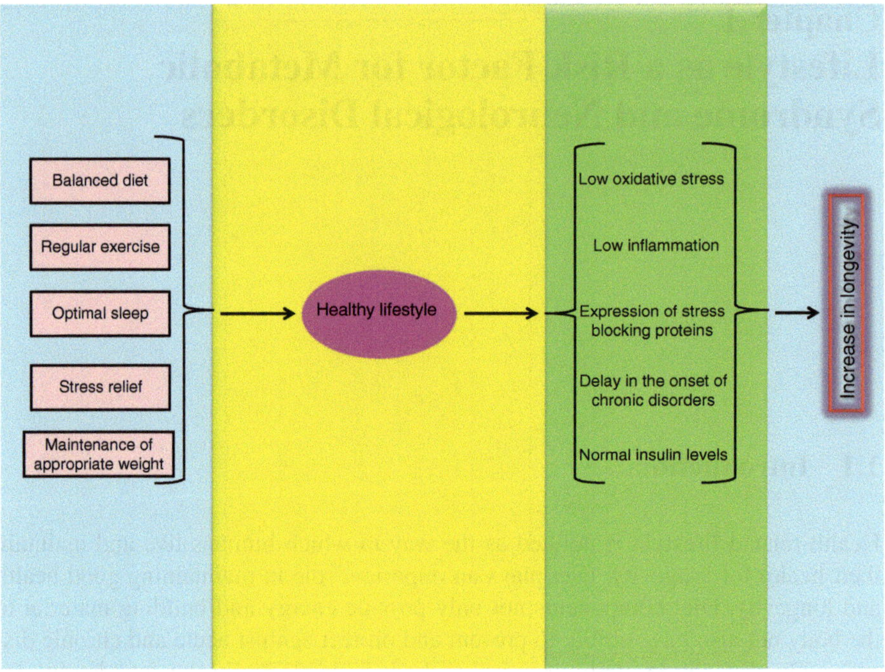


**Fig. 1.1** Factors modulating healthy lifestyle and their effect on longevity

daily calories from carbohydrates, whereas one-third of calories are derived from fat. The contribution of protein to daily energy in today's Western diet is about 15 %. In addition in the present-day American diet, the ratio between n-6 to n-3 fatty acid is about 20:1. High-fat diet with high n-6 to n-3 fatty acid ratio (20:1), aging, family history, and toxic environment are major risk factors for neurological disorders (Fig. 1.2). Thus, a shift has occurred from Paleolithic diet, which was high in fruits, vegetables, lean meats, and seafood, to Western diet containing processed foods high in sodium and hydrogenated fats and low in fiber. Since human diet consists of complicated mixtures of proteins, carbohydrates, fats, and both micro- and macronutrients, it is extremely difficult to identify and dissect out the contributions of any single component to nutrition and health. Recently, comprehensive approaches, such as the analyses of dietary patterns and/or diet quality indices and scores, focusing on more than a single nutrient have been used to investigate the association between diet and neurodegenerative diseases, such as AD and PD (Jew et al. 2009). These studies have demonstrated that Western diet, which is enriched in "red meat, alcohol, and soft drinks containing high-fructose corn syrup" (Fig. 1.3), is associated with a higher risk for developing chronic visceral diseases (obesity, cardiovascular diseases, diabetes, MetS, arthritis, and cancer) as well as neurodegenerative diseases (AD and PD). In contrast, the Paleolithic diet, on which human beings have evolved and lived for most of their existence, had n-6 to n-3 ratio of 1:1. This diet was high in fiber and rich in fruits, vegetables, lean meat, and fish (Fig. 1.4)

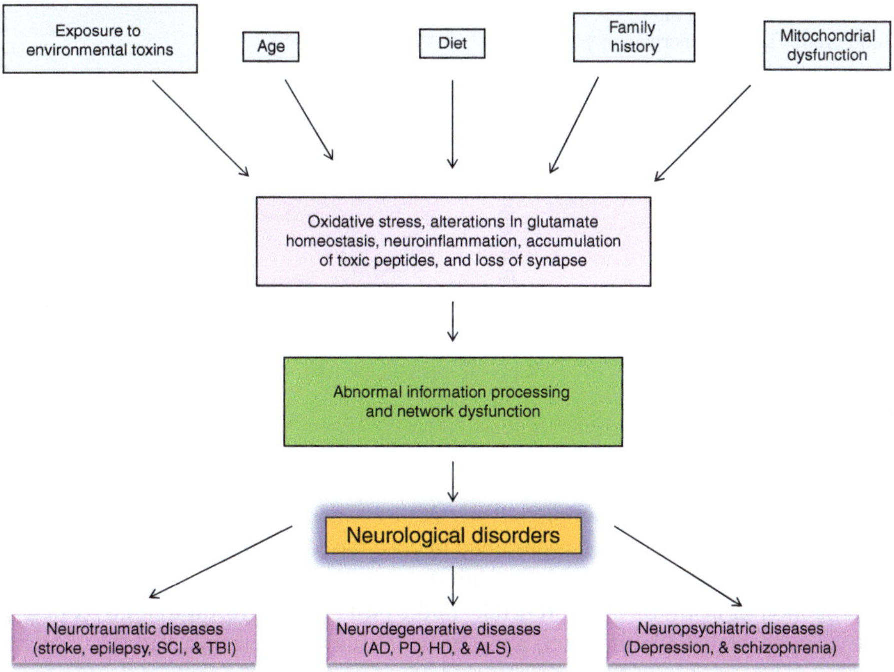


**Fig. 1.2** Factors modulating onset of neurodegenerative diseases

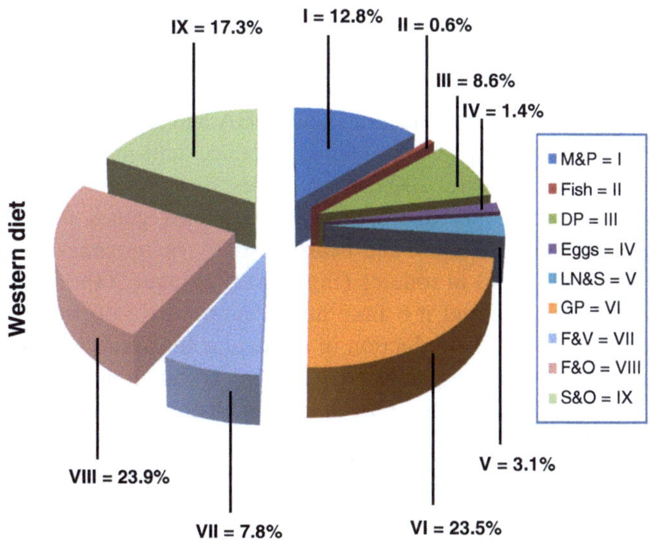


**Fig. 1.3** Contribution of food groups to daily energy intake in the present-day Western diet. Data are summarized from Hiza and Bente (2007). Meat and protein (M & P); dairy products (D & P); legumes, nuts, and soy (LN&S); grain products (GP); fruit and vegetables (F & V); fat and oils (F & O); and sugar and sweeteners (S & S)

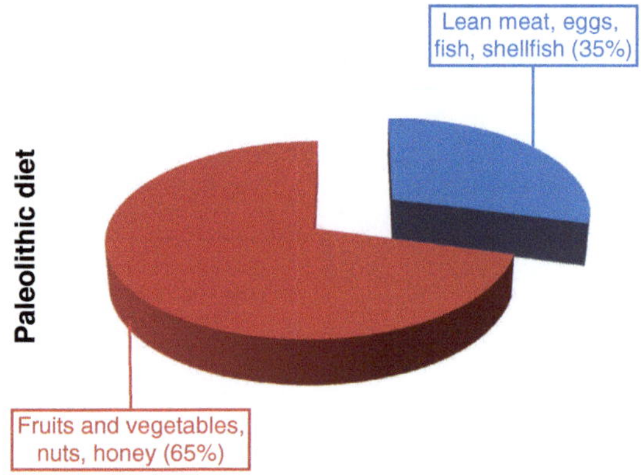


**Fig. 1.4** Contribution of food groups to daily energy intake in the Paleolithic diet. Data are summarized from Eaton and Cordain (1997)

(Cordain et al. 2005; Simopoulos 2002, 2008). Changes in eating habits, natural versus processed food, which is enriched in corn-based livestock and increase in the consumption of vegetable oil along with decrease in consumption of seafood, and introduction of fast food and soft drinks in the past 50 years result in high oxidative stress, increase in inflammation, and abnormal immune responses in the visceral organs and brain tissue. An adequate ratio of n-6 and n-3 fatty acids (1:1) promotes a healthier balance between n-6 and n-3 PUFA-derived lipid mediators and supports and maintains optimal visceral and neural membrane functions (Farooqui 2009). The generation of n-6 fatty acid (arachidonic acid, ARA)-derived prostaglandins (PGs), leukotrienes (LTs), and thromboxanes (TXs) produces oxidative stress and proinflammatory responses (Fig. 1.5) (Farooqui 2010, 2011), whereas n-3 fatty acid (docosahexaenoic and eicosapentaenoic acid, DHA and EPA)-derived docosanoids (neuroprotectins, resolvins, and maresins) produce anti-inflammatory and pro-resolutionary effects (Farooqui 2009, 2011). Thus, diet rich in DHA and EPA supports cognitive processes in humans and upregulates genes that facilitate the maintenance of synaptic function and neuroplasticity (neurogenesis, synaptogenesis, cortical reorganization) in rodents (Fig. 1.5) (Farooqui 2009). In contrast, diet enriched in saturated fats and n-6 fatty acids has negative effects on age-related cognitive decline through the reduction in molecular substrates that support cognitive function and increase the risk of neurological dysfunction through the generation of PGs, LTs, and TXs from n-6 fatty acids in both humans and animals (Farooqui 2009). Another difference between modern American diet and our ancestors' diets is the energy density. At some typical outlets the average energy density of the entire diet is 65 % and 145 % higher than the average British and traditional African diets respectively. This increase in energy density is against human weight regulatory mechanisms. Based on this information, it is suggested that the high energy densities of many fast foods challenge human appetite control systems with conditions for which human genome was not designed. The consumption of high-energy-density food in Western diet not only promotes weight gain and obesity but also results in

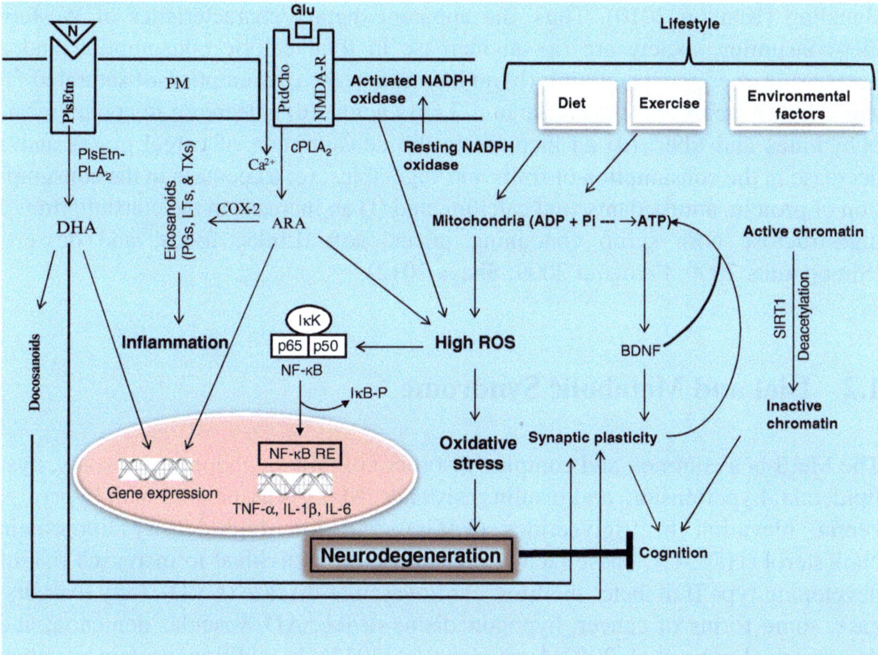


**Fig. 1.5** Effect of lifestyle on signal transduction processes associated with neurodegeneration and cognition. Plasma membrane (PM); *N*-methyl-D-aspartate receptor (NMDA-R); glutamate (Glu); neurotransmitter (N); phosphatidylcholine (PtdCho); ethanolamine plasmalogen (PlsEtn); plasmalogen-selective phospholipase $A_2$ (PlsEtn-PLA$_2$); cytosolic phospholipase $A_2$ (cPLA$_2$); cyclooxygenase (COX-2); arachidonic acid (ARA); reactive oxygen species (ROS); nuclear factor-κB (NF-κB); nuclear factor-κB response element (NF-κB-RE); inhibitory subunit of NF-κB (IκB); brain-derived neurotrophic factor (BDNF); sirtuin1 (SIRT1); tumor necrosis factor-α (TNF-α); interleukin-1β (IL-1β); interleukin-6 (IL-6); prostaglandins (PGs); leukotrienes (LTs); thromboxanes (TXs); and nicotinamide adenine dinucleotide phosphate oxidase (NADPH oxidase)

insulin resistance, inflammation, and oxidative stress in brain and visceral tissues. These processes result in short lifespan and accelerate the onset of aging-associated disorders, including diabetes, MetS, cancer, and neurodegenerative disorders. By contrast, a dietary regimen of moderate calorie restriction with adequate nutrient intake not only delays aging but may delay or attenuate age-related diseases such as diabetes, cancer, and AD (Farooqui and Farooqui 2009). Caloric restriction reduces insulin resistance, and it is the only proven mechanism to expand lifespan. Conversely, insulin resistance increases with age and obesity. Both these are risk factors for MetS, diabetes, and late-onset AD. Caloric restriction also activates stress pathways that increase resistance of organism to subsequent stress or nutritional limitation, an effect known as hormesis. Energy-sensing pathways are linked to the aging process and are regulated by insulin/IGF-1, sirtuins, TOR, and AMPK signaling (Kenyon 2010). Thus, the apparent dietary characteristics of Western diet-consuming society are (a) an increase in total caloric consumption and a decrease in energy expenditure; (b) an increase in the consumption of saturated fat and n-6 fatty acids and a decrease in n-3 fatty acids; (c) a decrease in complex carbohydrates and fiber; (d) an increase in the consumption of cereal grains and a decrease in the consumption of fruits and vegetables; (e) a decrease in the consumption of protein, antioxidants, and calcium; and (f) an increase in the consumption of high-fructose corn syrup containing juices, soft drinks, foods, and desserts (Simopoulos 2009; Farooqui 2009; Shaya 2012).

## 1.2 Diet and Metabolic Syndrome

The MetS is a common and complex disorder combining abdominal obesity, dyslipidemia, hypertension, and insulin resistance. MetS is associated with hyperglycemia, elevation in triglycerides (TAGs), and low high-density lipoprotein cholesterol (HDL-C). These factors predispose the individual to increased risk of developing type II diabetes mellitus, cardiovascular disease (CVD), fatty liver disease, some forms of cancer, hypogonadism, stroke, AD, vascular dementia, and depression (Lakka et al. 2002; Farooqui et al. 2012). In addition, protein-enriched "Western diet" has especially been associated with an increased risk of type II diabetes, confirmed by several cohort studies (Song et al. 2004; Villegas et al. 2006). Pathophysiological abnormalities that contribute to the development of the MetS include impaired mitochondrial oxidative phosphorylation and mitochondrial biogenesis, dampened insulin metabolic signaling, endothelial dysfunction, and associated myocardial functional abnormalities (Ren et al. 2010). Subjects with MetS also have autonomic nervous system dysfunction characterized by predominance of the sympathetic nervous system in many organs, including heart, kidneys, vasculature, adipose tissue, and muscles. The prevalence of the MetS varies between different populations. It is high in developed nations because of decrease in physical activity and increased consumption of energy-laden foods (high fat, high carbohydrate along with soft drinks), which thus lead to high rates of obesity. The prevalence of MetS in the United States is about 34–40 % for both sexes with variations according to ethnicity, age, and race. Lifestyle factors including smoking, lack of exercise, poor dietary habits as well as low socioeconomic status are associated with the development of MetS (Ford et al. 2002). The prevalence of MetS is increasing dramatically with age. Since the mean age of the US population is rising, further increase in MetS incidence appears inevitable. The associated health care costs are projected to increase to unsustainable levels in 10–20 years, not only in the United States but in Europe and Asia as well (Simmons et al. 2010).

## 1.3 Pathogenesis of Metabolic Syndrome

Although the etiology of MetS is not clear, high-fat diet with abundant protein and presence of fructose corn syrup in soft drinks, lack of exercise, obesity, and family history are considered important risk factors (Fig. 1.5) (Grundy et al. 2005). Above-mentioned diet is hyperenergetic. Dietary fats favor their metabolic processing, saving glucose, which additionally spares amino acids. It is well known that glucose entry into most cells is controlled by insulin. The excess of glucose promotes hyperinsulinemia and insulin resistance, which is characterized by the failure of peripheral tissues to appropriately regulate glucose homeostasis in response to insulin (Hegarty et al. 2003; Parillo and Riccardi 2004; Shulman 2002; Hwu et al. 2009). The available systems of energy disposal are unable to cope with the excess of substrates, since they are geared for saving not for spendthrift, which promotes and facilitates an unbearable overload of the storage mechanisms. Adipose tissue is an important energy sink, which stores the energy that cannot be used otherwise. However, adipose tissue growth also has limits, and the excess of energy induces inflammation, promoted by the ineffective intervention of the immune system. However, even under this acute situation, the presence of excess glucose remains favors its final conversion to fat. In addition to energy storage, adipose tissue plays an active role in many homeostatic processes including energy expenditure, appetite regulation, and glucose regulation. Fat tissue is critical for thyroid function, immune response, bone health maintenance, reproduction, and blood clotting. The adipose tissue is an active endocrine organ secreting free fatty acids, leptin, adiponectin, adipsin, complement factor 3, interlukin-6 (IL-6), TNF-α, angiotensinogen, and plasminogen activation inhibitor-1 (PAI-1), among others. Abnormal signaling and the deficiency of above-mentioned hormones result in deleterious effects (Kissebah et al. 1982; Xing et al. 1998). The increased expression of inflammatory adipokines and cytokines and deranged substrate handling by adipose tissue produce most of the MetS traits: insulin resistance (requirement of higher amounts of insulin to maintain euglycemia), obesity (accumulation of excess body fat, which manifests as increased weight or waist circumference), diabetes, hypertension, hyperlipidemia, and their compounded combined effects. Thus, a maintained excess of energy in the diet may induce difficulties in the disposal of glucose, eliciting inflammation and ultimately developing MetS. Elevation in the circulating lipids helps in raising arterial tension (Ayyobi and Brunzell 2003; Hwu et al. 2009). The induction of insulin resistance is also supported by abnormalities in β-cell secretory product, circulating insulin antagonists, and target tissue defect in insulin action. The target tissues predominantly include skeletal muscle, liver, and adipose tissue (Adediran et al. 2006). In addition, there is compelling evidence that small size at birth in full-term pregnancies is linked with the subsequent development of the major features of the MetS, namely, glucose intolerance, insulin resistance, type II diabetes, hypertension, dyslipidemia, and increased mortality from cardiovascular disease (Levitt et al. 2000). Levels of adiponectin and the ratio between leptin and adiponectin have also been suggested as surrogate markers of insulin resistance (Finucane et al. 2009).

It is proposed that the leptin–adiponectin ratio may be a stable indicator in non-fasted subjects since the fluctuations in their levels are modest (Finucane et al. 2009).

Type II diabetes is a component of MetS. Type II diabetes involves dysfunction of multiple organ systems, including impaired insulin action in muscle and adipose, defective control of hepatic glucose production, and insulin deficiency caused by loss of β-cell mass and function (Muoio and Newgard 2008). In the brain, diabetes produces atrophy and electrophysiological changes that may result in deficits in learning, memory, attention, executive function, and psychomotor efficiency (Wrighten et al. 2009). The diabetes-mediated changes in the CNS may, over time, lead to an acceleration in brain aging and increase the risk of age-related neurodegenerative diseases such as Alzheimer disease (Biessels et al. 2002; Farooqui et al. 2012). Recent epidemiological studies indicate that increased dairy and meat consumption in the Western diet may be responsible for the development of type II diabetes. This pathological condition is caused by complex set of interactions between genetic and environmental factors. Recent studies have indicated that human type II diabetes is a constellation of disorders associated with polymorphisms in a wide array of genes, with each individual gene accounting for <1 % of disease risk (Ridderstråle and Groop 2009). The presence of leucine in meat and dairy products may contribute to the pathogenesis of type II diabetes (Melnik 2012). Although the molecular mechanism associated with pathogenesis of leucine-mediated is not fully understood, it is proposed that leucine induces type II diabetes and obesity through the overstimulation of mammalian target of rapamycin complex 1 (mTORC1). The downstream target of mTORC1, the kinase S6K1, initiates insulin resistance through the phosphorylation of insulin receptor substrate-1, thereby increasing the metabolic burden of β cells. In addition, leucine-induced mTORC1–S6K1 signaling plays an important role in adipogenesis, thus increasing the risk of obesity-mediated insulin resistance (Melnik 2012). High consumption of leucine-rich proteins explains the increase in mTORC1-dependent insulin secretion, increase in β-cell growth, and β-cell proliferation promoting an early onset of replicative β-cell senescence with subsequent induction of β-cell apoptosis (Melnik 2012). These processes are hallmarks of type II diabetes and insulin resistance. In contrast, the antidiabetic drug metformin antagonizes leucine-mediated mTORC1 signaling. Plant-derived polyphenols and flavonoids have been identified as natural inhibitors of mTORC1. They exert antidiabetic and anti-obesity effects (Melnik 2012; Farooqui 2012).

A new hypothesis to explain the pathogenesis MetS has also been proposed (Hallfrisch 1990; Nakagawa et al. 2005, 2006). According to this hypothesis, MetS is a consequence of a high intake in carbohydrates, food with a high glycemic index, and fructose and relatively low intake of cholesterol and saturated fat. Fructose is absorbed in the intestine via specific transporters [glucose transporter 5 (GLUT5), GLUT2, and possibly SLC2A9] and metabolized largely (50–75 %) by the liver, with the rest being primarily metabolized by the kidney and adipocytes (Froesch 1972). In contrast to glucose, consumption of fructose stimulates food intake (Miller et al. 2002; Cha et al. 2008; Lane and Cha 2009). This effect may be due to the different mechanisms by which fructose enters glycolysis. Fructose enters the glycolytic pathway at the triose level, after the generation of fructose-1-phosphate via fructokinase. Because fructokinase consumes ATP at a greater rate than does phosphofructokinase-1 (one of the regulatory steps of glycolysis), fructose can also decrease intracellular ATP while increasing AMP levels (Cha et al. 2008; Lane and Cha 2009). In addition unlike glucose, fructose does not stimulate insulin release from the pancreas, because the pancreas lacks the GLUT5 transporter for fructose (Curry 1989). Insulin signals the secretion of the hormone leptin from adipocytes, which inhibits food intake (Havel 2002). Since pancreatic insulin secretion is not sensitive to fructose, the ingestion of fructose in high-fructose corn syrup may contribute to overeating and obesity in rodents (Teff et al. 2002). Once in the cell, fructose is phosphorylated to fructose-1-phosphate by fructokinase, which is cleaved by aldolase to form precursors for triacylglycerol (TAG) synthesis (Mayes 1993). These two enzymes are regulated neither by insulin nor by the energy status of the cell. As a consequence most fructose in portal blood is rapidly transformed into triose phosphate and ultimately converted to triacylglycerol (TAG), fatty acyl coenzyme A, and diacylglycerol (Stanhope and Havel 2008).

Acute fructose feeding also decreases leptin secretion in humans and produces leptin resistance (Teff et al. 2004). In fructose-fed rats, the anorectic effects of intraperitoneally administered leptin can be nearly abolished. This corresponds to a significant decrease in hypothalamic signal transducer and activator of transcription 3 (STAT3) phosphorylation in response to fructose (Shapiro et al. 2008). High-fructose diet also produces marked changes in the brain as well as liver (Park 2001). Fructose feeding promotes insulin resistance, which may interfere with memory formation (Convit 2005; Messier 2005; Dou et al. 2005). Liver metabolizes fructose and converts it into lipids (Havel 2005) and produces a large increase TAG (Basciano et al. 2005; Havel 2005). In liver, TAG interferes with insulin signaling (Kim et al. 2007). It is suggested that a high-fructose diet is analogous to a high-fat diet in many metabolic ways (Havel 2005).

In visceral and brain tissues, high fructose consumption not only leads to the formation of increased levels of fructose end products (glyceraldehydes, and dihydroxyacetone phosphate), but also results in the generation of non-enzymatic advanced glycation products (AGEs) through Maillard reaction. AGEs not only react with amino group of proteins or lipids inducing molecular rearrangements in membrane components but also produce their effect by interacting with receptor for advanced glycation end products (RAGE) (Neeper et al. 1992; Ahmed 2005). These receptors are involved in multiple physiological and pathological functions, including inflammation, oxidative stress, neurodegeneration, maintenance of homeostasis, tumorigenesis, promotion of neurite outgrowth, cell survival, cell migration, and neuronal differentiation (Fages et al. 2000; Ramasamy et al. 2005; Meneghini et al. 2010). Unlike glucose, which is utilized by all organs, fructose is solely metabolized in the liver. The enzyme that metabolizes fructose (fructokinase) is not regulated by negative feedback, so all fructose that enters the liver cells is rapidly phosphorylated by ATP leading to ATP depletion, a finding which has been consistently observed in in vivo animal models of diabetes and diabetic humans. ATP depletion activates enzymes of purine metabolism (AMP deaminase-1), which degrades adenine nucleotides to uric acid via xanthine oxidoreductase with the development of hyperuricemia (Hallfrisch 1990; Nakagawa et al. 2005). Serum uric acid levels have been associated with cardiovascular risk factors like hypertension and the MetS in adolescents (Alper et al. 2005; Baker et al. 2005).

Accumulation of AGEs in the vasculature occurs in patients with dialysis-related amyloidosis (Miyata et al. 1996) and diabetes (Schmidt et al. 1995). The progressive accumulation of AGEs in patients with diabetes supports the view that endogenous clearance mechanisms fail to function effectively at sites of AGE deposition. Increased methylglyoxal generated by nonenzymic reactions of carbohydrates and oxidized lipids reacts to amino and sulfhydryl group of proteins to form AGEs that impair protein function is found under hyperglycemic conditions in diabetes. Accumulated AGEs have capability of altering cellular properties by several mechanisms. Thus, direct effects of AGEs in the extracellular space include cross-linking with neighboring unrelated macromolecules (Brownlee 1995). Nonenzymic glycation has been shown to alter properties of FGF, as well as vitronectin and laminin (Brownlee 1995). However, as mentioned above, AGE engagement of cellular receptors may provide a biological amplification mechanism whereby glycoxidized adducts can modulate properties of multiple cells (Schmidt et al. 2001).

A substantial body of epidemiological and experimental evidence suggests that serum uric acid is an important, independent risk factor for cardiovascular, cerebrovascular, and renal diseases especially in patients with hypertension, heart failure, and diabetes. Elevated serum uric acid is highly predictive of mortality in patients with heart failure or coronary artery disease and of cardiovascular events in patients with diabetes (Alderman and Aiyer 2004). In addition, patients with hypertension and hyperuricemia have a three- to five-fold increased risk of experiencing coronary artery disease or cerebrovascular disease (stroke) compared with patients with normal uric acid levels. Although the mechanisms by which uric acid plays a pathogenetic role in cardiovascular and cerebrovascular diseases are unclear, it is shown that hyperuricemia is associated with deleterious effects on endothelial dysfunction, oxidative metabolism, platelet adhesiveness, hemorheology, and aggregation (Alderman and Aiyer 2004; Pasalic et al. 2012).

High insulin levels have been shown to reduce renal uric acid excretion leading to hyperuricemia, which results from a combination of stimulated formation and reduced elimination in uric acid. In addition to producing uric acid, fructose enhances the production of tumor necrosis factor-α (TNF-α) (Kanuri et al. 2011), a potent proinflammatory cytokine that induces insulin resistance and lipoprotein production. Fructose not only evokes alteration of post-receptor insulin signaling and decreases insulin receptor activation and phosphorylation of IRS-1 but also causes oxidative stress and mitochondrial dysfunction (Nagai et al. 2009).

Mounting evidence suggests that MetS is a proinflammatory state, which contributes to the accelerated atherogenesis, cardiovascular diseases, stroke, AD, and depression (Fig. 1.6) (Hansson 2005; Farooqui et al. 2012). Interleukin-6 (IL-6) and TNF-α are key adipocytokines that contribute to the pathogenesis of atherothrombotic disease through a variety of mechanisms (Kristiansen and Mandrup-Poulsen 2005).

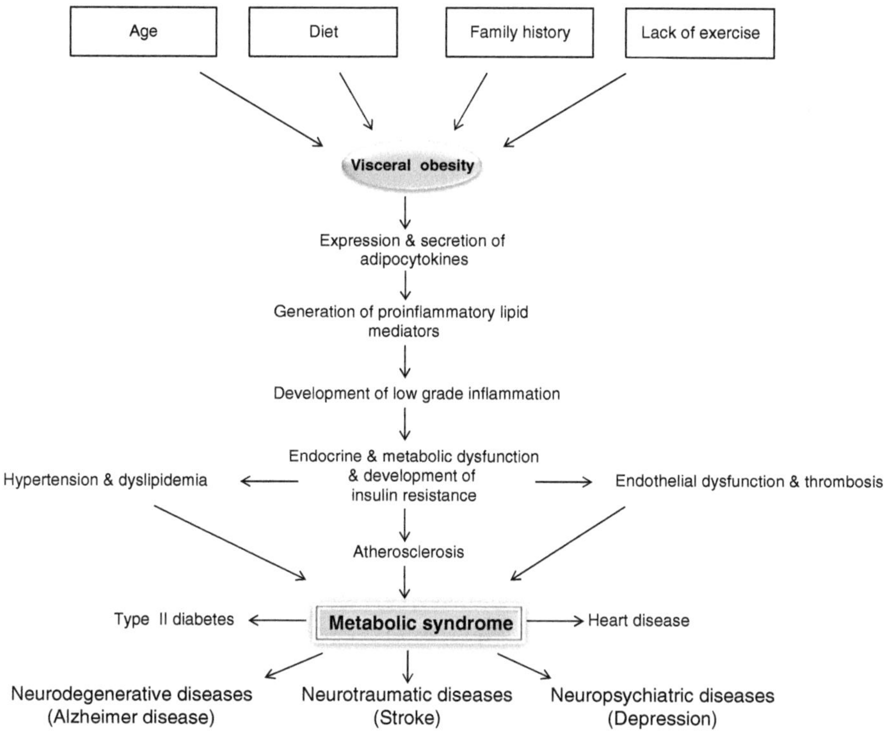


**Fig. 1.6** Events associated with the development of metabolic syndrome and neurological disorders

IL-6 amplifies the inflammatory response-stimulating hepatic production of acute-phase proteins including C-reactive protein and fibrinogen and modulates cell adhesion and chemotaxis (Kristiansen et al. 2005; Yudkin et al. 2000; Heinrich et al. 1990; Xing et al. 1998). Collectively, these studies suggest that fructose is extremely a proinflammatory sugar that facilitates AGEs formation and speeds up the aging process. It not only promotes the kind of dangerous growth of fat cells around the vital organs that are the hallmark of diabetes and heart disease, but also generates large quantities of uric acid responsible with hypertension and stroke.

## 1.4 Exercise and MetS

Exercise and healthy diet are major preventive therapies against insulin resistance, chronic inflammation, and obesity, the major components of the MetS (Lakka and Laaksonen 2007). Exercise initiates the maintenance of good cardiorespiratory and muscular fitness by increasing energy consumption, improving insulin sensitivity, increasing blood flow, and reducing inflammation. Although the molecular mechanisms underlying these benefits are not fully understood, it is well known that exercise not only improves the dyslipidemic profile by raising high-density lipoprotein cholesterol and lowering triglycerides in overweight and obese adults with characteristics of the MetS (Lakka and Laaksonen 2007), but also increases expression of GLUT4 and other proteins involved in insulin signaling and glucose metabolism (Houmard et al. 1993). These changes are due to alterations in the hormone production, blood flow, and activity of the nervous system. In addition, exercise modulates the expression of certain genes and proteins, such as heat shock proteins in the skeletal muscle. These proteins not only function to maintain homeostasis, facilitate repair from injury, and provide protection against future insults, but also promote permanent beneficial adaptations, which may include increase in activators of mitochondrial biogenesis such as peroxisome proliferator-activated receptor γ coactivator-1α (PGC-1α) and increase in activity of oxidative enzymes such as citrate synthase and succinate dehydrogenase (Carter et al. 2001; Fernström et al. 2004). In addition, heat shock protein peptides promote the production of anti-inflammatory cytokines, indicating immunoregulatory potential of these proteins. Therefore, the presence of immune responses to heat shock proteins in inflammatory diseases is considered as an attempt of the immune system to correct the inflammatory condition. These changes may decrease the likelihood of developing the diabetes and MetS, especially in high-risk groups (Lakka and Laaksonen 2007; Morton et al. 2009). It is also proposed that as a non-pharmacological intervention, exercise mediates upregulation of heat shock proteins, boosts the production of human growth hormone, and blocks maladapted pathways, which may not only provide protection against protein misfolding diseases, but also promote preservation of muscle function during aging (Morton et al. 2009).

It is well known that energy consumption and storage depends on the lifestyle. Active lifestyle controls weight; reduces risk of type II diabetes, MetS, heart disease, and some cancers; strengthens bones and muscles; and increases longevity. In contrast, a sedentary lifestyle leads in a positive energy balance, which results in accumulation of body fat. Under these circumstances, adipose tissue initiates the expression and secretion of adipocytokines (tumor necrosis factor-α (TNF-α), plasminogen activator inhibitor, and resistin) into the circulation. Adipocytokines not only induce low-grade inflammation but also impair glucose transport via the inhibition of insulin signal transduction. Insulin-induced activation of the insulin receptor (IR), phosphatidylinositol 3-kinase (PtdIns 3K), and Akt is retarded by the activation of IκB kinase (IKK) and degradation of IκB in the muscle tissue (De Alvaro et al. 2004). Furthermore, IKKβ silencing prevents TNF-α-induced impairments in insulin action on Akt phosphorylation and glucose uptake (Austin et al. 2008). Sedentary lifestyle produces a decrease in resting energy metabolic capacity due to atrophy of skeletal muscles, which are major energy-consuming tissues in the body. Muscle atrophy may result in loss of muscle fibers (Larsson and Ansved 1995) not only due to apoptotic cell death in the muscle (Sakuma and Yamaguchi 2010) but also due to decrease in differentiation of satellite cells (Conboy and Rando 2005) as well as decrease in protein synthesis and increase in protein degradation (Combaret et al. 2009). The activity of enzymes associated with aerobic metabolism and glucose uptake in muscle is also decreased in the sedentary lifestyle specifically in the middle age. In contrast, daily exercise produces a persistent increase in insulin action in the skeletal muscle of insulin-resistant individuals. The molecular mechanism(s) associated with increased glucose uptake with exercise has been attributed not only to the increased expression and/or activity of key signaling proteins involved in the regulation of glucose uptake and metabolism in skeletal muscle, but also to the upregulation of genes encoding energy metabolism as well as oxidative phosphorylation and lipid metabolism (Short 2009). Mounting evidence supports the view that exercise-mediated improvements in insulin sensitivity are related to changes in the expression and/or activity of proteins involved with insulin signal transduction in skeletal muscle such as the AMP-activated protein kinase (AMPK) and the protein kinase B (Akt) substrate AS160 (Hawley and Lessard 2008). In addition, exercise increases oxidative capacity of the skeletal muscle by upregulating lipid oxidation and the expression of proteins involved in mitochondrial biogenesis. This may be another mechanism by which exercise improves insulin sensitivity (Hawley and Lessard 2008).

In addition, exercise not only upregulates receptors, such as LXRα, PPARγ, scavenger receptor B1 (SRB1), and LDLR, and enzymes such as CYPs, LCAT, and paraoxonase 1, but also increases the expression of various transporters such as ABCA1, ABCG1, and apo A1 (Luoma 2010). Regular exercise improves oxygen uptake and weight control, upregulates mechanisms protecting arteries from atherosclerosis, and decreases mortality (Luoma 2010). Collective evidence suggests that exercise along with the consumption of healthy diet (fruits, vegetables, and n-3 fatty acids) not only improves the muscle function, increases production of nitric oxide, reduces systemic inflammation, blocks insulin resistance, decreases blood pressure, and increases levels of circulating endothelial progenitor cells and improves endothelial dysfunction leading to reduction in risk of developing MetS, diabetes, and heart diseases (Libby et al. 2002; Walther et al. 2008; Caulin-Glaser et al. 2005).

## 1.5 Effect of Exercise on Signaling in the Brain

Aerobic exercise training has been reported to increase the size of the anterior hippocampus, leading to improvements in spatial memory (Erickson et al. 2011). Exercise training increases hippocampal volume by 2%, effectively reversing age-related loss in volume by 1 to 2 y. At the molecular level exercise upregulates the expression of genes associated with the actions of brain-derived neurotrophic factor (BDNF) and insulin-like growth factor (IGF) systems. These genes modulate synaptic plasticity (Molteni et al. 2002a, b).

It is also shown that increase in hippocampal volume is associated with greater serum levels of BDNF, a mediator of neurogenesis in the dentate gyrus. Caudate nucleus and thalamus volumes are not affected by the aerobic exercise (Erickson et al. 2011). In addition, exercise also upregulates genes related to the *N*-methyl-D-aspartate receptor (NMDA-R) function and downregulates genes associated with the GABAergic system. In addition, genes for the expression of proteins associated with synapsin I and II, synaptotagmin, and syntaxin, calcium/calmodulin protein kinase II (CaMKII), mitogen-activated protein kinase (MAPK/ERK, I and II), protein kinase C (PKC-δ), or transcription factor cAMP response element binding protein (CREB) are also increased (Vaynman et al. 2003; Molteni et al. 2002a, b; Ding et al. 2006a, b). These findings support the view that exercise mediates its effects through the activation of BDNF-mediated intracellular signal pathways involving protein kinases (MAPK/ERK and CaMK) (Vaynman et al. 2003; Molteni et al. 2002a, b). Blocking BDNF signaling abolishes the effects of exercise on neuroplasticity and learning and memory (Vaynman et al. 2004). Based on these results, it is proposed that cross talk between IGF-1 and BDNF can support cellular processes associated with learning and memory and neural cell survival.

## 1.6 Pathogenesis of Neurological Disorders

Neurological disorders include neurotraumatic diseases (stroke, spinal cord injury, and traumatic brain injury), neurodegenerative diseases (AD, PD, HD, and ALS), and neuropsychological disorders (depression, dementia, schizophrenia, bipolar disorders). The most important risk factors for neurological disorders are old age, race/ethnicity, a positive family history, unhealthy lifestyle, exposure to environmental toxins, and endogenous factors. The onset of stroke, AD, PD, and ALS is often subtle and usually occurs in mid- to late life, and their progression depends on genetic and environmental factors (Graeber and Moran 2002). Depending on the brain regions where neurons degenerate, individuals experience physical disabilities and cognitive impairments. Currently, stroke is the most common single cause of disability. AD, which is characterized by the loss of neurons in nucleus basalis and hippocampus, is the leading cause of dementia among elderly (Farooqui 2010). Oxidative and nitrosative stress and neuroinflammation play a major role in neurodegeneration in neurotraumatic and neurodegenerative diseases. Persistent presence of oxidative and nitrosative stress and neuroinflammation induces the accumulation of damaged proteins, DNA, and membrane fragments. In addition, neurodegenerative diseases are accompanied by the accumulation of disease-specific proteins, such as accumulation of Aβ and its aggregates in the cerebral cortex and hippocampal region in AD, α-synuclein in the brain stem in PD, and huntingtin in striatal medium spiny neurons in HD. Furthermore, neurological diseases are accompanied by abnormalities in signal transduction processes along with elevated levels of lipid mediators (Farooqui et al. 2007; Farooqui 2010). Mounting evidence suggests that neurological disorders are multifactorial diseases that are caused by complex interactions among genetic factors, environmental factors, aging, and lifestyle. Environmental and dietary risk factors, such as heavy metals, hormones, cholesterol, high-fat diet, high alcohol intake, diet deficient in n-3 fatty acids, antioxidants and vitamins, and reduced levels of physical activity (exercise) may promote the onset and progression of neurological diseases.

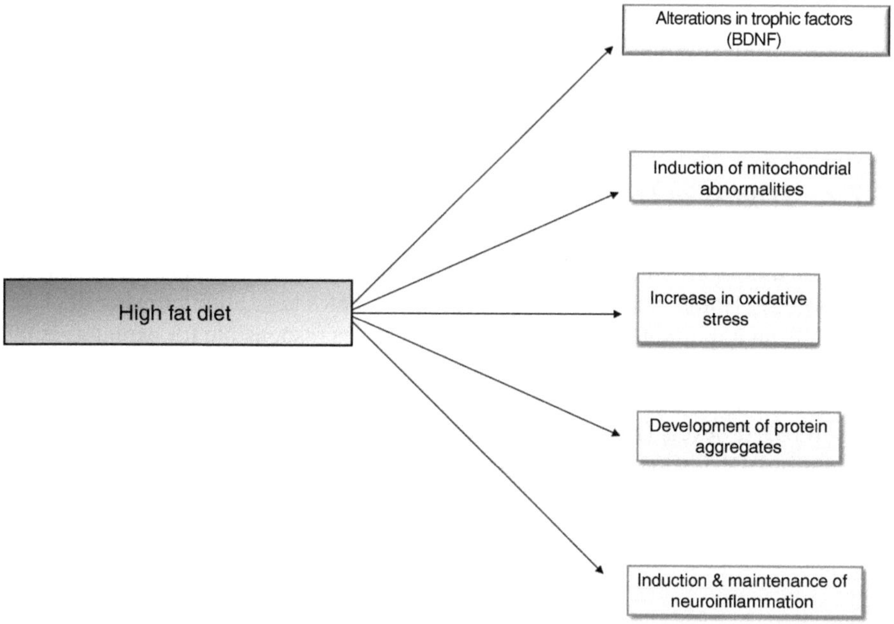


**Fig. 1.7** Effects of consumption of high-fat diet on neurochemical processes in the brain

## 1.7 Effect of High-Fat Diet on Neurological Disorders

Consumption and maintenance of rodents on high-fat diet produces many neurochemical changes in the brain, including increase in free fatty acids, induction of mitochondrial abnormalities, formation of protein aggregates (Aβ, neurofibrillary tangles, Lewy Bodies, and lipofuscin ceroid), alterations in trophic factors, neuroinflammation, and oxidative stress (Fig. 1.7). These changes induce functional impairments in the brain. The severity of brain impairments is highly dependent on the areas (brain region) and extent of the damage in the brain. Neurochemical changes with brain damage are closely associated with the pathogenesis of MetS and neurological disorders.

### *1.7.1 High-Fat Diet and Formation of Protein Aggregates*

Although human studies on link among diabetes, cognitive deterioration, and neuropathological changes have been controversial, many rodent studies show the ability of a high-fat and high-sucrose diet to promote the development of extracellular β-amyloid deposits (Beeri et al. 2005; Li et al. 2007). High-fat-diet-induced cognitive impairment is tied to altered hippocampal morphology/plasticity in normal rats (Granholm et al. 2008; Stranahan et al. 2008). Using normal C57BL/6 mice, it is also reported that a high-fat/high-cholesterol diet induces loss of working memory correlated with striking neuroinflammatory changes and increases in APP processing (Thirumangalakudi et al. 2008).

### 1.7.2 High-Fat Diet and Changes in BDNF Signaling

BDNF is a neurotrophic factor involved in the regulation of synaptic function and morphology (Horch 2004). High-fat diets and diabetes have ability to alter the level of BDNF, an important class of signaling molecules responsible for axon targeting, neuron growth, maturation of synapses during development, and synaptic plasticity, supporting the view that high-fat diet promotes brain disturbances through alterations in BDNF levels (Kanoski et al. 2007; Wu et al. 2004). As stated above, these changes in BDNF signaling are linked to deleterious changes in MAPK and CREB signaling, which are necessary for neuronal homeostasis (Kanoski et al. 2007; Wu et al. 2004; Jovanovic et al. 1996).

### 1.7.3 High-Fat Diet and Mitochondrial Abnormalities

High-fat diet not only downregulates the expression of genes involved in oxidative phosphorylation, but also decreases the expression of genes involved in mitochondria biogenesis (Nisoli et al. 2007). These processes are accompanied by an increase in fatty acid oxidation, as an alternative source for ATP generation (Nisoli et al. 2007; Handschin and Spiegelman 2006). Long-term upregulation of fatty oxidation in visceral tissues (muscle and liver) and in the brain may have adverse effects on cellular homeostasis through the elevation in oxidative stress and reduction in ATP production.

### 1.7.4 High-Fat Diet and Induction of Oxidative Stress and Inflammation

In visceral organs as well as in the brain, high-fat diet induces oxidative stress, which is characterized by a marked increase in superoxide anions ($O_2^{\bullet-}$); hydroxyl ($^{\bullet}OH$), alkoxyl ($RO^{\bullet-}$), and peroxyl radicals ($ROO^{\bullet}$); singlet oxygen; and hydrogen peroxide ($H_2O_2$). The major sources of oxidative stress are mitochondrial respiratory chain, xanthine/xanthine oxidase, myeloperoxidase in cytoplasm, oxidation of arachidonic acid (ARA) by cyclooxygenase (COX), lipoxygenase (LOX), and epoxygenase (EPOX) in cytoplasm, and NADPH oxidase in plasma membranes (Farooqui 2010). Under physiological conditions the antioxidant defense systems (superoxide dismutase, catalase, transferrin, and glutathione peroxide, glutathione, and vitamin C) within the body can easily neutralize the amount of reactive oxygen species (ROS) produced through three cellular antioxidative defense systems. However, disruption of tight control by high ROS levels results in the severe oxidative stress, which is closely associated with neural cell injury (Farooqui 2010). Low levels of ROS are needed not only for fundamental cellular functions, such as growth and adaptation responses, but also for optimal functioning of the immune system, which is involved in the defense against the penetrating microorganisms. However, generation of high ROS in neural cells facilitates the translocation of transcription factor (NF-κB) from cytoplasm to the nucleus, where it facilitates the expression of proinflammatory enzymes (phospholipases, cyclooxygenases, and nitric oxide synthases), proinflammatory cytokines and chemokines, growth factors, cell cycle regulatory molecules, and adhesion molecules (Fig. 1.5). As stated above, NF-κB is involved in the initiation and orchestration of inflammatory pathways through the production of TNF-α, IL-1β, and adhesion molecules. Although it is well known that high-fat diet-derived ROS interact and damage proteins, nucleic acids, and lipids, relative susceptibility of various brain regions to ROS-mediated damage and resulting neurobehavioral changes remains unknown (Farooqui 2010). Collective evidence suggests that consumption of high-fat diet enriched in saturated and n-6 fatty acids promotes insulin resistance, neuroinflammation, and oxidative stress, whereas consumption of diet enriched in monounsaturated and longer chain n-3 fatty acids not only retards neuroinflammation and oxidative stress, but also improves insulin sensitivity (Farooqui 2009, 2010). Less-refined and less-starchy carbohydrates (such as fresh fruit, whole grains, green leafy vegetables, nuts, legumes) in contrast to fruit juice, refined grains, and starchy vegetables have direct and independent beneficial effects on both glucose and insulin metabolism leading to less incidences of neurological disorders (stroke, AD, and PD) and heart disease (Kurth et al. 2006; Bernal-Pacheco and Román 2007; Galimanis et al. 2009).

The majority of neurodegenerative diseases are sporadic, but some neurodegenerative diseases are caused by the consumption of toxic diet. For example, Guam-type amyotrophic lateral sclerosis/Parkinsonism dementia (ALS/PDC) is caused by the presence of β-methylaminoalanine (BMAA), a neurotoxic, non-lipophilic, and nonprotein amino acid found in *Cycas circinalis*. This indigenous plant commonly ingested as a food or medicine by the Chamorros of Guam (Murch et al. 2004; Ince and Codd 2005). Recent studies have indicated that BMAA is produced by symbiotic cyanobacteria within specialized roots of the cycads (Ince and Codd 2005; Bradley and Mash 2009). BMAA is not restricted to Guam but is rather broadly present in the environment. High concentrations of BMAA have been found to be associated with cyanobacterial blooms outside of Guam, in marine (Banack et al. 2007) and freshwater environments (Metcalf et al. 2008). Recently, the occurrence and bioaccumulation of BMAA has been detected in a temperate aquatic ecosystem (Baltic Sea), suggesting a potential widespread human exposure (Jonasson et al. 2010).

After oral consumption, most of BMAA (80 %) passes from the digestive tract into the blood stream. BMAA then crosses the blood–brain barrier (BBB) by binding through large neutral amino acid carriers. At the physiological concentrations, blood bicarbonate ions react with BMAA to form a β-carbamate (Weiss and Choi 1988), which competes with glutamate for binding to NMDA receptors (Rakonczay et al. 1991; Matsuoka et al. 1993), AMPA receptors (Rakonczay et al. 1991; Matsuoka et al. 1993), and metabotropic type of glutamate receptors (Copani et al. 1991). Activation of the various glutamate receptors results in $Ca^{2+}$ influx causing depolarization and activation of $Ca^{2+}$-dependent enzymes (Brownson et al. 2002; Cucchiaroni et al. 2010). BMAA also inhibits the cystine/glutamate antiporter system $Xc^{-}$ (Liu et al. 2009) leading to retardation of cystine uptake, which results in glutathione depletion. This may contribute to increase in oxidative stress. Increases in intracellular $Ca^{2+}$ concentrations disrupt normal mitochondrial function and facilitate the release of ROS into the cytoplasm, thereby contributing marked increases in ROS levels (Lobner et al. 2007; Cucchiaroni et al. 2010). In addition, mitochondrial dysfunction also promotes the release of cytochrome c resulting in the induction of apoptosis (Cucchiaroni et al. 2010; Chiu et al. 2011). It is recently shown that BMAA is also present in brain tissues of North American patients who had died of AD (average concentration 95 μg/g, 0.8 mM) suggesting a possible etiological role for BMAA in non-Guamanian neurodegenerative diseases (Ince and Codd 2005; Bradley and Mash 2009). Cyanobacteria are ubiquitous throughout the world, so it is possible that all humans are exposed to low amounts of cyanobacterial BMAA, that protein-bound BMAA in human brains is a reservoir for chronic neurotoxicity, and that cyanobacterial BMAA is a major cause of progressive neurodegenerative diseases including ALS worldwide (Ince and Codd 2005; Bradley and Mash 2009).

Intoxication with 1-methyl-4-phenyl-1,2,3,6-tetrahydropyridine (MPTP) causes a severe and irreversible Parkinsonian syndrome that is similar but not identical to PD in pathology and progression. In addition, exposure to certain insecticides and herbicides, such as rotenone, also produces a Parkinson-like syndrome (Brown et al. 2006; Kamel and Hoppin 2004; Keifer and Firestone 2007). Similarly, another environmental toxin, paraquat (a commonly used herbicide), has also been implicated with the onset of PD. The neurotoxicity of these chemicals is accompanied by the blockade of electron flow from NADH dehydrogenase to coenzyme Q. The agents with the ability to improve mitochondrial respiration and ATP production have been shown to exert beneficial effects in PD patients as well as in the animal models of PD (AbdulWahid Arif and Ahmad Khan 2010).

In aged human brain, elevation in plasma fatty acids due to consumption of high-fat diet not only causes the activation of redox cycling of the copper–albumin complex and excessive lipid peroxidation but also leads to the accumulation of AGEs through the Maillard reaction. As stated above, AGEs interact with their receptors, RAGE, and stimulate them (Luchsinger and Mayeux 2004; Ahmed 2005; Ghosh et al. 2007). The deleterious effects of AGEs in brain and visceral tissues are attributed to their prooxidant and inflammatory actions. The binding of AGEs with RAGE not only results in the activation of the mitogen-activated protein kinases (MAPKs) and the phosphatidylinositol 3-kinase (PtdIns 3K) pathways but also an activation of NADPH oxidases, generation of superoxide, and activation of NF-κB, which migrates to the nucleus and induces the induction of inflammatory cytokines

(TNF-α, IL-1β, IL-6, and vascular cell adhesion molecule 1). Using double transgenic PDAPP-J20 mice overproducing either wild-type RAGE (APP/RAGE) or a mutant RAGE protein that inhibits the endogenous receptor function (i.e., a dominant negative RAGE receptor, APP/dnRAGE), it is shown that RAGE signaling contributes to Aβ-mediated neuronal perturbation through the involvement of NF-κB and MAPK signaling. This process leads to increase in synaptic transmission deficits, as well as cognitive impairment (Arancio et al. 2004).

Mounting evidence suggests that dietary lifestyle influences brain function. Thus, the consumption of high-fat "Western diet" rich in n-6 fatty acids and cholesterol elevates saturated fatty acids and is associated with the development of Alzheimer-like cognitive impairment (Pasinetti and Eberstein 2008) and favors the onset of stroke. These processes may have negative effects on age-related cognitive decline and mild cognitive impairment (MCI). Studies on animal models of AD have also shown that diet enriched in n-6 fatty acids and cholesterol can cause increase in amyloidosis and alterations in synaptic plasticity and behavior. These are essential features of Alzheimer pathology. In addition, high-fat and cholesterol-enriched diet and excessive sucrose intake enhance tau pathology in Tg mice (Cao et al. 2007). Hypercholesterolemia can cause tau hyperphosphorylation in the brain, which has been shown in apo E-deficient mice fed a high-cholesterol diet (Rahman et al. 2005).

## 1.8 Effect of Exercise on the Brain

As mentioned above, regular physical exercise ameliorates age-related neuronal loss and produces positive effect on neurological disorders (Trejo et al. 2002). In the brain, exercise induces both acute and long-term beneficial alterations, such as increased levels of various neurotrophic factors and enhanced cognition. Although the molecular mechanism associated with exercise-mediated alterations in the brain is not fully understood, it is well known that physical exercise increases the expression of IGF-1 and BDNF in the brain. These neurotrophins play a critical role in the induction of synaptic plasticity and learning and memory (Fig. 1.8) (Carro et al. 2001; Vaynman et al. 2004, 2006). BDNF is also known to promote recovery after neurological insult. BDNF mediates its effects through interactions with two types of receptors, namely, the tyrosine kinase (Trk) B receptor and the pan-neurotrophin receptor p75 (p75NTR) (Reichardt 2006). Pro-BDNF preferentially interacts with the p75NTR, whereas mBDNF selectively binds and activates the TrkB (Chao and Bothwell 2002; Ibanez 2002). BDNF acts through the protein tyrosine kinase receptor (TrkB) (Soppet et al. 1991) and regulates a number of processes including memory formation, learning and behavior, synaptic plasticity (Monteggia et al. 2004), synaptic efficacy, and neuronal connectivity. It not only facilitates the development of immature neurons but also enhances the survival of adult neurons (Tyler et al. 2002). In contrast, endogenous pro-BDNF-mediated activation of p75NTR is associated with long-term depression (LTD) in the hippocampus (Rosch et al. 2005;

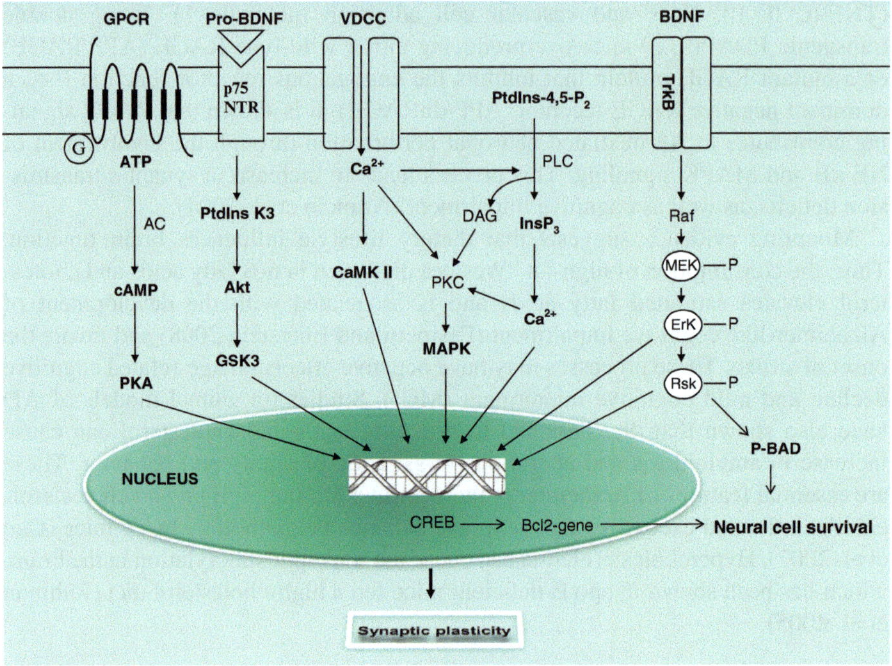


**Fig. 1.8** Diagram showing receptor-mediated signal transduction process involving BDNF and seven transmembrane-spanning G protein-coupled receptor (GPCR). Adenylyl cyclase (AC); cyclic AMP (cAMP); protein kinase A (PKA); cAMP response element binding protein (CREB); activating transcription factor 1 (ATF1); mitogen-activated kinase (MAPK); calmodulin-dependent protein kinase II (CaMKII); phospholipase C (PLC); diacylglycerol (DAG); phosphatidylinositol trisphosphate ($PtdInsP_3$); protein kinase C (PKC); phosphatidylinositol 3-kinases (PtdIns 3K); serine/threonine protein kinase (Akt); glycogen synthase kinase 3 (GSK3); brain-derived neurotrophic factor (BDNF); tyrosine kinase receptor (TrkB); extracellular signal-regulated kinase (MEK); extracellular signal-regulated kinase (Erk); Raf proto-oncoproteins (Raf); and ribosomal S6 kinase (Rsk)

Wu 2005) and induces apoptosis in peripheral neurons (Teng et al. 2005). Recent studies indicate that BDNF regulates energy homeostasis through the inclusion of dietary factors that initiate repair and recovery from neural injury (Farooqui 2010).

Exercise not only stimulates BDNF expression but also modulates signaling of proteins downstream to BDNF activation that plays an important role in synaptic function, such as synapsin I, and phosphorylated calcium/calmodulin protein kinase II and phosphorylated mitogen-activated protein kinase II (Ding et al. 2006a, b). Exercise also stimulates the expression of several key intermediates of the PtdIns 3K/Akt pathway, which contribute to neuronal survival through the phosphorylation of a number of target proteins, such as GSK-3β, endothelial nitric oxide synthase (eNOS), the proapoptotic Bcl-2 family member BAD, caspase-9, and the ubiquitin ligase murine double minute 2 (mdm2). Phosphorylation of BAD targets it to 14-3-3 protein where it is sequestered, thereby blocking its proapoptotic role (Fayard et al. 2005;

Chen and Russo-Neustdt 2007) (Fig. 1.8). Additionally, cAMP/PKA-mediated activation of synapsin I phosphorylation facilitates regenerative growth of neurons and promotes neuronal survival. Inhibition of IGF-1 receptor retards the exercise-induced increases in signal transduction. These results explain the molecular mechanisms by which IGF-1 modulates the BDNF system and mediates exercise-mediated synaptic and cognitive plasticity. BDNF not only facilitates long-term potentiation but also increases the activities of free radical scavenging enzymes and hence protects neurons against oxidative stress (Pelleymounter et al. 1996). Thus, interplay between IGF-1 and BDNF provides protection to neurons in brain (Hoyer 2004; Carro and Torres-Aleman 2004). In addition, exercise also upregulates the expression of the mitochondrial uncoupling protein 2, an energy-balancing factor involved in ATP formation and free radical management (Vaynman et al. 2006). These observations support the view that in brain physical exercise-mediated release of BDNF not only regulates hippocampal synaptic plasticity, but also controls body weight and energy status by reducing body weight, normalizing glucose levels, and increasing insulin sensitivity (Tsuchida et al. 2002; Gomez-Pinilla 2011).

## 1.9 Exercise-Mediated Neuroprotection in Neurological Disorders

Moderate physical exercise has preventive and therapeutic effects on several clinical and neurological disorders (Deslandes et al. 2009; Wichi et al. 2009). It is well known that aging and age-related neurological disorders are accompanied by the deterioration of the hippocampus in aging adult brain. This process induces memory impairment in late adulthood (Raz et al. 2005; Jack et al. 2010; Hillman et al. 2008). Exercise not only delays the deterioration of hippocampus but also enhances learning and memory formation along with improvement in memory retention, which is mediated by a variety of biochemical mechanisms and accompanied by increased cell proliferation and survival in the hippocampus of rodents (van Praag et al. 2005; Erickson et al. 2011). At the molecular level, effects are mediated, in part, by increased production and secretion of BDNF and its receptor tyrosine kinase TrkB (Vaynman et al. 2003, 2004; Wu et al. 2004; Li et al. 2008). In addition, increased blood flow volume to the brain during physical exercise also attenuates cognitive decline via mitigation of cerebrovascular risk, including the contribution of small vessel disease to dementia. In addition, it is becoming increasingly evident that exercise increases the expression of synaptic plasticity genes (Stranahan et al. 2010), gene products such as synapsin I and synaptophysin (Vaynman et al. 2004, 2006), and various neuroplasticity-related transcription factors such as cyclic adenosine monophosphate response element binding and intracellular kinases (Gomez-Pinilla et al. 2008). Exercise also enhances hippocampal dendritic length, dendritic spine complexity, and neurogenesis (Eadie et al. 2005; Farmer et al. 2004). Finally, exercise also stimulates long-term potentiation, which is a primary neurophysiologic substrate in learning (O'Callaghan et al. 2007). Furthermore, exercise not only modulates insulin signaling, neuronal activity, and energy metabolism in neural cells, but also controls glucose metabolism in visceral cells. MAP kinase and Wnt signaling pathways are closely associated with physical activity and learning and memory (Reichardt 2006). Wnt signaling also supports hippocampal plasticity and learning and memory along with neurogenesis and LTP (O'Callaghan et al. 2007). Thus, mounting evidence suggests that exercise not only protects brain tissue from deterioration, but also boosts cognitive and psychological health in older adults. However, participation in exercise among older adults is particularly poor. According to the 2007 National Health Interview Survey from the Centers for Disease Control and Prevention (CDC) Healthy People 2010 Database, only 14 % of adults 65–74 years old and 7 % over 75 years of age have been reported to perform regular exercise (www.cdc.gov/nchs/healthy_people.htm).

### *1.9.1 Effect of Exercise on Alzheimer Disease*

Studies on the effect of exercise in animal models of AD indicate that subjecting transgenic mice (mice expressing the skeletal muscle-specific mutant PS2 gene) to treadmill exercise for 3 months not only results in a reduction of Aβ-42 deposits, but also produces improvement in behavioral function, thereby restoring normal concentrations of total cholesterol, high-density lipoprotein cholesterol, and low-density lipoprotein cholesterol and triglyceride (Cho et al. 2003). Similarly, studies on the effect of exercise on treadmill for 16 weeks in the NSE/APPsw Tg mice indicate that exercise not only decreases levels of Aβ-42 peptides and retards processes associated with apoptotic cell death (the release of cytochrome c, activities of caspase-9, caspase-3, and Bax) but also inhibits the induction of glucose transporter 1 (GLUT1) and BDNF (Um et al. 2008). In addition, exercise provides neuroprotection by inducing the expression of superoxide dismutase-1 (SOD-1), catalase and Bcl-2, heat shock protein-70 (HSP-70), and glucose-regulated protein-78 (GRP-78). Furthermore, exercise training also contributes to the restoration of normal levels of serum total cholesterol, insulin, and glucose. All above-mentioned parameters suppress oxidative stress and excitotoxic injury. Collectively these studies suggest that exercise training represents a practical therapeutic strategy for human subjects suffering from AD (Um et al. 2008).

Balance dysfunction, gait disturbances, and falls are common problems in later stages of AD and dementia compared with older people in general (Manckoundia et al. 2006). Although, in normal older people, exercise reduces falls and improves their mood, but in AD patients exercise induces marginal effects on fall reduction and mood improvement (Cho et al. 2003). Studies on the effect of aerobic exercise in mild to moderate AD patients have been controversial. Some studies have indicated that exercise has no effect on global cognition and quality of life except for depression (Yu et al. 2013). Other studies indicate that exercise improves cognitive functions, such as tasks mediated by the hippocampus, and results in major changes in plasticity in the hippocampus. Interestingly, exercise-induced plasticity is also pronounced in *apo E ε4* carriers which express a risk factor for late-onset AD that may modulate the effect of treatments (Foster et al. 2011). Based on magnetic resonance imaging (MRI) studies, it is suggested that exercise may reduce hypertrophy in the hippocampus and promote the production of BDNF that enhances neurogenesis and plays a key role in positive cognitive effects (Foster et al. 2011).

### *1.9.2 Effect of Exercise on Parkinson Disease*

Studies on the effect of exercise in Parkinsonian animal models have revealed that exercise not only protects dopaminergic neurons from the toxic effects of neurotoxins, but also improves cognition in the toxin treated animals. This effect is mediated by brain neurotrophic factors and neuroplasticity. In these animal models, immobilization has the opposite effect (Ahlskog 2011). Astrocytes are known to contribute to functional recovery/maintenance against brain damage (Farooqui 2010). Studies on the effect of physical exercise on animal model of PD have indicated that physical exercise has no effect on tyrosine hydroxylase immunoreactivity in striatum, but it restores striatal GFAP expression (Dutra et al. 2012). Although the molecular mechanism associated with beneficial effects of exercise is fully understood, it is becoming increasingly important that exercise exerts its effect through various mechanisms such as modulation of neurotransmission, plasticity, neurogenesis, homeostasis, and neurotrophic factors. In humans, exercise also increases serum BDNF, which can cross the blood–brain barrier and induce its neuroprotective effect. PD risk in humans is significantly reduced by midlife exercise, documented in large prospective studies. In addition, PD patients who exercise improve cognitive scores and decrease risk of dementia (Ahlskog 2011).

### *1.9.3 Effect of Exercise on Depression*

Studies on the effect of exercise in animal models of depression indicate chronic stress induces depression-like behavior, reduces blood vessel density, and decreases neurogenesis in hippocampus (Kiuchi et al. 2012). Regular exercise training improves depression-like behavior and increases neurogenesis in the hippocampus (Kiuchi et al. 2012). It is suggested that exercise-mediated antidepressant action may be related to hippocampal microvasculature. In depressed human subjects, exercise reduces depressive symptoms associated with chronic diseases not only through its antidepressant effects (Herring et al. 2012) but also by improving depression by inducing changes in levels of endorphin, serotonin, and norepinephrine levels. In addition it is becoming increasingly evident that endocannabinoid system is closely associated with the etiology of depression. Thus, endocannabinoids

[*N*-arachidonoylglycerol (anandamide, AEA) and 2-arachidonoylglycerol (2-AG)], which are synthesized on demand in the brain, spinal cord, and visceral tissues, are agonist for the cannabinoid CB1 receptor. They not only alter cognitive and emotional behaviors and neurogenesis, but also modulate the levels of BDNF (Gorzalka and Hill 2010). These behaviors and molecules are also influenced by physical exercise, and it is speculated that the endocannabinoid system represents a crucial signaling system mediating the beneficial antidepressant effects of exercise. In addition to its putative antidepressant effects, the endocannabinoids may acutely influence mood through their effects on pain perception (Pertwee 2001) and facilitation of dopamine release in the nucleus accumbens (Maldonado et al. 2006; Cheer et al. 2007). Based on above observations, it is hypothesized that endocannabinoids may be linked with the so-called runners high, an intense but transient positive emotion during exercise (Keeney et al. 2008; Fuss and Gass 2010). The mechanism by which exercise modulates endocannabinoid levels is not fully understood. However, it is known that stress not only stimulates endocannabinoid signaling but also facilitates the release of cortisol (Hill and McEwen 2010). Thus, systemic administration of the glucocorticoid corticosterone results in an increase in the tissue content of AEA within several brain structures (amygdala, hippocampus, hypothalamus) in rats, while 2-AG is less responsive (as it increased only in the hypothalamus). The rapid increase in AEA suggests that glucocorticoids act through a nongenomic pathway (Hill et al. 2010), compatible with acute adaptations to a physiological stressful condition such as intensive physical exercise. It has been hypothesized that increase in cortisol secretion, triggered by long intense exercise, may stimulate AEA production (Hill et al. 2010).

## 1.10 Conclusion

MetS is a cluster of metabolic abnormalities including abdominal obesity linked to an excess of visceral fat, impaired fasting glucose, insulin resistance, hypertension, endothelial cell dysfunction, and dyslipidemia. Among above abnormalities, insulin resistance is the most important parameter. Multiple mechanisms and abnormalities in insulin signaling contribute to the development of insulin resistance. High levels of n-6 FFA and fructose consumption have been correlated with various parameters of the MetS, including hyperuricemia, hyperinsulinemia, hypertension, insulin resistance, leptin resistance, obesity, and dyslipidemia.

Fructose metabolism in the liver increases uric acid, which may be not only responsible for endothelial cell dysfunction, impaired insulin-induced vasodilation, and a consequent failure to increase muscle blood flow after a meal. These processes may lead to muscle insulin resistance. In addition, the metabolism of fructose in liver cells can cause the formation of AGEs and ROS, which can activate NF-κB, causing inflammation-linked insulin resistance. Modifications of the changes in body composition with aging by diet and exercise training could delay the onset of insulin resistance. MetS is promoted by sedentary lifestyle (high-fat diet and lack of exercise) and overnutrition resulting into sustained and moderate positive energy balance leading to increase in visceral fat accumulation and elevated portal delivery of FFA to the liver. At the molecular level, MetS is accompanied not only by the dysregulation in the expression of adipocytokines (TNF-α and IL-6), but also by alterations in levels of leptin, a peptide hormone secreted largely by adipose tissue that controls food intake and energy expenditure (Zurlo et al. 1990). Circulating levels of leptin parallel fat cell stores, which is increased with overfeeding and decreased with starvation. Leptin acts through leptin receptors, which are located in the brain, adipocytes, and endothelial cells. Leptin receptors are engaged signal transduction pathways associated with the STAT3 and phosphoinositide 3-kinase pathways. Leptin modulates TNF-α and IL-6 production and macrophage activation. Both TNF-α and IL-6 alter insulin sensitivity by triggering different key steps in the insulin signaling pathway. The overproduction of TNF-α and IL-6 in adipose tissue plays an important role in the pathogenesis of insulin resistance. These changes modulate immune response, generation of lipid mediators, and inflammation leading to alterations in the hypothalamic "bodyweight/appetite/satiety set point" inducing the initiation and development of the MetS. Neurological diseases are accompanied by excitotoxicity, oxidative stress, and neuroinflammation as common mechanisms of cell death. In addition, neurodegenerative diseases are accompanied by the accumulation of misfolded proteins, mitochondrial and proteasomal dysfunction, loss of synapses, and premature and slow death of certain neuronal populations in brain tissue. The most important risk factors for stroke and neurodegenerative diseases are old age, race/ethnicity, a positive family history, unhealthy lifestyle, and endogenous factors. In addition to the above-mentioned risk factors, potentially modifiable factors such as hypertension, hyperlipidemia, and environmental exposures have been identified as risk factors for neurological disorders. Although molecular mechanism associated with relationship between MetS and neurological disorders is not fully understood, it is proposed that impairment of endothelial cell function, abnormality in essential fatty acid metabolism, and alterations in lipid mediators along with abnormal insulin/leptin signaling may contribute to the relationship between metabolic syndrome and neurological disorders, such as stroke, AD, and depression. An active and stimulating lifestyle in late life as well as an optimal control of vascular and other chronic diseases both at middle age and late life can be two possible intervention strategies to prevent or postpone the onset of metabolic syndrome and neurological disorders.

## References


AbdulWahid Arif I, Ahmad Khan H (2010) Environmental toxins and Parkinson's disease: putative roles of impaired electron transport chain and oxidative stress. Toxicol Ind Health 26:121–128

Adediran OS, Jimoh AK, Ogbera AO (2006) Metabolic syndrome: The pathogenesis of its predictors. Postgrad Doctor Caribbean. 22:35–45

Ahlskog JE (2011) Does vigorous exercise have a neuroprotective effect in Parkinson disease? Neurology 77:288–294

Ahmed N (2005) Advanced glycation endproducts—role in pathology of diabetic complications. Diabetes Res Clin Pract 67:3–21

Alderman M, Aiyer KJ (2004) Uric acid: role in cardiovascular disease and effects of losartan. Curr Med Res Opin 20:369–379

Alper AB Jr, Chen W, Yau L, Srinivasan SR, Berenson GS, Hamm LL (2005) Childhood uric acid predicts adult blood pressure: the Bogalusa Heart Study. Hypertension 45:34–38

Arancio O, Zhang HP, Chen X, Lin C, Trinchese F, Puzzo D, Liu S, Hegde A, Yan SF, Stern A, Luddy JS, Lue LF, Walker DG, Roher A, Buttini M, Mucke L, Li W, Schmidt AM, Kindy M, Hyslop PA, Stern DM, Du Yan SS (2004) RAGE potentiates Aβ-induced perturbation of neuronal function in transgenic mice. EMBO J 23:4096–4105

Austin RL, Rune A, Bouzakri K, Zierath JR, Krook A (2008) SiRNA-mediated reduction of inhibitor of nuclear factor-κB kinase prevents tumor necrosis factor-α-induced insulin resistance in human skeletal muscle. Diabetes 57:2066–2073

Ayyobi AF, Brunzell JD (2003) Lipoprotein distribution in the metabolic syndrome, type 2 diabetes mellitus, and familial combined hyperlipidemia. Am J Cardiol 2003(92):27J–33J

Baker JF, Krishnan E, Chen L, Schumacher HR (2005) Serum uric acid and cardiovascular disease: recent developments, and where do they leave us? Am J Med 118:816–826

Banack SA, Johnson HE, Cheng R, Cox PA (2007) Production of the neurotoxin BMAA by a marine cyanobacterium. Mar Drugs 5:180–196

Basciano H, Federico L, Adeli K (2005) Fructose, insulin resistance, and metabolic dyslipidemia. Nutr Metab (Lond) 2:5

Beeri MS, Silverman JM, Davis KL, Marin D, Grossman HZ, Schmeidler J, Purohit DP, Perl DP, Davidson M, Mohs RC, Haroutunian V (2005) Type 2 diabetes is negatively associated with Alzheimer's disease neuropathology. J Gerontol A Biol Sci Med Sci 60:471–475

Bernal-Pacheco O, Román GC (2007) Environmental vascular risk factors: new perspectives for stroke prevention. J Neurol Sci 262:60–70

Biessels GJ, van der Heide LP, Kamal A, Bleys RL, Gispen WH (2002) Ageing and diabetes: implications for brain function. Eur J Pharmacol 441:1–14

Bradley WG, Mash DC (2009) Beyond Guam: the cyanobacteria/BMAA hypothesis of the cause of ALS and other neurodegenerative diseases. Amyotroph Lateral Scler 10(Suppl 2):7–20

Brown TP, Rumsby PC, Capleton AC, Rushton L, Levy LS (2006) Pesticides and Parkinson's disease—is there a link? Environ Health Perspect 114:156–164

Brownlee M (1995) Advanced glycosylation in diabetes and aging. Annu Rev Med 46:223–234

Brownson DM, Mabry TJ, Leslie SW (2002) The cycad neurotoxic amino acid, β-*N*-methylamino-l-alanine (BMAA), elevates intracellular calcium levels in dissociated rat brain cells. J Ethnopharmacol 82:159–167

Cao D, Lu H, Lewis TL, Li L (2007) Intake of sucrose-sweetened water induces insulin resistance and exacerbates memory deficits and amyloidosis in a transgenic mouse model of Alzheimer disease. J Biol Chem 282:36275–36282

Carro E, Torres-Aleman I (2004) The role of insulin and insulin-like growth factor I in the molecular and cellular mechanisms underlying the pathology of Alzheimer's disease. Eur J Pharmacol 490:127–133

Carro E, Trejo JL, Busiguina S, Torres-Aleman I (2001) Circulating insulin-like growth factor I mediates the protective effects of physical exercise. J Neurosci 21:5678–5684

Carter SL, Rennie CD, Hamilton SJ, Tarnopolsky MA (2001) Changes in skeletal muscle in males and females following endurance training. Can J Physiol Pharmacol 79:386–392

Caulin-Glaser T, Falko J, Hindman L, La Londe M, Snow R (2005) Cardiac rehabilitation is associated with an improvement in C-reactive protein levels in both men and women with cardiovascular disease. J Cardiopulm Rehabil 25:332–336

Cha SH, Wolfgang M, Tokutake Y, Chohnan S, Lane MD (2008) Differential effects of central fructose and glucose on hypothalamic malonyl-CoA and food intake. Proc Natl Acad Sci USA 105:16871–16875

Chao MV, Bothwell M (2002) Neurotrophins: to cleave or not to cleave. Neuron 33:9–12

Cheer JF, Wassum KM, Sombers LA, Heien ML, Ariansen JL, Aragona BJ, Phillips PE, Wightman RM (2007) Phasic dopamine release evoked by abused substances requires cannabinoid receptor activation. J Neurosci 27:791–795

Chen MJ, Russo-Neustdt AA (2007) Running exercise- and antidepressant-induced increases in growth and survival-associated signaling molecules are IGF-dependent. Growth Factors 25:118–131

Chiu AS, Gehringer MM, Welch JH, Neilan BA (2011) Does α-amino-β-methylaminopropionic acid (BMAA) play a role in neurodegeneration? Int J Environ Res Public Health 8:3728–3746

Cho JY, Hwang DY, Kang TS, Shin DH, Hwang JH, Lim CH, Lee SH, Lim HJ, Min SH, Seo SJ, Song YS, Nam KT, Lee KS, Cho JS, Kim YK (2003) Use of NSE/PS2m-transgenic mice in the study of the protective effect of exercise on Alzheimer's disease. J Sports Sci 21:943–951

Combaret L, Dardevet D, Béchet D, Taillandier D, Mosoni L, Attaix D (2009) Skeletal muscle proteolysis in aging. Curr Opin Clin Nutr Metab Care 12:37–41

Conboy IM, Rando TA (2005) Aging, stem cells and tissue regeneration: lessons from muscle. Cell Cycle 4:407–410

Convit A (2005) Links between cognitive impairment in insulin resistance: an explanatory model. Neurobiol Aging 26(Suppl 1):31–35

Copani A, Canonico PL, Catania MV, Aronica E, Bruno V, Ratti E, van Amsterdam FT, Gaviraghi G, Nicoletti F (1991) Interaction between beta-*N*-methylamino-L-alanine and excitatory amino acid receptors in brain slices and neuronal cultures. Brain Res 558:79–86

Cordain L, Eaton SB, Sebastian A, Mann N, Lindeberg S, Watkins BA, O'Keefe JH, Brand-Miller J (2005) Origins and evolution of the Western diet: health implications for the 21st century. Am J Clin Nutr 81:341–354

Cucchiaroni ML, Viscomi MT, Bernardi G, Molinari M, Guatteo E, Mercuri NB (2010) Metabotropic glutamate receptor 1 mediates the electrophysiological and toxic actions of the cycad derivative {beta}-*N*-Methylamino-L-alanine on substantia nigra pars compacta DAergic neurons. J Neurosci 30:5176–5188

Curry DL (1989) Effects of mannose and fructose on the synthesis and secretion of insulin. Pancreas 4:2–9

De Alvaro C, Teruel T, Hernandez R, Lorenzo M (2004) Tumor necrosis factor α produces insulin resistance in skeletal muscle by activation of inhibitor κB kinase in a p38 MAPK-dependent manner. J Biol Chem 279:17070–17078

Deslandes A, Moraes H, Ferreira C, Veiga H, Silveira H, Mouta R, Pompeu FA, Coutinho ES, Laks J (2009) Exercise and mental health: many reasons to move. Neuropsychobiology 59:191–198

Ding Q, Vaynman S, Akhavan M, Ying Z, Gomez-Pinilla F (2006a) Insulin-like growth factor I interfaces with brain-derived neurotrophic factor-mediated synaptic plasticity to modulate aspects of exercise-induced cognitive function. Neuroscience 140:823–833

Ding Q, Vaynman S, Souda P, Whitelegge JP, Gomez-Pinilla F (2006b) Exercise affects energy metabolism and neural plasticity-related proteins in the hippocampus as revealed by proteomic analysis. Eur J Neurosci 24:1265–1276

Dou JT, Chen M, Dufour F, Alkon DL, Zhao WQ (2005) Insulin receptor signaling in long-term memory consolidation following spatial learning. Learn Mem 12:646–655

Dutra MF, Jaeger M, Ilha J, Kalil-Gaspar PI, Marcuzzo S, Achaval M (2012) Exercise improves motor deficits and alters striatal GFAP expression in a 6-OHDA-induced rat model of Parkinson's disease. Neurol Sci 33(5):1137–1144

Eaton SB, Cordain L (1997) Evolutionary aspects of diet: old genes, new fuels. World Rev Nutr Diet 81:26–37

Eadie BD, Redila VA, Christie BR (2005) Voluntary exercise alters the cytoarchitecture of the adult dentate gyrus by increasing cellular proliferation, dendritic complexity, and spine density. J Comp Neurol 486:39–47

Erickson KI, Voss MW, Prakash RS, Basak C, Szabo A, Chaddock L, Kim JS, Heo S, Alves H, White SM, Wojcicki TR, Mailey E, Vieira VJ, Martin SA, Pence BD, Woods JA, McAuley E, Kramer AF (2011) Exercise training increases size of hippocampus and improves memory. Proc Natl Acad Sci USA 108:3017–3022

Fages C, Nolo R, Huttunen HJ, Eskelinen E, Rauvala H (2000) Regulation of cell migration by amphoterin. J Cell Sci 113:611–620

Farmer J, Zhao X, van Praag H, Wodtke K, Gage FH, Christie BR (2004) Effects of voluntary exercise on synaptic plasticity and gene expression in the dentate gyrus of adult male Sprague-Dawley rats in vivo. Neuroscience 124:71–79

Farooqui AA (2009) Beneficial effects of fish oil on human brain. Springer, New York

Farooqui AA (2010) Neurochemical aspects of neurodegenerative and neurotraumatic diseases. Springer, New York

Farooqui AA (2011) Lipid mediators and their metabolism in the brain. Springer, New York

Farooqui AA (2012) Phytochemicals, signal transduction, and neurological disorders. Springer, New York

Farooqui T, Farooqui AA (2009) Aging: an important factor for the pathogenesis of neurodegenerative diseases. Mech Ageing Dev 130:203–215

Farooqui AA, Horrocks LA, Farooqui T (2007) Modulation of inflammation in brain: a matter of fat. J Neurochem 101:577–599

Farooqui AA, Farooqui T, Panza F, Frisardi V (2012) Metabolic syndrome as a risk factor for neurological disorders. Cell Mol Life Sci 69:741–762

Fayard E, Tintignac LA, Baudry A, Hemmings BA (2005) Protein kinase B/Akt at a glance. J Cell Sci 118:5675–5678

Fernström M, Tonkonogi M, Sahlin K (2004) Effects of acute and chronic endurance exercise on mitochondrial uncoupling in human skeletal muscle. J Physiol 554:755–763

Finucane FM, Luan J, Wareham NJ, Sharp SJ, O'Rahilly S, Balkau B, Flyvbjerg A, Walker M, Hojlund K, Nolan JJ, Savage DB (2009) Correlation of the leptin:adiponectin ratio with measures of insulin resistance in nondiabetic individuals. Diabetologia 52:2345–2349

Ford ES, Giles WH, Dietz WH (2002) Prevalence of the metabolic syndrome among US adults: findings from the third National Health and Nutrition Examination Survey. JAMA 287:356–359

Foster PP, Rosenblatt KP, Kuljiš RO (2011) Exercise-induced cognitive plasticity, implications for mild cognitive impairment and Alzheimer's disease. Front Neurol 2:28

Froesch ER (1972) Fructose metabolism in adipose tissue. Acta Med Scand Suppl 542:37–46

Fuss J, Gass P (2010) Endocannabinoids and voluntary activity in mice: runner's high and long-term consequences in emotional behaviors. Exp Neurol 224:103–105

Galimanis A, Mono ML, Arnold M, Nedeltchev K, Mattle HP (2009) Lifestyle and stroke risk: a review. Curr Opin Neurol 22:60–68

Ghosh S, Novak EM, Innis SM (2007) Cardiac proinflammatory pathways are altered with different dietary n-6 linoleic to n-3 α-linolenic acid ratios in normal, fat-fed pigs. Am J Physiol Heart Circ Physiol 293:H2919–H2927

Gomez-Pinilla F (2011) The combined effects of exercise and foods in preventing neurological and cognitive disorders. Prev Med 52(Suppl 1):S75–S80

Gomez-Pinilla F, Vaynman S, Ying Z (2008) Brain-derived neurotrophic factor functions as a metabotrophin to mediate the effects of exercise on cognition. Eur J Neurosci 28:2278–2287

Gorzalka BB, Hill MN (2010) Putative role of endocannabinoid signaling in the etiology of depression and actions of antidepressants. Prog Neuropsychopharmacol Biol Psychiatry 35:1575–1585

Graeber MB, Moran LB (2002) Mechanisms of cell death in neurodegenerative diseases: fashion, fiction, and facts. Brain Pathol 12:385–390

Granholm AC, Bimonte-Nelson HA, Moore AB, Nelson ME, Freeman LR, Sambamurti K (2008) Effects of a saturated fat and high cholesterol diet on memory and hippocampal morphology in the middle-aged rat. J Alzheimers Dis 14:133–145

Grundy SM, Cleeman JI, Daniels SR, Donato KA, Eckel RH, Franklin BA, Gordon DJ, Krauss RM, Savage PJ, Smith SC Jr, Spertus JA, Costa F, American Heart Association; National Heart, Lung, and Blood Institute (2005) Diagnosis and management of the metabolic syndrome: an American Heart Association/National Heart, Lung, and Blood Institute Scientific Statement. Circulation 112:2735–2752

Hallfrisch J (1990) Metabolic effects of dietary fructose. FASEB J 4:2652–2660

Handschin C, Spiegelman BM (2006) PGC-1 coactivators, energy homeostasis, and metabolism. Endocr Rev 27:728–735

Hansson GK (2005) Inflammation, atherosclerosis, and coronary artery disease. N Engl J Med 352:1685–1695

Havel PJ (2002) Control of energy homeostasis and insulin action by adipocyte hormones: leptin, acylation stimulating protein, and adiponectin. Curr Opin Lipidol 13:51–59

Havel PJ (2005) Dietary fructose: implications for dysregulation of energy homeostasis and lipid/carbohydrate metabolism. Nutr Rev 63:133–157

Hawley JA, Lessard SJ (2008) Exercise training-induced improvements in insulin action. Acta Physiol (Oxf) 192:127–135

Hegarty BD, Furler SM, Ye J, Cooney GJ, Kraegen EW (2003) The role of intramuscular lipid in insulin resistance. Acta Physiol Scand 178:373–383

Heinrich PC, Castell JV, Andus T (1990) Interleukin-6 and the acute phase response. Biochem J 265:621–636

Herring MP, Puetz TW, O'Connor PJ, Dishman RK (2012) Effect of exercise training on depressive symptoms among patients with a chronic illness: a systematic review and meta-analysis of randomized controlled trials. Arch Intern Med 172:101–111

Hill MN, McEwen BS (2010) Involvement of the endocannabinoid system in the neurobehavioural effects of stress and glucocorticoids. Prog Neuropsychopharmacol Biol Psychiatry 34:791–797

Hill MN, Karatsoreos IN, Hillard CJ, McEwen BS (2010) Rapid elevations in limbic endocannabinoid content by glucocorticoid hormones in vivo. Psychoneuroendocrinology 35:1333–1338

Hillman CH, Erickson KI, Kramer AF (2008) Be smart, exercise your heart: exercise effects on brain and cognition. Nat Rev Neurosci 9:58–65

Hiza HAB, Bente L (2007) Nutrient content of the U.S. food supply, 1909–2004: a summary report. Home Economics Research Report No. 57. Center for Nutrition Policy and Promotion, U.S. Department of Agriculture, Alexandria, VA

Horch HW (2004) Local effects of BDNF on dendritic growth. Rev Neurosci 15:117–129

Houmard JA, Shinebarger MH, Dolan PL, Leggettfrazier N, Bruner RK, Mccammon MR, Israel RG, Dohm GL (1993) Exercise training increases GLUT-4 protein concentration in previously sedentary middle-aged men. Am J Physiol 264:E896–E901

Hoyer S (2004) Glucose metabolism and insulin receptor signal transduction in Alzheimer disease. Eur J Pharmacol 490:115–125

Hwu CM, Liou TL, Hsiao LC, Lin MW (2009) Prehypertension is associated with insulin resistance. Q J Med 102:705–711

Ibanez CF (2002) Jekyll-Hyde neurotrophins: the story of proNGF. Trends Neurosci 25:284–286

Ince PG, Codd GA (2005) Return of the cycad hypothesis—does the amyotrophic lateral sclerosis/parkinsonism dementia complex (ALS/PDC) of Guam have new implications for global health? Neuropathol Appl Neurobiol 31:345–353

Jack CR Jr, Wiste HJ, Vemuri P, Weigand SD, Senjem ML, Zeng G, Bernstein MA, Gunter JL, Pankratz VS, Aisen PS, Weiner MW, Petersen RC, Shaw LM, Trojanowski JQ, Knopman DS, Alzheimer's Disease Neuroimaging Initiative (2010) Brain beta-amyloid measures and magnetic resonance imaging atrophy both predict time-to-progression from mild cognitive impairment to Alzheimer's disease. Brain 133:3336–3348

Jew S, AbuMweis SS, Jones PJ (2009) Evolution of the human diet: linking our ancestral diet to modern functional foods as a means of chronic disease prevention. J Med Food 12:925–934

Jonasson S, Eriksson J, Berntzon L, Spacil Z, Ilag LL, Ronnevi LO, Rasmussen U, Bergman B (2010) Transfer of a cyanobacterial neurotoxin within a temperate aquatic ecosystem suggests pathways for human exposure. Proc Natl Acad Sci USA 107:9252–9257

Jovanovic JN, Benfenati F, Siow YL, Sihra TS, Sanghera TS, Pelech SL, Greengard P, Czernik AJ (1996) Neurotrophins stimulate phosphorylation of synapsin I by MAP kinase and regulate synapsin I actin interactions. Proc Natl Acad Sci USA 93:3679–3683

Kamel F, Hoppin JA (2004) Association of pesticide exposure with neurologic dysfunction and disease. Environ Health Perspect 112:950–958

Kanoski SE, Meisel RL, Mullins AJ, Davidson TL (2007) The effects of energy-rich diets on discrimination reversal learning and on BDNF in the hippocampus and prefrontal cortex of the rat. Behav Brain Res 182:57–66

Kanuri G, Spruss A, Wagnerberger S, Bischoff SC, Bergheim I (2011) Role of tumor necrosis factor-α (TNF-α) in the onset of fructose- induced nonalcoholic fatty liver disease in mice. J Nutr Biochem 22:527–534

Keeney BK, Raichlen DA, Meek TH, Wijeratne RS, Middleton KM, Gerdeman GL, Garland T Jr (2008) Differential response to a selective cannabinoid receptor antagonist (SR:141716 rimonabant) in female mice from lines selectively bred for high voluntary wheel-running behaviour. Behav Pharmacol 19:812–820

Keifer MC, Firestone J (2007) Neurotoxicity of pesticides. J Agromedicine 12:17–25

Kenyon CJ (2010) The genetics of ageing. Nature 464:504–512

Kim DS, Jeong SK, Kim HR, Kim DS, Chae SW, Chae HJ (2007) Effects of triglyceride on ER stress and insulin resistance. Biochem Biophys Res Commun 363:140–145

Kissebah AH, Vydelingum N, Murray M, Evans DJ, Hartz AJ, Kalkhoff RK, Adams PW (1982) Relation of body fat distribution to metabolic complications of obesity. J Clin Endocrinol Metab 54:254–260

Kiuchi T, Lee H, Mikami T (2012) Regular exercise cures depression-like behavior via VEGF-Flk-1 signaling in chronically stressed mice. Neuroscience 207:208–217

Kristiansen OP, Mandrup-Poulsen T (2005) Interleukin-6 and diabetes: the good, the bad, or the indifferent? Diabetes 54(Suppl 2):S114–S115

Kurth T, Moore SC, Gaziano JM, Kase CS, Stampfer MJ, Berger K, Buring JE (2006) Healthy lifestyle and the risk of stroke in women. Arch Intern Med 166:1403–1409

Lakka TA, Laaksonen DE (2007) Physical activity in prevention and treatment of the metabolic syndrome. Appl Physiol Nutr Metab 32:76–88

Lakka HM, Laaksonen DE, Lakka TA, Niskanen LK, Kumpusalo E, Tuomilehto J, Salonen JT (2002) The metabolic syndrome and total and cardiovascular disease mortality in middle-aged men. JAMA 288:2709–2716

Lane MD, Cha SH (2009) Effect of glucose and fructose on food intake via malonyl-CoA signaling in the brain. Biochem Biophys Res Commun 382:1–5

Larsson L, Ansved T (1995) Effects of ageing on the motor unit. Prog Neurobiol 45:397–458

Levitt NS, Lambert EV, Woods D, Hales CN, Andrew R, Seckl JR (2000) Impaired glucose tolerance and elevated blood pressure in low birth weight, nonobese, young south African adults: early programming of cortisol axis. J Clin Endocrinol Metab 85:4611–4618

Li ZG, Zhang W, Sima AA (2007) Alzheimer-like changes in rat models of spontaneous diabetes. Diabetes 56:1817–1824

Li Y, Luikart BW, Birnbaum S, Chen J, Kwon CH, Kernie SG, Bassel-Duby R, Parada LF (2008) TrkB regulates hippocampal neurogenesis and governs sensitivity to antidepressive treatment. Neuron 59:399–412

Libby P, Ridker PM, Maseri A (2002) Inflammation and atherosclerosis. Circulation 105:1135–1143

Liu X, Rush T, Zapata J, Lobner D (2009) [beta]-*N*-methylamino-L-alanine induces oxidative stress and glutamate release through action on system $Xc^-$. Exp Neurol 217:429–433

Lobner D, Piana PMT, Salous AK, Peoples RW (2007) [beta]-*N*-methylamino-L-alanine enhances neurotoxicity through multiple mechanisms. Neurobiol Dis 25:360–366

Luchsinger JA, Mayeux R (2004) Dietary factors and Alzheimer's disease. Lancet Neurol 3:579–587

Luoma PV (2010) Gene activation regresses atherosclerosis, promotes health, and enhances longevity. Lipids Health Dis 9:67

Maldonado R, Valverde O, Berrendero F (2006) Involvement of the endocannabinoid system in drug addiction. Trends Neurosci 29:225–232

Manckoundia P, Pfitzenmeyer P, d'Athis P, Dubost V, Mourey F (2006) Impact of cognitive task on the posture of elderly subjects with Alzheimer's disease compared to healthy elderly subjects. Mov Disord 21:236–241

Matsuoka Y, Rakonczay Z, Giacobini E, Naritoku D (1993) L-beta-methylaminoalanine-induced behavioral changes in rats. Pharmacol Biochem Behav 44:727–734

Mayes PA (1993) Intermediary metabolism of fructose. Am J Clin Nutr 58:S754–S765

Melnik BC (2012) Leucine signaling in the pathogenesis of type 2 diabetes and obesity. World J Diabetes 3:38–53

Meneghini V, Francese MT, Carraro L, Grilli M (2010) A novel role for the receptor for advanced glycation end-products in neural progenitor cells derived from adult subventricular zone. Mol Cell Neurosci 45:139–150

Messier C (2005) Impact of impaired glucose tolerance and type 2 diabetes on cognitive aging. Neurobiol Aging 26(Suppl 1):26–30

Metcalf JS, Banack SA, Lindsay J, Morrison LF, Cox PA, Codd GA (2008) Co-occurrence of beta-N-methylamino-L-alanine, a neurotoxic amino acid with other cyanobacterial toxins in British waterbodies, 1990–2004. Environ Microbiol 10:702–708

Miller CC, Martin RJ, Whitney ML, Edwards GL (2002) Intracerebroventricular injection of fructose stimulates feeding in rats. Nutr Neurosci 5:359–362

Miyata T, Hori O, Zhang J, Yan SD, Ferran L, Iida Y, Schmidt AM (1996) RAGE mediates the interaction of AGE-beta-2-microglobulin with human mononuclear phagocytes via an oxidant-sensitive pathway: implications for the pathogenesis of dialysis-related amyloidosis. J Clin Invest 98:1088–1094

Molteni R, Barnard RJ, Ying Z, Roberts CK, Gomez-Pinilla F (2002a) A high-fat, refined sugar diet reduces hippocampal brain-derived neurotrophic factor, neuronal plasticity, and learning. Neuroscience 112:803–814

Molteni R, Ying Z, Gomez-Pinilla F (2002b) Differential effects of acute and chronic exercise on plasticity related genes in the rat hippocampus revealed by microarray. Eur J Neurosci 16:1107–1116

Monteggia LM, Barrot M, Powell CM, Berton O, Galanis V, Gemelli T, Meuth S, Nagy A, Greene RW, Nestler EJ (2004) Essential role of brain-derived neurotrophic factor in adult hippocampal function. Proc Natl Acad Sci USA 101:10827–10832

Morton JP, Kayani AC, McArdle A, Drust B (2009) The exercise-induced stress response of skeletal muscle, with specific emphasis on humans. Sports Med 39:643–662

Muoio DM, Newgard CB (2008) Molecular and metabolic mechanisms of insulin resistance and β-cell failure in type 2 diabetes. Nat Rev Mol Cell Biol 9:193–205

Murch SJ, Cox PA, Banack SA (2004) A mechanism for slow release of biomagnified cyanobacterial neurotoxins and neurodegenerative disease in Guam. Proc Natl Acad Sci USA 101:12228–12231

Nagai Y, Yonemitsu S, Erion DM, Iwasaki T, Stark R, Weismann D, Dong J, Zhang D, Jurczak MJ, Löffler MG, Cresswell J, Yu XX, Murray SF, Bhanot S, Monia BP, Bogan JS, Samuel V, Shulman GI (2009) The role of peroxisome proliferator-activated receptor gamma coactivator-1 beta in the pathogenesis of fructose-induced insulin resistance. Cell Metab 9:252–264

Nakagawa T, Tuttle KR, Short RA, Johnson RJ (2005) Hypothesis: fructose-induced hyperuricemia as a causal mechanism for the epidemic of the metabolic syndrome. Nat Clin Pract Nephrol 1:80–86

Nakagawa T, Hu H, Zharikov S, Tuttle KR, Short RA, Glushakova O, Ouyang X, Feig DI, Block ER, Herrera-Acosta J, Patel JM, Johnson RJ (2006) A causal role for uric acid in fructose-induced metabolic syndrome. Am J Physiol Renal Physiol 290:F625–F631

Neeper M, Schmidt AM, Brett J, Yan SD, Wang F, Pan YC, Elliston K, Stern D, Shaw A (1992) Cloning and expression of a cell surface receptor for advanced glycosylation end products of proteins. J Biol Chem 267:14998–15004

Nisoli E, Clementi E, Carruba MO, Moncada S (2007) Defective mitochondrial biogenesis: a hallmark of the high cardiovascular risk in the metabolic syndrome? Circ Res 100:795–806
O'Callaghan RM, Ohle R, Kelly AM (2007) The effects of forced exercise on hippocampal plasticity in the rat: a comparison of LTP, spatial- and non-spatial learning. Behav Brain Res 176:362–366
Parillo M, Riccardi G (2004) Diet composition and the risk of type 2 diabetes: epidemiological and clinical evidence. Br J Nutr 92:7–19
Park CR (2001) Cognitive effects of insulin in the central nervous system. Neurosci Biobehav Rev 25:311–323
Pasalic D, Marinkovic N, Feher-Turkovic L (2012) Uric acid as one of the important factors in multifactorial disorders—facts and controversies. Biochem Med (Zagreb) 22:63–75
Pasinetti GM, Eberstein JA (2008) Metabolic syndrome and the role of dietary lifestyles in Alzheimer's disease. J Neurochem 106:1503–1514
Pelleymounter MA, Cullen MJ, Baker MB, Gollub M, Wellman C (1996) The effects of intrahippocampal BDNF and NGF on spatial learning in aged Long Evans rats. Mol Chem Neuropathol 29:211–226
Pertwee RG (2001) Cannabinoid receptors and pain. Prog Neurobiol 63:569–611
Rahman A, Akterin S, Flores-Morales A, Crisby M, Kivipelto M, Schultzberg M, Cedazo-Mínguez A (2005) High cholesterol diet induces tau hyperphosphorylation in apolipoprotein E deficient mice. FEBS Lett 579:6411–6416
Rakonczay Z, Matsuoka Y, Giacobini E (1991) Effects of L-beta-*N*-methylamino-L-alanine (L-BMAA) on the cortical cholinergic and glutamatergic systems of the rat. J Neurosci Res 29:121–126
Ramasamy R, Vannucci SJ, Yan SS, Herold K, Yan SF, Schmidt AM (2005) Advanced glycation end products and RAGE: a common thread in aging, diabetes, neurodegeneration, and inflammation. Glycobiology 15:16R–28R
Raz N, Lindenberger U, Rodrigue KM, Kennedy KM, Head D, Williamson A, Dahle C, Gerstorf D, Acker JD (2005) Regional brain changes in aging healthy adults: general trends, individual differences and modifiers. Cereb Cortex 15:1676–1689
Reichardt LF (2006) Neurotrophin-regulated signalling pathways. Philos Trans R Soc Lond B Biol Sci 361:1545–1564
Ren J, Pulakat L, Whaley-Connell A, Sowers JR (2010) Mitochondrial biogenesis in the metabolic syndrome and cardiovascular disease. J Mol Med (Berl) 88:993–1001
Ridderstråle M, Groop L (2009) Genetic dissection of type 2 diabetes. Mol Cell Endocrinol 297:10–17
Rosch H, Schweigreiter R, Bonhoeffer T, Barde YA, Korte M (2005) The neurotrophin receptor p75NTR modulates long-term depression and regulates the expression of AMPA receptor subunits in the hippocampus. Proc Natl Acad Sci USA 102:7362–7367
Sakuma K, Yamaguchi A (2010) The functional role of calcineurin in hypertrophy, regeneration, and disorders of skeletal muscle. J Biomed Biotechnol. 2010:721219
Schmidt A-M, Yan S-D, Stern DM (1995) The dark side of glucose. Nat Med 1:1002–1004
Schmidt AM, Yan SD, Yan SF, Stern DM (2001) The multiligand receptor RAGE as a progression factor amplifying immune and inflammatory responses. J Clin Invest 108:949–955
Shapiro A, Mu W, Roncal C, Cheng KY, Johnson RJ, Scarpace PJ (2008) Fructose-induced leptin resistance exacerbates weight gain in response to subsequent high-fat feeding. Am J Physiol Regul Integr Comp Physiol 295:R1370–R1375
Shaya GE (2012) Contributions of increased dietary linoleic acid and fructose to the metabolic syndrome. J Obes Weight Loss Ther 2:114
Short KR (2009) Introduction to symposium proceedings: the emerging interplay among muscle mitochondrial function, nutrition and disease. Am J Clin Nutr 89:453S–454S
Shulman GI (2002) Cellular mechanisms of insulin resistance. J Clin Invest 106:171–176
Simmons RK, Alberti KG, Gale EA, Colagiuri S, Tuomilehto J, Qiao Q, Ramachandran A, Tajima N, Brajkovich Mirchov I, Ben-Nakhi A, Reaven G, Hama Sambo B, Mendis S, Roglic G

(2010) The metabolic syndrome: useful concept or clinical tool? Report of a WHO Expert Consultation. Diabetologia 53:600–605
Simopoulos AP (2002) The importance of the ratio of omega-6/omega-3 essential fatty acids. Biomed Pharmacother 56:365–379
Simopoulos AP (2008) The importance of the omega-6/omega-3 fatty acid ratio in cardiovascular disease and other chronic diseases. Exp Biol Med (Maywood) 233:674–688
Simopoulos AP (2009) Evolutionary aspects of the dietary omega-6:omega-3 fatty acid ratio: medical implications. World Rev Nutr Diet 100:1–21
Song Y, Manson JE, Buring JE, Liu S (2004) A prospective study of red meat consumption and type 2 diabetes in middle-aged and elderly women: the women's health study. Diabetes Care 27:2108–2115
Soppet D, Escandon E, Maragos J, Middlemas DS, Reid SW, Blair J, Burton LE, Stanton BR, Kaplan DR, Hunter T, Nikolics K, Parada LF (1991) The neurotrophic factors brain-derived neurotrophic factor and neurotrophin-3 are ligands for the trkB tyrosine kinase receptor. Cell 65:895–903
Stanhope KL, Havel PJ (2008) Fructose consumption: potential mechanisms for its effects to increase visceral adiposity and induce dyslipidemia and insulin resistance. Curr Opin Lipidol 19:16–24
Stranahan AM, Norman ED, Lee K, Cutler RG, Telljohann RS, Egan JM, Mattson MP (2008) Diet-induced insulin resistance impairs hippocampal synaptic plasticity and cognition in middle-aged rats. Hippocampus 18:1085–1088
Stranahan AM, Lee K, Becker KG, Zhang Y, Mandsley S, Martin B, Cutler RG, Mattson MP (2010) Hippocampal gene expression patterns underlying the enhancement of memory by running in aged mice. Neurobiol Aging 31:1937–1949
Teff K, Elliot S, Tschoep MR (2002) Consuming high fructose meals reduces 24 hour plasma insulin and leptin concentrations, does not suppress circulating ghrelin, and increases postprandial and fasting triglycerides in women. Diabetes 52:A408
Teff KL, Elliott SS, Tschop M, Kieffer TJ, Rader D, Heiman M, Townsend RR, Keim NL, D'Alessio D, Havel PJ (2004) Dietary fructose reduces circulating insulin and leptin, attenuates postprandial suppression of ghrelin, and increases triglycerides in women. J Clin Endocrinol Metab 89:2963–2972
Teng HK, Teng KK, Lee R, Wright S, Tevar S, Almeida RD, Kermani P, Torkin R, Chen ZY, Lee FS, Kraemer RT, Nykjaer A, Hempstead BL (2005) ProBDNF induces neuronal apoptosis via activation of a receptor complex of p75NTR and sortilin. J Neurosci 25:5455–5463
Thirumangalakudi L, Prakasam A, Zhang R, Bimonte-Nelson H, Sambamurti K, Kindy MS, Bhat NR (2008) High cholesterol-induced neuroinflammation and amyloid precursor protein processing correlate with loss of working memory in mice. J Neurochem 106:475–485
Trejo JL, Carro E, Nunez A, Torres-Aleman I (2002) Sedentary life impairs self-reparative processes in the brain: the role of serum insulin-like growth factor-I. Rev Neurosci 13:365–374
Tsuchida A, Nonomura T, Nakagawa T, Itakura Y, Ono-Kishino M, Yamanaka M, Sugaru E, Taiji M, Noguchi H (2002) Brain-derived neurotrophic factor ameliorates lipid metabolism in diabetic mice. Diabetes Obes Metab 4:262–269
Tyler WJ, Alonso M, Bramham CR, Pozzo-Miller LD (2002) From acquisition to consolidation: on the role of brain-derived neurotrophic factor signaling in hippocampal-dependent learning. Learn Mem 9:224–237
Um HS, Kang EB, Leem YH, Cho IH, Yang CH, Chae KR, Hwang DY, Cho JY (2008) Exercise training acts as a therapeutic strategy for reduction of the pathogenic phenotypes for Alzheimer's disease in an NSE/APPsw-transgenic model. J Mol Med 22:529–539
van Praag H, Shubert T, Zhao C, Gage FH (2005) Exercise enhances learning and hippocampal neurogenesis in aged mice. J Neurosci 25:8680–8685
Vaynman S, Ying Z, Gomez-Pinilla F (2003) Interplay between BDNF and signal transduction modulators in the regulation of the effects of exercise on synaptic-plasticity. Neuroscience 122:647–657

Vaynman S, Ying Z, Gomez-Pinilla F (2004) Hippocampal BDNF mediates the efficacy of exercise on synaptic plasticity and cognition. Eur J Neurosci 20:2580–2590

Vaynman S, Ying Z, Wu A, Gomez-Pinilla F (2006) Coupling energy metabolism with a mechanism to support brain-derived neurotrophic factor-mediated synaptic plasticity. Neuroscience 139:1221–1234

Villegas R, Shu XO, Gao YT, Yang G, Cai H, Li H, Zheng W (2006) The association of meat intake and the risk of type 2 diabetes may be modified by body weight. Int J Med Sci 3:152–159

Walther C, Möbius-Winkler S, Linke A, Bruegel M, Thiery J, Schuler G, Halbrecht R (2008) Regular exercise training compared with percutaneous intervention leads to a reduction of inflammatory markers and cardiovascular events in patients with coronary artery disease. Eur J Cardiovasc Prev Rehabil 15:107–112

Weiss JH, Choi DW (1988) Beta-*N*-methylamino-L-alanine neurotoxicity: requirement for bicarbonate as a cofactor. Science 241:973–975

Wichi RB, Angelis K, Jones L, Irigoyen MC (2009) A brief review of chronic exercise intervention to prevent autonomic nervous system changes during the aging process. Clinics 64:253–258

Wrighten SA, Piroli GG, Grillo CA, Reagan LP (2009) A look inside the diabetic brain: contributors to diabetes-induced brain aging. Biochim Biophys Acta 1792:444–453

Wu D (2005) Neuroprotection in experimental stroke with targeted neurotrophins. NeuroRx 2:120–128

Wu A, Ying Z, Gomez-Pinilla F (2004) The interplay between oxidative stress and brain-derived neurotrophic factor modulates the outcome of a saturated fat diet on synaptic plasticity and cognition. Eur J Neurosci 18:1699–1707

Xing Z, Gauldie J, Cox G, Baumann H, Jordana M, Lei XF, Achong MK (1998) IL-6 is an antiinflammatory cytokine required for controlling local or systemic acute inflammatory responses. J Clin Invest 101:311–320

Yu F, Nelson NW, Savik K, Wyman JF, Dysken M, Bonas UG (2013) Affecting cognition and quality of life via aerobic exercise in Alzheimer's disease. West J Nurs Res 35(1):24–38

Yudkin JS, Kumari M, Humphries SE, Mohamed-Ali V (2000) Inflammation, obesity, stress and coronary heart disease: is interleukin-6 the link? Atherosclerosis 148:209–214

Zurlo F, Larson K, Bogardus C, Ravussin E (1990) Skeletal muscle metabolism is a major determinant of resting energy expenditure. J Clin Invest 86:1423–1427

# Chapter 2
# Glucose- and Fructose-Induced Toxicity in the Liver and Brain

## 2.1 Introduction

Fructose is a monosaccharide found in many fruits and honey. In sucrose (table sugar), fructose is linked with glucose through an alpha 1–4 glycoside bond. Fructose and glucose have the same chemical formula ($C_6H_{12}O_6$). However, fructose is a five-carbon sugar with a keto group on the second carbon, while glucose is a six-carbon sugar with an aldehyde group on the first carbon.

Both glucose and fructose are reducing sugars, but fructose is a ketose and glucose is an aldose. Unlike glucose, which is used by all cells in the body to generate energy and constitutes the nearly exclusive energy fuel for the brain, fructose is mainly metabolized by the liver.

In 1960, the food industry developed technologies for converting corn starch into high-fructose corn syrup (HFCS), which contains high levels of fructose (Marshall and Kooi 1957). For the food industry, the use of sugar created several technical problems. Sugar is not only expensive but is also hydrolyzed under acidic conditions (Salomonsson 2005), producing changes in the sweetness and flavor characteristics of the product. In addition, sugar has to be dissolved in water before its use. In contrast, HFCS is an inexpensive liquid, which is stable in acidic condition in foods and beverages. Between 1970 and 1990, the consumption of HFCS increased over 1,000 % (Bray et al. 2004). Humans have consumed fructose (16–20 g/day) for thousands of years, largely in fresh fruits and honey. Introduction of HFCS, which contains around 55 % free fructose, 42 % free glucose, and 3 % other sugars, has resulted in significant increase in consumption of fructose leading to typical daily consumptions amounting to 85–100 g/day. These days HFCS is used as a replacement to sucrose in processed foods and beverages such as jams, jellies, dairy products, baked desserts, cereals, canned fruits, candies, juices, sodas, and sports drinks (Bray et al. 2004). Because of low cost, sweetness, and increased shelf life of the above-mentioned products, the use of HFCS-containing foods has increased worldwide. In the USA, estimated annual per capita fructose consumption rose from 64 g/day during the 1970s to 81 g/day at the end of the 1990s (Elliott et al. 2002), with

A.A. Farooqui, *Metabolic Syndrome: An Important Risk Factor for Stroke, Alzheimer Disease, and Depression*, DOI 10.1007/978-1-4614-7318-3_2,

an additional increase in fructose intake (2.5 g/day) resulting from increased fruit and vegetable consumption. Based on this information, it is postulated that increased consumption of HFCS-containing food and soft drinks may be closely associated with stimulation of lipogenesis, high plasma levels of triacylglycerols (TAGs), obesity, and cardiovascular disease (CVD) (Bray et al. 2004). In addition, when ingested in large amounts as part of a hypercaloric diet, fructose can cause hepatic insulin resistance, leptin resistance, increase in total and visceral fat mass, and accumulation of ectopic fat in the liver and skeletal muscle (Shapiro et al. 2008). In the long run, these effects may be closely associated with the pathophysiology of the metabolic syndrome (MetS). However, consumption of fructose in moderate amounts does not produce above mentioned effects due to decrease in glycemic response to glucose loads, and improvement in glucose tolerance (Tappy 2012). Thus, consumption of fructose in moderate amounts does not produce any changes in insulin signaling and on levels of TAG (Moore et al. 2001). Like humans, in rodents the intake of large quantities of fructose-containing food and drinks or HFCS induces insulin resistance (Morino et al. 2006), increases levels of plasma TAGs and VLDL-TAGs, and causes high blood pressure due to increased production of uric acid (Nakagawa et al. 2006; Parks et al. 2008). Similarly, consumption of a high-fat diet also results in increased levels of blood TAGs and glucose, and it is proposed that an overexpression of hepatic 1,6-bisphosphatase, which is involved in gluconeogenesis, may be responsible for hyperglycemia in type II diabetes, a metabolic condition characterized by an elevated blood glucose concentration that results from inadequate insulin action in insulin-sensitive tissues and from abnormal insulin secretion (Visinoni et al. 2008). Growing evidence indicates that a high-fructose diet produces insulin resistance not only by significantly reducing the protein expression of insulin receptor, insulin receptor substrate-1, but also by altering expression of Akt and GLUT4. Fructose feeding also produces ventricular dilatation, ventricular hypertrophy, decrease in ventricular contractile function, infiltration of inflammatory cells in the heart, and hepatic steatosis (Patel et al. 2009; Chang et al. 2007). In the liver, fructose feeding causes both microvesicular and macrovesicular steatosis with periportal fibrosis and lobular inflammation (Kawasaki et al. 2009). Consequently, the administration of fructose-rich diet has been used as a model for investigating the development of the MetS (Rayssiguier et al. 2006; Panchal and Brown 2011).

## 2.2 Metabolism of Glucose and Fructose in the Liver

Glucose is utilized by all body tissues to produce energy. The uptake of glucose is regulated by a variety of glucose transporters that differ in their tissue distribution, regulation, and kinetics (Thorens and Mueckler 2010). In addition, consumption of glucose triggers the release of insulin, which further promotes glucose uptake by increasing the translocation of glucose transporter 4 (GLUT4) in striated muscle (skeletal and cardiac) and adipose tissues to further facilitate its uptake. After its absorption into the liver and other tissues, glucose either is stored as glycogen or undergoes glycolysis (Kelley et al. 1988; Shulman et al. 1990). Glucose is synthesized in the liver via a combination of two biochemical processes: gluconeogenesis and glycogenolysis. The abnormal levels of endogenous glucose synthesis in patients with type II diabetes are predominately due to an increase in the rate of gluconeogenesis compared with nondiabetic subjects, as glycogenolysis is largely unchanged (Magnusson et al. 1992).

After its absorption through the gut, about 15–30 % glucose enters the liver through the portal vein. Glucose enters the hepatocytes by the membrane transporter GLUT2. Once in the cell, glucose is phosphorylated by glucokinase into glucose-6-phosphate. This metabolite is transformed into fructose 1,6-bisphosphate through the action of phosphofructokinase, a multisubunit allosteric enzyme (Fig. 2.1a). Fructose 1,6-bisphosphate is metabolized to triose phosphate and pyruvate by aldolase A. Pyruvate is then decarboxylated to acetyl coenzyme A and enters the tricarboxylic acid cycle (TCA cycle), a reaction sequence for energy production. The net equations for glucose breakdown (glycolysis and TCA cycle) are as follows:

1. $C_6H_{12}O_6 + 2\ \mathrm{ADP} + 2\ [\mathrm{P}]\mathrm{i} + 2\ \mathrm{NAD}^+ \rightarrow 2\ \mathrm{pyruvate} + 2\ \mathrm{ATP} + 2\ \mathrm{NADH}.$

2. $\mathrm{Acetyl\ CoA} + 3\ \mathrm{NAD}^+ + \mathrm{Q} + \mathrm{GDP} + [\mathrm{P}]\mathrm{i} + 2\ H_2O \rightarrow \mathrm{CoA\text{-}SH} + 3\ \mathrm{NADH} +$
   $3\ H^+ + \mathrm{QH2} + \mathrm{GTP} + 2\ CO_2.$

The energy is then used to synthesize ATP and NADH, which are used by the cell to run cellular processes efficiently. In the absence of oxygen, the glucose cannot be metabolized as thoroughly, and so its metabolism yields lactic acid and fewer molecules of ATP and NADH. Intracellular ATP and citrate exert a negative feedback on phosphofructokinase, so that hepatic glucose catabolism is tuned to the energy status of the liver cells. Insulin regulates glucose metabolism not only by modulating the expression of glucokinase but also activities of key glycolytic enzymes. This allows much of the glucose arriving via the portal vein to bypass the liver and reach the systemic circulation. Thus, following the consumption of glucose-sweetened beverages with meals, plasma glucose and insulin concentrations are increased by 4–5 mM and 70–100 μU/ml, respectively (Stanhope et al. 2008). This way catabolism of glucose by hepatocytes and other visceral cells produces energy for cell survival (Tappy 2012).

Brain tissue represents only 2 % of the total body weight. It utilizes and metabolizes 20–30 % of total glucose (Sieber et al. 1989). The brain oxidizes approximately 120 g of glucose per day (Sieber and Traystman 1992). The majority of brain glucose is converted to ATP energy for the maintenance of normal brain functions including cognition. During hypoglycemia other tissues cease to utilize glucose all together in order to increase glucose availability to the brain (Lehninger et al. 2005). GLUT1 (45 kDa) is a key regulator of glucose transport into and out of the brain across the blood–brain barrier (BBB) acting to maintain glucose homeostasis in the

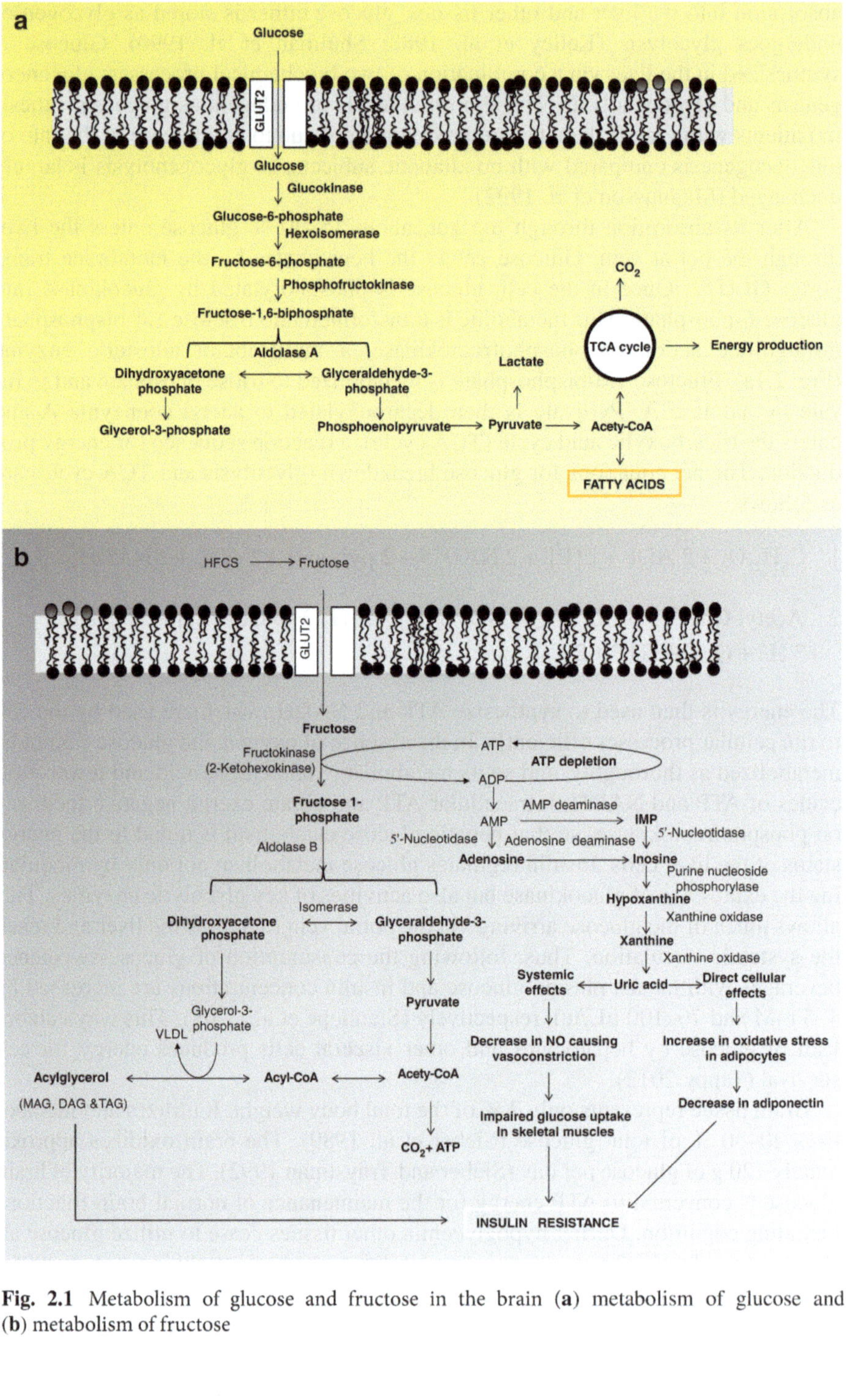


**Fig. 2.1** Metabolism of glucose and fructose in the brain (**a**) metabolism of glucose and (**b**) metabolism of fructose

brain. It is also highly expressed in proliferating cells of the early developing embryo, cardiac muscle, human erythrocytes, and astrocytes (Morgello et al. 1995). GLUT1 is found in the cells forming blood–tissue barriers (Mann et al. 2003), such as the choroid plexus and blood–spinal cord barrier, and the ependymal lining of the cerebral ventricles (Vannucci 1994). GLUT1 in the BBB and GLUT3 in the neurons are not sensitive to insulin that is why brain glucose metabolism is so widely viewed as being independent of insulin. However, GLUT4 is insulin responsive and is expressed particularly in brain regions involved in memory and cognition such as the hippocampus (Watson and Craft 2004).

Like visceral tissues, in the brain, glucose is either metabolized through glycolysis or stored in astrocytes as glycogen. During aerobic cellular respiration, the process of glycolysis in the brain yields pyruvate, which is converted into acetate during transition reaction. As stated above, acetate is used in the citric acid cycle for the production of ATP through the electron transport chain within the mitochondria.

Absorption of fructose from the intestine into the portal blood is facilitated by glucose transporter 5 (GLUT5) at the brush border and basolateral membranes of the jejunum. Fructose through the portal vein enters the liver, where like glucose it is transported into the hepatocytes by GLUT2. In the hepatocytes, fructose is phosphorylated by fructokinase (2-ketohexokinase) producing fructose 1-phosphate. Fructokinase exists in two alternatively spliced isoforms known as fructokinase C and fructokinase A (Hayward and Bonthron 1998). Fructokinase C is expressed primarily in the liver, kidney, and intestines, whereas fructokinase A is more ubiquitous (Diggle et al. 2009). Although, both fructokinases can metabolize fructose, fructokinase C is considered as the primary enzyme involved in fructose metabolism due to its lower $K_m$ (Asipu et al. 2003). Because of its higher $K_m$, fructokinase A is not considered presently to be actively involved in fructose metabolism but rather has been postulated to act on as yet unknown substrates (Asipu et al. 2003). In contrast to hexokinase, the fructokinase pathway of fructose metabolism bypasses phosphofructokinase, an enzyme that tightly regulates glycolysis. Thus, while glucose metabolism is *negatively* regulated by phosphofructokinase, fructose can *continuously* enter the glycolytic pathway. Fructose 1-phosphate is transformed into dihydroxyacetone phosphate and glyceraldehyde-3-phosphate by the action of aldolase B (Fig. 2.1b) (Jianghai et al. 2012). Some of these three-carbon molecules are either converted into glucose through gluconeogenesis or transformed into triacylglycerol (TAG). In the extrahepatic metabolism, fructose bypasses fructokinase and allows carbon from fructose to enter glycolysis downstream of this enzyme. The three-carbon molecules can eventually be used for the synthesis of glycerol and fatty acids, which through esterification can form TAGs. Another important aspect of fructose metabolism is that unlike other sugars, it raises uric acid levels in humans and rodent's sera (Stirpe et al. 1970; Stavric et al. 1976) (see below).

The liver and gut are major sites for the metabolism of fructose. As stated above, fructose enters the glycolytic pathway at the triose level, after the generation of fructose 1-phosphate via fructokinase, which is not regulated by energy status. In addition, fructokinase consumes ATP at a greater rate than does phosphofructokinase-1; fructose decreases intracellular ATP while increasing AMP levels (Cha et al. 2008; Lane and Cha 2009). It is demonstrated that fructose-induced depletion of ATP is similar to ATP depletion in ischemia. This effect not only produces transient arrest in protein synthesis (Maenpaa et al. 1968) but also increases the production of inflammatory proteins, induces endothelial dysfunction, and promotes oxidative stress (Cirillo et al. 2009). These processes lead to greater fructose uptake by the liver with little of the consumed fructose reaching the circulation. In addition, ATP depletion also produces intracellular phosphate depletion and AMP generation, with stimulation of AMP deaminase and the stepwise degradation of AMP to purine products including uric acid (van den Berghe et al. 1977). As uric acid rises inside the cells, it spills out into the circulation. Accumulating evidence suggests that fructose is distinct from glucose in its ability to cause intracellular phosphate depletion, ATP depletion, and uric acid generation in the liver (Lanaspa et al. 2012). Approximately 70 % of the uric acid is excreted through the kidneys and 30 % of the uric acid is excreted through the gastrointestinal tract. Overproduction of uric acid through over-consumption of fructose and purine-rich foods may result in hyperuricemia. Coffee consumption not only decreases serum uric acid levels, but also lowers the risk of developing gout (Choi et al. 2007) through various mechanisms (Kela et al. 1980; Petrie et al. 2004), including the inhibition of xanthine oxidase (Kela et al. 1980; Petrie et al. 2004), lowering of insulin levels (Wu et al. 2005), and increase in insulin sensitivity in rats (Arnlov et al. 2004).

Glucose stimulates the release of insulin from the pancreas. Glucose-induced insulin secretion is regulated by (a) transport of glucose into the β cells through the translocation of the GLUT2; (b) generation of ATP through the oxidation of glucose; (c) elevation of ATP/ADP ratio induces closure of cell-surface ATP-sensitive $K^+$ ($K_{ATP}$) channels, leading to cell membrane depolarization and activation of voltage-gated $Ca^{2+}$ channels; (d) extracellular $Ca^{2+}$ influx into the β cell; and (e) a rise in cytosolic $Ca^{2+}$ triggering the exocytosis of insulin granules (Gong and Muzumdar 2012).

Unlike glucose, in the hepatocytes fructose is largely transported via GLUT2 and it does not stimulate insulin release from the pancreas, because the pancreas lacks the GLUT5 transporter for fructose (Curry 1989). As stated above, fructose is rapidly phosphorylated by fructokinase, yielding fructose 1-phosphate. Unlike phosphofructokinase, which is involved in the metabolism of glucose, fructokinase is not inhibited by ATP and thus may not be involved in maintaining the cellular energy states (Samuel 2011). Consumption of fructose-enriched diet results in the generation of glycerol-3-phosphate (G-3-P), which causes fat to become fixed in fat tissue. Fructose tricks our body into gaining weight by turning off our body's appetite control system. Fructose does not suppress ghrelin (the "hunger hormone") and doesn't stimulate leptin (the "satiety hormone"), which together results in feeling hungry all the time, even though one has eaten. Thus, consumption of diet enriched in fructose contributes to overeating and obesity in rodents due to increase in leptin secretion (Froesch 1972; Teff et al. 2002; Havel 2002). Consumption of fructose also influences activities of mitochondrial enzymes. Thus, fructose consumption also results in an increase flux of acetyl CoA through the TCA cycle with a concomitant increase in cellular energy status (increased ATP/ADP ratio and NADH/NAD$^+$ ratio). A high NADH/NAD$^+$ ratio in the mitochondria results in substrate inhibition of isocitrate dehydrogenase in the TCA cycle. This inhibition not only increases the export of citrate to the cytosol, but also activates acetyl-CoA carboxylase and increases the synthesis of malonyl-CoA, a precursor of fatty acid synthesis (Locke et al. 2008; Cox et al. 2012). Elevated cytosolic concentrations of malonyl CoA is known to inhibit the carnitine shuttle via carnitine palmitoyltransferase (CPT), resulting in reduced entry of fatty acids into the mitochondria and decrease in fatty acid oxidation (Locke et al. 2008; Cox et al. 2012). In animals and humans, L-carnitine is biosynthesized primarily in the liver and kidneys from the amino acids lysine or methionine with vitamin C (ascorbic acid) being essential to the synthesis of L-carnitine (Steiber et al. 2004). L-carnitine is necessary for the transport long-chain fatty acids from the cytosol to sites of β-oxidation in the mitochondria. Acetate is a by-product of β-oxidation and provides usable energy via the citric acid cycle. Under normal conditions, the availability of carnitine is not a limiting step in β-oxidation. However, an elevation in carnitine content may increase the rate of fatty acid oxidation causing a reduction in glucose utilization, preserving muscle glycogen content, and producing maximal rates of ATP production (Owen and Sunram-Lea 2011). High plasma fatty acid levels in obesity may require more carnitine for efficient β-oxidation. Chronic high-fat feeding not only produces severe age-related decrease in carnitine homeostasis, but also coincides with reduction in hepatic expression of carnitine regulatory genes and increase in sequestration of carnitine in the skeletal muscle acylcarnitine pool. Whole body carnitine homeostasis is regulated at multiple levels, including dietary intake, intestinal absorption, de novo biosynthesis, and renal reabsorption (Noland et al. 2009). In neuronal cells, the L-carnitine shuttle mediates translocation of the acetyl moiety from the mitochondria into the cytosol and promotes the synthesis of acetylcholine and of acetylcarnitine (Nalecz and Nalecz 1996). The neurochemical effects of acetylcarnitine include modulation of brain energy and phospholipid metabolism, cellular macromolecules (such as neurotrophic factors and neurohormones), synaptic morphology, and synaptic transmission of multiple neurotransmitters (Furlong 1996).

Fructose is a lipogenic sugar. The consumption of fructose or high-fructose corn syrup increases the deposition of fat in the liver due to elevation in fatty acyl coenzyme A and diacylglycerol (DAG) (Stanhope and Havel 2008). DAG promotes hepatic insulin resistance (Morino et al. 2005), possibly through the activation of novel-PKC (Samuel et al. 2007). Novel-PKC decreases tyrosine phosphorylation and/or increases serine phosphorylation of the insulin receptor and IRS-1, resulting in decreased insulin sensitivity, increased hepatic glucose production, and increased fasting glucose and insulin concentrations, and impaired glucose tolerance. In addition, fructose has been shown to induce increase in the synthesis of ceramide, TAG, and uric acid along with increase in the expression of the factor of transcription SREBP-1c, the principal inducer of hepatic lipogenesis (Fig. 2.2) (Matsuzaka et al. 2004; Vila et al. 2008; Seneff et al. 2011). This effect of fructose on SREBP-1c requires peroxisome proliferator-activated receptor γ coactivator-1β (PGC-1β). Fructose also activates the hepatic transcription factors carbohydrate-responsive

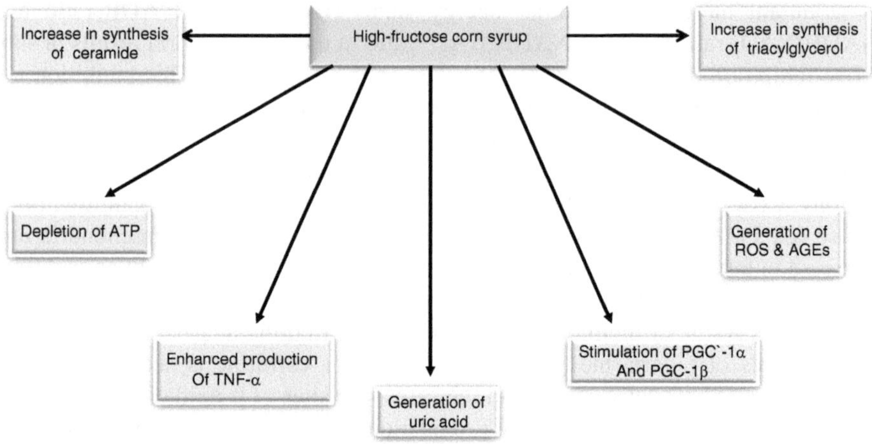


**Fig. 2.2** Harmful effects of high-fructose corn syrup

element-binding protein (ChREBP), which upregulates the expression of hepatic fatty acid synthase and acetyl-CoA carboxylase (Koo et al. 2008). Fructose administration also produces greater postprandial hypertriglyceridemia than glucose, and it can also result in higher apolipoprotein B (apo B) levels (Swarbrick et al. 2008). Apo B assembles atherogenic lipoproteins and promotes the development of atherosclerosis (Olofsson and Boren 2005). Therefore, long-term consumption of diets containing 25 % of energy from fructose produces a lipoprotein profile that is associated with the development of atherosclerosis. The consumption of fructose stimulates food intake (Miller et al. 2002; Cha et al. 2008; Lane and Cha 2009). In addition, a definite relationship has been shown to occur between MetS and hyperhomocysteinemia, which is associated with initiation of cardiovascular and cerebrovascular diseases. Rats fed a fructose-enriched diet have a 72 % higher homocysteine levels after 5 weeks compared to chow-fed controls (Oron-Herman et al. 2003). Elevation in homocysteine levels has also been reported to occur in vascular diseases, coronary artery disease, and in patients with type II diabetes (Okada et al. 1999).

Fructose feeding is also an effective way for inducing fatty liver (Ouyang et al. 2008). It is proposed that metabolic disturbances caused by high-fructose consumption underlie the induction of insulin resistance in both humans and animal models (Stanhope and Havel 2009). Insulin resistance is manifested by the failure of endogenous insulin to hinder hepatic gluconeogenesis and to induce glucose peripheral uptake. Such dysfunction is compensated by hyperinsulinemia and associated with alteration in glucose and lipid homeostasis leading to muscular fat storage as well as visceral fat deposition (Lonardo et al. 2005). Peripheral insulin resistance is promoted by interplay among three major organs, namely, the liver, adipose tissue, and skeletal muscles, with alterations in insulin signaling that involves changes in insulin receptors and post-receptor defects at the molecular level (Farooqui et al. 2012).

These include alterations in the relative abundance of the two insulin receptor isoforms, decrease in insulin affinity to its receptors, improper insulin receptor kinase activity, decrease in autophosphorylation, and abnormalities in glucose transporter translocation and activation (De Fronzo 1997). Fructose-induced insulin-resistant state is characterized by a profound metabolic dyslipidemia, which appears to result from hepatic and intestinal overproduction of atherogenic lipoprotein particles. In addition, fructose feeding also decreases leptin secretion in humans and produces leptin resistance (Teff et al. 2004). In rats, high-fructose diet not only produces hepatic leptin resistance through the enhancement in the amount of suppressor of cytokine 3 (SOCS3) but also decreases serine/threonine phosphorylation of key proteins in leptin signaling (Vila et al. 2008; Li et al. 2010).

At the molecular level, fructose stimulates the expression of protein tyrosine phosphatase 1B (Li et al. 2010). Fructose-fed rats exhibit impaired c-Jun $NH_2$-terminal kinase (JNK) and mitogen-activated protein kinase signaling and increase the expression of FOXO1 due to SOCS3 expression (Vila et al. 2008). In turn, these changes lead to decrease in peroxisome proliferator-activated receptor alpha (PPARα), reduction in fatty acid oxidation, and accumulation of TAG in the liver. Conversely, the activation of PPARα reverses leptin resistance in fructose-fed rats (Li et al. 2010). In addition, fructose feeding also increases hepatic ceramide levels, suggesting that incomplete fatty acid oxidation due to PPARα impairments provides substrate for ceramide synthesis (Vila et al. 2008). This may also result in activation of protein phosphatase 2A and thus contributing to deficiencies in leptin signaling and exacerbate metabolic disease (Vila et al. 2008). Collectively, these studies indicate that exposure of the liver to increased quantities of fructose leads to induction of insulin resistance, rapid stimulation of lipogenesis, and accumulation of TAG in rodent (Seneff et al. 2011). The adipocytes store TAG in the form of lipid droplets inducing adipocyte hypertrophy. Adipocyte hypertrophy is accompanied by the accumulation of multinucleated macrophages, which attempt to protect the contents of dysfunctional adipocytes from further glycation, fructation, and oxidation (Seneff et al. 2011). The exposure of macrophages to advanced glycation end products (AGEs) results in formation of dysfunctional macrophages, which enter into the artery wall and initiate the formation of plaque leading thrombosis (Seneff et al. 2011). Thus, emerging evidence from recent epidemiological and biochemical studies clearly suggests that the liver metabolizes fructose and produces a large increase in TAG (Basciano et al. 2005; Havel 2005; Stanhope and Havel 2009). In the liver, TAG interferes with insulin signaling (Kim et al. 2007). Based on the above information, it is proposed that a high-fructose diet is analogous to a high-fat diet in many metabolic ways (Havel 2005). It is also reported that alterations in metabolic, endocrine, and adipose tissue functions induced by the consumption of a fructose-rich diet (FRD) to an adult female rat can be prevented, at least partly, by pretreatment with progesterone (P4) (Castrogiovanni et al. 2012). The molecular mechanism involved in this process is not fully understood. However, pretreatment with P4 has been reported to modulate the FRD-induced up-regulation of leptin production and down-regulation of plasminogen activator inhibitor-1(PAI-1) production

(Castrogiovanni et al. 2012). Thus, growing evidence supports the concept that P4 exerts a dual effect upon fat tissue metabolism: It enhances in vivo lipogenic activity, induced by administration of an FRD, and inhibits the PAI-1 overproduction caused by this diet.

## 2.3 Metabolism of Glucose and Fructose in the Brain

Glucose is a major circulating sugar and primary fuel for the animal brain for energy production. Glucose enters in neural cells by the membrane transporter GLUT1 and GLUT4. In neural cells glucose is metabolized by a mechanism, which is similar to liver. The hypothalamus plays an important role in the control of food intake and energy expenditure (Cha et al. 2008). Insulin receptor-expressing neurons in the rat are localized in important hypothalamic and hindbrain areas that modulate glucose homeostasis, energy intake and expenditure, and neuroendocrine and autonomic functions (Unger et al. 1991). Activation of insulin receptors in the hypothalamus causes activation of PtdIns 3K and MAPK signaling pathways (Unger et al. 1991). Activation of insulin receptor produces short-term changes in neuronal activity (Spanswick et al. 2000) or prolonged changes in gene transcription and neuronal plasticity (Levin et al. 2006).

Malonyl CoA, an intermediate in fatty acid synthesis, serves as an indicator of energy status in the hypothalamic neurons. Thus, increase in hypothalamic malonyl CoA indicates energy surplus, causing a decrease in food intake and an increase in energy expenditure. In contrast, a decrease in hypothalamic malonyl CoA signals an energy deficit, inducing an increase in appetite and a decrease in body energy expenditure (Fig. 2.3). The cellular levels of malonyl CoA are determined not only by its rate of synthesis, which is catalyzed by acetyl-CoA carboxylase (ACC), but also by the rate of removal of malonyl CoA, by fatty acid synthase. Physiologically, the levels of hypothalamic malonyl CoA are determined through phosphorylation/dephosphorylation of ACC by 5′-AMP-dependent kinase (AMPK) in response to changes in the AMP/ATP ratio, an indicator of energy status. Thus, changes of hypothalamic malonyl CoA during feeding and fasting cycles are closely associated with changes in the phosphorylation state and activity of ACC mediated through AMPK-dependent mechanism. AMPK acts as a "fuel sensor" that increases fatty acid β-oxidation during periods of increase in energy demand or decreases fatty acid β-oxidation during periods of low demand, secondary to respective decreases and increases in ACC activity and malonyl CoA levels (Dyck and Lopaschuk 2006). Metabolism of glucose in the brain produces ATP through the glycolytic pathway. This increase in ATP produces a decrease in AMP level (Lane and Cha 2009; Cha et al. 2008). Because AMP is an activator of AMPK, a drop in AMP level may result in dephosphorylation and inactivation of AMPK. Since acetyl-CoA carboxylase (ACC) is a substrate of AMPK, lowering AMP increases the catalytic activity of ACC and thereby the level of its reaction product, malonyl CoA, which signals the anorexigenic–orexigenic neuropeptide system to suppress food intake. In addition, the hypothalamus contains an atypical and brain-specific carnitine palmitoyltransferase-1

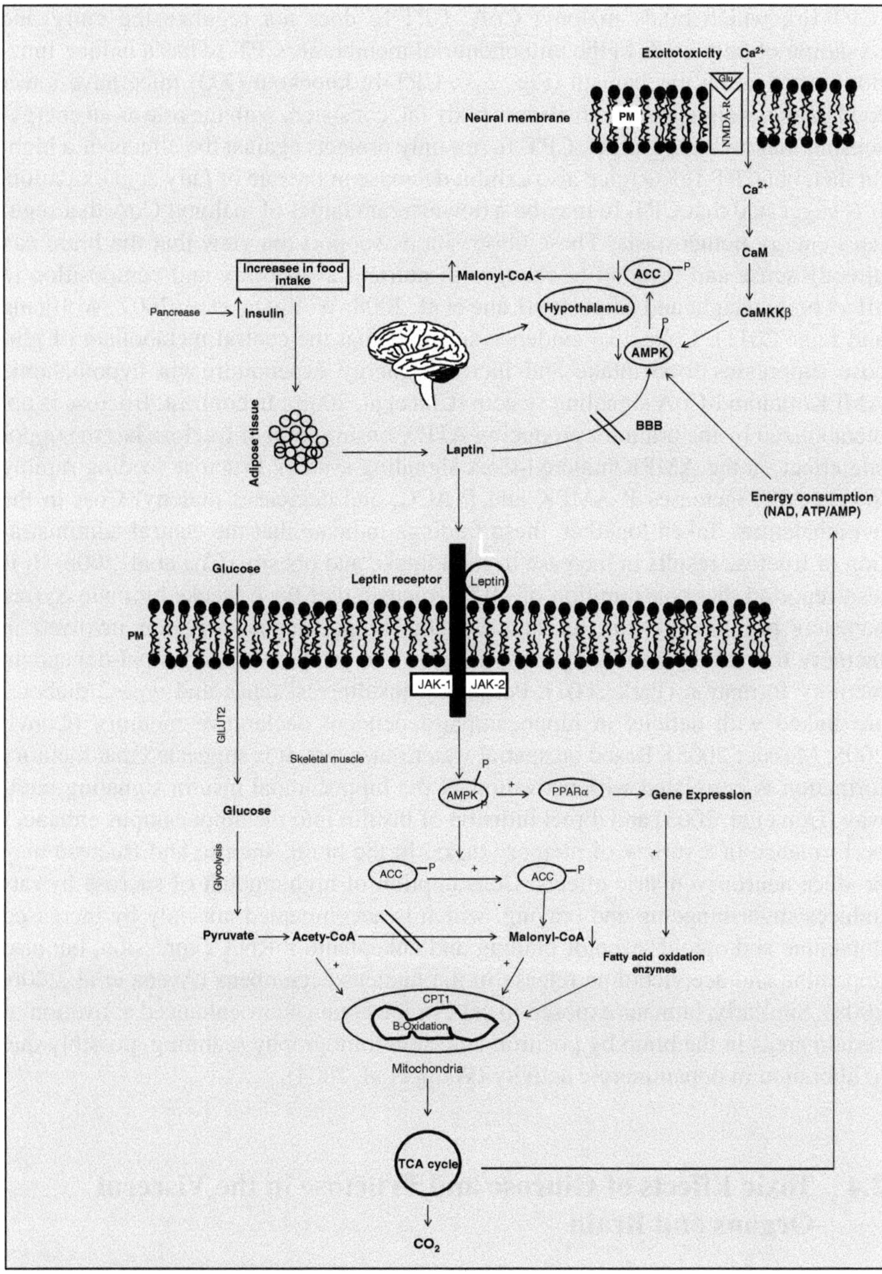


**Fig. 2.3** Hypothetical model showing the effect of malonyl CoA on the hypothalamus and skeletal muscles. Glucose transporter 2 (GLUT2); Janus kinase 1 (JAK1); Janus kinase 2 (JAK2); acetyl-CoA carboxylase (ACC); 5′-AMP-dependent kinase (AMPK); carnitine palmitoyl-CoA transferase-1c (CPT-1c); peroxisome proliferator-activated receptor (PPAR); and calmodulin (CaM). Toxicity of glucose and fructose and generation of advanced glycation end products (AGEs). Reactive oxygen species (ROS); Janus kinase (JAK); protein kinase C (PKC); and mitogen-activated protein kinase (MARK)

(CPT-1c), which binds malonyl CoA. CPT-1c does not regulate the entry and oxidation of fatty acids in the mitochondrial membrane. CPT-1c has a unique function or activation mechanism (Fig. 2.3). CPT-1c knockout (KO) mice have lower food intake, weigh less, and have less body fat, consistent with the role as an energy-sensing malonyl-CoA target. CPT-1c not only protects against the effects of a high-fat diet, but CPT-1cKO mice also exhibit decrease in the rate of fatty acid oxidation. It is suggested that CPT-1c may be a downstream target of malonyl CoA that regulates energy homeostasis. These observations support the view that the brain can directly sense and respond to changes in nutrient availability and composition to affect body weight and adiposity (Lane et al. 2008; Wolfgang et al. 2007; Wolfgang and Lane 2011). Collective evidence suggests that the central metabolism of glucose suppresses food intake and increase energy expenditure via hypothalamic AMPK/malonyl-CoA signaling system (Cha et al. 2008). In contrast, fructose is not metabolized by the brain for producing ATP. Consumption of fructose has the opposite effect on the AMPK/malonyl-CoA signaling system. Fructose feeding rapidly lowers ATP, increases P-AMPK and P-ACC, and decreases malonyl CoA in the hypothalamus. Taken together, these findings indicate that the central administration of fructose results in increase in food intake and obesity (Cha et al. 2008). It is also reported that consumption of 60 % fructose diet for 6 weeks by male Syrian hamsters produces insulin resistance in the hippocampus, a region involved in memory formation. Insulin signaling is known to promote hippocampal-dependent memory formation (Park 2001). Peripheral insulin resistance and type 2 diabetes are linked with deficits in hippocampal-dependent declarative memory (Convit 2005; Messier 2005). Based on spatial water maze test, it is suggested that memory formation is correlated with activation of the hippocampal insulin signaling pathway (Dou et al. 2005) and direct infusion of insulin into the hippocampus enhances performance in a variety of memory tasks. In the brain, sucrose and fructose may produce neuropsychiatric effects. Consumption of high amount of sucrose by rats induces sugar bingeing and craving, which is accompanied not only by increased dopamine and opioid receptor binding and enkephalin mRNA expression, but also dopamine and acetylcholine release in the nucleus accumbens (Avena et al. 2006, 2008). Similarly, humans exposed to cake or ice cream show enhanced activation of certain areas in the brain by positron emission tomography scanning, possibly due to alteration in dopaminergic activity (Wang et al. 2004).

## 2.4 Toxic Effects of Glucose and Fructose in the Visceral Organs and Brain

Diabetes is a complex condition characterized by high levels of blood glucose due to defects in insulin production, insulin action, or both. Diabetic patients often develop secondary complications. Mostly these secondary complications arise in various tissues due to abnormal insulin-independent glucose uptake. Pathophysiological complications in diabetic patients are mainly caused by increased sorbitol production, oxidative–nitrosative stress, endogenous antioxidant depletion, enhanced lipid peroxidation, metabolic changes, and altered hormonal responses (Brownlee 2005). However, the major mechanism that precedes all the above-mentioned pathways is increase in intracellular glucose level and its downstream metabolism. Hyperglycemia contributes to the generation of superoxides, which inhibit glyceraldehyde-3-phosphate dehydrogenase (GAPDH) activity in vivo by modifying the enzyme with polymers of ADP-ribose (Du et al. 2003). Specific inhibitors of poly(ADP-ribose) polymerase (PARP) prevent the inhibition of GAPDH. Under normal conditions, PARP resides in the nucleus in an inactive form. Under hyperglycemic conditions, increased production of ROS produces DNA damage and activates PARP in the nucleus. PARP splits the $NAD^+$ molecule into nicotinic acid and ADP-ribose. PARP then proceeds to make polymers of ADP-ribose, which accumulate on GAPDH and other nuclear proteins. GAPDH is commonly thought to reside exclusively in the cytosol. However, it normally shuttles in and out of the nucleus, where it plays a critical role in DNA repair (Sawa et al. 1997; Du et al. 2003) (Fig. 2.4). In diabetes and MetS, alterations in glucose metabolism, due to insulin resistance, result in increase in generation of methylglyoxal (MG) (Beisswenger et al. 2003; Thornalley 1993a). As stated above, under normal conditions, glucose is metabolized to glyceraldehyde-3-phosphate (G3P) via glycolysis. G3P is then converted to 1,3-diphosphoglycerate by glyceraldehyde-3-phosphate dehydrogenase (GAPDH), which is upregulated, and G3P is further metabolized to form pyruvate. During insulin resistance, abnormalities in insulin function may result in downregulation of GAPDH, slowing glucose metabolism and increasing metabolism via the polyol pathway (Thornalley 1993a; Alexander et al. 1988). The polyol pathway involves aldo-keto reductase enzymes which utilize a wide variety of carbonyl compounds as substrates and reduce them into sugar alcohols (polyols) through the participation of nicotinic acid adenine dinucleotide phosphate (NADPH). For example, glucose is transformed into sorbitol by the enzyme aldose reductase, and sorbitol is then oxidized to fructose by the enzyme sorbitol dehydrogenase (SDH), with $NAD^+$ as a cofactor (Giacco and Brownlee 2010).

The hyperglycemia and metabolic abnormalities in diabetes and MetS also result in mitochondrial superoxide overproduction not only in endothelial cells of both large and small vessels but also in the myocardium. This increased superoxide production causes the activation of five major pathways, which are associated with in the pathogenesis of complications in diabetes and MetS. These pathways include polyol pathway flux, increased production of AGEs, increased expression of the receptor for AGEs (RAGE) and its activating ligands, activation of protein kinase C (PKC) isoforms, and overactivity of the hexosamine pathway (Fig. 2.4) (Giacco and Brownlee 2010). Persistent and excessive activation of PKC isoforms (β and δ) in cultured vascular cells enhances de novo synthesis of DAG from glucose via triose phosphate, whose availability is increased because increased ROS inhibit activity of the glycolytic enzyme GAPDH, raising intracellular levels of the DAG precursor triose phosphate (Inoguchi et al. 1992; Scivittaro et al. 2000).

Healthy aging is accompanied by significant decrease in glucose uptake and metabolism in specific cortical regions, an effect that is more pronounced in the elderly

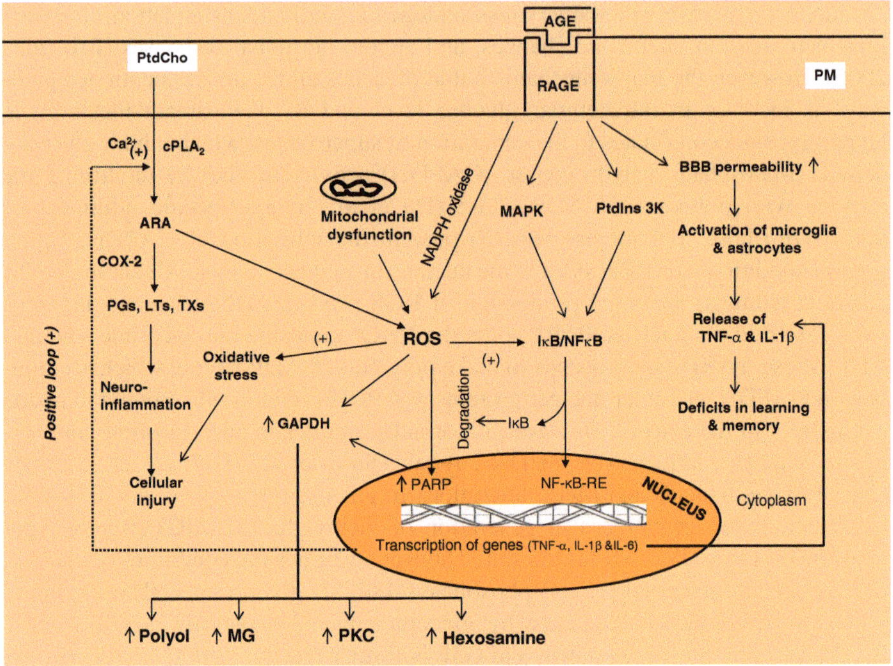


**Fig. 2.4** Interactions between AGE and AGE receptors (RAGE) along with induction of ROS formation. Plasma membrane (PM); advanced glycation products (AGEs); receptor for advanced glycation end products (RAGE); blood–brain barrier (BBB); phosphatidylcholine (PtdCho); cytosolic phospholipase $A_2$ ($cPLA_2$); cyclooxygenase (COX-2); arachidonic acid (ARA); reactive oxygen species (ROS); nuclear factor-kappaB (NF-κB); nuclear factor-kappaB response element (NF-κB-RE); inhibitory subunit of NF-κB (IκB); tumor necrosis factor-α (TNF-α); interleukin-1β (IL-1β); interleukin-6 (IL-6); prostaglandins (PGs); leukotrienes (LTs); thromboxanes (TXs); methylglyoxal (MG); mitogen-activated protein kinase (MAPK); phosphatidylinositol 3-kinase (PtdIns 3K); protein kinase C (PKC); nicotinamide adenine dinucleotide phosphate oxidase (NADP oxidase); poly(ADP-ribose) polymerase (PARP); and glyceraldehyde-3-phosphate dehydrogenase (GAPDH)

patients with Alzheimer disease (AD) (Kalpouzos et al. 2009a, b). Decrease in utilization and metabolism of glucose has also been reported to occur in asymptomatic ApoE4 carriers as young as 20–30 years old in the same brain regions as clinically affected AD patients (Reiman et al. 2004; Barberger-Gateau et al. 2011). Alterations in the above parameters are the earliest sign of brain hypometabolism abnormality that occur in the living persons that are at risk of developing AD. This suggestion is supported by various clinical and experimental models, studies of family history and genetic susceptibility to AD, postmortem brain analysis, and from in vitro and animal models (Cunnane et al. 2011). An important issue that has to be addressed is to establish whether brain hypometabolism contributes to the development and/or progression of AD or whether these metabolic changes in brain glucose metabolism are caused by earlier neurodegenerative processes that reduce the demand for glucose in the affected brain areas (Cunnane et al. 2011).

Although, glucose is the main cerebral energy substrate, but during fasting or low carbohydrate intake, glucose is replaced by ketone bodies as a primary replacement fuel (Owen et al. 1967). It is interesting to note that in the elderly and in ApoE4 carriers, increase in blood ketone bodies and their uptake by the brain counterbalances lower brain glucose uptake (Barberger-Gateau et al. 2011; Freemantle et al. 2009). High levels of blood ketone bodies are known to facilitate cognitive performance in ApoE4 noncarriers but not in the carriers (Reger et al. 2004). This suggestion is supported by a larger scale study with 152 subjects, who were diagnosed with mild to moderate AD (Henderson et al. 2009).

An excess of fructose and glucose in the bloodstream is known to produce extensive glycation damage to vulnerable proteins (McPherson et al. 1988; Kikuchi et al. 2003). Glycation (fructated or glucated) is a nonenzymic and posttranslational modification of proteins initiated by the primary addition of a sugar aldehyde or ketone to the amino groups of proteins. Glycation not only induces impaired function, but glycated proteins become highly susceptible to oxidative damage. Glycated proteins are also resistant to degradation by lysosomal enzymes. For example, glycated hemoglobin (hemoglobin damaged by glucose exposure) is used as marker for diabetes, and glycated LDL are poorly recognized by lipoprotein receptors and scavenger receptors (Zimmermann et al. 2001). Over time, these proteins and their debris accumulate in the blood serum and along the arterial walls. These damaged proteins are collectively known as AGEs (Brownlee 2001; Giacco and Brownlee 2010). AGEs play a critical role in aging, diabetes, atherosclerosis and cardiovascular diseases, and in neurodegenerative diseases. Hyperglycemia is closely linked with accelerated AGEs formation (Méndez et al. 2010). The major AGEs in vivo are formed from highly reactive intermediate carbonyl groups, known as α-dicarbonyls or oxoaldehydes, including 3-deoxyglucosone, glyoxal, and MG (Brownlee 2001; Kim et al. 2005). Some of these AGEs are also formed from increased fatty acid oxidation in arterial endothelial cells (Wautier and Schmidt 2004).

MG is a highly reactive metabolite of glucose. It is produced through enzymic and nonenzymic pathways. MG is primarily formed by the degradation of the glycolytic intermediates, dihydroxyacetone phosphate, and glyceraldehyde-3-phosphate (Thornalley 1993b). MG levels rise in hyperglycemia and diabetes (Brownlee 2001). The polyol pathway transforms excess of glucose into sorbitol, which in turn produces a variety of intracellular changes in vascular tissues including conversion of glucose to fructose by sorbitol dehydrogenase (Fig. 2.5). In the polyol pathway, aldose reductase has a low affinity for glucose. At the normal glucose concentration (nondiabetic subjects), this reaction is very slow and a very small percentage of glucose is utilized through this pathway. However, in diabetic subjects high levels of glucose (hyperglycemia) induce an overproduction of ROS via the mitochondrial electron transport chain, leading to the activation of aldose reductase and the subsequent elevation in conversion of glucose to sorbitol. This reaction is catalyzed by aldose reductase (Madonna and De Caterina 2011). American diet, which is high in fructose- and fat-enriched processed food products, contributes to the formation of MG (Vasdev et al. 2006).

Nonenzymically, MG is spontaneously formed in a process of glycolysis from dihydroxyacetone phosphate as a by-product during the formation of

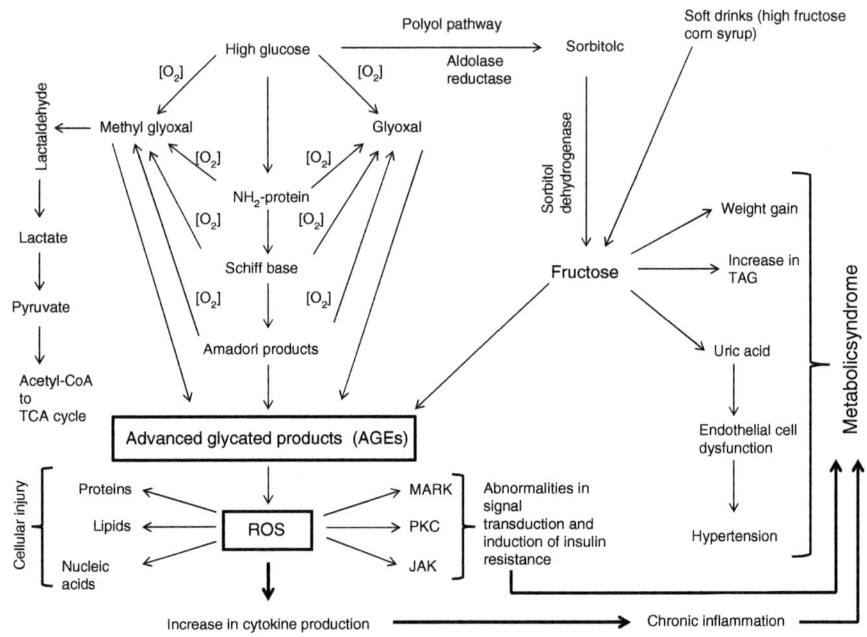


**Fig. 2.5** Glucose and fructose toxicity and generation of AGEs. Mitogen-activated protein kinase (MAPK); reactive oxygen species (ROS); protein kinase C (PKC); c-Jun N-terminal kinase (JNK); and triacylglycerol (TAG)

glyceraldehydes-3-phosphate in the vascular endothelial cells and smooth muscle cells (Wu 2006). It should be noted that triose phosphate pool, in turn, is regulated by cellular levels of fructose, not by glucose. Interestingly, recent studies have shown that plasma levels of fructose are more involved in MG overproduction than glucose, which is mainly metabolized by the over-activated polyol pathway (Wang et al. 2008) (Fig. 2.5). The enzyme, glyoxalase 1 (GLO1), converts MG into the less reactive substance, D-lactate (Thornalley 2003). In addition, in vitro, overexpression of GLO1 prevents MG accumulation (Shinohara et al. 1998); conversely, GLO1 inhibition increases MG accumulation and decreases cellular viability (Kuhla et al. 2006). It is becoming increasingly evident that expression of GLO1 is associated with behavioral changes in rodents and humans. Thus in mice, GLO1 expression is not only associated with anxiety-like behavior (Hovatta et al. 2005; Loos et al. 2009; Benton et al. 2011; Distler et al. 2012) but also with depression (Benton et al. 2011) and neuropathic pain (Jack et al. 2011; Bierhaus et al. 2012). Accumulating evidence indicates that GLO1 increases anxiety by reducing levels of MG, which acts as a GABA receptor agonist and reduces GABAA receptor activation. A single intraperitoneal injection of low doses of MG produces anxiolytic effect within minutes. At high doses, acute MG administration produces sedative effects, including locomotor depression, ataxia, hypothermia, and lethargy (Distler et al. 2012). Acute MG administration also induces hyperalgesia equivalent to that caused by diabetes in mice (Bierhaus et al. 2012). Together, these studies indicate that MG mediates multiple behavioral phenotypes and that MG's effects are opposite to those of GLO1 overexpression (Distler et al. 2012).

It is also shown that MG also lowers hydrogen sulfide ($H_2S$) concentrations in cultured vascular smooth muscle cells both directly by scavenging $H_2S$ and indirectly by downregulating cystathionine γ-lyase expression, suggesting an important interaction between MG and $H_2S$ (Chang et al. 2010). Hyperglycemia increases the formation of MG, and this process may be an important molecular mechanism linking diabetes to endothelial damage, a systemic pathological condition which can be broadly defined as an imbalance between vasodilating and vasoconstricting substances produced by the endothelium (Deanfield et al. 2005). MG not only modifies lysine, arginine, and cysteine residues in peptides or proteins to yield irreversible advanced glycosylation end products (AGEs), but also increases the generation of reactive oxygen species (ROS) and oxidative stress in endothelial cells (Brownlee 2001; Kim et al. 2005; Madonna and De Caterina 2011). Accumulating evidence suggests that increase in MG is linked to the development of insulin resistance associated with type II diabetes mellitus and MetS (Dhar et al. 2011), as well as hypertension (Vasdev et al. 1998; Wang et al. 2005).

During AGE formation, glucose or fructose attaches itself with the α-amino group of either the amino terminus of proteins or lysine residues via nucleophilic attack: The product is known as a Schiff base (Fig. 2.5), which undergoes an Amadori rearrangement forming ketoamines (Singh et al. 2001). These ketoamines, synonymously termed Amadori products, either undergo through an oxidative or nonoxidative pathway forming irreversible AGE modifications (Singh et al. 2001). Collective evidence suggests that hyperglycemia in diabetes and MetS causes excessive glycation of proteins found in serum (e.g., albumin, hemoglobin, and LDL) and in the vessel wall (e.g., collagen, fibronectin) (Singh et al. 2001). These modified circulating proteins can then bind to AGE receptors and activate them, thereby initiating vascular pathological changes in the vessel wall. AGEs not only contribute to oxidative stress, but also play an important role in initiating endothelial cell dysfunction (Funk et al. 2012). Intracellular production of AGE precursors can damage visceral and neural cells by three general mechanisms: (a) AGE-mediated modification of intracellular proteins causes alterations in their functions, (b) AGE may produce alterations in extracellular matrix components and with matrix receptors (integrins) on the surface of cells, and (c) plasma proteins modified by AGE precursors may bind to AGE receptors (RAGE) on cells (macrophages, vascular endothelial cells, and vascular smooth muscle cells) mediating the production of ROS, which in turn activates the pleiotropic transcription factor, nuclear factor-kappaB (NF-κB), causing multiple pathological changes in gene expression (Goldin et al. 2006).

Although molecular mechanism of RAGE receptor signaling is not fully understood, RAGE ligation is known to stimulate endothelial ROS production (Yan et al. 1994), and the NADPH oxidase inhibitor diphenyliodonium (DPI) significantly blunts this induction (Wautier and Guillausseau 2001). In addition, AGE/RAGE interactions also upregulate the expression of proinflammatory gene expression (VCAM-1, E-selectin, cytokines) in endothelial cells (Schmidt et al. 1995) and NADPH oxidase activation (Wautier and Guillausseau 2001; Higai et al. 2006).

AGEs and their precursors can diffuse out of the cell and modify extracellular matrix molecules nearby (McLellan et al. 1994), which changes signaling between the matrix and the cell and causes cellular dysfunction (Charonis et al. 1990).

The macrophages exposed to AGEs become dysfunctional and, on entry into the artery wall, contribute to plaque formation and thrombosis (Seneff et al. 2011). As stated above, AGEs mediate their effects by interacting with RAGE, a group of cell surface receptor, which belongs to the immunoglobulin superfamily (Neeper et al. 1992; Ahmed 2005) (Fig. 2.4). RAGE is a type I transmembrane protein composed of three extracellular immunoglobulin-like domains (V, C1, and C2), a single transmembrane domain and a short cytoplasmic tail thought to be important in signal transduction (Hudson et al. 2008). RAGEs are expressed in the brain in neurons, microglia, and astrocytes (Lue et al. 2001; Sasaki et al. 2001; Yan et al. 1996). Aβ is a specific ligand for RAGE, which interacts with the N-terminal domain of RAGE (Chaney et al. 2005). RAGE expression is elevated in AD pathology-enriched brain regions, including the hippocampus and inferior frontal cortex, when compared to the cerebellum where AD pathology is limited. The interactions of RAGE with its diverse ligands (AGEs, Aβ, amphoterin, and S100B protein) mediate multiple physiological and pathological functions, including inflammation, oxidative stress, neurodegeneration, maintenance of homeostasis, tumorigenesis, promotion of neurite outgrowth, cell survival, cell migration, and neuronal differentiation (Fages et al. 2000; Ramasamy et al. 2005; Meneghini et al. 2010). These processes involve NF-κB, MAPKs, PtdIns 3K/Akt, Rho GTPases, JAK/STAT, and Src family kinases (Arumugam et al. 2004; Fang et al. 2010; Kislinger et al. 1999). Blocking the activation of RAGE with neutralizing antibody in the presence of retinoic acid disrupted the progression of normal neuronal differentiation. RAGEs are expressed on multiple cell types and AGE binding with RAGE stimulates mitogen-activated protein kinases (MAPKs), the phosphatidylinositol 3-kinase (PtdIns 3K), JAK/STAT (signal transducers and activators of transcription), and the Rho GTPases Rac-1 and Cdc42 pathways leading to the activation of NF-κB, which translocates to the nucleus where it activates the transcription of genes for cytokines, growth factors, and adhesive molecules, such as tumor necrosis factor-α (TNF-α), interleukin-6 (IL-6), well-known inflammation promoters, and vascular cell adhesion molecule 1 (VCAM1) (Xing et al. 1998; Kalousova et al. 2002; Lukic et al. 2008; Wautier et al. 2001). In addition, AGE–RAGE interaction activates NADPH oxidase (a complex of enzymes which produces superoxide) and when this complex is upregulated, it increases intracellular oxidative stress and promotes cellular injury (Fig. 2.4). These processes result in the downregulation of superoxide dismutase isoforms, glutathione peroxidase, and heme oxygenase-2 in various body tissues (Roberts et al. 2006) not only leading to the elevation in plasma levels of malondialdehyde but also in impairment of vasodilatory response to acetylcholine.

Fructose-derived AGEs not only promote the formation of cross-linkages between key molecules in the basement membrane but also interact with specific receptors on the cell surfaces resulting in abnormal intracellular signaling and disrupting cell function (Fig. 2.4) (Suárez et al. 1989; Seneff et al. 2011). Unlike glucose, which is utilized by all organs, fructose is solely metabolized in the

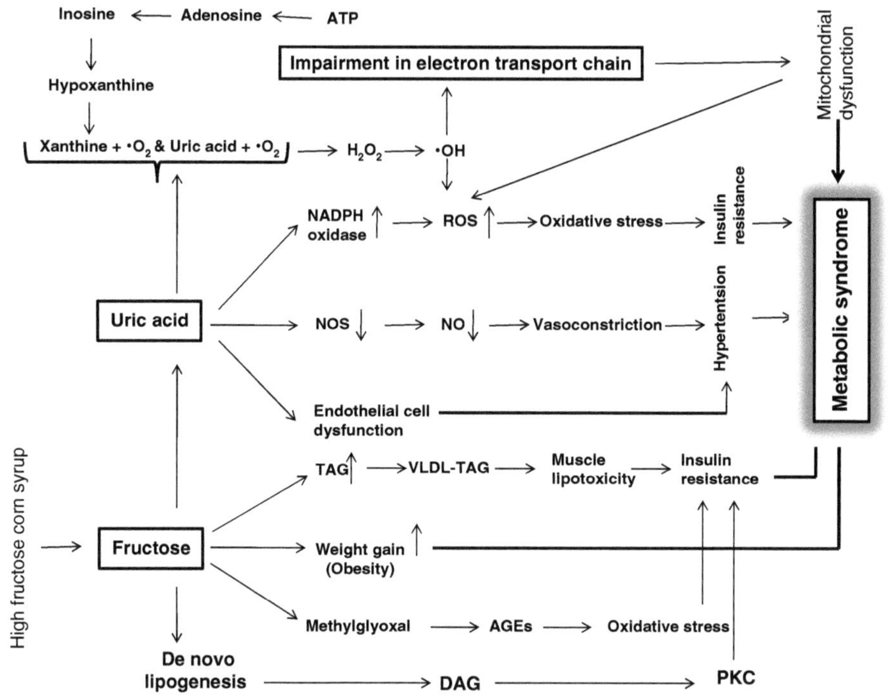


**Fig. 2.6** Biochemical effects of fructose-derived uric acid. Superoxide anions ($O_2^{\bullet-}$); hydroxyl ($^{\bullet}OH$); nicotinamide adenine dinucleotide phosphate oxidase (NADP oxidase); *upward arrow* indicates increase and *downward arrow* indicates decrease

liver. The enzyme that metabolizes fructose (fructokinase) is not regulated by negative feedback, so all fructose that enters the liver cells is rapidly phosphorylated by ATP leading to ATP depletion, a finding, which has been consistently observed in in vivo animal models of diabetes and humans. ATP depletion activates enzymes of purine metabolism (AMP deaminase-1), which degrade adenine nucleotides to uric acid via xanthine oxidoreductase with the development of hyperuricemia (Hallfrisch 1990; Nakagawa et al. 2005) (Fig. 2.6). Thus, uric acid is synthesized in the liver from purine compounds provided by the diet or by the endogenous pathway of purine synthesis de novo. Serum uric acid levels, even within the normal range, are associated with other cardiovascular risk factors and predict cardiovascular events in adults (Gao et al. 2007; Gagliardi et al. 2009). Biologically, uric acid plays an important role in worsening of insulin resistance in animal models of diabetes by inhibiting the bioavailability of nitric oxide, which is essential for insulin-stimulated glucose uptake (Khosla et al. 2005). Serum uric acid levels are also associated with its deleterious effects on endothelial function, platelet adhesion and aggregation, or oxidative metabolism leading to hypertension and cardiovascular disease (Alper et al. 2005) (Fig. 2.7). In addition, increase in uric acid is also associated with elevated circulating levels of systemic inflammatory mediators, such as monocyte chemoattractant protein-1, NF-κB, interleukin-1β, interleukin-6, and tumor necrosis

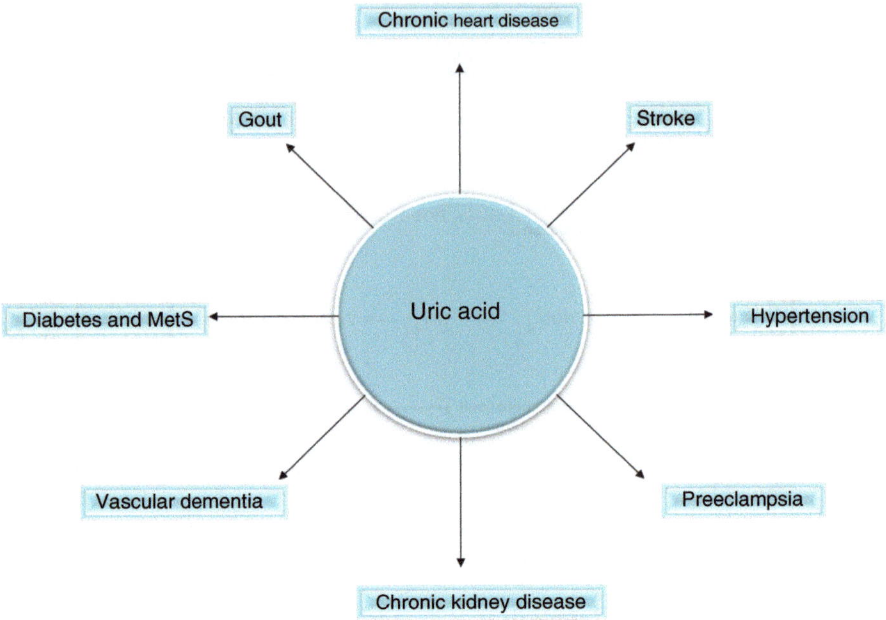


**Fig. 2.7** Pathological conditions associated with elevated levels of uric acid

factor-α, and vascular smooth muscle proliferation (Johnson et al. 2003; Kanellis et al. 2004).

There are several proposed neurohormonal links between weight loss and blood pressure. Body mass index is an important mediator of the weight–blood pressure relationship. There is a mostly linear relationship between weight and blood pressure; as weight is regained, the blood pressure benefit is mostly lost. Physical activity, but more so physical fitness (the physiological benefit obtained from physical activity), has a dose-dependent blood pressure benefit but reaches a plateau at which there is no further benefit (Frisoli et al. 2011). However, even just a modest physical activity can have a meaningful blood pressure effect. High blood pressure is accompanied by sustained arterial pressure produced by a combination of acquired or genetic metabolic defects involved in blood pressure regulation. Blood pressure is regulated by environmental factors such as diet and lifestyle (Deshmukh et al. 1998). Although, uric acid increases blood pressure through the inhibition of nitric oxide synthase, but several other possibilities have been proposed. These possibilities include increase in sympathetic nervous system activity (Farah et al. 2006; Verma et al. 1999), insulin resistance, increased oxidative stress, elevation in circulating catecholamines (Tran et al. 2009), decrease in nitric oxide (NO) bioavailability, enhancement in renin–angiotensin system activity and angiotensin II levels (a potent vasoconstrictor) (Tran et al. 2009), increase in sodium reabsorption (De Fronzo 1981), impaired endothelium-dependent relaxation (Katakam et al. 1998),

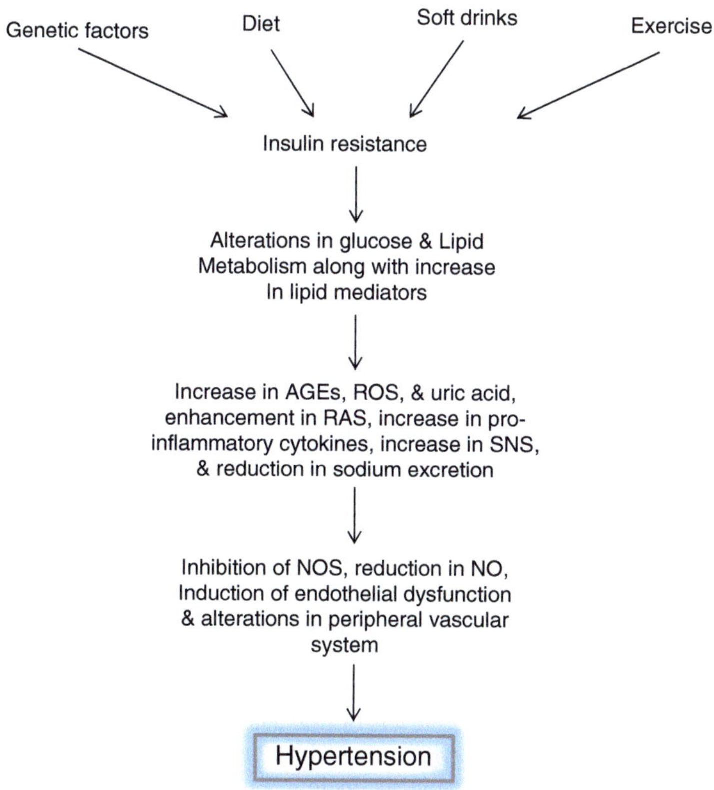


**Fig. 2.8** Factors that modulate hypertension. Advanced glycation products (AGEs); reactive oxygen species (ROS); renin–angiotensin system (RAS); sympathetic nervous system (SNS); nitric oxide synthase (NOS); and nitric oxide (NO)

and increase in secretion of endothelin-1 (ET-1) (Juan et al. 1998) (Fig. 2.8). Many of these factors may also contribute to an increased vascular tone and impaired endothelial function. Furthermore, it is also reported that hyperinsulinemia itself may cause hypertension as long-term insulin administration causes an increased blood pressure in rats (Meehan et al. 1994), which is reversed upon discontinuation of the insulin (Hsieh and Huang 1993). Hypertension impairs functional hyperemia, the process by which brain activity and blood flow are coordinated. This impairment is induced by dysregulation of vasoactive mediators such as nitric oxide (NO) and endothelin-1, by oxidative stress, by structural alteration of the blood vessels, and by inadequate cerebral autoregulation (Iadecola and Davisson 2008), and all of these processes have been linked to insulin resistance (Crafts 2009).

In addition to high blood pressure, increase in uric acid levels also promotes gout (Yoo et al. 2009). In vitro, uric acid increases NADPH oxidase activity and oxidative stress in adipocytes, leading to increase in p38 MAP kinase activity and insulin resistance (Sautin et al. 2007; Lee et al. 2009). Lowering uric acid in fructose-fed rats improves components of the MetS, including a reduction in blood pressure, serum triglycerides, hyperinsulinemia, and weight gain supporting the view that uric acid may play a causal role in fructose-mediated MetS in animal models (Nakagawa et al. 2006). Thus, intake of fructose- and purine-rich foods raises uric acid levels and may play a role in the epidemic of MetS that is occurring around the globe (Cirillo et al. 2006, 2009). Uric acid-mediated derangement of renal function is an independent predictor not only of cardiovascular disease outcome but also gout (Baker et al. 2005). In addition, several studies have indicated that uric acid is also associated with diabetic nephropathy along with production of cytokines and induction of inflammatory cascades. All these processes may contribute to progression of microvascular disease and thereby resulting in renal injury in diabetic nephropathy, a major cause of morbidity and mortality in patients with diabetes (Jalal et al. 2010, 2011).

Uric acid also functions as an antioxidant. The antioxidant effects of uric acid are protective in several neurological diseases, including multiple sclerosis and Parkinson disease. However, several observational studies have indicated that high levels of serum uric acid increase the risk of stroke. The current data show elevation in uric acid in high-fat diet associated with hypertriglyceridemia and hypercholesterolemia suggesting that an increased level of uric acid may be a risk factor for the presence of silent brain infarction (SBI). So serum uric acid level might be a good serum marker of underlying SBI or future stroke (Heo and Lee 2010) (Fig. 2.7).

The exact mechanism by which hypertriglyceridemia, hypercholesterolemia, along with generation of elevated levels of uric acid may promote SBI is unknown yet. However, there are many ways through which generation of uric acid may affect the brain. The proposed mechanisms include enhanced inflammatory activity, increased action of free radicals, and reduction in insulin transport into the brain in the subjects with MetS (Lusis et al. 2008). Considering that the majority of SBI are caused by small vessel occlusion, the risk factors of lacunar infarctions can be the plausible suspects of SBI in MetS. Conversely, uric acid can also function as a pro-oxidant, either by generating radicals during its degradation or by stimulating NADPH oxidase (Sautin et al. 2007). Uric acid can also stimulate innate immunity through the effects of microcrystalline uric acid on the function of dendritic cells and T cells (Shi et al. 2006).

## 2.5 Conclusion

Although, glucose is metabolized by liver and brain tissues for energy production, but fructose is primarily metabolized by the liver. Unlike glucose, fructose is not utilized by the brain and does not stimulate insulin secretion, due to its hepatic metabolism and the low level of expression of the fructose transporter GLUT5 in pancreatic β cells. Metabolism of fructose differs from glucose not only in the involvement of different transporters but also in the first three enzymes involved in its metabolism. The enzyme fructokinase uses ATP to transform fructose into fructose

1-phosphate. Unlike enzymes associated with glucose metabolism (glucokinase and phosphofructokinase), which are regulated by ATP, the fructokinase is not regulated by ATP. This results in production of fructose-1-phosphate, depletion of ATP, and generation of AMP, which is then metabolized to uric acid. Uric acid causes proinflammatory effects on vascular cells and adipocytes. It also functions as an antioxidant. Uric acid also stimulates innate immunity through the effects of microcrystalline uric acid on the function of dendritic cells and T cells. In the liver, high-fructose flux also leads to enhanced accumulation of hepatic TAG resulting in impaired glucose and lipid metabolism and increase risk of MetS and cardiovascular diseases. Fructose produces damaging effects in hepatocytes, tubular cells, adipocytes, and intestinal epithelial cells because it is highly reactive as a reducing agent and a precursor of AGE. The liver promotes removal of high levels of fructose aggressively from the bloodstream to prevent the damaging effects of glycation/fructation on serum lipids and proteins. Glycated/fructated proteins not only show impaired functions but are also more susceptible to oxidative damage. Glycated/fructated proteins are ultimately converted into toxic AGEs. In humans, metabolism of high levels of fructose and generation of AGE are associated with increased total energy intake, body weight, visceral adiposity, insulin resistance, and dyslipidemia. These abnormalities contribute to the pathogenesis of obesity, diabetes, MetS, fatty liver, cardiovascular diseases, and neurological disorders. Accumulating evidence suggests that fructose is distinct from glucose in its ability to cause intracellular phosphate depletion, ATP depletion, and uric acid generation in the liver. High levels of intracellular uric acid, AGEs, and ROS can induce inflammatory effects and oxidative stress not only in vascular cells and adipocytes but also in the liver and brain.

## References


Ahmed N (2005) Advanced glycation endproducts—role in pathology of diabetic complications. Diabetes Res Clin Pract 67:3–21

Alexander MC, Lomanto M, Nasrin N, Ramaika C (1988) Insulin stimulates glyceraldehyde-3-phosphate dehydrogenase gene expression through cis-acting DNA sequences. Proc Natl Acad Sci USA 85:5092–5096

Alper AB Jr, Chen W, Yau L, Srinivasan SR, Berenson GS, Hamm LL (2005) Childhood uric acid predicts adult blood pressure: the Bogalusa Heart Study. Hypertension 45:34–38

Arnlov J, Vessby B, Riserus U (2004) Coffee consumption and insulin sensitivity. JAMA 291:1199–1201

Arumugam T, Simeone DM, Schmidt AM, Logsdon CD (2004) S100P stimulates cell proliferation and survival via receptor for activated glycation end products (RAGE). J Biol Chem 279: 5059–5065

Asipu A, Hayward BE, O'Reilly J, Bonthron DT (2003) Properties of normal and mutant recombinant human ketohexokinases and implications for the pathogenesis of essential fructosuria. Diabetes 52:2426–2432

Avena NM, Rada P, Hoebel BG (2006) Sugar bingeing in rats. Curr Protoc Neurosci Chapter 9:Unit9.23C

Avena NM, Rada P, Hoebel BG (2008) Evidence for sugar addiction: behavioral and neurochemical effects of intermittent, excessive sugar intake. Neurosci Biobehav Rev 32:20–39

Baker JF, Krishnan E, Chen L, Schumacher HR (2005) Serum uric acid and cardiovascular disease: recent developments, and where do they leave us? Am J Med 118:816–826

Barberger-Gateau, Samieri C, Féart C, Plourde M (2011) Dietary omega 3 polyunsaturated fatty acids and Alzheimer's disease: interaction with apolipoprotein E genotype. Curr Alzheimer Res 8:479–491

Basciano H, Federico L, Adeli K (2005) Fructose, insulin resistance, and metabolic dyslipidemia. Nutr Metab (Lond) 2:5

Beisswenger PJ, Howell SK, Smith K, Szwergold BS (2003) Glyceraldehyde-3-phosphate dehydrogenase activity as an independent modifier of methylglyoxal levels in diabetes. Biochim Biophys Acta 1637:98–106

Benton CS, Miller BH, Skwerer S, Suzuki O, Schultz LE, Cameron MD, Marron JS, Pletcher MT, Wiltshire T (2011) Evaluating genetic markers and neurobiochemical analytes for fluoxetine response using a panel of mouse inbred strains. Psychopharmacology (Berl) 221:297–315

Bierhaus A, Fleming T, Stoyanov S, Leffler A, Babes A, Neacsu C, Sauer SK, Eberhardt M, Schnölzer M, Lasitschka F, Neuhuber WL, Kichko TI, Konrade I, Elvert R, Mier W, Pirags V, Lukic IK, Morcos M, Dehmer T, Rabbani N, Thornalley PJ, Edelstein D, Nau C, Forbes J, Humpert PM, Schwaninger M, Ziegler D, Stern DM, Cooper ME, Haberkorn U, Brownlee M, Reeh PW, Nawroth PP (2012) Methylglyoxal modification of Na(v)1.8 facilitates nociceptive neuron firing and causes hyperalgesia in diabetic neuropathy. Nat Med 18:926–933

Bray GA, Nielsen SJ, Popkin BM (2004) Consumption of high-fructose corn syrup in beverages may play a role in the epidemic of obesity. Am J Clin Nutr 79:537–543

Brownlee M (2001) Biochemistry and molecular cell biology of diabetic complications. Nature 414:813–820

Brownlee M (2005) The pathobiology of diabetic complications: a unifying mechanism. Diabetes 54:1615–1625

Castrogiovanni D, Alzamendi A, Ongaro L, Giovambattista A, Gaillard RC, Spinedi E (2012) Fructose rich diet-induced high plasminogen activator inhibitor-1 (PAI-1) production in the adult female rat: protective effect of progesterone. Nutrients 4:1137–1150

Cha SH, Wolfgang M, Tokutake Y, Chohnan S, Lane MD (2008) Differential effects of central fructose and glucose on hypothalamic malonyl-CoA and food intake. Proc Natl Acad Sci USA 105:16871–16875

Chaney MO, Stine WB, Kokjohn TA, Kuo YM, Esh C, Rathman A, Luehr DC, Schmidt AM, Stern D, Yan SD, Roher AE (2005) RAGE and amyloid β interactions: atomic force microscopy and molecular modeling. Biochim Biophys Acta 1741:199–205

Chang KC, Liang JT, Tseng CD, Wu C-D, Hsu K-L, Wu M-S, Lin Y-T, Tseng T (2007) Aminoguanidine prevents fructose-induced deterioration in left ventricular-arterial coupling in Wistar rats. Br J Pharmacol 151:341–346

Chang T, Untereiner A, Liu J, Wu L (2010) Interaction of methylglyoxal and hydrogen sulfide in rat vascular smooth muscle cells. Antioxid Redox Signal 12:1093–1100

Charonis AS, Reger LA, Dege JE, Kouzi-Koliakos K, Furcht LT, Wohlhueter RM, Tsilibary EC (1990) Laminin alterations after in vitro nonenzymatic glycosylation. Diabetes 39:807–814

Choi HK, Willett W, Curhan G (2007) Coffee consumption and risk of incident gout in men: a prospective study. Arthritis Rheum 56:2049–2055

Cirillo P, Sato W, Reungjui S, Heinig M, Gersch M, Sautin Y, Nakagawa T, Johnson RJ (2006) Uric acid, the metabolic syndrome, and renal disease. J Am Soc Nephrol 17:S165–S168

Cirillo P, Gersch M, Mu W, Scherer PM, Kim KM, Gesualdo L, Henderson GN, Johnson RJ, Sautin YY (2009) Ketohexokinase-dependent proinflammatory effect of fructose in human proximal tubular cells. J Am Soc Nephrol 20:545–553

Convit A (2005) Links between cognitive impairment in insulin resistance: an explanatory model. Neurobiol Aging 26(Suppl 1):31–35

Cox CL, Stanhope KL, Schwarz JM, Graham JL, Hatcher B, Griffen SC, Bremer AA, Berglund L, McGahan JP, Havel PJ, Keim NL (2012) Consumption of fructose-sweetened beverages for 10 weeks reduces net fat oxidation and energy expenditure in overweight/obese men and women. Eur J Clin Nutr 66:201–208

Craft S (2009) The role of metabolic disorders in Alzheimer disease and vascular dementia: two roads converged. Arch Neurol 66:300–305

Cunnane S, Nugent S, Roy M, Courchesne-Loyer A, Croteau E, Tremblay S, Castellano A, Pifferi F, Bocti C, Paquet N, Begdouri H, Bentourkia M, Turcotte E, Allard M, Barberger-Gateau P, Fulop T, Rapoport SI (2011) Brain fuel metabolism, aging, and Alzheimer's disease. Nutrition 27:3–20

Curry DL (1989) Effects of mannose and fructose on the synthesis and secretion of insulin. Pancreas 4:2–9

De Fronzo RA (1981) The effect of insulin on renal sodium metabolism. A review with clinical implications. Diabetologia 21:165–171

De Fronzo RA (1997) Pathogenesis of type 2 diabetes: metabolic and molecular implications for identifying diabetes genes. Diabetes Rev 5:177–269

Deanfield J, Donald A, Ferri C, Giannattasio C, Halcox J, Halligan S, Lerman A, Mancia G, Oliver JJ, Pessina AC, Rizzoni D, Rossi GP, Salvetti A, Schiffrin EL, Taddei S, Webb DJ, Working Group on Endothelin and Endothelial Factors of the European Society of Hypertension (2005) Endothelial function and dysfunction. Part I: methodological issues for assessment in the different vascular beds: a statement by the working group on endothelin and endothelial factors of the European society of hypertension. J Hypertens 23:7–17

Deshmukh R, Smith A, Lilly LS (1998) Hypertension. In: Lilly LS (ed) Pathophysiology of heart disease, 2nd edn. Lippincott Williams and Wilkins, Philadelphia, PA, pp 267–288

Dhar A, Dhar I, Jiang B, Desai KM, Wu L (2011) Chronic methylglyoxal infusion by minipump causes pancreatic beta cell dysfunction and induces type 2 diabetes in Sprague-Dawley rats. Diabetes 60:899–908

Diggle CP, Shires M, Leitch D, Brooke D, Carr IM, Markham AF, Hayward BE, Asipu A, Bonthron DT (2009) Ketohexokinase: expression and localization of the principal fructose-metabolizing enzyme. J Histochem Cytochem 57:763–774

Distler MG, Plant LD, Sokoloff G, Hawk AJ, Aneas I, Wuenschell GE, Termini J, Meredith SC, Nobrega MA, Palmer AA (2012) Glyoxalase 1 increases anxiety by reducing GABAA receptor agonist methylglyoxal. J Clin Invest 122:2306–2315

Dou JT, Chen M, Dufour F, Alkon DL, Zhao WQ (2005) Insulin receptor signaling in long-term memory consolidation following spatial learning. Learn Mem 12:646–655

Du X, Matsumura T, Edelstein D, Rossetti L, Zsengeller Z, Szabo C, Brownlee M (2003) Inhibition of GAPDH activity by poly(ADP-ribose) polymerase activates three major pathways of hyperglycemic damage in endothelial cells. J Clin Invest 112:1049–1057

Dyck JR, Lopaschuk GD (2006) AMPK alterations in cardiac physiology and pathology: enemy or ally? J Physiol 574:95–112

Elliott SS, Keim NL, Stern JS, Teff K, Havel PJ (2002) Fructose, weight gain, and the insulin resistance syndrome. Am J Clin Nutr 76:911–922

Fages C, Nolo R, Huttunen HJ, Eskelinen E, Rauvala H (2000) Regulation of cell migration by amphoterin. J Cell Sci 113:611–620

Fang F, Lue LF, Yan S, Xu H, Luddy JS, Chen D, Walker DG, Stern DM, Yan S, Schmidt AM, Chen JX, Yan SS (2010) RAGE-dependent signaling in microglia contributes to neuroinflammation, Abeta accumulation, and impaired learning/memory in a mouse model of Alzheimer's disease. FASEB J 24:1043–1055

Farah V, Elased KM, Chen Y, Key MP, Cunha TS, Ingoyen M, Morris M (2006) Nocturnal hypertension in mice consuming a high fructose diet. Auton Neurosci 130:41–50

Farooqui AA, Farooqui T, Panza F, Frisardi V (2012) Metabolic syndrome as a risk factor for neurological disorders. Cell Mol Life Sci 69:741–762

Freemantle E, Vandal M, Tremblay-Mercier J, Plourde M, Poirier J, Cunnane S (2009) Metabolic response to a ketogenic breakfast in the healthy elderly. J Nutr Health Aging 13:293

Frisoli TM, Schmieder RE, Grodzicki T, Messerli FH (2011) Beyond salt: lifestyle modifications and blood pressure. Eur Heart J 32:3081–3087

Froesch ER (1972) Fructose metabolism in adipose tissue. Acta Med Scand Suppl 542:37–46

Funk SD, Yurdogul A Jr, Orr W (2012) Hyperglycemia and endothelial dysfunction in atherosclerosis: lessons from type 1 diabetes. Int J Vasc Med 2012:569654

Furlong J (1996) Acetyl-L-carnitine: metabolism and applications in clinical practice. Altern Med Rev 1:85–93

Gagliardi ACM, Miname MH, Santos RD (2009) Uric acid: a marker of increased cardiovascular risk. Atherosclerosis 202:11–17

Gao X, Qi L, Qiao N, Choi HK, Curhan G, Tucker KL, Ascherio A (2007) Intake of added sugar and sugar-sweetened drink and serum uric acid concentration in US men and women. Hypertension 50:306–312

Giacco F, Brownlee M (2010) Oxidative stress and diabetic complications. Circ Res 107:1058–1070

Goldin A, Beckman JA, Schmidt AM, Creager MA (2006) Advanced glycation end products: sparking the development of diabetic vascular injury. Circulation 114:597–605

Gong Z, Muzumdar RH (2012) Pancreatic function, Type 2 diabetes, and metabolism in aging. Int J Endocrinol PMC 3362843

Hallfrisch J (1990) Metabolic effects of dietary fructose. FASEB J 4:2652–2660

Havel PJ (2002) Control of energy homeostasis and insulin action by adipocyte hormones: leptin, acylation stimulating protein, and adiponectin. Curr Opin Lipidol 13:51–59

Havel PJ (2005) Dietary fructose: implications for dysregulation of energy homeostasis and lipid/carbohydrate metabolism. Nutr Rev 63:133–157

Hayward BE, Bonthron DT (1998) Structure and alternative splicing of the ketohexokinase gene. Eur J Biochem 257:85–91

Henderson ST, Vogel JL, Barr LJ, Garvin F, Jones JJ, Costantini LC (2009) Study of the ketogenic agent AC-1202 in mild to moderate Alzheimer's disease: a randomized, double-blind, placebo-controlled, multicenter trial. Nutr Metab (Lond) 6:31

Heo SH, Lee SH (2010) High levels of serum uric acid are associated with silent brain infarction. J Neurol Sci 297:6–10

Higai K, Shimamura A, Matsumoto K (2006) Amadori-modified glycated albumin predominantly induces E-selectin expression on human umbilical vein endothelial cells through NADPH oxidase activation. Clin Chim Acta 367:137–143

Hovatta I, Tennant RS, Helton R, Marr RA, Singer O, Redwine JM, Ellison JA, Schadt EE, Verma IM, Lockhart DJ, Barlow C (2005) Glyoxalase 1 and glutathione reductase 1 regulate anxiety in mice. Nature 438:662–666

Hsieh PS, Huang WC (1993) Chemical sympathectomy attenuates hyperinsulinemia-induced hypertension in conscious rats. Nutr Metab Cardiovasc Dis 3:173–178

Hudson BI, Carter AM, Harja E, Kalea AZ, Arriero M, Yang H, Grant PJ, Schmidt AM (2008) Identification, classification, and expression of RAGE gene splice variants. FASEB J 22:1572–1580

Iadecola C, Davisson RL (2008) Hypertension and cerebrovascular dysfunction. Cell Metab 7:476–484

Inoguchi T, Battan R, Handler E, Sportsman JR, Heath W, King GL (1992) Preferential elevation of protein kinase C isoform beta II and diacylglycerol levels in the aorta and heart of diabetic rats: differential reversibility to glycemic control by islet cell transplantation. Proc Natl Acad Sci USA 89:11059–11063

Jack MM, Ryals JM, Wright DE (2011) Characterisation of glyoxalase I in a streptozocin-induced mouse model of diabetes with painful and insensate neuropathy. Diabetologia 54:2174–2182

Jalal DI, Rivard CJ, Johnson RJ, Maahs DM, McFann K, Rewers M, Snell-Bergeon JK (2010) Serum uric acid levels predict the development of albuminuria over 6 years in patients with type 1 diabetes: findings from the Coronary Artery Calcification in Type 1 Diabetes study. Nephrol Dial Transplant 25:1865–1869

Jalal DI, Maahs DM, Hovind P, Nakagawa T (2011) Uric acid as a mediator of diabetic nephropathy. Semin Nephrol 31:459–465

Jianghai L, Wang R, Desai K, Wu L (2012) Upregulation of aldolase B and overproduction of methylglyoxal in vascular tissues from rats with metabolic syndrome. Cardiovasc Res 92:494–503

Johnson R, Kang D, Feig D, Kivlighn S, Kanellis J, Watanabe S, Tuttle KR, Rodriguez-Iturbe B, Herrera-Acosta J, Mazzali M (2003) Is there a pathogenetic role for uric acid in hypertension and cardiovascular and renal disease? Hypertension 41:1183–1189

Juan CC, Fang VS, Hsu YP, Huang DB, Hsia DB, Yu PC, Kwok CF, Ho LT (1998) Overexpression of vascular endothelin-1 and endothelin-A receptors in a fructose-induced hypertensive rat model. J Hypertens 16:1775–1782

Kalousova M, Skrha J, Zima T (2002) Advanced glycation end-products and advanced oxidation protein products in patients with diabetes mellitus. Physiol Res 51:597–604

Kalpouzos G, Eustache F, de la Sayette V, Viader F, Chetelat G, Desgranges B (2009a) Working memory and FDG-PET dissociate early and late onset alzheimer disease patients. J Neurol 252:548–558

Kalpouzos G, Chetelat G, Baron JC, Landeau B, Mevel K, Godeau C, Barré L, Constans JM, Viader F, Eustache F, Desgranges B (2009b) Voxel-based mapping of brain gray matter volume and glucose metabolism profiles in normal aging. Neurobiol Aging 30:112–124

Kanellis J, Feig D, Johnson R (2004) Does asymptomatic hyperuricaemia contribute to the development of renal and cardiovascular disease? An old controversy renewed. Nephrology (Carlton) 9:394–399

Katakam PV, Ujhelyi MR, Hoenig ME, Miller AW (1998) Endothelial dysfunction precedes hypertension in diet-induced insulin resistance. Am J Physiol 275:R788–R792

Kawasaki T, Igarashi K, Koeda T, Sugimoto K, Nakagawa K, Hayashi S, Yamaji R, Inui H, Fukusato T, Yamanouchi T (2009) Rats fed fructose-enriched diets have characteristics of non-alcoholic hepatic steatosis. J Nutr 139:2067–2071

Kela U, Vijayvargiya R, Trivedi CP (1980) Inhibitory effects of methylxanthines on the activity of xanthine oxidase. Life Sci 27:2109–2119

Kelley D, Mitrakou A, Marsh H, Schwenk F, Benn J, Sonnenberg G, Arcangeli M, Aoki T, Sorensen J, Berger M (1988) Skeletal muscle glycolysis, oxidation, and storage of an oral glucose load. J Clin Invest 81:1563–1571

Khosla UM, Zharikov S, Finch JL, Nakagawa T, Roncal C, Mu W, Krotova K, Block ER, Prabhakar S, Johnson RJ (2005) Hyperuricemia induces endothelial dysfunction. Kidney Int 67:1739–1742

Kikuchi S, Shinpo K, Takeuchi M, Yamagishi S, Makita Z, Sasaki N, Tashiro K (2003) Glycation—a sweet tempter for neuronal death. Brain Res Brain Res Rev 41:306–323

Kim W, Hudson BI, Moser B, Guo J, Rong LL, Lu Y, Qu W, Lalla E, Lerner S, Chen Y, Yan SS, D'Agati V, Naka Y, Ramasamy R, Herold K, Yan SF, Schmidt AM (2005) Receptor for advanced glycation end products and its ligands: a journey from the complications of diabetes to its pathogenesis. Ann N Y Acad Sci 1043:553–561

Kim DS, Jeong SK, Kim HR, Kim DS, Chae SW, Chae HJ (2007) Effects of triglyceride on ER stress and insulin resistance. Biochem Biophys Res Commun 363:140–145

Kislinger T, Fu C, Huber B, Qu W, Taguchi A, Du Yan S, Hofmann M, Yan SF, Pischetsrieder M, Schmidt AM (1999) N(epsilon)-(carboxymethyl)lysine adducts of proteins are ligands for receptor for advanced glycation end products that activate cell signaling pathways and modulate gene expression. J Biol Chem 274:31740–31749

Koo HY, Wallig MA, Chung BH, Nara TY, Cho BH, Nakamura MT (2008) Dietary fructose induces a wide range of genes with distinct shift in carbohydrate and lipid metabolism in fed and fasted rat liver. Biochim Biophys Acta 1782:341–348

Kuhla B, Luth HJ, Haferburg D, Weick M, Reichenbach A, Arendt T, Münch G (2006) Pathological effects of glyoxalase I inhibition in SH-SY5Y neuroblastoma cells. J Neurosci Res 83:1591–1600

Lanaspa MA, Sanchez-Lozada LG, Cicerchi C, Li N, Roncal-Jimenez CA, Ishimoto T, Le M, Garcia GE, Thomas JB, Rivard CJ, Andres-Hernando A, Hunter B, Schreiner G, Rodriguez-Iturbe B, Sautin YY, Johnson RJ (2012) Uric acid stimulates fructokinase and accelerates fructose metabolism in the development of fatty liver. PLoS One 7:e47948

Lane MD, Cha SH (2009) Effect of glucose and fructose on food intake via malonyl-CoA signaling in the brain. Biochem Biophys Res Commun 382:1–5

Lane MD, Wolfgang M, Cha SH, Dai Y (2008) Regulation of food intake and energy expenditure by hypothalamic malonyl-CoA. Int J Obes (Lond) 32(Suppl 4):S49–S54
Lee MY, Martin AS, Mehta PK, Dikalova AE, Garrido AM, Lyons E (2009) Mechanisms of vascular smooth muscle NADPH oxidase 1 (Nox1) contribution to injury-induced neointimal formation. Arterioscler Thromb Vasc Biol 29:103–110
Lehninger A, Nelson D, Cox M (2005) Lehninger principles of biochemistry. W. H. Freeman, New York
Levin BE, Kang L, Sanders NM, Dunn-Meynell AA (2006) Role of neuronal glucosensing in the regulation of energy homeostasis. Diabetes 55(Suppl 2):S122–S130
Li JM, Li YC, Kong LD, Hu QH (2010) Curcumin inhibits hepatic protein-tyrosine phosphatase 1B and prevents hypertriglyceridemia and hepatic steatosis in fructose-fed rats. Hepatology 51:1555–1566
Locke GA, Cheng D, Witmer MR, Tamura JK, Haque T, Carney RF, Rendina AR, Marcinkeviciene J (2008) Differential activation of recombinant human acetyl-CoA carboxylases 1 and 2 by citrate. Arch Biochem Biophys 475:72–79
Lonardo A, Lombardini S, Ricchi M, Scaglioni F, Loria P (2005) Review article: hepatic steatosis and insulin resistance. Aliment Pharmacol Ther 22:64–70
Loos M, van der Sluis S, Bochdanovits Z, van Zutphen IJ, Pattij T, Stiedl O, Neuro-BSIK Mouse Phenomics consortium, Smit AB, Spijker S (2009) Activity and impulsive action are controlled by different genetic and environmental factors. Genes Brain Behav 8:817–828
Lue LF, Walker DG, Brachova L, Beach TG, Roger J, Schmidt AM, Stern DM, Yan SD (2001) Involvement of microglial receptor for advanced glycation endproducts (RAGE) in Alzheimer's disease: identification of a cellular activation mechanism. Exp Neurol 171:29–45
Lukic IK, Humpert PM, Nawroth PP, Bierhaus A (2008) The RAGE pathway: activation and perpetuation in the pathogenesis of diabetic neuropathy. Ann N Y Acad Sci 1126:76–80
Lusis AJ, Attie AD, Reue K (2008) Metabolic syndrome: from epidemiology to systems biology. Nat Rev Genet 9:819–830
Madonna R, De Caterina R (2011) Cellular and molecular mechanisms of vascular injury in diabetes—Part II: cellular mechanisms and therapeutic targets. Vascul Pharmacol 54:75–79
Maenpaa PH, Raivio KO, Kekomaki MP (1968) Liver adenine nucleotides: fructose-induced depletion and its effect on protein synthesis. Science 161:1253–1254
Magnusson I, Rothman D, Katz L, Shulman R, Shulman G (1992) Increased rate of gluconeogenesis in type II diabetes mellitus. A 13C nuclear magnetic resonance study. J Clin Invest 90:1323–1327
Mann GE, Yudilevich DL, Sobrevia L (2003) Regulation of amino acid and glucose transporters in endothelial and smooth muscle cells. Physiol Rev 83:183–252
Marshall RO, Kooi ER (1957) Enzymatic conversion of D-glucose to D-fructose. Science 125:648–649
Matsuzaka T, Shimano H, Yahagi N, Amemiya-Kudo M, Okazaki H, Tamura Y, Iizuka Y, Ohashi K, Tomita S, Sekiya M, Hasty A, Nakagawa Y, Sone H, Toyoshima H, Ishibashi S, Osuga J, Yamada N (2004) Insulin-independent induction of sterol regulatory element-binding protein-1c expression in the livers of streptozotocin-treated mice. Diabetes 53:560–569
McLellan AC, Thornalley PJ, Benn J, Sonksen PH (1994) Glyoxalase system in clinical diabetes mellitus and correlation with diabetic complications. Clin Sci (Lond) 87:21–29
McPherson JD, Shilton BH, Walton DJ (1988) Role of fructose in glycation and cross-linking of proteins. Biochemistry 27:1901–1907
Meehan WP, Buchanan TA, Hsueh W (1994) Chronic insulin administration elevates blood pressure in rats. Hypertension 23:1012–1017
Méndez JD, Xie J, Aguilar-Hernández M, Méndez-Valenzuela V (2010) Trends in advanced glycation end products research in diabetes mellitus and its complications. Mol Cell Biochem 341:33–41
Meneghini V, Francese MT, Carraro L, Grilli M (2010) A novel role for the receptor for advanced glycation end-products in neural progenitor cells derived from adult subventricular zone. Mol Cell Neurosci 45:139–150

Messier C (2005) Impact of impaired glucose tolerance and type 2 diabetes on cognitive aging. Neurobiol Aging 26(Suppl 1):26–30
Miller CC, Martin RJ, Whitney ML, Edwards GL (2002) Intracerebroventricular injection of fructose stimulates feeding in rats. Nutr Neurosci 5:359–362
Moore MC, Davis SN, Mann SL, Cherrington AD (2001) Acute fructose administration improves oral glucose tolerance in adults with type 2 diabetes. Diabetes Care 24:1882–1887
Morgello S, Uson RR, Schwartz EJ, Haber RS (1995) The human blood-brain barrier glucose transporter (glut1) is a glucose transporter of gray matter astrocytes. Glia 14:43–54
Morino K, Petersen KF, Dufour S, Befroy D, Frattini J, Shatzkes N, Neschen S, White MF, Bilz S, Sono S, Pypaert M, Shulman GI (2005) Reduced mitochondrial density and increased IRS-1 serine phosphorylation in muscle of insulin-resistant offspring of type 2 diabetic parents. J Clin Invest 115:3587–3593
Morino K, Petersen KF, Shulman GL (2006) Molecular mechanisms of insulin resistance in humans and their potential links with mitochondrial dysfunction. Diabetes 55(Suppl 2):S9–S15
Nakagawa T, Tuttle KR, Short RA, Johnson RJ (2005) Hypothesis: fructose-induced hyperuricemia as a causal mechanism for the epidemic of the metabolic syndrome. Nat Clin Pract Nephrol 1:80–86
Nakagawa T, Hu H, Zharikov S, Tuttle KR, Short RA, Glushakova O, Ouyang X, Feig DI, Block ER, Herrera-Acosta J, Patel JM, Johnson RJ (2006) A causal role for uric acid in fructose-induced metabolic syndrome. Am J Physiol Renal Physiol 290:F625–F631
Nalecz KA, Nalecz MJ (1996) Carnitine—a known compound, a novel function in neural cells. Acta Neurobiol Exp (Wars) 56:597–609
Neeper M, Schmidt AM, Brett J, Yan SD, Wang F, Pan YC, Elliston K, Stern D, Shaw A (1992) Cloning and expression of a cell surface receptor for advanced glycosylation end products of proteins. J Biol Chem 267:14998–15004
Noland RC, Koves TR, Seiler SE, Lum H, Lust RM, Ilkayeva O, Stevens RD, Hegardt FG, Muoio DM (2009) Carnitine insufficiency caused by aging and overnutrition compromises mitochondrial performance and metabolic control. J Biol Chem 284:22840–22852
Okada E, Oida K, Tada H, Asazuma K, Eguchi K, Tohda G, Kosaka S, Takahashi S, Miyamori I (1999) Hyperhomocysteinemia is a risk factor for coronary arteriosclerosis in Japanese patients with type 2 diabetes. Diabetes Care 22:484–490
Olofsson SO, Boren J (2005) Apolipoprotein B: a clinically important apolipoprotein which assembles atherogenic lipoproteins and promotes the development of atherosclerosis. J Intern Med 258:395–410
Oron-Herman M, Rosenthal T, Sela BA (2003) Hyperhomocysteinemia as a component of syndrome X. Metabolism 52:1491–1495
Ouyang X, Cirillo P, Sautin Y, McCall S, Bruchette JL, Diehl AM, Johnson RJ, Abdelmalek MF (2008) Fructose consumption as a risk factor for non-alcoholic fatty liver disease. J Hepatol 48:993–999
Owen L, Sunram-Lea SI (2011) Metabolic agents that enhance ATP can improve cognitive functioning: a review of the evidence for glucose, oxygen, pyruvate, creatine, and L-carnitine. Nutrients 3:735–755
Owen OE, Morgan AP, Kemp HG, Sullivan JM, Herrera MG, Cahill GF Jr (1967) Brain metabolism during fasting. J Clin Invest 46:1589–1595
Panchal SK, Brown L (2011) Rodent models for metabolic syndrome research. J Biomed Biotechnol 2011:351982
Park CR (2001) Cognitive effects of insulin in the central nervous system. Neurosci Biobehav Rev 25:311–323
Parks EJ, Skokan LE, Timlin MT, Dingfelder CS (2008) Dietary sugars stimulate fatty acid synthesis in adults. J Nutr 138:1039–1046
Patel J, Iyer A, Brown L (2009) Evaluation of the chronic complications of diabetes in a high fructose diet in rats. Indian J Biochem Biophys 46:66–72
Petrie HJ, Chown SE, Belfie LM, Duncan AM, McLaren DH, Conquer JA, Graham TE (2004) Caffeine ingestion increases the insulin response to an oral-glucose-tolerance test in obese men before and after weight loss. Am J Clin Nutr 80:22–28

Ramasamy R, Vannucci SJ, Yan SS, Herold K, Yan SF, Schmidt AM (2005) Advanced glycation end products and RAGE: a common thread in aging, diabetes, neurodegeneration, and inflammation. Glycobiology 15:16R–28R

Rayssiguier Y, Gueux E, Nowacki W, Rock E, Mazur A (2006) High fructose consumption combined with low dietary magnesium intake may increase the incidence of the metabolic syndrome by inducing inflammation. Magnes Res 19:237–243

Reger MA, Henderson ST, Hale C, Cholerton B, Baker LD, Watson GS, Hyde K, Chapman D, Craft S (2004) Effects of [beta]-hydroxybutyrate on cognition in memory-impaired adults. Neurobiol Aging 25:311–314

Reiman EM, Chen K, Alexander GE, Caselli RJ, Bandy D, Osborne D, Saunders AM, Hardy J (2004) Functional brain abnormalities in young adults at genetic risk for late-onset alzheimer's dementia. Proc Natl Acad Sci USA 101:284–289

Roberts CK, Barnard RJ, Sindhu RK, Jurczak M, Ehdaie A, Vaziri ND (2006) Oxidative stress and dysregulation of NAD(P)H oxidase and antioxidant enzymes in diet-induced metabolic syndrome. Metabolism 55:928–934

Salomonsson I (2005) Shelf life: sucrose hydrolysis. Danisco Sugar A/S, 2005, Copenhagen, Denmark. www.danisco.com/cms/resources/file/eb241b041a6ed65/Shelf%20life.pdf. Accessed 15 Mar 2007

Samuel VT (2011) Fructose induced lipogenesis: from sugar to fat to insulin resistance. Trends Endocrinol Metab 22:60–65

Samuel VT, Liu ZX, Wang A, Beddow SA, Geisler JG, Kahn M, Zhang X, Monia BP, Bhanot S, Shulman GI (2007) Inhibition of protein kinase Cepsilon prevents hepatic insulin resistance in nonalcoholic fatty liver disease. J Clin Invest 117:739–745

Sasaki N, Toki S, Chowei H, Saito T, Nakano N, Hayashi Y, Takeuchi M, Makita Z (2001) Immunohistochemical distribution of the receptor for advanced glycation end products in neurons and astrocytes in Alzheimer's disease. Brain Res 888:256–262

Sautin YY, Nakagawa T, Zharikov S, Johnson RJ (2007) Adverse effects of the classic antioxidant uric acid in adipocytes: NADPH oxidase-mediated oxidative/nitrosative stress. Am J Physiol Cell Physiol 293:C584–C596

Sawa A, Khan AA, Hester LD, Snyder SH (1997) Glyceraldehyde-3-phosphate dehydrogenase: nuclear translocation participates in neuronal and nonneuronal cell death. Proc Natl Acad Sci USA 94:11669–11674

Schmidt AM, Hori O, Chen JX, Li JF, Crandall J, Zhang J, Cao R, Yan SD, Brett J, Stern D (1995) Advanced glycation endproducts interacting with their endothelial receptor induce expression of vascular cell adhesion molecule-1 (VCAM-1) in cultured human endothelial cells and in mice: a potential mechanism for the accelerated vasculopathy of diabetes. J Clin Invest 96:1395–1403

Scivittaro V, Ganz MB, Weiss MF (2000) AGEs induce oxidative stress and activate protein kinase C-beta (II) in neonatal mesangial cells. Am J Physiol 278:F676–F683

Seneff S, Wainwright G, Mascitelli L (2011) Is the metabolic syndrome caused by a high fructose, and relatively low fat, low cholesterol diet? Arch Med Sci 7:8–20

Shapiro A, Mu W, Roncal C, Cheng KY, Johnson RJ, Scarpace PJ (2008) Fructose-induced leptin resistance exacerbates weight gain in response to subsequent high-fat feeding. Am J Physiol Regul Integr Comp Physiol 295:R1370–R1375

Shi Y, Galusha SA, Rock KL (2006) Cutting edge: elimination of an endogenous adjuvant reduces the activation of CD8 T lymphocytes to transplanted cells and in an autoimmune diabetes model. J Immunol 176:3905–3908

Shinohara M, Thornalley PJ, Giardino I, Beisswenger P, Thorpe SR, Onorato J, Brownlee M (1998) Overexpression of glyoxalase-I in bovine endothelial cells inhibits intracellular advanced glycation endproduct formation and prevents hyperglycemia-induced increases in macromolecular endocytosis. J Clin Invest 101:1142–1147

Shulman GI, Rothman DL, Jue T, Stein P, DeFronzo RA, Shulman RG (1990) Quantitation of muscle glycogen synthesis in normal subjects and subjects with non-insulin-dependent diabetes by 13C nuclear magnetic resonance spectroscopy. N Engl J Med 322:223–228

Sieber FE, Traystman RJ (1992) Special issues: glucose and the brain. Crit Care Med 20:104–114

Sieber FE, Derrer SA, Saudek CD, Traystman RJ (1989) Effect of hypoglycemia on cerebral metabolism and carbon dioxide responsivity. Am J Physiol 256:H697–H706
Singh R, Barden A, Mori T, Beilin L (2001) Advanced glycation end-products: a review. Diabetologia 44:129–146
Spanswick D, Smith MA, Mirshamsi S, Routh VH, Ashford ML (2000) Insulin activates ATP-sensitive K+channels in hypothalamic neurons of lean, but not obese rats. Nat Neurosci 3:757–758
Stanhope KL, Havel PJ (2008) Fructose consumption: potential mechanisms for its effects to increase visceral adiposity and induce dyslipidemia and insulin resistance. Curr Opin Lipidol 19:16–24
Stanhope KL, Havel PJ (2009) Fructose consumption: considerations for future research on its effects on adipose distribution, lipid metabolism, and insulin sensitivity in humans. J Nutr 139:1236S–1241S
Stanhope KL, Griffen SC, Bair BR, Swarbrick MM, Keirn WL, Havel PJ (2008) Twenty-four-hour endocrine and metabolic profiles following consumption of high-fructose corn syrup-, sucrose-, fructose-, and glucose-sweetened beverages with meals. Am J Clin Nutr 87:1194–1203
Stavric B, Johnson WJ, Clayman S, Gadd RE, Chartrand A (1976) Effect of fructose administration on serum urate levels in the uricase inhibited rat. Experientia 32:373–374
Steiber A, Kerner J, Hoppel C (2004) Carnitine: a nutritional, biosynthetic, and functional perspective. Mol Aspects Med 25:455–473
Stirpe F, Della Corte E, Bonetti E, Abbondanza A, Abbati A, De Stefano F (1970) Fructose-induced hyperuricaemia. Lancet 2:1310–1311
Suárez G, Rajaram R, Oronsky AL, Gawinowicz MA (1989) Nonenzymatic glycation of bovine serum albumin by fructose (fructation). Comparison with the Maillard reaction initiated by glucose. J Biol Chem 264:3674–3679
Swarbrick MM, Stanhope KL, Elliott SS, Graham JL, Krauss RM, Christiansen MP, Griffen SC, Keim NL, Havel PJ (2008) Consumption of fructose-sweetened beverages for 10 weeks increases postprandial triacylglycerol and apolipoprotein-B concentrations in overweight and obese women. Br J Nutr 100:947–952
Tappy L (2012) Q&A: 'toxic' effects of sugar: should we be afraid of fructose? BMC Biol 10:42
Teff K, Elliot S, Tschoep MR (2002) Consuming high fructose meals reduces 24 hour plasma insulin and leptin concentrations, does not suppress circulating ghrelin, and increases postprandial and fasting triglycerides in women. Diabetes 52:A408
Teff KL, Elliott SS, Tschop M, Kieffer TJ, Rader D, Heiman M, Townsend RR, Keim NL, D'Alessio D, Havel PJ (2004) Dietary fructose reduces circulating insulin and leptin, attenuates postprandial suppression of ghrelin, and increases triglycerides in women. J Clin Endocrinol Metab 89:2963–2972
Thorens B, Mueckler M (2010) Glucose transporters in the 21st Century. Am J Physiol Endocrinol Metab 298:E141–E145
Thornalley PJ (1993a) Modification of the glyoxalase system in disease processes and prospects for therapeutic strategies. Biochem Soc Trans 21:531–534
Thornalley PJ (1993b) The glyoxalase system in health and disease. Mol Aspects Med 14:287–371
Thornalley PJ (2003) Glyoxalase I—structure, function and a critical role in the enzymatic defence against glycation. Biochem Soc Trans 31:1343–1348
Tran LT, MacLeod KM, McNeill JH (2009) Chronic etanercept treatment prevents the development of hypertension in fructose-fed rats. Mol Cell Biochem 330:219–228
Unger JW, Moss AM, Livingston JN (1991) Immunohistochemical localization of insulin receptors and phosphotyrosine in the brainstem of the adult rat. Neuroscience 42:853–861
van den Berghe G, Bronfman M, Vanneste R, Hers HG (1977) The mechanism of adenosine triphosphate depletion in the liver after a load of fructose. A kinetic study of liver adenylate deaminase. Biochem J 162:601–609
Vannucci SJ (1994) Developmental expression of glut1 and glut3 glucose transporters in rat brain. J Neurochem 62:240–246
Vasdev S, Ford CA, Longerich L, Parai S, Gadag V, Wadhawan S (1998) Aldehyde induced hypertension in rats: prevention by N-acetyl cysteine. Artery 23:10–36

Vasdev S, Gill VD, Singal PK (2006) Modulation of oxidative stress-induced changes in hypertension and atherosclerosis by antioxidants. Exp Clin Cardiol 11:206–216

Verma S, Bhanot S, McNeill JH (1999) Sympathectomy prevents fructose-induced hyperinsulinemia and hypertension. Eur J Pharmacol 373:R1–R4

Vila L, Roglans N, Alegret M, Sanchez RM, Vazquez-Carrera M, Laguna JC (2008) Suppressor of cytokine signaling-3 (SOCS-3) and a deficit of serine/threonine (Ser/Thr) phosphoproteins involved in leptin transduction mediate the effect of fructose on rat liver lipid metabolism. Hepatology 48:1506–1516

Visinoni S, Fam BC, Blair A, Rantzau C, Lamont BJ, Bouwman R, Watt MJ, Proietto J, Favaloro JM, Andrikkopoulos S (2008) Increased glucose production in mice overexpressing human fructose-1,6-bisphosphatase in the liver. Am J Physiol Endocrinol Metab 295:E1132–E1141

Wang GJ, Volkow ND, Thanos PK, Fowler JS (2004) Similarity between obesity and drug addiction as assessed by neurofunctional imaging: a concept review. J Addict Dis 23:39–53

Wang X, Desai K, Chang T, Wu L (2005) Vascular methylglyoxal metabolism and the development of hypertension. J Hypertens 23:1565–1573

Wang X, Jia X, Chang T, Desai K, Wu L (2008) Attenuation of hypertension development by scavenging methylglyoxal in fructose-treated rats. J Hypertens 26:765–772

Watson GS, Craft S (2004) Modulation of memory by insulin and glucose: neuropsychological observations in Alzheimer's disease. Eur J Pharmacol 490:97–113

Wautier JL, Guillausseau PJ (2001) Advanced glycation end products, their receptors and diabetic angiopathy. Diabetes Metab 27:535–542

Wautier JL, Schmidt AM (2004) Protein glycation: a firm link to endothelial cell dysfunction. Circ Res 95:233–238

Wautier MP, Chappey O, Corda S, Stern DM, Schmidt AM, Wautier JL (2001) Activation of NADPH oxidase by AGE links oxidant stress to altered gene expression via RAGE. Am J Physiol Endocrinol Metab 280:E685–E694

Wolfgang MJ, Lane MD (2011) Hypothalamic malonyl-CoA and CPT1c in the treatment of obesity. FEBS J 278:552–558

Wolfgang MJ, Cha SH, Sidhaye A, Chohnan S, Cline G, Shulman GI, Lane MD (2007) Regulation of hypothalamic malonyl-CoA by central glucose and leptin. Proc Natl Acad Sci USA 104:19285–19290

Wu L (2006) Is methylglyoxal a causative factor for hypertension development? Can J Physiol Pharmacol 84:129–139

Wu T, Willett WC, Hankinson SE, Giovannucci E (2005) Caffeinated coffee, decaffeinated coffee, and caffeine in relation to plasma C-peptide levels, a marker of insulin secretion, in U.S. women. Diabetes Care 28:1390–1396

Xing Z, Gauldie J, Cox G, Baumann H, Jordana M, Lei XF, Achong MK (1998) IL-6 is an antiinflammatory cytokine required for controlling local or systemic acute inflammatory responses. J Clin Invest 101:311–320

Yan SD, Schmidt AM, Anderson GM, Zhang J, Brett J, Zou YS, Pinsky D, Stern D (1994) Enhanced cellular oxidant stress by the interaction of advanced glycation end products with their receptors/binding proteins. J Biol Chem 269:9889–9897

Yan SD, Chen X, Fu J, Chen M, Zhu H, Roher A, Slattery T, Zhao L, Nagashima M, Moser J, Migheli A, Nawroth P, Stern D, Schmidt AM (1996) RAGE and amyloid-β peptide neurotoxicity in Alzheimer's disease. Nature 382:685–691

Yoo HG, Lee SI, Chae HJ, Park SJ, Lee YC, Yoo WH (2009) Prevalence of insulin resistance and metabolic syndrome in patients with gouty arthritis. Rheumatol Int 20:231–235

Zimmermann R, Panzenböck U, Wintersperger A, Levak-Frank S, Graier W, Glatter O, Fritz G, Kostner GM, Zechner R (2001) Lipoprotein lipase mediates the uptake of glycated LDL in fibroblasts, endothelial cells, and macrophages. Diabetes 50:1643–1653

# Chapter 3
# Essential Fatty Acid Metabolism in Metabolic Syndrome and Neurological Disorders

## 3.1 Introduction

Fatty acids are important building blocks of complex lipids. They contain hydrocarbon chains with a carboxyl group at one end and a methyl group at the other. The biological reactivity of fatty acids is defined not only by the length of the carbon chain but also by both the number and position of any double bonds present. Saturated fatty acids do not contain double bonds within the acyl chain. In contrast, unsaturated fatty acids contain at least one double bond. When two or more double bonds are present, unsaturated fatty acids are referred as polyunsaturated fatty acids (PUFAs) (Calder 2008, 2009). Dietary fat is a mixture of fatty acids, with saturated, monounsaturated (MUFA), and PUFA. Fatty acids not only provide diversity and chemical specificity to the complex lipids found in neural and nonneural membranes but are also associated with energy supply and regulation of a wide variety of cellular processes, including gene expression (Farooqui et al. 2000). Essential fatty acids are fatty acids that humans and other animals cannot synthesize and must obtain them from diet because these fatty acids are required not only for maintaining the orientation of internal membrane domains for interactions among intracellular organelles but also for providing an appropriate environment for the optimal functioning of receptors, membrane-bound enzymes, transporters, and ion channels (Farooqui et al. 2000). In addition, these fatty acids also provide and modulate membrane fluidity. Two families of essential fatty acids, namely, n-6 (ω-6) and n-3 (ω-3), are known to occur in human diet. The n-6 fatty acids are abundantly found in meat and vegetable oils. The n-6 fatty acids are synthesized from their precursor, linoleic acid (LA; 18:2n-6), through different steps involving a series of enzymes called desaturases and elongases located in the endoplasmic reticulum of hepatocytes. The most biologically active n-6 fatty acid is arachidonic acid (ARA; 20:4n-6). This n-6 fatty acid is a major component of mammalian cell membrane phospholipids. It accounts for up to 25 % of all phospholipid fatty acids (Zhou and Nilsson 2001). A major function of ARA is to serve as a precursor for prostaglandins (PGs) and thromboxanes (TXs) via the cyclooxygenase pathway or to leukotrienes (LTs),

A.A. Farooqui, *Metabolic Syndrome: An Important Risk Factor for Stroke, Alzheimer Disease, and Depression*, DOI 10.1007/978-1-4614-7318-3_3,

lipoxins (LXs), hydroperoxyeicosatetraenoic acids (HPETE), and hydroxyeicosatetraenoic acids (HETE) via the lipoxygenase pathway (Farooqui 2011). Certain PGs, TXs, and LTs have proinflammatory actions, whereas LXs produce anti-inflammatory effects. In addition, essential fatty acids and their long-chain metabolites modulate the activities of angiotensin converting and HMG-CoA reductase enzymes, enhance acetylcholine levels in the brain, increase the synthesis of endothelial nitric oxide, augment diuresis, and enhance insulin action (Das 2008). Essential fatty acids and their long-chain metabolites react with nitric oxide (NO) to yield respective nitroalkene derivatives that exert cell-signaling actions via ligation and activation of peroxisome proliferator-activated receptors (PPARs) (Das 2008). Thus, essential fatty acids and their derivatives have varied biological actions that may have relevance to their involvement in several physiological and pathological processes. Free fatty acid metabolism responds to varying metabolic states partially by induction of enzymes that promote either catabolic or anabolic processes. Two major classes of transcriptional regulators are associated with fatty acid metabolism, the PPARs and the sterol regulatory element-binding proteins (SREBPs). In general, PPARγ and SREBP-1c regulate processes involved in lipogenesis, whereas lipolytic enzymes are induced by PPARα (Kersten 2002). PPARα plays an important role in the regulation of cellular uptake and activation and β-oxidation of fatty acids. The natural, preferentially binding ligands of PPARα are long-chain unsaturated fatty acids including ARA, LA, and oleic acid, but saturated fatty acids like palmitic acid can also act as ligands (Gottlicher et al. 1992). SREBP family of transcription factors governs transcriptional activation of a large number of genes involved in the regulation of lipid metabolism, including lipogenesis, cholesterol transport, and synthesis (Foufelle and Ferre 2002). Of interest is the high expression of SREBP-1c in liver and adipose tissue (Horton et al. 2002) and its detection in pancreatic β cells (Kakuma et al. 2000).

ARA is also a ligand for nuclear receptors which take part in the subcellular control of metabolic pathways. Covalent modification of proteins by fatty acid acylation enables their incorporation into membranes. The degree of fatty acid unsaturation in membrane phospholipids dictates the biophysical properties of the membrane, which in turn influences many critical membrane-associated functions including the activities of membrane-bound proteins such as ion channels and endo- and exocytosis (Spector and Yorek 1985; Horrocks and Farooqui 2004). Depletion of ARA from membranes may alter the physical and functional properties of membranes (Ariyama et al. 2010). ARA also functions as a transcriptional regulator by modulating signal transduction at the cell surface, by altering membrane fluidity, or cell-surface interactions by acylating membrane proteins (Fig. 3.1) (Clarke 2001; Farooqui et al. 2000; Farooqui 2009).

n-3 family of fatty acids consists of α-linolenic acid (ALA; 18:3n-3), eicosapentaenoic acid (EPA), and docosahexaenoic acids (DHA). ALA is converted to eicosapentaenoic acid (EPA) and DHA (Farooqui 2009). However, this conversion is ineffective in general and appears to get progressively more ineffective with age.

n-3 PUFAs are one of the major components of the lipid bilayer of neural cell membranes. Diets deficient in DHA affect electroretinogram response amplitudes in animal models and young infants (Weisinger et al. 1996; Farooqui 2009). DHA

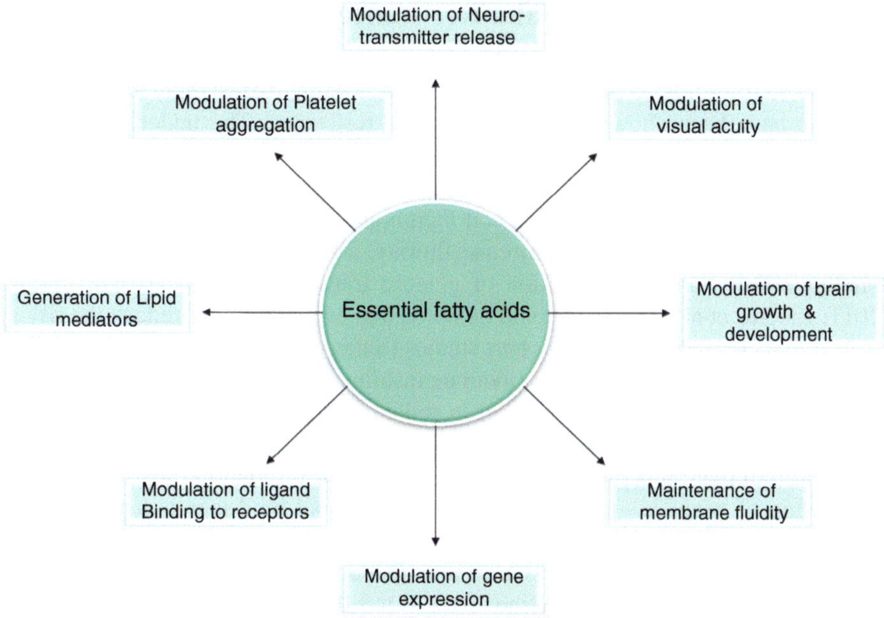


**Fig. 3.1** Role of essential fatty acids in the visceral and brain tissues

in particular is widely distributed throughout the brain in gray matter and myelin and is found in high concentration in the outer segment of retinal photoreceptors (Farooqui 2009). By comparison, EPA is present in very low amounts in the brain and retina (Farooqui 2009). EPA supplementation improves mood and alleviates the intensity of psychiatric symptoms in schizophrenia and depression (Stahl et al. 2008). In general, n-6 and n-3 fatty acids produce opposing effects in the context of oxidative stress and inflammation (Farooqui 2009).

LA competes with ALA for endogenous conversion to the long-chain derivatives EPA and DHA, and LA also inhibits incorporation of DHA and EPA into tissues. Thus, high-LA levels in the diet generally result in low n-3 fatty acid status. The proportions of ARA and DHA in neural membrane vary considerably in various subclasses of phospholipids. ARA is distributed rather evenly in gray and white matter and among the different types of neural cells. In contrast, DHA is highly enriched not only in neuronal membranes but also in synaptic membranes. PtdIns and PtdH contain high levels of arachidonoyl groups (Farooqui and Horrocks 2007). Rat brain contains 2.42, 0.58, 1.65, and 6.92 μmol ARA/g brain in PtdCho, PtdSer, PtdIns, and PtdEtn, respectively. Similarly, rat brain contains 1.69, 3.72, 0.14, and 11.08 DHA/g brain in PtdCho, PtdSer, PtdIns, and PtdEtn, respectively (Rapoport 2008). It is becoming increasingly evident that the degree of unsaturation in fatty acids not only determines membrane order, packing, and fluidity (physicochemical characteristics) but also contributes to the pathogenesis of a variety of disease states, including MetS, diabetes, obesity, neurotraumatic diseases, neurodegenerative diseases, cardiovascular diseases, and various types of cancers (Farooqui 2009).

## 3.2 Fatty Acids and Insulin Resistance

Effects of fatty acids on insulin resistance are becoming increasingly evident. Thus, MUFAs have been shown to improve insulin resistance. The underlying mechanisms associated with MUFA-induced improvement of insulin resistance are not fully understood. However, it is suggested that MUFA may contribute to insulin resistance through their effects on cell membrane fatty acid composition (Riserus et al. 2009), modulation of membrane fluidity, ion permeability, insulin receptor binding/affinity, and up-regulation of glucose transporters (Thomas and Pfeiffer 2011; Riserus et al. 2009). It should be noted that MUFAs do not reduce the risk of type II diabetes in prospective cohort studies (Salas-Salvado et al. 2011).

PUFAs have been reported to modulate insulin resistance depending on the consumption of PUFA type (n-3 or n-6). In rodents, supplementation with n-3 PUFA from fish oil, but not with n-6 PUFA, reduces and reverses insulin resistance not only through their anti-inflammatory effects but also through the regulation of the expression of genes associated with carbohydrate and lipid metabolism. n-6 fatty acids do not produce insulin resistance reversal effect in rodents. In humans, n-6 PUFA may improve insulin resistance and also decrease risk of diabetes (Riserus et al. 2009). However, results on the effect of n-3 PUFA are inconsistent (McAuley and Mann 2006; Riserus et al. 2009). Some PUFAs suppress lipogenic gene expression in vitro, partly by PPAR gamma-mediated effects, which may be related to beneficial effects of n-6 PUFA consumption in humans (Riserus et al. 2009).

In contrast, saturated fatty acids are key contributors of insulin resistance not only by regulating enzyme activities involved in lipid metabolism and modulating inflammatory gene expression but also controlling the transcription factors that modulate glucose and lipid metabolism (Staiger et al. 2005; Rioux and Legrand 2007). Saturated fatty acids have been shown to induce hypothalamic inflammation by activating signal transduction though TLR4, which leads to endoplasmic reticulum stress, in situ expression of inflammatory cytokines, and, eventually, apoptosis of neurons, all contributing to the development of adipostatic hormone resistance and anomalous expression of the neurotransmitters involved in the regulation of energy homeostasis (Milanski et al. 2009; Moraes et al. 2009). Based on available information, it is suggested that desaturation indexes are also linked to adiposity and insulin resistance. For example, the ratios of palmitic and stearic acid (18:0) to palmitoleic (16:1) and oleic acid (18:1), respectively, have been correlated with adiposity and insulin resistance, while the ratio of dihomo-gamma-linoleic acid (20:3n-6) to arachidonic acid (20:4n-6) is negatively associated with obesity and insulin resistance (Jumpertz et al. 2012).

## 3.3 Metabolism of ARA in the Liver

ARA is consumed in the diet in meats, eggs, and vegetable oil. It is also synthesized in the liver from LA, the most abundant dietary PUFA. The biosynthesis of ARA from LA also requires a series of desaturation and elongation steps catalyzed by

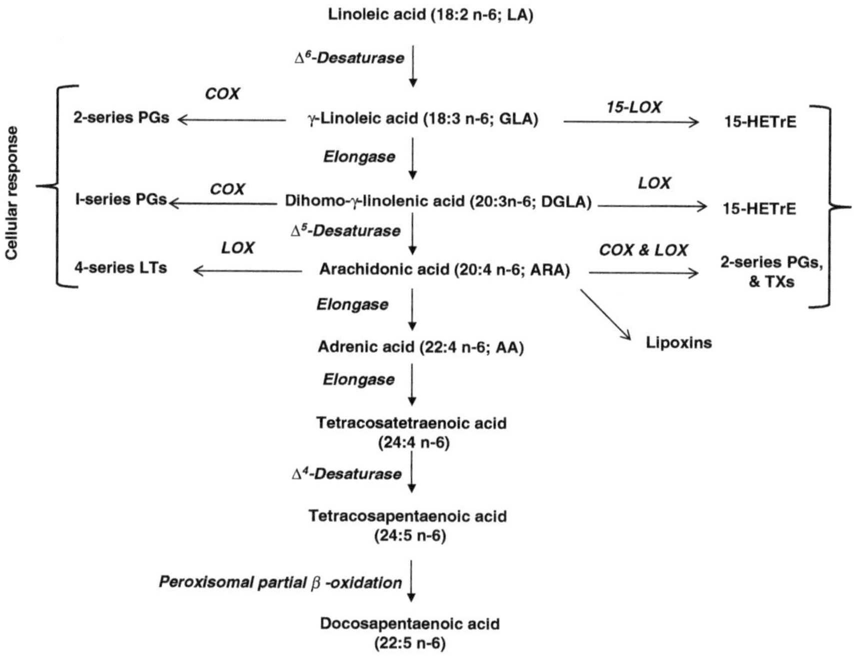


**Fig. 3.2** Synthesis of arachidonic acid from linoleic acid and transformation of arachidonic acid in eicosanoids. Linoleic acid (LA); γ-linolenic acid (GLA, 18:3); dihomo-GLA (DGLA, 20:3), arachidonic acid (ARA); 15-(S)-hydroxy-8,11,13-eicosatrienoic acid (15-HETrE)

elongases and $\Delta^6$- and $\Delta^5$-desaturases (Fig. 3.2). LA is converted to γ-linolenic acid (GLA, 18:3) by the enzyme $\Delta^6$-desaturase and GLA is elongated to form dihomo-GLA (DGLA, 20:3). DGLA is transformed into ARA (20:4) by the enzyme $\Delta^5$-desaturase. Fatty acid $\Delta^6$-desaturation is a rate-limiting step in the conversion of LA to ARA. Genetic variation in human desaturase genes affects fatty acid metabolism, plasma lipid profiles, and risk of disease development (Brenna et al. 2009; Lattka et al. 2010; Merino et al. 2010). ARA is transported to other cell types via serum albumin or lipoproteins (Simopoulos 1999). In liver, many nutritional, hormonal, and physiological factors modulate ARA biosynthesis (Bezard et al. 1994). Dietary fatty acids exert a great influence and are often inhibitory. Thus, dietary ALA inhibits $\Delta^6$-desaturation of LA. The desaturation products ARA, EPA, and DHA inhibit $\Delta^6$-desaturation of LA and $\Delta^5$-desaturation of DGLA (dihomo-gamma-linolenic acid). With regard to hormones, insulin and thyroxin are necessary to $\Delta^6$- and $\Delta^5$-desaturation activities, whereas other hormones (glucagon, epinephrine, ACTH, glucocorticoids) inhibit desaturation (Bezard et al. 1994). $\Delta^6$-Desaturation is also modulated by vitamin B6. This vitamin is a cofactor for $\Delta^6$-desaturase as well as trans-sulfuration pathway of homocysteine. Low levels of this vitamin may contribute to heart disease by increasing homocysteine levels (Cabrini et al. 2005). Both DGLA and ARA are substrates for cyclooxygenases (COX). Through a series of free radical reactions, COX metabolizes DGLA and ARA to form various bioactive metabolites, namely, the 1 and the 2 series of prostaglandins ($PGs_1$ and $PGs_2$), respectively (Fig. 3.2). Unlike $PGs_2$, which produces proinflammatory effects, $PGs_1$ actually induces anti-inflammatory and anticancer effects (Levin et al. 2002). For example, $PGE_1$, one form of $PGs_1$, inhibits vascular smooth muscle cell proliferation, reduces vascular cell adhesion, and attenuates the development of atherosclerosis (Fan and Chapkin 1998).

The mode by which eicosanoids mediate their biological homeostatic and inflammatory effects depends on binding to their respective G protein-coupled receptors (GPCRs). Specific GPCRs have been identified for all eicosanoids, where there are at least nine known eicosanoid receptor forms have been reported to occur in mouse and man (Farooqui 2011). Although most of the PG receptors are localized at the plasma membrane of platelets, vascular smooth muscle cells, neural cells, and mast cells, some are situated at the nuclear envelope (Ide et al. 2003; Farooqui 2011). Four of these receptor subtypes bind $PGE_2$ ($EP_1$–$EP_4$), two bind $PGD_2$ ($DP_1$ and $DP_2$), and more specific receptors bind $PGF_{2\alpha}$, $PGI_2$, and $TXA_2$ (FP, IP, and TP) (Harizi et al. 2008). $PGE_2$ and $PGI_2$ are the predominant proinflammatory eicosanoids, and, through their activation of $EP_2$ and IP, respectively, they can increase vascular permeability and leukocyte infiltration.

## 3.4 Metabolism of ARA in Metabolic Syndrome

It is well known that MetS is linked with obesity, which contributes to insulin resistance through elevation of circulating levels of free fatty acids derived from the adipocytes (Chien et al. 2011). These free fatty acids not only inhibit glucose uptake, glycogen synthesis, and glycolysis inducing hyperglycemia but also enhance the metabolism of fatty acids particularly ARA, which is a precursor for reactive oxygen species (ROS). ROS not only damage cells by oxidizing DNA, protein, and lipid but also activate a variety of stress-sensitive intracellular signaling pathways, such as NF-κB, $p38^{MAPK}$, and JNK/SAPK (Klaunig et al. 2010). Activation of these pathways results in increased expression of numerous gene products that may cause cellular damage and play a major role in insulin resistance (Wei et al. 2008; Davis et al. 2009). Accordingly, it is proposed that there is a link among the FFA-induced increased ROS generation, activation of stress-sensitive pathways, and eventual development of insulin resistance (Cusi et al. 2000).

Consumption of high-fat diets is a central risk factor for MetS and obesity (Bessesen 2008). Adverse effects of high-fat diet on metabolic homeostasis are linked to adipose tissue physiology and are highly influenced by gender (Macotela et al. 2009). The imbalance between caloric intake and energy expenditure leads to hyperplasia and hypertrophy of adipocytes depending on the type as much as the amount of dietary lipids (Funaki 2009). Specific dietary fatty acids can directly activate Toll-like receptors, which are better known as components of the innate immune system recognizing bacteria-derived fatty acids (Shi et al. 2006; Suganami et al. 2007). The resulting immune response facilitates the systemic activation of

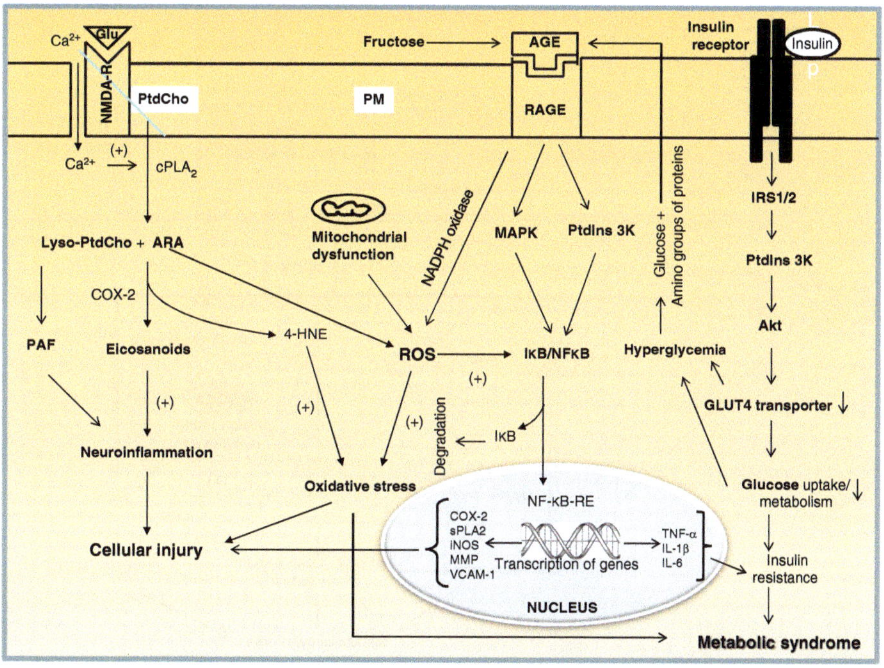


**Fig. 3.3** Hypothetical diagram showing interactions among NMDA receptor, RAGE receptor, and insulin receptor. Plasma membrane (PM); *N*-methyl-D-aspartate receptor (NMDA-R); glutamate (Glu); phosphatidylcholine (PtdCho); lysophosphatidylcholine (lyso-PtdCho); platelet-activating factor (PAF); cytosolic phospholipase $A_2$ ($cPLA_2$); cyclooxygenase (COX-2); arachidonic acid (ARA); ROS; nuclear factor-κB (NF-κB); nuclear factor-κB response element (NF-κB-RE); inhibitory subunit of NF-κB (IκB); tumor necrosis factor-α (TNF-α); interleukin-1β (IL-1β); interleukin-6 (IL-6); nicotinamide adenine dinucleotide phosphate oxidase (NADPH oxidase); secretory phospholipase A2 (sPLA2); inducible nitric oxide synthase (iNOS); matrix metalloproteinases (MMP); vascular cell adhesion molecule 1 (VCAM-1); mitogen-activated protein kinase (MARK); and glucose transporter type 4 (GLUT4)

proinflammatory pathways including NF-κB, TNF-α, or IL-6 (Shi et al. 2006; Suganami et al. 2007). This chain of events leads to insulin resistance, setting in motion the pathogenesis of the MetS (Tschop and Thomas 2006). Accumulating evidence suggests that excessive intake of total fat (>37 % of daily energy intake (Vessby et al. 2001), irrespective of fatty acid composition, may worsen insulin resistance, likely via a combination of factors that include interference with binding of insulin to its receptors and accumulation of triglycerides in skeletal muscle (Thomas and Pfeiffer 2011). However, if total fat intake remains <30 %, the type of fatty acids consumed appears to differently influence IR (Vessby et al. 2001).

MetS subjects exhibit oxidative stress through the production of advanced glycated end product (AGE) and ROS (Figs. 3.3 and 3.4). AGEs are synthesize from nonenzymic oxidation of high glucose levels. ROS is formed from enzymic and nonenzymic oxidation of ARA (Farooqui 2009). ROS activates nuclear factor-κB

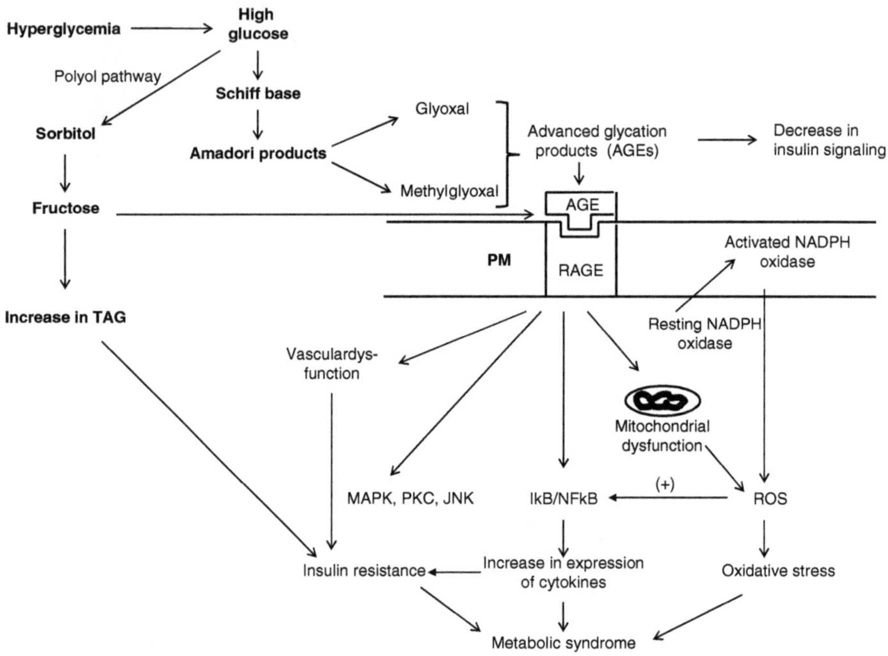


**Fig. 3.4** Generation of AGE from glucose and fructose. Reactive oxygen species (ROS); microtubule affinity-regulating kinase (MARK); protein kinase C (PKC); Janus kinase (JAK); and triacylglycerol (TAG)

(NF-κB), a transcription factor that resides in the cytoplasm of immune cells (particularly B and T lymphocytes) as well as neural cells. NF-κB is a dimeric transcription factor composed of five members, p50, RelA/p65, c-Rel, RelB, and p52, that can diversely combine to form the active transcriptional dimer. NF-κB controls the expression of genes that regulate a broad range of biological processes in the brain such as synaptic plasticity, neurogenesis, and differentiation (Hayden and Ghosh 2008). NF-κB activity is regulated by the IκB protein. The activity of IκB protein is controlled through phosphorylation by upstream IκB kinases (IKKs). In its latent form NF-κB is bound to an inhibitory protein (IκB), which is present in cytoplasm. After activation by ROS, IκB undergoes phosphorylation at serine residues by a ubiquitin-dependent protein kinase IκB and finally degraded by the proteasome. NF-κB rapidly migrates into the cell nucleus where it upregulates the expression of many genes, such as genes involved in cell cycle regulation through actions on the cyclin-dependent kinase system (Joyce et al. 2001), proinflammatory cytokines (TNF-α, IL-2, IL-6, and IL-8), and apoptosis (Baldwin 1996; Piette et al. 1997) (Fig. 3.3).

Inflammation is an intrinsically beneficial mechanism that leads to removal of offending factors and restoration of tissue structure and physiological function. The acute phase of inflammation involves rapid influx of blood granulocytes, typically neutrophils, followed swiftly by monocytes that mature into inflammatory macrophages that subsequently proliferate and thereby modulate the functions of resident tissue macrophages. ARA-derived PGs play a key role in the generation of the inflammatory response and their biosynthesis is significantly increased in inflamed tissue. Once the initiating noxious stimulus is removed via phagocytosis, the inflammatory reaction goes through resolution of inflammation. During the resolution of inflammation, granulocytes are eliminated and macrophages and lymphocytes return to normal pre-inflammatory numbers and phenotypes. It should be noted that failure of inflammation to resolve may predispose to autoimmunity, MetS, and neurological disorders (Farooqui 2011).

Under physiological conditions, ARA is released by phospholipase $A_2$ and metabolized to low levels of eicosanoids (PGs, LTs, TXs, LX), which can cross neural cell membrane and act on neighboring cells by interacting with EP, LT, TX, LX, and ALX receptors (Phillis et al. 2006; Farooqui 2011) and modulate signal transduction processes. High levels of ARA-derived eicosanoids promote inflammation and oxidative stress, whereas various LXs (LXA4, LXB4, 15-epi-LXA4, and 15-epi-LXB4) facilitate anti-inflammatory and pro-resolution responses (Chiang et al. 2006; Farooqui 2011). ARA-derived lipid nonenzymic lipid mediators include 4-hydroxynonenal (4-HNE), isoprostanes, isoketal, and isofurans. These ARA-derived nonenzymic lipid mediators induce prothrombotic, proaggregatory, and proinflammatory effects throughout the body (Farooqui 2011).

In diabetes and MetS patients, production of eicosanoids and ROS is markedly elevated through increase in ARA oxidation by cyclooxygenases and lipoxygenases, activation of NADPH oxidase, and mitochondrial dysfunction (Farooqui et al. 2012). In addition, enzymes, such as xanthine/xanthine oxidase, myeloperoxidase, and cytochrome P450, also generate ROS in the cytoplasm. Low levels of ROS are important for maintaining physiological glucose homeostasis, whereas prolonged exposure to high ROS levels results in chronic oxidative stress and impaired glucose homeostasis (Wang and Hai 2011). Importantly, pancreatic islets are particularly vulnerable to oxidative stress as they express low levels of antioxidant proteins (Tiedge et al. 1997). In addition, increased ROS levels are observed in animal models of type II diabetes (i.e., db/db mice, GK rats, and ZDF rats) and human type II diabetes patients (Ihara et al. 1999; Tanaka et al. 1999). Oxidative stress-mediated low-grade chronic inflammation through the generation of proinflammatory mediators (eicosanoids, platelet-activating factor, and cytokines) induces insulin resistance, which is influenced by genetically modified environmental factors, including decreased physical activity and poor nutrition (n-6 fatty acid-enriched diet along with consumption of soft drinks) (Pickup et al. 2004). The release of these proinflammatory cytokines (TNF-1$\alpha$, IL-6, and IL-1$\beta$) and other mediators (ceramide and diacylglycerols) results in the local recruitment of monocytes within the adipose tissues (Farooqui et al. 2012). The differentiation of the monocytes into macrophages is accompanied by an increase in the release of inflammatory factors and chemokines not only locally within the adipose tissue but also systemically, such that the inflammatory response is propagated to various tissues, especially in insulin-sensitive organs such as the liver and skeletal muscle, thus contributing to overall insulin resistance (Shoelson et al. 2006). There is also evidence to suggest that abovementioned proinflammatory stimuli stimulate endothelial cells and increase the production and expression of soluble adhesion molecules, such as intercellular adhesion molecule-1, vascular cell adhesion molecule 1, and E-selectin (Ceriello et al. 1996). Moreover, it is also shown that elevated levels of intercellular adhesion molecule-1 and E-selectin in diabetes significantly predict future risk of heart disease and stroke (Ridker et al. 1998; Farooqui et al. 2012). CD40 ligand is a trimeric, transmembrane protein of the TNF family. Together with its receptor, it is an important contributor to the inflammatory processes that lead to atherosclerosis and thrombosis. Both CD40 and CD40 ligand have been associated with human atheroma (Schonbeck and Libby 2001). CD40 and CD40 ligand are closely associated with the pathogenesis of cardiovascular disease and stroke. Decrease in high levels of soluble CD40 ligand not only reduces heart disease in diabetic patients but also decreases the risk of stroke (Lim et al. 2004; Kalofoutis et al. 2007; Farooqui et al. 2012).

Among nonenzymic oxidation products of ARA, 4-HNE freely diffuses across membranes and reacts with proteins and DNA. Enhanced synthesis of 4-HNE is a characteristic feature of many inflammatory processes, such as ischemia and MetS (Mattson 2009). Brain and visceral tissues neutralize 4-HNE either spontaneously or through glutathione-S-transferase-catalyzed conjugation with reduced glutathione to yield glutathionyl-HNE (GS-HNE), which induces peritonitis in mice, resulting in the recruitment of leukocytes and in vivo generation of proinflammatory lipid mediators, including PGs and LTs. The inflammatory response induced by GS-HNE can be mitigated by $RvD_1$ in vivo, leading to dampening of inflammation. These results support the notion that GSH-dependent metabolism of reactive lipid peroxidation products, such as 4-HNE, can result in local production of new proinflammatory mediators with novel bioactions (Spite et al. 2009; Farooqui 2011).

## 3.5 Metabolism of n-3 Fatty Acids in the Liver

As stated above, DHA is a major component of brain gray matter and retina in most mammalian species and is considered essential for normal neurological and cellular developments (Farooqui 2009). DHA is obtained directly from the diet. When the diet does not provide an adequate supply of DHA, the liver has capacity to synthesize DHA from ALA in liver through a series of elongation and desaturation step (Scott and Bazan 1989) (Fig. 3.5). Earlier steps of DHA synthesis in the liver take place at the endoplasmic reticulum and consist of sequential alternating elongation and desaturation steps catalyzed by fatty acid elongase, $\Delta^6$- and $\Delta^5$-desaturase. Desaturases are the key enzymes for the synthesis of DHA. They share common regulatory features, including dependence of expression on insulin and induction by peroxisome proliferators. A key regulator of desaturase gene expression is sterol regulatory element-binding protein-1c (SREBP-1c), which mediates transcriptional activation of $\Delta^6$-desaturase gene by insulin (Nakamura and Nava 2002). It is proposed that $\Delta^6$-desaturase step is the rate-limiting step in the biosynthesis of DHA

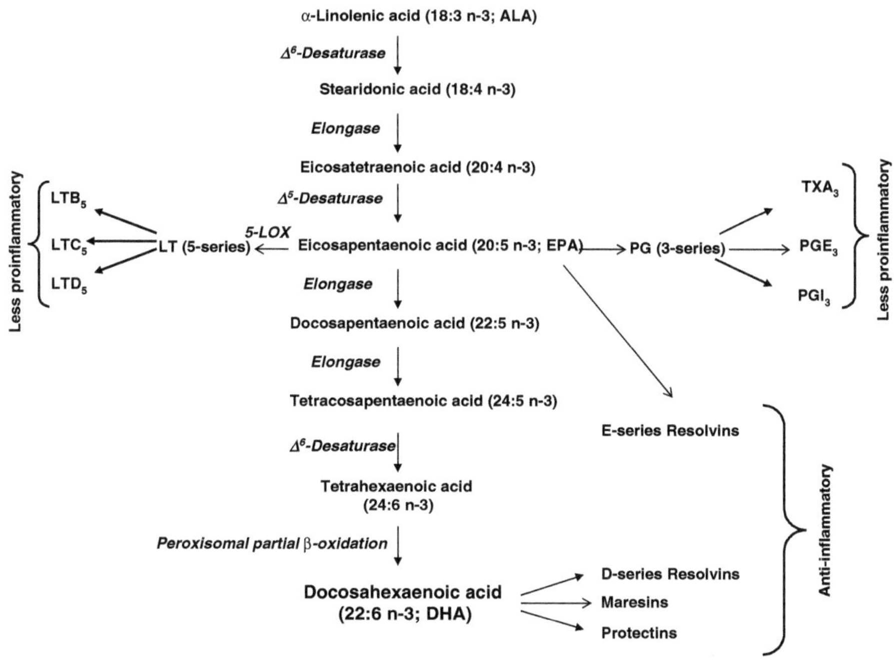


**Fig. 3.5** Synthesis of docosahexaenoic acid from **α**-linolenic acid (ALA) and generation of eicosanoids and docosanoids from eicosapentaenoic acid and docosahexaenoic acid

(Sprecher et al. 1999). Another pathway for the synthesis of n-3 (DHA) involves a separate channeled carnitine-dependent mitochondrial pathway (Infante and Huszagh 1997, 1998). In outer mitochondrial membrane, DHA synthesis starts with an 18:3n-3 carnitine complex. This complex is transported across the outer mitochondrial membrane, followed by binding of 18:3n-3 to a multifunctional enzyme complex. Two enzymes (elongases and desaturases) and α-tocopherol metabolites are required for the synthesis of DHA. Activities of these enzymes are significantly decreased in chronic diseases such as diabetes, hypertension, hyperlipidemia, and MetS (Das 2007). These studies support the proposal that DHA is synthesized through endoplasmic reticulum, mitochondrial, and peroxisome pathways. Later pathway involves channeled carnitine and separate n-3-specific desaturases. The activity of $\Delta^6$-desaturase is significantly lower in livers of rats fed with B6-deficient diet than in the pair-fed control group (approximately 64 %).

Enzymic oxidation of DHA is catalyzed by15-LOX. This enzyme catalyzes oxidation of DHA producing resolvins of the D series (RvDs) and neuroprotectin $D_1$/protectin $D_1$ ($NPD_1/PD_1$), which has been described in the vascular cells (Serhan and Chiang 2008) (Fig. 3.6). Resolvins and protectins are collectively called docosanoids. They not only antagonize the effects of eicosanoids but also modulate leukocyte trafficking as well as functions downregulating the expression of cytokines

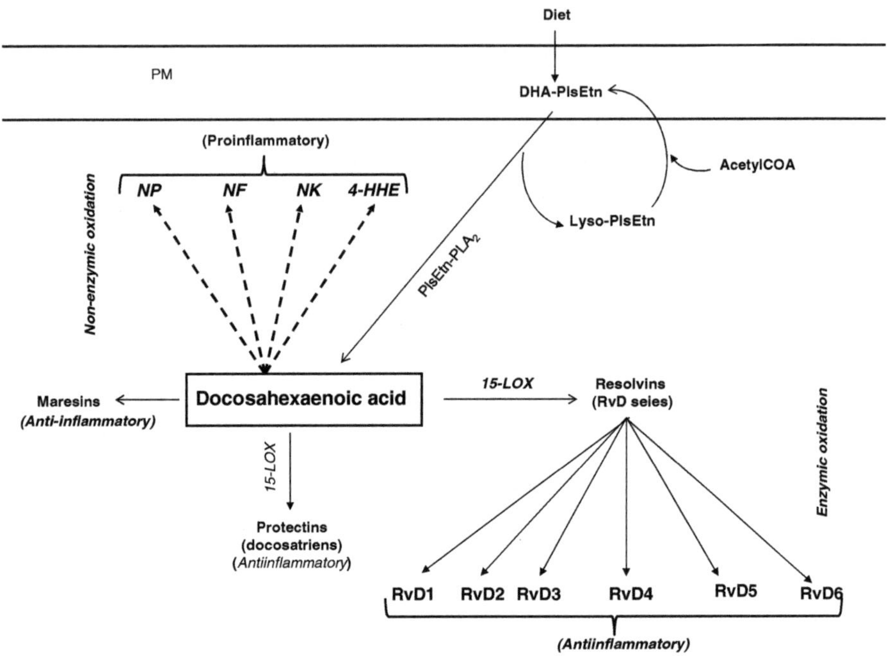


**Fig. 3.6** Generation of enzymic and nonenzymic lipid mediators from docosahexaenoic acid. Plasma membrane (PM); plasmalogen (PlsEtn); plasmalogen-selective $PLA_2$ (PlsEtn-$PLA_2$); lyso-ethanolamine plasmalogen (lyso-PlsEtn); neuroprostane (NP); neurofuran (NF); neuroketal (NK); 4-hydroxyhexenal (4-HHE); resolvin D (RvD); and 15-lipoxygenase (15-LOX)

(Farooqui 2009, 2011). The synthesis of aspirin-triggered forms of D series resolvins (AT-Rv) from DHA is catalyzed by sequential oxygenation initiated by aspirin-acetylated COX-2 enzyme (Fig. 3.6). These mediators not only antagonize the effects of ARA-derived PGs, LTs, and TXs, but also modulate leukocyte trafficking and downregulate the expression of cytokines in neural and nonneural cells. They possess potent anti-inflammatory, neuroprotective, and pro-resolving properties (Serhan 2005; Farooqui 2011). D series resolvins exert potent agonist actions on macrophages and vascular endothelial cells. They control the magnitude of the local inflammatory response. They act through specific receptors found in neural and nonneural cells. These receptors are called as resolvin D receptors ($resoDR_1$) (Serhan and Chiang 2008). These receptors have not been characterized in neural and nonneural tissues. Thus, detailed investigations on link between enzymic activities ($PLA_2$, COX, LOX, and EPOX and their downstream enzymes) and $resoDR_1$ and $resoER_1$ in neural and nonneural cells are urgently required. 15-LOX also converts DHA into protectin $D_1$ through epoxide intermediate with epoxy group at the 16(17) position (Fig. 3.7). The reaction sequence of biosynthesis for $PD_1$ via the epoxide intermediate distinguishes it from the formation of the double dioxygenation product 10S,17S-dihydroxy-DHA (Serhan 2010). The

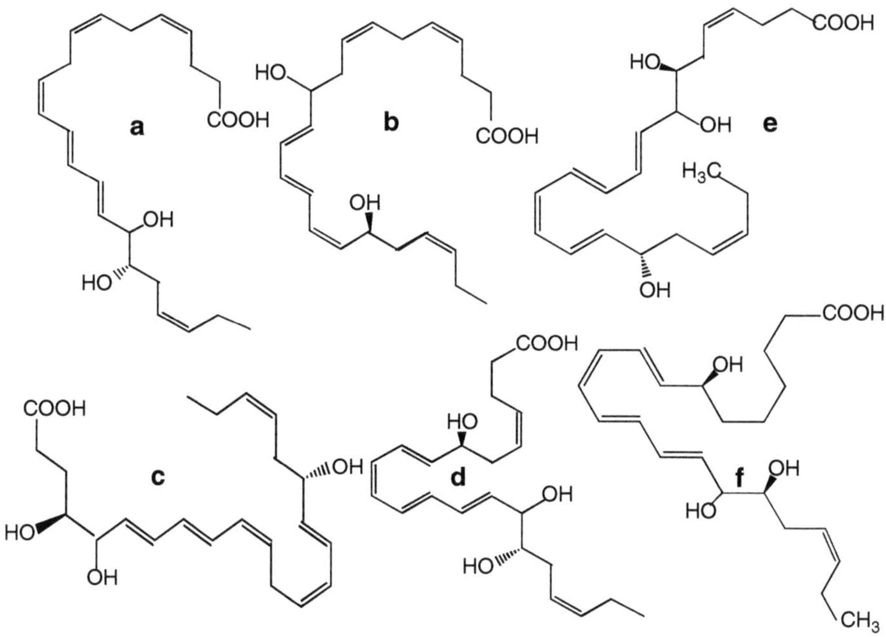


**Fig. 3.7** Chemical structures of lipid mediators derived from docosahexaenoic acid. 10, 17 S-Docosatriene (**a**); 16,17 S-docosatrienes (**b**); 4S, 5,17 S-resolvin (**c**); 7S,16,17 S-resolvin (**d**); resolving $D_1$ (**e**); and resolving $D_2$ (**f**)

occurrence of $PD_1$ has also been reported in brain, where it is named neuroprotectin $D_1$ (10R, 17S-dihydroxy-docosa-4Z,7Z,11E,13E,15Z,19Z-hexaenoic acid, $NPD_1$) (Marcheselli et al. 2003; Mukherjee et al. 2004).

Increased consumption of DHA alleviates obesity-induced insulin resistance in obese mice (González-Périz et al. 2009). These beneficial effects of DHA are associated with upregulation of genes involved in insulin sensitivity (PPARγ), glucose transport (GLUT2 and GLUT4), and insulin receptor signaling (IRS-1 and IRS-2) in both adipose tissue and liver (González-Périz et al. 2009). In addition, DHA upregulates the expression and the generation of the potent anti-inflammatory and insulin-sensitizing adipokine and adiponectin and promotes the phosphorylation of AMPK. These beneficial responses are accompanied by the decrease in formation of ARA-derived eicosanoids ($PGE_2$ and 5-HETE) and a concomitant increase in the formation of protective DHA-derived lipid mediators (protectins and resolvins), which facilitate the insulin-sensitizing and antisteatotic effects exerted by DHA (González-Périz et al. 2009). In addition, dietary supplementation of DHA increases GLUT1 expression in the rat brain endothelial cells and astrocytes (Pifferi et al. 2007), supporting a positive correlation between DHA level in brain membranes and glucose transporter expression in the brain (Brenna and Diau 2007). Expressions of brain genes encoding for proteins of the mitochondrial respiratory chain are

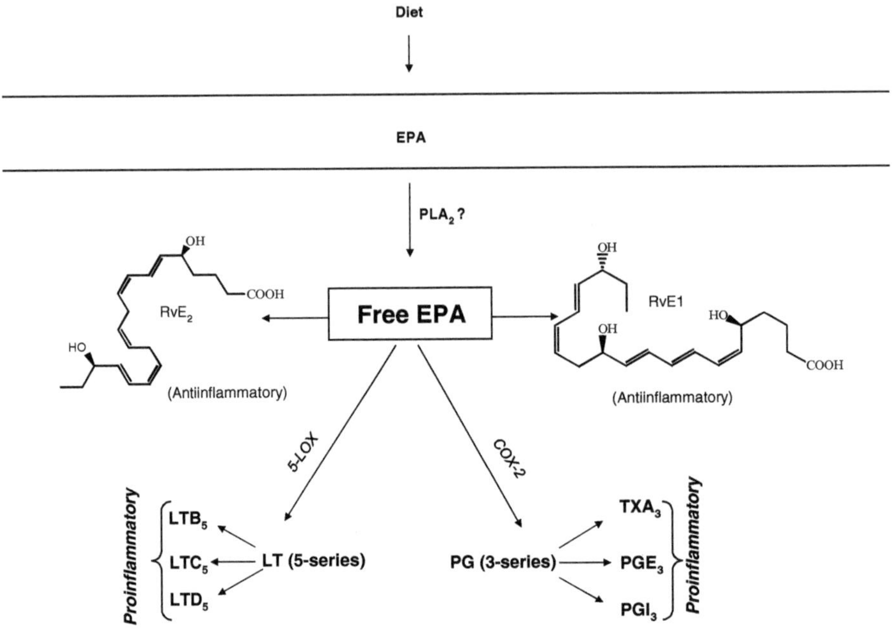


**Fig. 3.8** Generation of lipid mediators from eicosapentaenoic acid. Eicosapentaenoic acid (EPA); phospholipase $A_2$ ($PLA_2$); cyclooxygenase-2 (COX-2); 5-lipoxygenase (5-LOX); 5-series leukotrienes ($LTB_5$, $LTC_5$, and $LTD_5$) and 3-series prostaglandins ($PGE_3$ and $PGI_3$), and 3-series thromboxanes ($TX_3$)

significantly increased in animals fed a diet supplemented with DHA (Cunnane et al. 2009). Thus, emerging evidence suggests that DHA plays an important role in brain glucose/brain energy metabolism.

Like ARA, EPA is oxidized by COXs and LOXs and generates 3-series of prostaglandins ($PGE_3$) and thromboxanes ($TX_3$) and the 5-series of leukotrienes ($LTB_5$) (Fig. 3.8). These eicosanoids have different biological properties than the corresponding analogs produced by the oxidation of ARA. For example, $TXA_3$ is less active than $TXA_2$ in aggregating platelets and constricting blood vessels (Calder and Grimble 2002; Yin et al. 2007; Calder 2009; Farooqui 2011). $PGE_3$, $TX_3$, and $LTB_5$ antagonize the proinflammatory effects of $PGE_2$, $TX_2$, and $LTB_2$ and less suitable for the activation of EP receptors. In addition to forming less active lipid mediator, EPA mediates its effects through the modulation of anti-inflammatory cytokines secretion (Schmitz and Ecker 2008). Thus, studies on the effects of PGs ($PGE_2$ vs. $PGE_3$) and LTs ($LTB_4$ vs. $LTB_5$) on endothelium permeability and mononuclear adhesion and migration across endothelial cell cultures indicate that $PGE_3$ produces more pronounced effect on trans-endothelial Evans blue-albumin (EBA) permeability than $PGE_2$, and these effects are antagonized by $EP_1$ and $EP_2$ antagonists (Moreno 2009). $LTB_4$ and $LTB_5$ produce very little effect on EBA extravasation. Accumulating evidence suggests that the enhancement of endothelial permeability in the presence of polymorphonuclear (PMN) cells is modulated by interplay between EPA- and ARA-derived LTs and PGs (Yin et al. 2007; Moreno 2009).

EPA is a precursor of E series resolvins, including resolvin $E_1$ ($RvE_1$; (5S,12R,18R)-trihydroxy-6Z,8E,10E,14Z,16E-eicosapentaenoic acid) and resolvin $E_2$ (15*S*,18*R*-dihydroxy-EPE) (24, 25) (Fig. 3.8). In vasculature, $RvE_1$ and $RvE_2$ exert potent anti-inflammation/pro-resolution effects in vivo (Arita et al. 2006) by interacting with specific G protein-coupled receptors (see below). $RvE_1$ suppresses the activation of NF-κB by tumor necrosis factor-α (TNF-α) through binding to human polymorphonuclear leukocyte (PMN) (Arita et al. 2007). Intrathecal $RvE_1$ injections retard spontaneous pain and heat and mechanical hypersensitivity induced by intrathecal capsaicin and TNF-α. $RvE_1$ mediates its anti-inflammatory activity not only by decreasing neutrophil infiltration, paw edema, but also by inhibiting the expression of proinflammatory cytokine (Xu et al. 2010).

## 3.6 Metabolism of n-3 Fatty Acids in Metabolic Syndrome

It is well known that insulin resistance, low-grade inflammation, hypertension, and endothelial cell dysfunction are closely associated with the pathophysiology of MetS. n-3 FFAs inhibit inflammation, lower blood pressure, improve vascular endothelial cell function, and increase insulin sensitivity (Farooqui 2009). Meta-analysis studies on humans have confirmed that n-3 FFAs not only decrease insulin resistance in type II diabetes and MetS patients but also reduce levels of triacylglycerols. It is also shown that saturated fat, monounsaturated fat, and trans fat are significantly associated with the risk of developing MetS. In contrast, n-3 FFAs have been shown to play a protective role in the risk of MetS (Breslow 2006; Ascherio et al. 1995). n-3 fatty acids not only induce the expression and the production of the potent anti-inflammatory, antisteatotic, and insulin-sensitizing adipokine, adiponectin, but also induce the phosphorylation of AMPK (González-Périz et al. 2009). These beneficial responses correlate with a decrease in the formation of n-6 fatty acid-derived eicosanoids ($PGE_2$ and 5-HETE) and a concomitant increase in the generation of protective n-3 fatty acid-derived lipid mediators (protectins and resolvins), which mimic the insulin-sensitizing and antisteatotic effects exerted by n-3 fatty acids. Interestingly, the effects of resolvins and protectins appeared to be more potent than their n-3 precursors (González-Périz et al. 2009). Thus, intraperitoneal injections of nanogram doses of an EPA-derived metabolite ($RvE_1$) produce significant insulin-sensitizing effects by inducing adiponectin, glucose transporter-4 (GLUT4), and insulin receptor signaling-1 (IRS-1) expression in adipose tissue and induce significant protection against hepatic steatosis (González-Périz et al. 2009). Similarly, in leptin receptor-deficient (db/db) obese and diabetic mice, nanogram doses of DHA-derived $RvD_1$ not only improve glucose tolerance but also decrease fasting blood glucose and increase insulin-stimulated Akt phosphorylation while reducing the formation of crown-like structures rich in inflammatory macrophages in adipose tissue (Hellmann et al. 2011). Recent studies have also shown that ARA-derived $LXA_4$ also produces beneficial effects in an experimental model of age-associated adipose inflammation (Börgeson et al. 2012). Based on these results, it can be proposed that EPA- and DHA-derived $RvE_1$ and $RvD_1$- and ARA-derived $LXA_4$ can be used as templates for designing novel therapies to combat the progression of metabolic complications associated with MetS. In addition, n-3 fatty acids do not produce any side effects and more importantly do not interact adversely with the common drug therapies used in the management and treatment of hypertension, dyslipidemia, type II diabetes, and obesity/MetS, and in some instances work synergistically, thereby providing additional cardiovascular and cerebrovascular benefits (Abeywardena and Patten 2011). Although the molecular mechanism n-3 FFA-mediated decrease in insulin resistance is not fully understood, it is recently proposed that polyunsaturated fatty acids/essential fatty acids (PUFAs/EFAs) modulate insulin resistance through several mechanisms including (a) increase in the number of insulin receptors due to increased membrane fluidity; (b) an increase in GLUT4 mRNA and protein level in adipocytes (Peyron-Caso et al. 2002); (c) generation of anti-inflammatory and antiatherosclerotic lipid mediators (lipoxins, resolvins, protectins, and maresins) (Das 2006, 2008, 2010; Merched et al. 2008); (d) downregulation of the expression of adhesion molecules (Weiss et al. 1995; Das 2000); (e) inhibition of proinflammatory cytokine (TNF-α, macrophage migration factor (MIF), high-mobility group box 1 (HMGB1), and IL-6) production (Blok et al. 1996; Das 2002); (f) upregulation of brain-derived neurotrophic factor (BDNF) production in the brain and gut along with inhibition of $PGI_2$ synthesis (Cole and Frautschy 2010; Boesmans et al. 2008; Santhanam et al. 2010); and (i) regulation of both the secretion and actions of various hypothalamic neurotransmitters and peptides that regulate appetite, satiety, food intake, and insulin secretion (Lavialle et al. 2010; Basselin et al. 2005). Insulin has been reported to activate $\Delta^6$- and $\Delta^5$-desaturases (Rimoldi et al. 2001) producing enhancement in PUFAs synthesis. Increase in PUFA synthesis potentiates the action of insulin. According to Das (2010) an adequate consumption of PUFAs (n-3 FFAs) from childhood to old age results in optimal incorporation of n-3 FFAs into the cell membrane of not only endothelial cells but also other visceral and neuronal tissues. This process may provide sufficient levels of lipid mediators (lipoxins, resolvins, protectins, and maresins), which protect visceral and brain tissues from the injurious effects of oxidative stress and low-grade chronic inflammation, processes that contribute to the pathogenesis of diabetes and MetS (Das 2010).

Increased consumption of DHA results in a decrease in the level of ARA in neural membrane phospholipids, thereby decreasing the level of substrate for the production of proinflammatory eicosanoids and cytokines (Fig. 3.9). The anti-inflammatory effects of DHA may also be due to downregulation of iNOS in microglial cell and peritoneal macrophages in a dose-dependent manner. This inhibition is accompanied not only by inhibition of the oxidative stress-sensitive NF-κB activation but also by upregulation of GSH (Farooqui 2009). Collectively, these studies suggest that because of differences in their mode of action, excess of DHA-enriched diet can interfere with

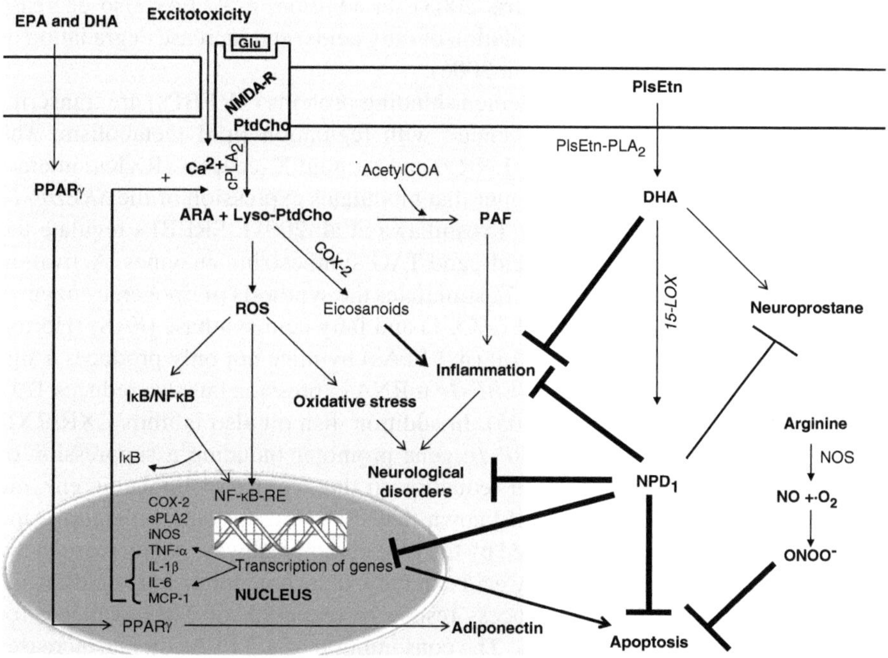


**Fig. 3.9** Effect of neuroprotectin D$_1$ (NPD$_1$) on inflammation. Oxidative stress and apoptosis. Plasma membrane (PM); *N*-methyl-D-aspartate receptor (NMDA-R); glutamate (Glu); phosphatidylcholine (PtdCho); lysophosphatidylcholine (lyso-PtdCho); platelet-activating factor (PAF); plasmalogen (PlsEtn); plasmalogen-selective PLA$_2$ (PlsEtn-PLA$_2$); cytosolic phospholipase A$_2$ (cPLA$_2$); cyclooxygenase (COX-2); 5-lipoxygenase (5-LOX); arachidonic acid (ARA); reactive oxygen species (ROS); nuclear factor-κB (NF-κB); nuclear factor-κB response element (NF-κB-RE); inhibitory subunit of NF-κB (IκB); tumor necrosis factor-α (TNF-α); interleukin-1β (IL-1β); interleukin-6 (IL-6); secretory phospholipase A$_2$ (sPLA$_2$); and inducible nitric oxide synthase (iNOS)

ARA metabolism through their short-lived lipid mediators, which act as local hormones or vice versa.

## 3.7 n-3 Fatty Acid-Mediated Reduction in Triacylglycerol Levels

It is well known that n-3 FFAs reduce the levels of TAG in blood. However, the mechanisms by which n-3 FFA reduces serum TAGs are not well understood. Based on preclinical and clinical studies, it is suggested that n-3 FFAs not only reduce hepatic VLDL-TAG synthesis/secretion, but also enhance TAG clearance from circulating VLDL particles by promoting increased lipoprotein lipase activity

(Harris et al. 2008; Park and Harris 2003). In addition, n-3 FFAs also decrease hepatic lipogenesis, increase β-oxidation of fatty acids, and increase degradation of apo B-100 (Harris and Bulchandani 2006).

PPARs and sterol regulatory element-binding proteins (SREBPs) are transcription factors that are closely associated with regulation lipid metabolism. The nuclear receptors liver X receptor (LXR)α and retinoid X receptor (RXR)α interact with each other to form a heterodimer that modulates expression of the *SREBP-1c* gene by binding to the *SREBP-1c* (Yoshikawa et al. 2001). SREBPs regulate the expression of cholesterol-, fatty acid-, and TAG-synthesizing enzymes. Activation of the transcription factor SREBP-1c stimulates the synthesis of lipogenic enzymes such as acetyl-CoA carboxylase-1 (ACC1) and fatty acid synthase (FAS) (Horton et al. 1998). Consumption of fish oil (n-3 FFAs) by mice not only produces a significant decrease in the hepatic *SREBP-1c* mRNA expression but also reduces TAG levels (Le Jossic-Corcos et al. 2005). In addition, fish oil also inhibits LXR/RXR heterodimer binding to the *SREBP-1c* gene promoter inducing a suppression of *SREBP-1c* mRNA expression and reduction in the activity of lipogenic enzyme (Yoshikawa et al. 2002). It is well known that FFAs are not only utilized for the synthesis of TAG, but also degraded by the β-oxidation pathway. An increase in the rate of β-oxidation of hepatic fatty acid can reduce the amount of fatty acids available for TAG synthesis. This process results in reduction of TAG available for incorporation into VLDL particles. The consumption of n-3 FFAs by rats increases in free fatty acid β-oxidation in peroxisomes and mitochondria (Harris and Bulchandani 2006). This process results in less availability of free fatty acids for TAG and VLDL synthesis. Similarly in human subjects consumption of n-3 FFAs also results in a faster rate of hepatic fatty acid oxidation (Dagnelie et al. 1994). Among n-3 FFAs, EPA interacts with all forms of PPAR (PPARα, β, and γ), and PPARα is known to contribute to n-3 fatty acid-mediated modulation of fatty acid β-oxidation (Jump 2002). It is also shown that n-3 FFAs may increase TAG removal from circulating VLDL and chylomicron particle, through increase in hydrolysis by LPL (Park and Harris 2003).

## 3.8 Fatty Acids in the Brain

Unlike earlier view that the brain does not use free fatty acids as an energy source, it is becoming increasingly evident that free fatty acids and specifically their intermediates may be involved in the central control of energy balance regulation (Lopez et al. 2007, 2008; Lam et al. 2005). Although the importance of FFAs in the pathogenesis of MetS in visceral organs is well known, very little is known about the uptake and metabolism of fatty acids in the brain of MetS patients (Eckel et al. 2005). Animal studies have indicated that hypothalamic sensing of circulating fatty acids is closely associated with the control of nutrient intake and, thus, energy balance. For example, it is shown that manipulation of central hypothalamic FFA metabolism normalizes energy and glucose homeostasis in overfed rats (Pocai et al. 2006).

Studies on mechanisms behind the complex hypothalamic regulatory circuitry that regulates energy expenditure and feeding behavior are difficult to address in humans in vivo due to ethical reasons, so most data in this regard have been obtained in small animals (Lam et al. 2005; Pocai et al. 2006; Morgan et al. 2004). Recently, positron emission tomography (PET), a noninvasive and quantitative procedure, has been used to study the fate of long-chain fatty acids (Arai et al. 1995), and it is reported that MetS patients have increased brain free fatty acid uptake compared to healthy subjects. Weight loss in MetS patients is able to partly reverse the increase in brain fatty acids. These studies not only support the possible role of brain free fatty acid in regulation in body weight but also suggest that catabolic effect of central free fatty acids is disturbed in MetS and this can be partially reversed after weight loss (Karmi et al. 2010)

## 3.9 Role of Essential Fatty Acids in the Brain

Fatty acid composition of neural membrane is regulated by diet. As stated earlier, ARA and DHA are enriched in brain, where they are required during pre- and postnatal development to build functionally critical brain circuits. Neurogenesis, synaptogenesis, dentrite formation, photoreceptor membrane biogenesis, and biogenesis of other neural membranes take place during early postnatal development. However, despite their abundance in neural membranes, these fatty acids cannot be synthesized de novo synthesis by the human brain (Farooqui 2009). They, or their precursors, must be ingested from dietary sources and transported to the brain (Horrocks and Farooqui 2004; Bazan et al. 2011). Present-day Western diet is characterized by an increase in saturated fat, n-6 PUFA, and trans fatty acid intake, as well as an overall decrease in n-3 PUFA intake (Simopoulos 2004, 2006; Cordain Eaton et al. 2005; Farooqui et al. 2007; Farooqui 2009; Simopoulos 2011). Due to the increased consumption of LA-rich vegetable oils, n-6 PUFA consumption has become progressively much higher than that of n-3 PUFA. Western diet now has a ratio of n-6 to n-3 fatty acids of about 20:1 and fatty acids now represent 28–42 % of total energy consumed by Western populations (Olivier et al. 2011; Eaton et al. 2010), whereas, in ancestral diet (the Paleolithic diet), fatty acid consumption was only approximately 20–30 % of total energy (Olivier et al. 2011; Eaton et al. 2010). This diet had an n-6 to n-3 ratio of about 4: 1. As a result of the increased consumption of LA-rich vegetable oils in the Western diet, n-6 PUFA consumption has become progressively much higher than that of n-3 PUFA (Simopoulos 2011). n-6 fatty acid-enriched diet promotes the pathogenesis of neurotraumatic, neurodegenerative, and neuropsychiatric diseases (Farooqui 2009). In contrast, diet enriched in n-3 fatty acids exerts cardioprotective, immunosuppressive, neuroprotective, and psychoprotective effects (Simopoulos 2006; Farooqui et al. 2007; Farooqui 2009). Essential fatty acids not only promote and maintain synaptic plasticity and induce neurogenesis in the brain circuits that also reduce oxidative stress and neuroinflammation through the generation of lipoxins, neuroprotectins, and resolvins (Bazan 2006;

Serhan 2005; Farooqui 2009). Low levels of ARA are involved in the regulation of long-term potentiation and in the retrograde signal generated by the activation of glutamate AMPA and NMDA receptors (Volterra et al. 1992; Williams et al. 1989). Furthermore, ARA is converted by different enzymes into cannabinoids and eicosanoids, these derivatives being involved in synaptic signaling and neuroinflammation (Sheinin et al. 2008; Farooqui 2009). In addition, ARA also exerts its own effects, including the inhibition of presynaptic and postsynaptic channels as well as the formation and recycling of synaptic vesicles (Marza et al. 2008). Additionally, DHA and EPA also facilitate optimal membrane fluidity and promote appropriate neurotrophic support (Farooqui 2009). DHA is a potent regulator of NF-κB via multiple mechanisms. Thus, DHA directly inhibits NF-κB activation in the brain (Komatsu et al. 2003; Farooqui 2009). DHA also produces a dose-dependent inhibitory effect on NO production. Furthermore, DHA inhibits the activation of NF-κB not only by upregulating intracellular glutathione but also by inhibiting the production of cytokines and expression of adhesion molecules (Farooqui 2009). Finally, DHA is a precursor for docosanoids (resolvins and neuroprotectins), which produce anti-inflammatory effects in the brain (Farooqui 2011). Mice fed a DHA-rich diet produce significantly higher levels of BDNF in the striatum, an area of the brain involved in the pathogenesis of Parkinson disease (Rapoport et al. 2007). Mice fed on DHA-deficient diet for 4 weeks after weaning have reduced levels of striatum DHA and BDNF compared with control mice (Miyazawa et al. 2010). Production of BDNF in the brain not only modulates neurogenesis but is also associated with the long-term maintenance of the function, shape, and plasticity of neurons, especially in high electrical areas, during aging (Barde 1994).

DHA also modulates CREB, a transcription factor that is situated at the distal end of the cAMP signaling cascade and is a major transcriptional regulator of numerous functions in adult neural cells, including learning and memory and neuronal survival (Fujioka et al. 2004). CREB is activated by phosphorylation via several mechanisms, including neurotrophic factor cascades. Immature neurons of the hippocampus express pCREB (Nakagawa et al. 2002). It is becoming increasingly evident that CREB and NF-κB regulate diverse sets of genes that in general appear to act in opposite ways. CREB is prominent in plasticity and growth, whereas NF-κB in oxidative inflammatory gene induction is generally associated with degeneration (Zou and Crews 2006). It is proposed that the intensity of completion between CREB and NF-κB may play a pivotal role in diabetes-induced impairments of hippocampal neurogenesis and is closely associated with cognitive deficits (Alvarez-Nolting et al. 2012). Treatment with DHA not only reverses alterations in NF-κB, pCREB, lipid peroxidation, and antioxidant defenses but also restores the decrease of neurogenesis in diabetes (Alvarez-Nolting et al. 2012).

In contrast as stated above, ARA initiates the production of ROS and promotes the expression of cytokines, especially IL-1β, the most potent proinflammatory cytokine that induces stress and anxiety-like behavior in rodents (Song et al. 2003, 2008). IL-1β also stimulates neurotransmitter systems, thereby increasing the turnover of noradrenaline, serotonin, and dopamine (Song 2002; Song et al. 2008). ARA containing cannabinoid agonists through presynaptic mechanisms modulate the release of several neurotransmitters implicated in the control of anxiety and depression. They suppress the outflow of glutamate not only in the hippocampus but also in periaqueductal gray and amygdala. Cannabinoids mediate inhibitory effect on corticolimbic release of norepinephrine, dopamine, serotonin, and anxiogenic neuropeptide corticotropin-releasing factor and cholecystokinin. ARA-containing cannabinoid agonists also interfere with GABAergic transmission in the amygdala, hippocampus, and prefrontal cortex. Accumulating evidence suggests that essential fatty acids play different role in the neuroendocrine-immune network, and the ratio of n-6 to n-3 fatty acid is an important dietary factor in reducing oxidative stress and inflammation and promoting neuroprotective and psychoprotective effects on human brain.

## 3.10 Metabolism of ARA in Neurological Disorders

Neurological disorders include neurotraumatic, neurodegenerative, and neuropsychiatric diseases. Neurotraumatic diseases include stroke, spinal cord injury SCI), and traumatic brain injury (TBI), whereas neurodegenerative diseases include Alzheimer's disease (AD), Parkinson disease (PD), Huntington disease (HD), and multiple sclerosis (MS) (Farooqui 2010). Neuropsychiatric diseases include depression, schizophrenia, bipolar disorders, autism, mood disorders, and attention-deficit/hyperactivity disorder (Farooqui 2009). Since all these diseases involve alterations in signal transduction and functional abnormalities in brain and spinal cord function, they are called as neurological disorders. Neurotraumatic, neurodegenerative, and neuropsychiatric diseases share excitotoxicity, oxidative stress, and neuroinflammation as a common mechanism of cell injury and death (Fig. 3.9). The onset of neurological diseases may occur when neurons fail to respond adaptively to age-related increase in oxidative and nitrosative stress and neuroinflammation (Farooqui 2010). Persistent presence of oxidative and nitrosative stress and neuroinflammation induces the accumulation of damaged proteins, DNA, and membrane fragments (Farooqui 2009, 2010). Abnormalities in signal transduction processes along with elevated levels of lipid mediators and disturbance in stress resistance mechanisms have also been reported to contribute to the pathophysiology of neurological disorders (Farooqui 2011). It is proposed that upregulation of interplay (cross talk) among excitotoxicity, oxidative stress, and neuroinflammation may be a common mechanism of brain damage in neurological disorders (Farooqui and Horrocks 2007; Farooqui 2010, 2011).

In neurological disorders the stimulation of isoforms of $PLA_2$, COX, and LOX results in the degradation of neural membrane phospholipids and accumulation of $PGE_2$, a major ARA-derived PG, which is closely associated with neuroinflammation and apoptotic and necrotic types of cell deaths (Phillis et al. 2006; Farooqui 2010).

In addition, lysophospholipids are metabolized to platelet-activating factor, a lipid mediator, which also promotes neuroinflammation. Thus, an uncontrolled ARA cascade sets the stage for the increased production of ROS (oxidative stress) as well as neuroinflammation. ROS-mediated damage to membrane proteins, ion channels, and receptors induces abnormalities in signal transduction processes associated with pathophysiology of neurological disorders. The action of 12-LOX on ARA produces lipid hydroperoxides, which inhibit reacylation of phospholipids and therefore interfere with restoration of normal phospholipid composition in neuronal membranes producing abnormalities in membrane fluidity and permeability (Farooqui and Horrocks 2007). Alterations in membrane fluidity and permeability not only increase influx of $Ca^{2+}$ and lead to stimulation of $Ca^{2+}$-dependent enzymes leading to intensification and upregulation of oxidative stress and neuroinflammation but also cause functional changes in the activity of receptors and other proteins embedded in the neural membranes. These processes may cause apoptotic and necrotic types of cell deaths in the brain and spinal cord. In brain ARA has been implicated in the pathogenesis of AD. Imaging of radiolabeled ARA in brains of human subjects has revealed increased ARA metabolism in patients with AD (Esposito et al. 2008). Furthermore, accumulation of ARA is known to induce the polymerization of tau protein (Wilson and Binder 1997) and induction of apoptosis in neurons (Farooqui and Horrocks 2007). In addition, ARA can also bind receptors at synapses or diffuse into the cell nucleus and interact with transcription factors to control transcription for cytokines. ARA-derived lipid nonenzymic lipid mediators include 4-HNE, isoprostanes, isoketal, and isofurans. ARA-derived enzymic and nonenzymic lipid mediators produce prothrombotic, proaggregatory, and proinflammatory effects (Farooqui 2011).

Studies on the ability of fatty acids to enhance or suppress AD like characteristics in COS-7 cell culture model system indicate that the presence of double bonds in all *cis* configuration at positions $\Delta^{13,16,19}$ (as occur in DPA, DHA, DTA, and $C_2H_4$-DHA chains) decreases the production of Aβ peptide. In contrast, double bonds with an all *cis* configuration either at positions $\Delta^{9,12}$ (in linolenic acid (LNA), αLNA, stearic acid) or at $\Delta^{8,11,14}$ (in EPA and ARA) increase Aβ levels supporting the view that fatty acid chains may contain templates that may control the production of Aβ in cell culture model (Amtul et al. 2011a) (Table 3.1). These results provide a structural basis for studies on the interactions of fatty acids with membrane-associated proteins, such as APP as well as Aβ production and degradation. These results are supported by studies on an early-onset AD transgenic mouse model expressing the double-mutant form of human APP, Swedish (K670N/M671L) and Indiana (V717F) (Amtul et al. 2011b). Mice fed with a high-protein, low-fat (18 % reduction), cholesterol-free diet enriched with oleic acid exhibit an increase in Aβ40/Aβ42 ratio, decrease in levels of β-site APP cleaving enzyme (BACE), and reduction in presenilin levels along with reduction in production of amyloid plaques in the brain. The decrease in BACE levels is accompanied by increased levels of a nonamyloidogenic soluble form of APP (sAPPα). Furthermore, the low-fat/+OA diet also produces an augmentation of

**Table 3.1** Modulation of Aβ deposition by fatty acids

| Fatty acid | Structure template | Effect | Toxic end products |
|---|---|---|---|
| Linoleic acid | 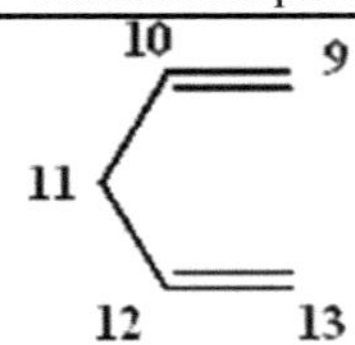  | Detrimental | |
| Arachidonic acid | 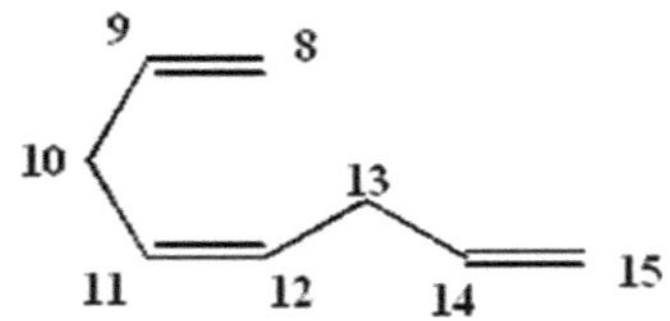  | Detrimental | PGs, LTs, TXs, LX (enzymic) 4-HNE, IsoP (nonenzymic) |
| Arachidonic acid metabolite | 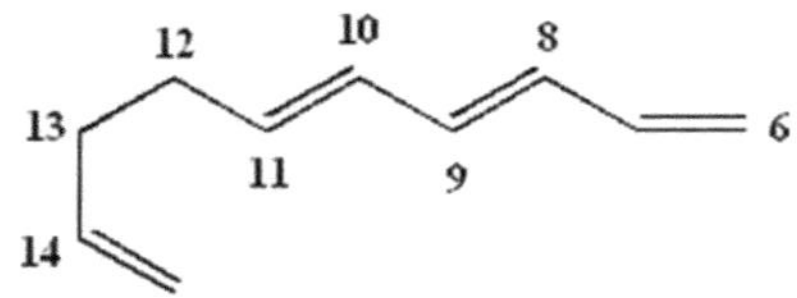  | Detrimental | PGs, LTs, TXs, LX (enzymic) 4-HNE, IsoP (nonenzymic) |
| Docosahexaenoic acid | 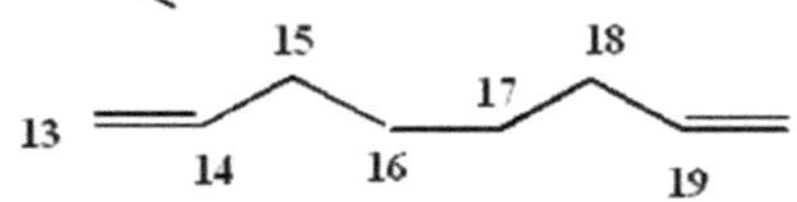  | Beneficial | PGE3, resolvins, and protectins (enzymic) 4-HNE, NeuroP (nonenzymic) |

Prostaglandins (PGs) and thromboxanes (TXs) via the cyclooxygenase pathway or to leukotrienes (LTs), lipoxins (LXs), hydroperoxyeicosatetraenoic acids (HPETE), and hydroxyeicosatetraenoic acids (HETE) insulin-degrading enzyme and insulin-like growth factor II (Amtul et al. 2011b). It is proposed that more studies on the template of fatty acids may provide important insights into the mechanism(s) that mediate their effects on Aβ metabolism as well as rationale for the use of specific naturally occurring fatty acids as novel AD therapeutics (Amtul et al. 2011a, b).

## 3.11 Metabolism of n-3 Fatty Acids in Neurological Disorders

Recent studies indicate that levels of DHA are reduced in temporal cortex and mid-frontal cortex from 37 subjects with Alzheimer's disease, compared to 17 closely matched control subjects (Astarita et al. 2010). As stated above, in mammals enzymic pathway for the synthesis of DHA is primarily localized to the liver (Marszalek and Lodish 2005; Rapoport and Igarashi 2009). Studies on lipid composition of liver samples from 14 patients and 9 control subjects indicate that liver tissue from AD patients contains reduced levels of DHA but elevated levels of tetracosahexaenoic acid and other n-3 fatty acids suggesting a defect in the conversion of tetracosahexaenoic acid into DHA, a process that occurs in liver peroxisomes (Fig. 3.2) (Astarita et al. 2010). Using an untargeted lipidomic methodology, it is also shown that the levels of nervonic acid (24:1n-9) and other monounsaturated fatty acids synthesized by stearoyl-CoA desaturase (SCD) are markedly increased in frozen brain samples from AD patients, compared to age-matched controls, and that these changes strongly correlate with cognitive impairment (Astarita et al. 2011). Although the functional consequences of elevated brain monounsaturated fatty acids levels are not fully understood at the present time, the importance of peripheral SCD activity is becoming increasingly evident in the development of metabolic diseases. Thus, increase in SCD activity is closely associated with the key aspects (insulin resistance, abdominal adiposity, and hyperlipidemia) of the MetS (Sampath and Ntambi 2008; Paillard et al. 2008; Warensjo et al. 2005), a condition which may be an important risk factor the pathogenesis of stroke, AD, and depression (Farooqui et al. 2012). Furthermore, animal studies also indicate that SCD-derived palmitoleate (16:1n-7) participates as a blood-borne messenger in the control of energy homeostasis and insulin resistance (Cao et al. 2008). These studies support the view that alterations in levels of free fatty acids play a key role in cognitive deficits in aging and AD (Astarita et al. 2010).

As stated above, metabolism of DHA generates anti-inflammatory lipid mediators, such as resolvins, protectins, and neuroprotectins, which directly or indirectly suppress the activity of NF-κB (Wada et al. 2003; Bazan 2006; Farooqui and Horrocks 2007; Phillis et al. 2006; Farooqui 2009). This results in neuroprotective effects of DHA. DHA also acts as a ligand for the PPARγ and the RXR receptors. In conjunction with these receptors, which act as transcription factors, DHA modulates various neurochemical processes that maintain cellular homeostasis by modulating lipid metabolism, neural cell differentiation, and apoptosis (Farooqui 2009). DHA and EPA also quench gene expression of COX-2, thereby diminishing the formation of proinflammatory eicosanoids. In addition, DHA and EPA scavenge of free radicals, which diminish inflammatory response. Finally, DHA increases brain levels of neuroprotective brain-derived neurotrophic factor and reduces the ARA and its oxidative metabolites. The cross talk among these molecular processes has distinct neuroprotective effects not only through the stabilization of neural membranes and modulation of ion channels and receptors but also through inhibition of inflammatory processes and generation of anti-inflammatory lipid mediators (Farooqui et al. 2007; Farooqui 2009).

The nonenzymic DHA-derived lipid mediators include 4-HHE, neuroprostanes, neuroketal, and neurofurans. These lipid mediators produce milder oxidative and inflammatory effects than ARA-derived nonenzymic lipid mediators (Farooqui 2011). Since DHA also plays important roles in modulation of gene expression and neurotransmission, involving serotonin, norepinephrine, and dopamine receptors, it is proposed that DHA deficit may play a role in etiology of mood and cognitive dysfunction in neurological diseases (Farooqui 2009; Farooqui et al. 2012). Beyond their structural importance in neural and retinal membranes, DHA and EPA are involved in the modulation of the immune system (Farooqui 2009, 2010). Enrichment of DHA in diet blocks neuroinflammation and oxidative stress in neurological disorders not only through its effects on physicochemical properties of neural cell membranes but also through modulation of genes and generation of resolvins, $NPD_1$, and maresins. $NPD_1$ attenuates NF-κB production and inhibits COX2 expression, reduces influx of polymorphonucleocytes, and counters apoptosis, thus promoting neural cell survival (Serhan et al. 2008; Bazan 2006). Furthermore, D series resolvins block TNF-α-induced IL-1β transcript in microglial cells and limit PMN infiltration into inflamed brains (Serhan et al. 2008; Bazan 2006). Accumulating evidence suggests that because of antioxidants, anti-inflammatory, antiexcitotoxic, and antiapoptotic properties and clinical safety, DHA can be used as an ideal prophylactic agent for the treatment of neurological disorders (Hogyes et al. 2003; Bazan 2006; Lukiw et al. 2005; Farooqui 2009; Grimm et al. 2011; Farooqui et al. 2012). APP processing is modulated by the presence of DHA in diet (Farooqui 2009; Farooqui et al. 2012). DHA affects APP processing in AD brain. Thus, it reduces amyloidogenic processing by inhibiting β- and γ-secretase activity (Grimm et al. 2011). Increase in ROS and RNS generation results in microglial cell activation in the brain. It not only facilitates β-secretase-mediated degradation of APP and increase in production of Aβ but also enhances the expression of proinflammatory cytokines—a potentially vicious cycle, which promotes apoptotic cell death in visceral circulatory system as well as in the brain (Farooqui 2010). NO produces differential regulation on β-secretase. At low concentration it increases β-secretase activity, whereas at modest to high concentrations, it suppresses β-secretase activity either transcriptionally or posttranslationally through S-nitrosylation. Thus, S-nitrosylation may represent a basic regulatory mechanism for maintaining β-secretase at physiological levels (Kwak et al. 2011). In contrast, as mentioned above, α-secretase in brain promotes neuroprotection and in visceral organs promotes insulin sensitivity and optimal mitochondrial and endothelial function. Based on these findings, it is proposed that DHA reduces Aβ generation not by a single predominant mechanism but rather by moderately modulating multiple cellular functions at once (Farooqui 2009; Grimm et al. 2011). DHA decreases amyloidogenic processing of APP and enhances the nonamyloidogenic pathway, suggesting a directed change in cellular properties. It is becoming increasingly evident that generation of Aβ requires a cascade of several steps, which are modulated by DHA. DHA is not as effective as γ-secretase inhibitors in reducing Aβ production. However, based on antioxidants, anti-inflammatory, antiexcitotoxic and anti-apoptotic properties and clinical safety of DHA, it is suggested that DHA can be used as a neuroprotective agent in AD, stroke, and depression. However, to date studies on the dietary intervention of n-3 fatty acids in humans have so far failed. I propose that long-term DHA supplementation (from childhood to old age) may not only restore signal transduction processes associated with behavioral deficits and learning activity but also produce several neuroendocrinological and immunological effects on brain tissue, which may be beneficial in neurotraumatic, neurodegenerative, and neuropsychiatric diseases (Farooqui 2010, 2011).

## 3.12 Conclusion

FFAs produce dual effects on insulin secretion depending on the duration of exposure. Acute exposure to FFA induces glucose-stimulated insulin secretion, whereas chronic exposure attenuates glucose sensitivity of pancreatic β cells. The coordinated control of these processes by lipid-sensing transcription factors and its relevance to β-cell dysfunction in type II diabetes mellitus and MetS is drawing a lot of attention. DHA and ARA are important components of neural membrane phospholipids. They are obtained either directly from the diet or synthesized from their dietary precursors, ALA and LA, in the liver through a series of elongation and desaturation steps, which take place in the endoplasmic reticulum. Increases in the ratio of n-6:n-3 PUFA may potentiate inflammatory processes and consequently predispose to MetS and neurological disorders. The change in ratio and increase in n-6 PUFA consumption is associated with the production of various eicosanoids, which may be involved in regulation of inflammation and immune responses. DHA and ARA compete for incorporation into the glycerophospholipids of neuronal membranes, and increased levels of DHA/EPA in neurons result in a decrease in the level of ARA in glycerophospholipids. ARA is released from phospholipids by $PLA_2$s and metabolized into eicosanoids (PGs, LTs, TXs, and LXs) by COX and LOX enzymes. High concentrations of eicosanoids modulate various metabolic activities. Eicosanoids produce platelet aggregation, hemorrhage, vasoconstriction, and vasodilation. ARA-derived eicosanoids not only produce proinflammatory effects but also have important homeostatic functions in regulating both the promotion and resolution of inflammation in the immune response. In contrast, DHA and EPA are transformed into docosanoids (resolvins, neuroprotectins, and maresins). Docosanoids produce antioxidant, anti-inflammatory, and antiapoptotic effects. Chronic diseases such as diabetes, obesity, MetS, and neurodegenerative diseases are associated with increased production of $PGE_2$, $LTB_4$, $TXA_2$, IL-1β, IL-6, and TNF-α. Studies in animal models of MetS and neurological disorders indicate that the consumption of DHA and EPA produces beneficial effects by inhibiting oxidative stress and inflammation and inducing the cell survival genes.

## References


Abeywardena MY, Patten GS (2011) Role of ω3 Longchain polyunsaturated fatty acids in reducing cardio-metabolic risk factors. Endocr Metab Immune Disord Drug Targets 11:232–246

Alvarez-Nolting R, Arnal E, Barcia JM, Miranda M, Romero FJ (2012) Protection by DHA of early hippocampal changes in diabetes: possible role of CREB and NF-κB. Neurochem Res 37:105–115

Amtul Z, Uhrig M, Rozmahel RF, Beyreuther K (2011a) Structural insight into the differential effects of omega-3 and omega-6 fatty acids on the production of Abeta peptides and amyloid plaques. J Biol Chem 286:6100–7107

Amtul Z, Westaway D, Cechetto DF, Rozmahel RF (2011b) Oleic acid ameliorates amyloidosis in cellular and mouse models of Alzheimer's disease. Brain Pathol 21:321–329

Arai T, Wakabayashi S, Channing MA, Dunn BB, Der MG, Bell JM, Herscovitch P, Eckelman WC, Rapoport SI, Chang MC (1995) Incorporation of [1-carbon-11]palmitate in monkey brain using PET. J Nucl Med 36:2261–2267

Arita M, Oh SF, Chonan T, Hong S, Elangovan S, Sun YP, Uddin J, Petasis NA, Serhan CN (2006) Metabolic inactivation of resolvin E1 and stabilization of its anti-inflammatory actions. J Biol Chem 281:22847–22854

Arita M, Ohira T, Sun YP, Elangovan S, Chiang N, Serhan CN (2007) Resolvin E1 selectively interacts with leukotriene B4 receptor BLT1 and ChemR23 to regulate inflammation. J Immunol 178:3912–3917

Ariyama H, Kono N, Matsuda S, Inoue T, Arai H (2010) Decrease in membrane phospholipid unsaturation induces unfolded protein response. J Biol Chem 16:22027–22030

Ascherio A, Rimm EB, Stampfer MJ, Giovannucci EL, Willett WC (1995) Dietary intake of marine n-3 fatty acids, fish intake, and the risk of coronary disease among men. N Engl J Med 332:977–982

Astarita G, Jung KM, Berchtold NC, Nguyen VQ, Gillen DL, Head E, Cotman CW, Piomelli D (2010) Deficient liver biosynthesis of docosahexaenoic acid correlates with cognitive impairment in Alzheimer's disease. PLoS One 5:e12538

Astarita G, Jung KM, Vasilevko V, Dipatrizio NV, Martin SK, Cribbs DH, Head E, Cotman CW, Piomelli D (2011) Elevated stearoyl-CoA desaturase in brains of patients with Alzheimer's disease. PLoS One 6:e24777

Baldwin AS Jr (1996) The NF-kappa B and I kappa B proteins: new discoveries and insights. Annu Rev Immunol 14:649–683

Barde YA (1994) Neurotrophins: a family of proteins supporting the survival of neurons. Prog Clin Biol Res 390:45–56

Basselin M, Chang L, Seemann R, Bell JM, Rapoport SI (2005) Chronic lithium administration to rats selectively modifies 5-HT2A/2C receptor-mediated brain signaling via arachidonic acid. Neuropsychopharmacol 30:461–472

Bazan NG (2006) The onset of brain injury and neurodegeneration triggers the synthesis of docosanoid neuroprotective signaling. Cell Mol Neurobiol 26:901–913

Bazan NG, Molina MF, Gordon WC (2011) Docosahexaenoic acid signalolipidomics in nutrition: significance in aging, neuroinflammation, macular degeneration, Alzheimer's, and other neurodegenerative diseases. Annu Rev Nutr 31:321–351

Bessesen DH (2008) Update on obesity. J Clin Endocrinol Metab 93:2027–2034

Bezard J, Blond JP, Bernard A, Clouet P (1994) The metabolism and availability of essential fatty acids in animal and human tissues. Reprod Nutr Dev 34:539–568

Blok WL, Katan MB, van der Meer JW (1996) Modulation of inflammation and cytokine production by dietary (n-3) fatty acids. J Nutr 126:1515–1533

Boesmans W, Gomes P, Janssens J, Tack J, Vanden Berghe P (2008) Brain-derived neurotrophic factor amplifies neurotransmitter responses and promotes synaptic communication in the enteric nervous system. Gut 57:314–322

Börgeson E, McGillicuddy FC, Harford KA, Corrigan N, Higgins DF, Maderna P, Roche HM, Godson C (2012) Lipoxin A4 attenuates adipose inflammation. FASEB J 26:4287–4294

Brenna JT, Salem N Jr, Sinclair AJ, Cunnane SC (2009) α-Linolenic acid supplementation and conversion to n-3 long-chain polyunsaturated fatty acids in humans. Prostaglandins Leukot Essent Fatty Acids 80:85–91

Breslow JL (2006) n-3 fatty acids and cardiovascular disease. Am J Clin Nutr 83(Suppl 6):1477S–1482S

Brenna JT, Diau GY (2007) The influence of dietary docosahexaenoic acid and arachidonic acid on central nervous system polyunsaturated fatty acid composition. Prostaglandins Leukot Essent Fatty Acids 77:247–250

Cabrini L, Bochicchio D, Bordoni A, Sassi S, Marchetti M, Maranesi M (2005) Correlation between dietary polyunsaturated fatty acids and plasma homocysteine concentration in vitamin B6-deficient rats. Nutr Metab Cardiovasc Dis 15:94–99

Calder PC, Grimble RF (2002) Polyunsaturated fatty acids, inflammation and immunity. Eur J Clin Nutr 56:S14–S19

Calder PC (2008) Polyunsaturated fatty acids, inflammatory processes and inflammatory bowel diseases. Mol Nutri Food Res 52:885–897

Calder PC (2009) Polyunsaturated fatty acids and inflammatory processes: New twists in an old tale. Biochimie 91:791–795

Cao H, Gerhold K, Mayers JR, Wiest MM, Watkins SM, Hotamisligil GS (2008) Identification of a lipokine, a lipid hormone linking adipose tissue to systemic metabolism. Cell 134:933–944

Ceriello A, Falleti E, Bortolotti N, Motz E, Cavarape A, Russo A, Gonano E, Bartoli E (1996) Increased circulating intercellular adhesion molecule-1 levels in type II diabetic patients: the possible role of metabolic control and oxidative stress. Metabolism 45:498–501

Chiang N, Serhan CN, Dahlén SE, Drazen JM, Hay DW, Rovati GE, Shimizu T, Yokomizo T, Brink C (2006) The lipoxin receptor ALX: potent ligand-specific and stereoselective actions in vivo. Pharmacol Rev 58:463–487

Chien KL, Chao CL, Kuo CH, Lin HJ, Liu PH, Chen PR, Hsu HC, Lee BC, Lee YT, Chen MF (2011) Plasma fatty acids and the risk of metabolic syndrome in ethnic Chinese adults in Taiwan. Lipids Health Dis 10:33

Clarke SD (2001) Polyunsaturated fatty acid regulation of gene transcription: a molecular mechanism to improve the metabolic syndrome. J Nutr 131:1129–1132

Cole GM, Frautschy SA (2010) DHA may prevent age-related dementia. J Nutr 140:869–874

Cordain Eaton SB, Sebastian A, Mann N, Lindeberg S, Watkins BA, O'Keefe JH, Brand-Miller J (2005) Origins and evolution of the Western diet: health implications for the 21st century. Am J Clin Nutri 81:341–354

Cunnane SC, Plourde M, Pifferi F, Begin M, Feart C, Barberger-Gateau P (2009) Fish, docosahexaenoic acid and Alzheimer's disease. Prog Lipid Res 48:239–256

Cusi K, Maezono K, Osman A, Pendergrass M, Patti ME, Mandel G, DeFronzo RA, Kahn CR, Mandarino LJ (2000) Insulin resistance differentially affects the PI 3-kinase- and MAP kinase-mediated signaling in human muscle. J Clin Invest 105:311–320

Dagnelie PC, Rietveld T, Swart GR, Stijnen T, van den Berg JW (1994) Effect of dietary fish oil on blood levels of free fatty acids, ketone bodies and triacylglycerol in humans. Lipids 29:41–45

Das UN (2000) Interaction(s) between essential fatty acids, eicosanoids, cytokines, growth factors and free radicals: relevance to new therapeutic strategies in rheumatoid arthritis and other collagen vascular diseases. Prostaglandins Leukot Essent Fatty Acids 44:201–210

Das UN (2002) The lipids that matter from infant nutrition to insulin resistance. Prostaglandins Leukot Essent Fatty Acids 67:1–12

Das UN (2006) Essential fatty acids: biochemistry, physiology, and pathology. Biotechnol J 1:420–439

Das UN (2007) A defect in the activity of $\Delta^6$ and $\Delta^5$-desaturases may be a factor in the initiation and progression of atherosclerosis. Prostaglandins Leukot Essent Fatty Acids 76:251–268

Das UN (2008) Essential fatty acids and their metabolites could function as endogenous HMG-CoA reductase and ACE enzyme inhibitors, anti-arrhythmic, anti-hypertensive, anti-atherosclerotic, anti-inflammatory, cytoprotective, and cardioprotective molecules. Lipids Health Dis 7:37

Das UN (2010) A defect in $\Delta^6$ and $\Delta^5$ desaturases may be a factor in the initiation and progression of insulin resistance, the metabolic syndrome and ischemic heart disease in South Asians. Lipids Health Dis 9:130

Davis JE, Gabler NK, Walker-Daniels J, Spurlock ME (2009) The c-Jun N-terminal kinase mediates the induction of oxidative stress and insulin resistance by palmitate and toll-like receptor 2 and 4 ligands in 3T3-L1 adipocytes. Horm Metab Res 41:523–530

Eaton SB, Konner MJ, Cordain L (2010) Diet-dependent acid load, Paleolithic nutrition, and evolutionary health promotion. Am J Clin Nutri 91:295–297

Eckel RH, Grundy SM, Zimmet PZ (2005) The metabolic syndrome. Lancet 365:1415–1428

Esposito G, Giovacchini G, Liow JS, Bhattacharjee AK, Greenstein D, Schapiro M, Hallett M, Herscovitch P, Eckelman WC, Carson RE, Rapoport SI (2008) Imaging neuroinflammation in Alzheimer's disease with radiolabeled arachidonic acid and PET. J Nucl Med 49:1414–1421

Fan YY, Chapkin RS (1998) Importance of dietary gamma-linolenic acid in human health and nutrition. J Nutr 128:1411–1414

Farooqui AA, Horrocks LA, Farooqui T (2000) Glycerophospholipids in brain: their metabolism, incorporation into membranes, functions, and involvement in neurological disorders. Chem Phys Lipids 106:1–29

Farooqui AA, Horrocks LA (2007) Glycerophospholipids in brain. Springer, New York

Farooqui AA, Ong WY, Horrocks LA, Chen P, Farooqui T (2007) Comparison of biochemical effects of statins and fish oil in brain: the battle of the titans. Brain Res Rev 956:443–471

Farooqui AA (2009) Beneficial effects of fish oil on human brain. Springer, New York

Farooqui AA (2010) Neurochemical aspects of neurotraumatic and neurodegenerative disease. Springer, New York

Farooqui AA (2011) Lipid mediators and their metabolism in the brain. Springer, New York

Farooqui AA, Farooqui T, Panza F, Frisardi V (2012) Metabolic syndrome as a risk factor for neurological disorders. Cell Mol Life Sci 69:741–762

Foufelle F, Ferre P (2002) New perspectives in the regulation of hepatic glycolytic and lipogenic genes by insulin and glucose: a role for the transcription factor sterol regulatory element binding protein-1c. Biochem J 366:377–391

Fujioka T, Fujioka A, Duman RS (2004) Activation of cAMP signaling facilitates the morphological maturation of newborn neurons in adult hippocampus. J Neurosci 24:319–328

Funaki M (2009) Saturated fatty acids and insulin resistance. J Med Invest 56:88–92

González-Périz A, Horrillo R, Ferré N, Gronert K, Dong B, Morán-Salvador E, Titos E, Martínez-Clemente M, López-Parra M, Arroyo V, Clària J (2009) Obesity-induced insulin resistance and hepatic steatosis are alleviated by omega-3 fatty acids: a role for resolvins and protectins. FASEB J 23:1946–1957

Gottlicher M, Widmark E, Li Q, Gustafsson JA (1992) Fatty acids activate a chimera of the clofibric acid-activated receptor and the glucocorticoid receptor. Proc Natl Acad Sci USA 89:4653–4657

Grimm MO, Kuchenbecker J, Grösgen S, Burg VK, Hundsdörfer B, Rothhaar TL, Friess P, de Wilde MC, Broersen LM, Penke B, Péter M, Vígh L, Grimm HS, Hartmann T (2011) Docosahexaenoic acid reduces amyloid beta production via multiple pleiotropic mechanisms. J Biol Chem 286:14028–14039

Harizi H, Corcuff JB, Gualde N (2008) Arachidonic-acid-derived eicosanoids: roles in biology and immunopathology. Trends Mol Med 14:461–469

Harris WS, Bulchandani D (2006) Why do omega-3 fatty acids lower serum triglycerides? Curr Opin Lipidol 17:387–393

Harris WS, Kris-Etherton PM, Harris KA (2008) Intakes of long-chain omega-3 fatty acid associated with reduced risk for death from coronary heart disease in healthy adults. Curr Atheroscler Rep 10:503–509

Hayden MS, Ghosh S (2008) Shared principles in NF-kappaB signaling. Cell 132:344–362

Hellmann J, Tang Y, Kosuri M, Bhatnagar A, Spite M (2011) Resolvin D1 decreases adipose tissue macrophage accumulation and improves insulin sensitivity in obese-diabetic mice. FASEB J 25:2399–2407

Hogyes E, Nyakas C, Kiliaan A, Farkas T, Penke B, Luiten PG (2003) Neuroprotective effect of developmental docosahexaenoic acid supplement against excitotoxic brain damage in infant rats. Neuroscience 119:999–1012

Horrocks LA, Farooqui AA (2004) Docosahexaenoic acid in the diet: its importance in maintenance and restoration of neural membrane function. Prostaglandins Leukot Essent Fat Acids 70:361–372

Horton JD, Shimomura I, Brown MS, Hammer RE, Goldstein JL, Shimano H (1998) Activation of cholesterol synthesis in preference to fatty acid synthesis in liver and adipose tissue of transgenic mice overproducing sterol regulatory element-binding protein-2. J Clin Investig 101:2331–2339

Horton JD, Goldstein JL, Brown MS (2002) SREBPs: activators of the complete program of cholesterol and fatty acid synthesis in the liver. J Clin Invest 109:1125–1131

Ide T, Egan K, Bell-Parikh LC, FitzGerald GA (2003) Activation of nuclear receptors by prostaglandins. Thrombosis Res 110:311–315

Ihara Y, Toyokuni S, Uchida K, Odaka H, Tanaka T, Ikeda H, Hiai H, Seino Y, Yamada Y (1999) Hyperglycemia causes oxidative stress in pancreatic beta-cells of GK rats, a model of type 2 diabetes. Diabetes 48:927–932

Infante JP, Huszagh VA (1997) On the molecular etiology of decreased arachidonic (20:4n-6), docosapentaenoic (22:5n-6) and docosahexaenoic (22:6n-3) acids in Zellweger syndrome and other peroxisomal disorders. Mol Cell Biochem 168:101–115

Infante JP, Huszagh VA (1998) Analysis of the putative role of 24-carbon polyunsaturated fatty acids in the biosynthesis of docosapentaenoic (22:5n-6) and docosahexaenoic (22:6n-3) acids. FEBS Lett 431:1–6

Joyce D, Albanese C, Steer J, Fu M, Bouzahzah B, Pestell RG (2001) NF-kappaB and cell-cycle regulation: the cyclin connection. Cytokine Growth Factor Rev 12:73–90

Jump DB (2002) The biochemistry of n-3 polyunsaturated fatty acids. J Biol Chem 277:8755–8758

Jumpertz R, Guijarro A, Pratley RE, Mason CC, Piomelli D, Krakoff J (2012) Associations of fatty acids in cerebrospinal fluid with peripheral glucose concentrations and energy metabolism. PLoS One 7:e41503

Kakuma T, Lee Y, Higa M, Wang Z, Pan W, Shimomura I, Unger RH (2000) Leptin, troglitazone, and the expression of sterol regulatory element binding proteins in liver and pancreatic islets. Proc Natl Acad Sci USA 97:8536–8541

Kalofoutis C, Piperi C, Kalofoutis A, Harris F, Phoenix D, Singh J (2007) Type II diabetes mellitus and cardiovascular risk factors: current therapeutic approaches. Exp Clin Cardiol 12:17–28

Karmi A, Iozzo P, Viljanen A, Hirvonen J, Fielding BA, Virtanen K, Oikonen V, Kemppainen J, Viljanen T, Guiducci L, Haaparanta-Solin M, Nagren K, Solin O, Nuutila P (2010) Increased brain fatty acid uptake in metabolic syndrome. Diabetes 99:2171–2177

Kersten S (2002) Effects of fatty acids on gene expression: role of peroxisome proliferator-activated receptor α, liver X receptor α and sterol regulatory element-binding protein-1 c. Proc Nutr Soc 61:371–374

Klaunig JE, Kamendulis LM, Hocevar BA (2010) Oxidative stress and oxidative damage in carcinogenesis. Toxicol Pathol 38:96–109

Komatsu W, Ishihara K, Murata M, Saito H, Shinohara K (2003) Docosahexaenoic acid suppresses nitric oxide production and inducible nitric oxide synthase expression in interferon-gamma plus lipopolysaccharide-stimulated murine macrophages by inhibiting the oxidative stress. Free Radic Biol Med 34:1006–1016

Kwak Y-D, Wang R, Li JJ, Zhang Y-W, Xu H, Liao F-F (2011) Differential regulation of BACE1 expression by oxidative and nitrosative signals. Mol Neurodegeber 6:17

Lam TK, Schwartz GJ, Rossetti L (2005) Hypothalamic sensing of fatty acids. Nat Neurosci 8:579–584

Lattka E, Illig T, Koletzko B, Heinrich J (2010) Genetic variants of the FADS1 FADS2 gene cluster as related to essential fatty acid metabolism. Curr Opin Lipidol 21:64–69

Lavialle M, Denis I, Guesnet P, Vancassel S (2010) Involvement of omega-3 fatty acids in emotional responses and hyperactive symptoms. J Nutr Biochem 21:899–905

Le Jossic-Corcos C, Gonthier C, Zaghini I, Logette E, Shechter I, Bournot P (2005) Hepatic farnesyl diphosphate synthase expression is suppressed by polyunsaturated fatty acids. Biochem J 385:787–794

Levin G, Duffin KL, Obukowicz MG, Hummert SL, Fujiwara H, Needleman P, Raz A (2002) Differential metabolism of dihomo-gamma-linolenic acid and arachidonic acid by cyclo-oxygenase-1 and cyclo-oxygenase-2: implications for cellular synthesis of prostaglandin E1 and prostaglandin E2. Biochem J 365:489–496

Lim HS, Blann AD, Lip GY (2004) Soluble CD40 ligand, soluble P-selectin, interleukin-6, and tissue factor in diabetes mellitus: Relationships to cardiovascular disease and risk factor intervention. Circulation 109:2524–2528

Lopez M, Lelliott CJ, Vidal-Puig A (2007) Hypothalamic fatty acid metabolism: a housekeeping pathway that regulates food intake. Bioessays 29:248–261

Lopez M, Lage R, Saha AK, Perez-Tilve D, Vazquez MJ, Varela L, Sangiao-Alvarellos S, Tovar S, Raghay K, Rodriguez-Cuenca S, Deoliveira RM, Castaneda T, Datta R, Dong JZ, Culler M, Sleeman MW, Alvarez CV, Gallego R, Lelliott CJ, Carling D, Tschop MH, Dieguez C, Vidal-Puig A (2008) Hypothalamic fatty acid metabolism mediates the orexigenic action of ghrelin. Cell Metab 7:389–399

Lukiw WJ, Cui J-G, Marcheselli VL, Bodker M, Botkjaer A, Gotlinger K, Serhan CN, Bazan NG (2005) A role for docosahexaenoic acid-derived neuroprotectin D1 in neural cell survival and Alzheimer disease. J Clin Invest 115:2774–2783

Macotela Y, Boucher J, Tran TT, Kahn CR (2009) Sex and depot differences in adipocyte insulin sensitivity and glucose metabolism. Diabetes 58:803–812

Marcheselli VL, Hong S, Lukiw WJ, Tian XH, Gronert K, Musto A, Hardy M, Gimenez JM, Chiang N, Serhan CN, Bazan NG (2003) Novel docosanoids inhibit brain ischemia-reperfusion-mediated leukocyte infiltration and pro-inflammatory gene expression. J Biol Chem 278:43807–43817

Marszalek JR, Lodish HF (2005) Docosahexaenoic acid, fatty acid-interacting proteins, and neuronal function: breastmilk and fish are good for you. Annu Rev Cell Dev Biol 21:633–657

Marza E, Long T, Saiardi A, Sumakovic M, Eimer S, Hall DH, Lesa GM (2008) Polyunsaturated fatty acids influence synaptojanin localization to regulate synaptic vesicle recycling. Mol Biol Cell 19:833–842

Mattson MP (2009) Roles of the lipid peroxidation product 4-hydroxynonenal in obesity, the metabolic syndrome, and associated vascular and neurodegenerative disorders. Exp Gerontol 44:625–633

McAuley K, Mann J (2006) Thematic review series: patient-oriented research. Nutritional determinants of insulin resistance. J Lipid Res 47:1668–1676

Merched AJ, Ko K, Gotlinger KH, Serhan CN, Chan L (2008) Atherosclerosis: evidence for impairment of resolution of vascular inflammation governed by specific lipid mediators. FASEB J 22:3595–3606

Merino DM, Ma DW, Mutch DM (2010) Genetic variation in lipid desaturases and its impact on the development of human disease. Lipids Health Dis 9:63

Milanski M, Degasperi G, Coope A, Morari J, Denis R, Cintra DE, Tsukumo DM, Anhe G, Amaral ME, Takahashi HK, Curi R, Oliveira HC, Carvalheira JB, Bordin S, Saad MJ, Velloso LA (2009) Saturated fatty acids produce an inflammatory response predominantly through the activation of TLR4 signaling in hypothalamus: implications for the pathogenesis of obesity. J Neurosci 29:359–370

Miyazawa D, Yasui Y, Yamada K, Ohara N, Okuyama H (2010) Regional differences of the mouse brain in response to an [alpha]-linolenic acid-restricted diet: Neurotrophin content and protein kinase activity. Life Sci 87:490–494

Moraes JC, Coope A, Morari J, Cintra DE, Roman EA, Pauli JR, Romanatto T, Carvalheira JB, Alexandre LR, Oliveira M, Saad J, Velloso LA (2009) High-fat diet induces apoptosis of hypothalamic neurons. PLoS One 4:e5045

Moreno JJ (2009) Differential effects of arachidonic and eicosapentaenoic acid-derived eicosanoids on polymorphonuclear transmigration across endothelial cell cultures. J Pharmacol Exp Ther 331:1111–1117

Morgan K, Obici S, Rossetti L (2004) Hypothalamic responses to long-chain fatty acids are nutritionally regulated. J Biol Chem 279:31139–31148

Mukherjee PK, Marcheselli VL, Serhan CN, Bazan NG (2004) Neuroprotectin D1: a docosahexaenoic acid-derived docosatriene protects human retinal pigment epithelial cells from oxidative stress. Proc Natl Acad Sci USA 101:8491–8496

Nakagawa S, Kim JE, Lee R, Malberg JE, Chen J, Steffen C, Zhang YJ, Nestler EJ, Duman RS (2002) Regulation of neurogenesis in adult mouse hippocampus by cAMP and the cAMP response element-binding protein. J Neurosci 22:3673–3682

Nakamura MT, Nava TY (2002) Gene regulation of mammalian desaturases. Biochem Soc Trans 30:1076–1079

Olivier MC, Vanessa L, Isabelle A (2011) Why and how meet n-3 PUFA dietary recommendations? Gastroenterol Res Pract 11 p. Article ID 364040

Paillard F, Catheline D, Duff FL, Bouriel M, Deugnier Y, Pouchard M, Daubert JC, Legrand P (2008) Plasma palmitoleic acid, a product of stearoyl-coA desaturase activity, is an independent marker of triglyceridemia and abdominal adiposity. Nutr Metab Cardiovasc Dis 18:436–440

Park Y, Harris WS (2003) Omega-3 fatty acid supplementation accelerates chylomicron triglyceride clearance. J Lipid Res 44:455–463

Peyron-Caso E, Fluteau-Nadler S, Kabir M, Guerre-Millo M, Quignard-Boulange A, Slama G, Rizkalla SW (2002) Regulation of glucose transport and transporter 4 (GLUT-4) in muscle and adipocytes of sucrose-fed rats: effects of N-3 poly- and monounsaturated fatty acids. Horm Metab Res 34:360–366

Phillis JW, Horrocks LA, Farooqui AA (2006) Cyclooxygenases, lipoxygenases, epoxygenases in CNS: their role and involvement in neurological disorders. Brain Res Rev 52:201–243

Pickup JC, Hil DP, Path FRC (2004) Inflammation and activated innate immunity in the pathogenesis of type 2 diabetes. Diabetes Care 27:813–823

Piette J, Piret B, Bonizzi G, Schoonbroodt S, Merville MP, Legrand-Poels S, Bours V (1997) Multiple redox regulation in NFkappaB transcription factor activation. Biol Chem 378:1237–1245

Pifferi F, Jouin M, Alessandri JM, Haedke U, Roux F, Perrière N, Denis J, Lavialle M, Guesnet P (2007) n-3 Fatty acids modulate brain glucose transport in endothelial cells of the blood–brain barrier. Prostaglandins Leukot Essent Fatty Acids 77(5–6):279–286

Pocai A, Lam TK, Obici S, Gutierrez-Juarez R, Muse ED, Arduini A, Rossetti L (2006) Restoration of hypothalamic lipid sensing normalizes energy and glucose homeostasis in overfed rats. J Clin Invest 116:1081–1091

Rapoport SI (2008) Arachidonic acid and the brain. J Nutr 138:2515–2520

Rapoport SI, Rao JS, Igarashi M (2007) Brain metabolism of nutritionally essential polyunsaturated fatty acids depends on both the diet and the liver. Prostaglandins Leukot Essent Fatty Acids 77:251–261

Rapoport SI, Igarashi M (2009) Can the rat liver maintain normal brain DHA metabolism in the absence of dietary DHA? Prostaglandins Leukot Essent Fatty Acids 81:119–123

Ridker PM, Hennekens CH, Roitman-Johnson B, Stampfer MJ, Allen J (1998) Plasma concentration of soluble intercellular adhesion molecule 1 and risks of future myocardial infarction in apparently healthy men. Lancet 351:88–92

Rimoldi OJ, Finarelli GS, Brenner RR (2001) Effects of diabetes and insulin on hepatic delta 6 desaturase gene expression. Biochem Biophys Res Commun 283:323–326

Rioux V, Legrand P (2007) Saturated fatty acids: simple molecular structures with complex cellular functions. Curr Opin Clin Nutr Metab Care 10:752–758

Riserus U, Willett WC, Hu FB (2009) Dietary fats and prevention of type 2 diabetes. Progr Lipid Res 48:44–51

Salas-Salvado J, Martinez-Gonzalez MA, Bullo M, Ros E (2011) The role of diet in the prevention of type 2 diabetes. Nutr Metabol Cardiovasc Dis 21(Suppl 2):B32–B48

Sampath H, Ntambi JM (2008) Role of stearoyl-CoA desaturase in human metabolic disease. Future Lipidol 3:163–173

Santhanam AV, Smith LA, Katusic ZS (2010) Brain-derived neurotrophic factor stimulates production of prostacyclin in cerebral arteries. Stroke 41:350–356

Schmitz G, Ecker J (2008) The opposing effects of n-3 and n-6 fatty acids. Prog Lipid Res 47:147–155

Schonbeck U, Libby P (2001) CD40 signaling and plaque instability. Circ Res 89:1092–1103

Scott BL, Bazan NG (1989) Membrane docosahexaenoate is supplied to the developing brain and retina by the liver. Proc Natl Acad Sci USA 86:2903–2907

Serhan CN (2005) Novel ω-3-derived local mediators in anti-inflammation and resolution. Pharmacol Ther 105:7–21

Serhan CN, Yacoubian S, Yang R (2008) Anti-inflammatory and proresolving lipid mediators. Annu Rev Pathol 3:279–312

Serhan CN (2010) Novel lipid mediators and resolution mechanisms in acute inflammation: to resolve or not? Am J Pathol 177:1576–1591

Serhan CN, Chiang N (2008) Endogenous pro-resolving and anti-inflammatory lipid mediators: a new pharmacologic genus. Br J Pharmacol 153(Suppl 1):S200–S215

Sheinin A, Talani G, Davis MI, Lovinger DM (2008) Endocannabinoid-and mGluR5-dependent short-term synaptic depression in an isolated neuron/bouton preparation from the hippocampal CA1 region. J Neurophysiol 100:1041–1052

Shi H, Kokoeva MV, Inouye K, Tzameli I, Yin H, Flier JS (2006) TLR4 links innate immunity and fatty acid-induced insulin resistance. J Clin Invest 116:3015–3025

Shoelson SE, Lee J, Goldfine AB (2006) Inflammation and insulin resistance. J Clin Invest 116:1793–1801

Simopoulos AP (1999) Essential fatty acids in health and chronic disease. Am J Clin Nutr 70:560S–569S

Simopoulos AP (2004) Omega-3 fatty acids and antioxidants in edible wild plants. Biol Res 37:263–277

Simopoulos AP (2006) Evolutionary aspects of diet, the omega-6/omega-3 ratio and genetic variation: nutritional implications for chronic diseases. Biomed Pharmacother 60:502–507

Simopoulos AP (2011) Evolutionary aspects of diet: the omega-6/omega-3 ratio and the brain. Mol Neurobiol 44:203–215

Song C (2002) Pro- and anti-inflammatory cytokines in Alzheimer's disease: new therapeutic options. Int J Neuropsychopharmacol 5:S14

Song C, Manku MS, Horrobin DF (2008) Long-chain polyunsaturated fatty acids modulate interleukin-1β-induced changes in behavior, monoaminergic neurotransmitters, and brain inflammation in rats. J Nutri 138:954–963

Song C, Li X, Leonard BE, Horrobin DF (2003) Effects of dietary n-3 or n-6 fatty acids on interleukin-1beta-induced anxiety, stress, and inflammatory responses in rats. J Lipid Res 44:1984–1991

Spector AA, Yorek MA (1985) Membrane lipid composition and cellular function. J Lipid Res 26:1015–1035

Spite M, Summers L, Porter TF, Srivastava S, Bhatnagar A, Serhan CN (2009) Resolvin D1 controls inflammation initiated by glutathione-lipid conjugates formed during oxidative stress. Br J Pharmacol 158:1062–1073

Sprecher H, Chen Q, Yin FQ (1999) Regulation of the biosynthesis of 22:5n-6 and 22:6n-3: a complex intracellular process. Lipids 34(Suppl):S153–S156

Stahl LA, Begg DP, Weisinger RS, Sinclair AJ (2008) The role of omega-3 fatty acids in mood disorders. Curr Opin Investig Drugs 9:57–64

Staiger H, Staiger K, Haas C, Weisser M, Machicao F, Haring HU (2005) Fatty acid-induced differential regulation of the genes encoding peroxisome proliferator-activated receptor-gamma coactivator-1alpha and -1beta in human skeletal muscle cells that have been differentiated in vitro. Diabetologia 48:2115–2118

Suganami T, Tanimoto-Koyama K, Nishida J, Itoh M, Yuan X, Mizuarai S, Kotani H, Yamaoka S, Miyake K, Aoe S, Kamei Y, Ogawa Y (2007) Role of the Toll-like receptor 4/NFkappaB pathway in saturated fatty acid-induced inflammatory changes in the interaction between adipocytes and macrophages. Arterioscler Thromb Vasc Biol 27:84–91

Tanaka Y, Gleason CE, Tran PO, Harmon JS, Robertson RP (1999) Prevention of glucose toxicity in HIT-T15 cells and Zucker diabetic fatty rats by antioxidants. Proc Natl Acad Sci USA 96:10857–10862

Thomas T, Pfeiffer AF (2011) Foods for the prevention of diabetes: how do they work? Diabetes Metab Res Rev 28:25–49

Tiedge M, Lortz S, Drinkgern J, Lenzens S (1997) Relation between antioxidant enzyme gene expression and antioxidative defense status of insulin-producing cells. Diabetes 46:1733–1742

Tschop M, Thomas G (2006) Fat fuels insulin resistance through Toll-like receptors. Nat Med 12:1359–1361

Vessby B, Uusitupa M, Hermansen K, Riccardi G, Rivellese AA, Tapsell LC, Nälsén C, Berglund L, Louheranta A, Rasmussen BM, Calvert GD, Maffetone A, Pedersen E, Gustafsson IB, Storlien LH (2001) Substituting dietary saturated for monounsaturated fat impairs insulin sensitivity in healthy men and women: the KANWU study. Diabetologia 44:312–319

Volterra A, Trotti D, Cassutti P, Tromba C, Galimberti R, Lecchi P, Racagni G (1992) A role for the arachidonic acid cascade in fast synaptic modulation: ion channels and transmitter uptake systems as target proteins. Adv Exp Med Biol 318:147–158

Wada M, DeLong CJ, Hong YH, Rieke CJ, Sidhu RS, Yuan C, Warnock M, Schmaier AH, Yokoyama C, Smyth EM, Wilson SJ, FitzGerald GA, Garavito RM, de Sui X, Wang X, Zhao X, Mao ZY, Wang XM, Liu ZL (2003) Neuroprotective effect of docosahexaenoic acid on glutamate-induced cytotoxicity in rat hippocampal cultures. Neuroreport 14:2457–2461

Wang X, Hai CX (2011) ROS acts as a double-edged sword in the pathogenesis of type 2 diabetes mellitus: is Nrf2 a potential target for the treatment? Mini Rev Med Chem 11:1082–1092

Warensjo E, Riserus U, Vessby B (2005) Fatty acid composition of serum lipids predicts the development of the metabolic syndrome in men. Diabetologia 48:1999–2005

Wei Y, Sowers JR, Clark SE, Li W, Ferrario CM, Stump CS (2008) Angiotensin II-induced skeletal muscle insulin resistance mediated by NF-kappaB activation via NADPH oxidase. Am J Physiol Endocrinol Metab 294:E345–E351

Weisinger HS, Vingrys AJ, Sinclair AJ (1996) Effect of dietary n-3 deficiency on the electroretinogram in the guinea pig. Ann Nutr Metab 40:91–98

Weiss JM, Pilarski KA, Weyl A, Peschen M, Schöpf E, Vestweber D, Vanscheidt W, Simon JC (1995) Prostaglandin E1 inhibits TNF alpha-induced T-cell adhesion to endothelial cells by selective down-modulation of ICAM-1 expression on endothelial cells. Exp Dermatol 4:302–307

Williams JH, Errington ML, Lynch MA, Bliss TV (1989) Arachidonic acid induces a long-term activity-dependent enhancement of synaptic transmission in the hippocampus. Nature 341:739–742

Wilson DM, Binder LI (1997) Free fatty acids stimulate the polymerization of tau and amyloid beta peptides. In vitro evidence for a common effector of pathogenesis in Alzheimer's disease. Am J Pathol 150:2181–2195

Xu ZZ, Zhang L, Liu T, Park JY, Berta T, Yang R, Serhan CN, Ji RR (2010) Resolvins RvE1 and RvD1 attenuate inflammatory pain via central and peripheral actions. Nat Med 16:592–597

Yin H, Brooks JD, Gao L, Poeter N, Morrow JD (2007) Identification of novel autoxidation products of the n-3 fatty acid eicosapentaenoic acid in vitro and in vivo. J Biol Chem 282:29890–29901

Yoshikawa T, Shimano H, Memiya-Kudo M, Yahagi N, Hasty AH, Matsuzaka T, Okazaki H, Tamura Y, Iizuka Y, Ohashi K, Osuga J, Harada K, Gotoda T, Kimura S, Ishibashi S, Yamada N (2001) Identification of liver X receptor-retinoid X receptor as an activator of the sterol regulatory element-binding protein 1c gene promoter. Mol Cell Biol 21:2991–3000

Yoshikawa T, Shimano H, Yahagi N, Ide T, Amemiya-Kudo M, Matsuzaka T, Nakakuki M, Tomita S, Okazaki H, Tamura Y, Iizuka Y, Ohashi K, Takahashi A, Sone H, Osuga Ji J, Gotoda T, Ishibashi S, Yamada N (2002) (LXR) binding to LXR response elements. J Biol Chem 277:1705–1711

Zhou L, Nilsson A (2001) Sources of eicosanoid precursor fatty acid pools in tissues. J Lipid Res 42:1521–1542

Zou J, Crews F (2006) CREB and NF-κB transcription factors regulate sensitivity to excitotoxic and oxidative stress induced neuronal cell death. Cell Mol Neurobiol 26:4–6

# Chapter 4
# Lipid Mediators in Metabolic Syndrome and Neurological Disorders

## 4.1 Introduction

Neural membranes are composed of glycerophospholipids (phosphatidylethanolamine (PtdEtn), plasmenylethanolamine (PlsEtn), phosphatidylcholine (PtdCho), phosphatidylinositol (PtdIns), phosphatidic acid (PtdH), and phosphatidylserine (PtdSer)), sphingolipids (sphingomyelins, cerebrosides, and gangliosides), cholesterol, and proteins. Among glycerophospholipids, PtdCho accounts for more than 50 % of all cell membrane glycerophospholipids and is the main bilayer-forming lipid. The distribution of lipids in two leaflets of the lipid bilayer is asymmetric (Ikeda et al. 2006; Yamaji-Hasegawa and Tsujimoto 2006). Sphingolipids are minor components of neural membranes. Many complex glycosphingolipids, including gangliosides, are not bilayer-forming lipids. However, they insert themselves in the lipid bilayer through their hydrophobic ceramide moieties. Thus, sphingolipids are mainly located on the external leaflet of the plasma membrane. Glycerophospholipids and sphingolipids contribute to the lipid asymmetry, whereas cholesterol and sphingolipids form lipid microdomains or lipid rafts, which float within the membrane and acts as molecular sorting machines and platforms for signal transduction pathways (Simons and Ikonen 2000; Zajchowski and Robbins 2002; Lucero and Robbins 2004). In addition, sphingolipids can influence the localization and functions of transporters and receptors, such as the AMPA receptor in neurons (Hou et al. 2008). In the brain, cholesterol not only serves as a precursor for the synthesis of oxysterols, steroid hormones, vitamin D, and lipoproteins but regulates activities of membrane-bound enzymes, receptors, and ion channels (Simons and Ikonen 2000). In the neural membrane, cholesterol not only provides neural plasticity and facilitates synaptic transmission but also modulates endocytosis and antigen expression.

Among glycerophospholipids, PtdEtn, PlsEtn, and PtdSer are enriched in docosahexaenoic acid (DHA, 22:6n-3) at the *sn*-2 position of the glycerol moiety, whereas PtdCho, PtdIns, and PtdH contain arachidonic acid (ARA, 20:4n-6) (Farooqui et al. 2000; Tillman and Cascio 2003). Cholesterol plays a dual role:

A.A. Farooqui, *Metabolic Syndrome: An Important Risk Factor for Stroke, Alzheimer Disease, and Depression*, DOI 10.1007/978-1-4614-7318-3_4,

regulation of protein–sphingolipid interactions through a fine-tuning of sphingolipid conformation (indirect effect), and facilitation of channel formation through direct binding to amyloidogenic proteins (Fantini and Yahi 2010). In addition to providing integrity and asymmetry, phospholipid, sphingolipid, and cholesterol act as storage depots for lipid mediators. Although the generation of physiological levels of lipid mediators is necessary for many neural cell function, such as modulation of cell cycle progression, cell activation, differentiation, proliferation, survival, growth arrest, and migration, generation of high amounts of lipid mediators due to enrichment of glycerophospholipid, sphingolipid, and cholesterol metabolism disrupts the neural and visceral membrane asymmetry resulting in insulin resistance, impaired glucose tolerance, and endothelial cell dysfunction in metabolic syndrome (MetS) and neurodegeneration in neurotraumatic and neurodegenerative diseases (Farooqui 2010, 2011; Farooqui et al. 2012).

## 4.2 Lipid Mediators in the Brain

Lipid mediators are a class of bioactive molecules that are synthesized from membrane glycerophospholipids, sphingolipids, and cholesterol in response to extracellular and intracellular stimuli. Lipid mediators are regarded as (local) hormones (Murakami 2011; Farooqui 2011). In the brain, the action of phospholipase $A_2$ ($PLA_2$) on membrane phospholipid results in the release of polyunsaturated fatty acids and lysophospholipids. Lysophospholipids are acetylated to platelet-activating factor (PAF) and its analogs (Farooqui 2011). Among polyunsaturated fatty acids (PUFAs), arachidonic acid (ARA) is a precursor for lipid mediators called as eicosanoids. Eicosanoids include prostaglandins (PGs), leukotrienes (LTs), thromboxanes (TXs), epoxyeicosatrienoic acids (EETs), and lipoxins (LXs) (Fig. 4.1). These lipid mediators are generated by the action of cyclooxygenases, lipoxygenases, and epoxygenases (Phillis et al. 2006). ARA can also be metabolized to 2-arachidonylglycerol and anandamide. ARA-derived lipid mediators modulate neural cell function by interacting with PG, LT, TX, LX, and cannabinoid receptors (Farooqui 2011). At the basal (physiological) levels, most ARA-derived lipid mediators act as local hormones or autacoids. They participate in signal transduction processes to modulate normal neural cell functions. Marked increases have been reported to occur in ARA-derived enzymic lipid mediators. It is proposed that generation of high levels of ARA-derived lipid mediated may be responsible for neurodegeneration in neurotraumatic and neurodegenerative diseases (Farooqui 2010, 2011). The nonenzymic lipid mediators of ARA metabolism include 4-hydroxynonenal (4-HNE), isoprostanes (IsoPs), isoketal (IsoK), and isofuran (IsoF) (Fig. 4.2). Levels of ARA-derived lipid mediators are markedly increased in neurotraumatic and neurodegenerative diseases. High levels of these lipid mediators produce their toxic effect not only by inhibiting the activities

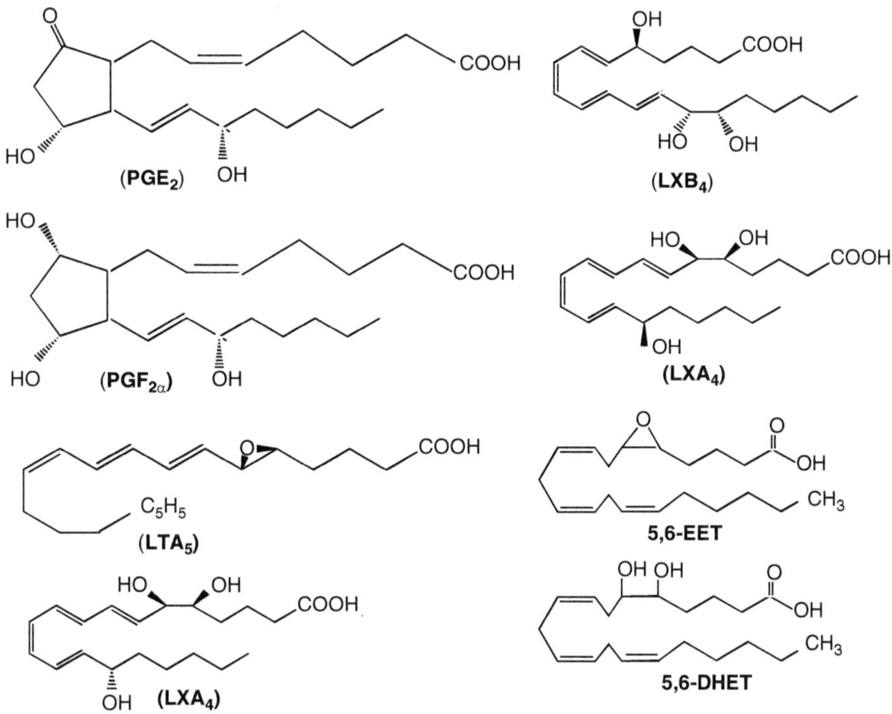


**Fig. 4.1** Chemical structures of enzymic lipid mediators of arachidonic acid metabolism

of various enzymes (phospholipases, protein kinases, adenylate cyclases, nitric oxide synthases, $Ca^{2+}$-ATPases, etc.) that are essential for neural cell function and survival but also by inducing mitochondrial stress and interacting with nucleic acids and producing abnormalities in DNA and RNA metabolism (Farooqui 2011).

Docosahexaenoic acid (DHA) is the precursor of docosanoids, which include resolvins, protectins, and maresins (Farooqui 2011, 2012a) (Fig. 4.3). Nonenzymic oxidation of DHA results in the formation of 4-hydroxyhexanal (4-HHE), neuroprostane, neuroketal, and neurofuran (Fig. 4.2). These lipid mediators also produce toxic effects in neural cells through their interactions with enzymes.

Sphingolipid-derived lipid mediators include ceramide, ceramide-1-phosphate, sphingosine, and sphingosine-1-phosphate. In the brain, sphingolipid-derived lipid mediators participate not only in modulating neural cell survival, migration, adhesion, and differentiation but also through induction and suppression of apoptosis, modulation of enzyme activities, and cell cycle arrest (Fig. 4.4). Several neurodegenerative and demyelinating diseases (Alzheimer's disease, amyotrophic lateral sclerosis, multiple sclerosis, and metachromatic leukodystrophy) are accompanied by dysregulation in sphingolipid metabolism, abnormal sphingolipid patterns and

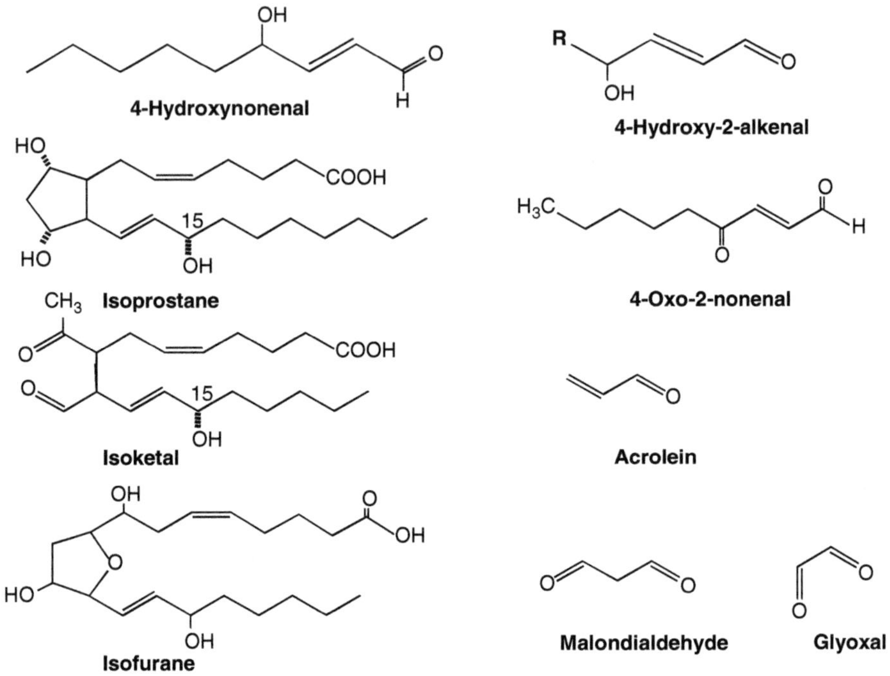


**Fig. 4.2** Chemical structures of nonenzymic lipid mediators of arachidonic acid metabolism

altered membrane organization, and increase in levels of sphingolipid-derived lipid mediators. It is proposed that changes in sphingolipid-derived lipid mediators may contribute to neurodegeneration in neurotraumatic and neurodegenerative diseases (Farooqui 2011).

The brain contains cytochrome P450-dependent oxygenases, which convert cholesterol into 24(*S*)-hydroxycholesterol, 25-hydroxycholesterol, and 27-hydroxycholesterol (Fig. 4.5). The major mechanism of brain cholesterol elimination is the conversion of cholesterol into 24(*S*)-hydroxycholesterol by CYP46A1, a neuron-specific cytochrome P450, which is enriched in the striatum and the cortex. Cholesterol is also oxidized to cholesterol oxides and converted into cholesterol ester via acyl-CoA:cholesterol acyltransferase (ACAT) (Björkhem et al. 1998). Hydroxysterols bind two different targets, liver X receptors (LXRs) and oxysterol-binding proteins (ORPs) (Russell 2000; Fu et al. 2001). They perturb rafts, regulate cholesterol homeostasis, modulate gene expression, induce apoptosis, modulate enzyme activities, and facilitate exocytosis.

Emerging evidence suggests that glycerophospholipid-, sphingolipid-, and cholesterol-derived lipid mediators modulate many processes, including growth, differentiation, adhesion, migration, and apoptosis. Their dysregulations contribute

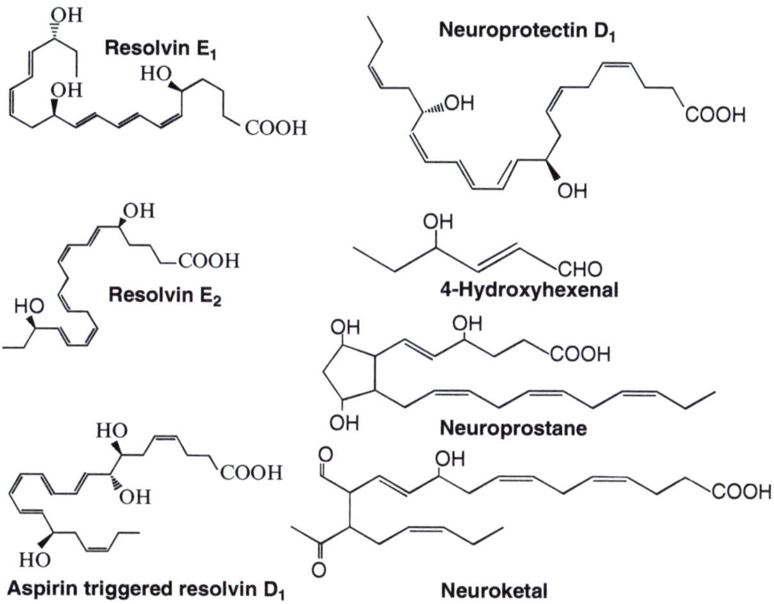


**Fig. 4.3** Chemical structures of enzymic and nonenzymic lipid mediators of docosahexaenoic acid metabolism

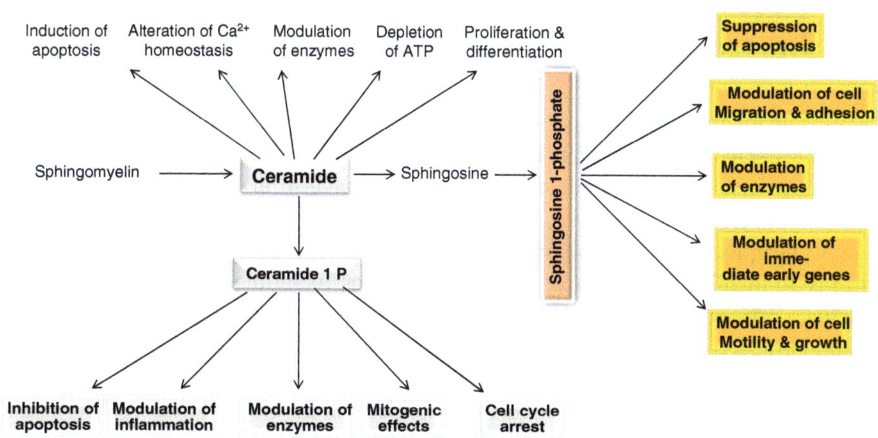


**Fig. 4.4** Sphingolipid-derived lipid mediators and their roles in the brain

to abnormal signal transduction processes, which are often linked to various visceral and neurological disorders including infertility, atherosclerosis, MetS, cancer, ischemia, and Alzheimer's disease (Farooqui 2011).

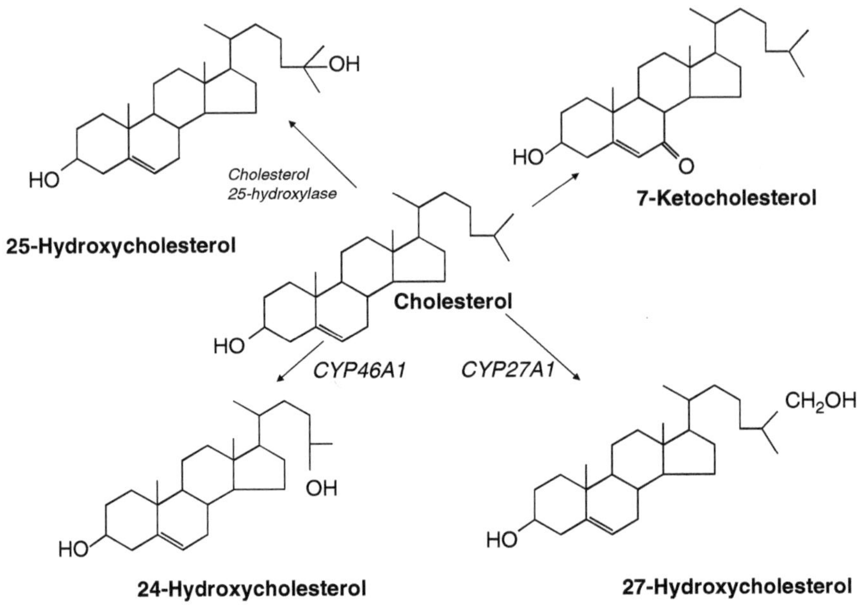


**Fig. 4.5** Chemical structures of cholesterol-derived lipid mediators found in the brain

## 4.3 Fatty Acids and Their Lipid Mediators in Metabolic Syndrome and Neurological Disorders

MetS is characterized by central adiposity, dyslipidemia, impaired glucose tolerance, insulin resistance, and hypertension (Fig. 4.6). Insulin resistance occurs when normal amounts of insulin are inadequate to produce a normal insulin response in cells. Insulin resistance is caused by defects in intracellular insulin signaling. Interactions of insulin with its receptors influence the functional integrity of the insulin receptor including its affinity to insulin. A number of other mechanisms including consumption of high-fructose corn syrup may also lead to insulin resistance in rodents. Fructose is metabolized to uric acid, which produces inhibition of endothelium-dependent vasodilation, to impair insulin signaling, but also induces oxidative stress through the production of advanced glycated products and causes lipotoxicity in the skeletal muscle. More studies are required to evaluate whether these mechanisms may be responsible for the development of insulin resistance in humans with years-long exposure to fructose.

Since properties of the cell membrane, especially its fluidity, depend on the lipid constitution of the membrane, presence of high levels of saturated fatty acids makes membranes more rigid (i.e., decrease in fluidity) leading to decrease in the number of insulin receptors and the affinity of insulin to its receptor (Ginsberg et al. 1982, 1991; Bruneau et al. 1987). This, in turn, induces insulin resistance (Das 2010). In contrast, elevation in the amount of PUFA in the cell membrane produces a

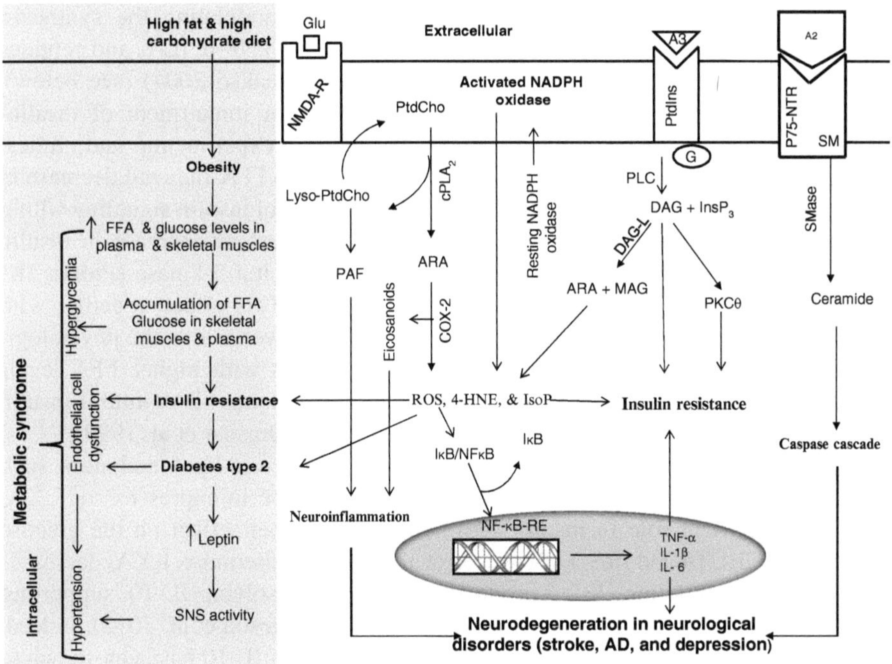


**Fig. 4.6** Hypothetical diagram showing the accumulation of free fatty acids in skeletal muscles, induction of insulin resistance, endothelial cell dysfunction, hypertension in metabolic syndrome by high-fat diet, onset of inflammation by phospholipid-derived lipid mediators, and induction of caspase cascade by ceramide. Glutamate (Glu); phosphatidylcholine (PtdCho); lyso-phosphatidylcholine (lyso-PtdCho); phosphatidylinositol (PtdIns); arachidonic acid (ARA); platelet-activating factor (PAF); diacylglycerol (DAG); monoacylglycerol (MAG); inositol 1,4,5-trisphosphate ($InsP_3$); phospholipase $A_2$ ($PLA_2$); phospholipase C (PLC); diacylglycerol lipase (DAG-L); cyclooxygenase-2 (COX-2); sphingomyelin (SM); sphingomyelinase (SMase); pan-neurotrophin receptor p75 (p75NTR); protein kinase Cθ (PKCθ); reactive oxygen species (ROS); 4-hydroxynonenal (4-HNE); and isoprostane (IsoP)

significant increase in the number of high-affinity sites for insulin (Ginsberg et al. 1982, 1991; Bruneau et al. 1987). Insulin resistance also affects the function of endothelial cells and cells of the vascular wall. This further increases CVD risk. Insulin not only optimizes glucose uptake by skeletal muscles, which are major peripheral users of glucose, but also reduces liver glucose production after a meal and reduces fatty acid release by adipose tissue. These three major functions are closely associated with glucose homeostasis. Increased delivery of free fatty acids (FFAs) to peripheral tissues is one of the factors linking obesity with insulin resistance. FFAs produce insulin resistance in the liver by blocking the ability of insulin to suppress glycogenolysis (Boden et al. 2002). This leads to an increase in hepatic glucose production while the patient is already in the hyperglycemic state. Hyperinsulinemia promotes the hepatic uptake of FFAs and the production of intracellular triglycerides, resulting in fatty liver disease.

Visceral fat also influences insulin resistance by modulating the synthesis, storage, and release of inflammatory mediators, including TNF-α, IL-6, and peptides such as leptin, resistin, and adiponectin (Boden and Laakso 2004) (see below). Evidence about the crucial role of FFA availability for impairment of insulin-stimulated muscle glucose uptake is supported by both experimental and clinical studies (Shulman 2000). The mechanism associated with FFA-induced decrease in insulin-stimulated glucose uptake involves the inhibition of insulin signaling within muscle cells. Lipid infusion has been reported to prevent an increase in insulin receptor substrate-1 (IRS-1)-associated phosphatidylinositol 3-kinase (PtdIns 3K) activity in response to insulin (Dresner et al. 1999). This effect, together with impaired Akt serine phosphorylation, has been reported even within the physiological range of plasma FFA concentrations and increases with higher FFA levels (Belfort et al. 2005). Collective evidence suggests that some FFAs inhibit insulin signaling by modifying the intramyocellular lipid pool (Dresner et al. 1999).

Recent studies have also indicated that some ARA-derived lipid mediators, such as $LXA_4$, also increase insulin sensitivity. Thus, increase in expression of $LXA_4$ plays an important role in insulin sensitivity through their effect on the glucose transporter GLUT4 and IRS-1 (Börgeson et al. 2012). Furthermore, $LXA_4$ decreases IL-6 secretion while increasing production of the pro-resolving IL-10, supporting the view that $LXA_4$ acts in a pro-resolving manner (Börgeson et al. 2012). Indeed, IL-10 inversely correlates with type II diabetes and MetS. IL-10 has been shown to inhibit IL-6-mediated insulin resistance, attenuate MCP-1 secretion, and promote GLUT4 and IRS-1 expression (Lumeng et al. 2007; Gonzalez-Periz and Claria 2010). Importantly, $LXA_4$-mediated reversal of insulin desensitization correlates with restored adipose Akt activation, which is necessary for translocation of the glucose-sensitizing GLUT4 receptor from the cytosol to the plasma membrane (Börgeson et al. 2012).

In contrast, neurological disorders are accompanied with site-specific premature and slow death of specific neuronal populations and loss of synapses in the brain and spinal cord that modulate thinking, skilled movements, decision making, cognition, and memory (Soto and Estrada 2008). At this stage very little is known about the effect of central adiposity, dyslipidemia, impaired glucose tolerance, insulin resistance, and hypertension in neurological disorders. The risk factors for MetS include high-fat and high-carbohydrate diet (Fig. 4.6) along with increased consumption of high-fructose corn syrup-containing soft drinks and lack of physical exercise, whereas risk factors for neurological disorders include age, family history, and exposure to toxic environment (Farooqui 2010).

Although the relationship between MetS and the risk of neurological disorders is still a matter of debate, MetS facilitates the carotid plaque formation and intima-media thickening in blood vessels, and amyloidoses, processes which make one susceptible to type II diabetes mellitus, stroke, Alzheimer's disease (AD), and depression (Farooqui et al. 2012) (Fig. 4.6). In addition, inflammation and oxidative stress, which are supported by lipid mediators derived from membrane phospholipids, sphingolipids, and cholesterol, have been increasingly recognized as critical steps in the pathogenesis of both metabolic syndrome as well as neurological disorders. Collective evidence suggests that elevated levels of phospholipid-, sphingolipid-, and cholesterol-derived lipid mediators, changes in $Ca^{2+}$ homeostasis, and increase in levels of inflammatory cytokine in neural cell may contribute to neurodegeneration in neurological disorders (Farooqui 2010). In contrast, dyslipidemia, impaired glucose tolerance, insulin resistance, and increase in levels of adipokines and chemokines may contribute to the pathogenesis of MetS (Farooqui et al. 2012). Clinicians recommend a low-fat diet for the treatment of MetS in order to facilitate weight loss, but epidemiological studies do not support a consistent relation between the intake of dietary fat and adiposity (Ludwig et al. 1999) because at the molecular level energy from fat is often replaced with energy from carbohydrates (Gross et al. 2004). A high-carbohydrate diet has been reported to produce the biochemical features of MetS (Garg et al. 1999) with adverse effects on lipid profile and glucose tolerance, worsening the metabolic abnormalities in those with or predisposed to MetS individuals (Riccardi and Rivellese 2000) (Fig. 4.6).

Long-chain fatty acids are common dietary nutrients. They are not only major energy source for visceral and neural cells but also precursors for synthesis of cellular complex lipids, which act as precursors for lipid mediators that modulate oxidative stress and inflammation (Prentki and Murthy Madiraju 2008; Farooqui et al. 2012). In addition, they modulate glucose metabolism and adipogenesis and regulate gene expression via various transcription factors, such as PPARs (Schroeder et al. 2008) and FOXO1 (Nahle et al. 2008; Buteau and Accili 2007; Su and Abumrad 2009). Visceral tissue (except for the liver and adipose tissue) and the brain synthesize fatty acids via de novo synthesis and depend on fatty acid intake through diet for their needs, emphasizing the physiological importance of fatty acids in diet. The amount and type of dietary fat play important roles in the development of obesity, insulin resistance, and neurological disorders (Riccardi et al. 2004). Studies on the effect of two types of high-fat diet (HFD), namely, vegetal fat diet and animal fat diet, identical in their lipid content (proportion) but different in their fat nature, indicate that both types of HFD promote weight gain than the control standard diet with a faster increase in male compared to female animals (El Akoum et al. 2011). High concentrations of saturated fatty acids in the animal fat diet group not only produce lipotoxic effects in visceral tissues (pancreas, liver, muscle, adipose tissue) and brain but also alter visceral cell and neural cell membrane dynamics (Manco et al. 2004; Farooqui 2010). These processes may retard the dimerization of insulin and leptin receptors on cell surface and retard insulin and leptin signaling. In addition, levels of circulating adiponectin as well as its adipose tissue mRNA expression levels are significantly reduced in the HFD-fed male mice. In contrast, its plasma levels remain unchanged in female animals. Adiponectin mRNA levels are significantly reduced in adipose tissue of both HFD-fed groups with a more marked decrease in the animal fat diet group compared to the vegetal fat diet-consuming group. Only the HFD-fed male mice developed diabetes. Levels of insulin are only increased in the animal fat diet-fed group in both sexes, and these changes are associated with increased resistin levels. Animals consuming vegetal fat diet do not develop any apparent metabolic alteration in females despite the increased weight gain (El Akoum et al. 2011).

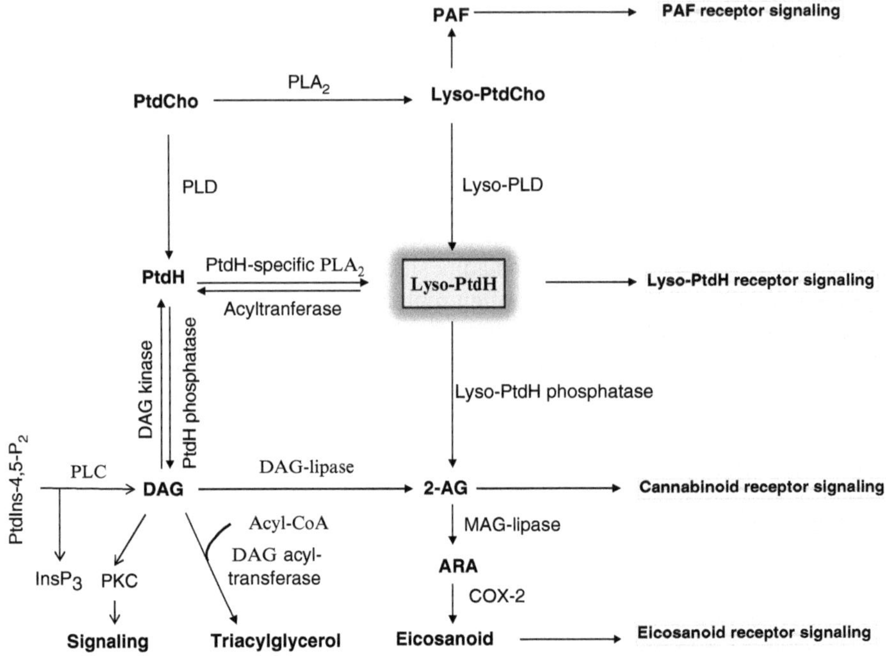


**Fig. 4.7** Generation of platelet-activating factor, diacylglycerol, lyso-PtdH, 2-arachidonylglycerol from phosphatidylinositol 4,5-bis phosphate (PtdIns 4,5-bis$P_2$), and PtdCho. Platelet-activating factor (PAF); phospholipase $A_2$ ($PLA_2$); phosphatidylcholine (PtdCho); lyso-phosphatidylcholine (lyso-PtdCho); phosphatidic acid (PtdH); lyso-phosphatidic acid (lyso-PtdH); diacylglycerol (DAG); arachidonic acid (ARA); 2-arachidonylglycerol (2-AG); and cyclooxygenase-2 (COX-2)

Diet high in saturated fat promotes obesity, insulin resistance, hypertension, diabetes, heart disease, and MetS, whereas diet enriched in ARA increases chances of neurological disorders in rodents and humans (Morrison et al. 2005) (Fig. 4.7). Diet high in olive (oleic acid) and fish oils (DHA and eicosapentaenoic acid, EPA), namely, Mediterranean diet, prevents or attenuates the development of obesity, diabetes, hypertension, heart disease, and neurological disorders (Morrison et al. 2005; Storlien et al. 1987) (Fig. 4.7). Studies on animal models of diabetes and neurological disorders indicate that depletion of n-3 fatty acids in diet can produce many symptoms of metabolic syndrome and increase risk factors for neurological disorders (Portois et al. 2009). The hypothetical link between PUFA and insulin resistance in both the brain and the peripheral tissues (muscles and liver) is represented by lipid rafts (microdomains). These rafts are involved in the intracellular trafficking of lipids and lipid-anchored protein. They can be partially modified by diet, particularly (but not exclusively) by dietary fatty acids (Edidin 2003). However, it is not clear whether dietary PUFAs incorporate into raft lipids or whether their low affinity to cholesterol disallows this and causes phase separation from rafts and displacement of raft proteins. Although there is increasing evidence supporting the view that membrane rafts and their modulation have an impact on the pathogenesis of type II diabetes, metabolic syndrome, and Alzheimer's disease, it is too early to affirm whether modulation of lipid rafts is responsible for the immunomodulatory effects of n-3 PUFAs. In addition to dietary fatty acids, gangliosides and cholesterol may also modulate stability of rafts in a number of tissues, and recent work highlights sphingolipids in membrane rafts as potential targets for the inhibition of tumor growth factor by n-3 PUFA (Yaqoob and Shaikh 2010). Fatty acids and ganglioside have been reported to contribute to the modulation of age-related cognitive decline in psychiatric disorders and AD and require clarification, particularly with respect to the contribution of lipid rafts, which serve as local organizing sites for the coordination of numerous signaling events by bringing receptors, effectors, and downstream signaling proteins into close proximity and promoting interactions (Simons and Toomre 2000). It is important to note that the localization of proteins (the glutamate receptors) in the synapse modulates neuronal plasticity and cognition. This localization is determined by the positioning of lipid raft domains at synapses (Hering et al. 2003). The roles of lipid rafts in metabolic disorder pathogenesis and in insulin resistance are only just emerging, but compelling evidence supports the growing importance of membrane rafts in pathological processes (Yaqoob and Shaikh 2010). Several aspects of brain metabolism clearly respond to insulin action, and although insulin and IGF-1 are supplied by circulation, a smaller proportion of insulin is produced in the brain itself (Plata-Salaman 1991). Moreover, insulin receptors (IRs) have been found in different brain areas with variable densities, in particular, in the olfactory bulb, hypothalamus, cerebral cortex, and hippocampus (Unger et al. 1989). Therefore, impairments of insulin and IGF-1 signaling lead to decreased energy metabolism and increase in oxidative stress. It is manifested by reduced glucose uptake and ATP production (de la Monte and Wands 2005). Reduced ATP adversely affects cellular homeostasis, membrane permeability, and fundamental processes required for synaptic maintenance and remodeling, which are needed for learning and establishing new memory. In addition to a metabolic function, insulin and IGF-1 modulate neuronal growth, survival, differentiation, migration, gene expression, protein synthesis, cytoskeletal assembly, synapse formation, and plasticity. In addition, they regulate growth, survival, and myelin production/maintenance in oligodendrocytes (D'Ercole et al. 1996).

The Mediterranean diet is now recognized as a dietary model, which not only retards cardiovascular disease and metabolic syndrome through the improvement of endothelial cell dysfunction (Angelico et al. 2012) but also partially blocks age-related diseases and neurodegeneration due to the presence of flavonoids in the Mediterranean diet (Farooqui 2012b). The Mediterranean diet not only improves endothelial function and significantly decreases waist circumference but also reduces plasma glucose and serum insulin in metabolic syndrome. In addition, a healthy n-3/n-6 PUFA ratio of the Mediterranean diet also plays an important role in improving cardiovascular and cerebrovascular functions (Farooqui 2009; Frisardi et al. 2010).

The hypolipidemic and anti-obesity effects of n-3 PUFA depend not only on the suppression of lipogenesis but also on the increase in fatty acid oxidation in several visceral tissues, including the liver (Teran-Garcia et al. 2007), adipose tissue (Flachs et al. 2005), and intestine (van Schothorst et al. 2009). This metabolic switch may reduce the accumulation of toxic fatty acid derivatives while protecting insulin signaling in the liver and muscle (Neschen et al. 2007). The effects of n-3 PUFAs and their active metabolites (González-Périz et al. 2009) are mediated by peroxisome proliferator-activated receptors (PPARα and PPARβ). It is reported that PPARγ, liver X receptor α, hepatic nuclear factor-4, sterol regulatory element-binding protein-1c (SREBP-1c), and carbohydrate-responsive element-binding protein may also contribute to the effects of n-3 PUFA (Neschen et al. 2006, 2007).

In contrast, Western diet contains high amounts of ARA, palmitic, and stearic acids. It is deficient in DHA and EPA (Farooqui 2009). The high intake of food enriched in ARA elevates levels of ARA-derived eicosanoids and upregulates the expression of proinflammatory cytokines (TNF-α and IL-1β). ARA-derived eicosanoids have prothrombotic, proaggregatory, vasoconstrictive, and proinflammatory properties. In addition, ARA is also a precursor of 4-hydroxynonal (4-HNE), a 9 carbon α, β-unsaturated aldehyde, and isoprostanes, which are generated by the lipid peroxidation of ARA (Fig. 4.2). Very little information is available on 4-HNE levels in visceral tissues of patients with MetS. However, based on detailed investigation, it is proposed that 4-HNE not only damages pancreatic beta cells, but impairs the ability of muscle and liver cells to respond to insulin (Mattson 2009). In addition, 4-HNE may facilitate atherosclerosis by modifying lipoproteins and may mediate cardiac cell damage by impairing metabolic enzymes (Mattson 2009) (see below). The levels of $F_2$-isoprostanes (IsoPs) and total HETEs and cholesterol oxidation products (such as 24- and 27-hydroxycholesterol), 8-OHdG, and allantoin (a product of oxidative damage to uric acid) are not changed in different risk categories of metabolic syndrome (Tsai et al. 2009; Seet et al. 2010). However, in individuals with type II diabetes, levels of urinary $F_2$-isoprostanes are increased only in those individuals who had type II diabetes for 7 years, indicating that oxidative damage in this condition is possibly a late event (Helmersson et al. 2004). Studies on the relationship between concentrations of plasma $F_2$-isoprostanes, isofurans, measures of obesity, and various cardiometabolic risk factors indicate that plasma $F_2$-isoprostanes and isofurans are elevated with increase in obesity and associated cardiometabolic risk factors, including CRP and blood pressure (Alkazemi et al. 2012). High concentrations of 4-HNE induce oxidative stress and apoptotic cell death (Farooqui 2011). In contrast, diet enriched in EPA and DHA generates docosanoids, which not only downregulates proinflammatory cytokines but also have anti-inflammatory, antithrombotic, antiarrhythmic, hypolipidemic, and vasodilatory effects. Lipid peroxidation of DHA results in the formation of 4-hydroxyhexanal (4-HHE). Like 4-HNE, 4-HHE also induces oxidative stress. Oxidative stress and inflammation are associated with the progression of insulin resistance and MetS (Urakawa et al. 2003). Some investigators have suggested that inflammation should be included as a component in the definition of MetS because it is such an important part of its pathophysiology (Haffner 2006). Hyperglycemia, increase in plasma FFA levels, and hyperinsulinemia have been linked to increased production of reactive oxygen species (ROS), which activate nuclear factor-κB (NF-κB), a proinflammatory transcription factor, that triggers a signaling cascade leading to a continued synthesis of oxidative species, induction of proinflammatory cytokines, and the low-grade chronic inflammation (Evans et al. 2002). Collective evidence suggests that levels of eicosanoids and docosanoids in the brain and visceral tissues are partly regulated by diet (Farooqui 2009). Accumulation of saturated fatty acids in myocytes, hepatocytes, cardiomyocytes, and pancreatic β cells and the metabolic products of n-6 fatty acids have been linked to insulin resistance and metabolic syndrome (Muoio and Newgard 2006). Incubation of muscle cells with saturated fatty acid (palmitic acid) results in reduction in glucose uptake, whereas incubation of these cells with unsaturated fatty acids (n-6 and n-3 fatty acids) increases glucose uptake (Dimopoulos et al. 2006). The molecular mechanism associated with this process is not fully understood. However, it is suggested that saturated fatty acids impair the insulin-dependent activation of Akt, which has been linked strongly to the hormonal regulation of glucose transport and glycogen synthesis (Hajduch et al. 2001). By inhibiting the condensation of serine with palmitoyl CoA to form 3-ketosphinganine, the inhibitor (myriocin) decreases ceramide synthesis without elevating intermediates such as sphinganine, which are upstream of ceramide formation and also have biologic effects (Merrill 2002). Myriocin treatment of rats infused with different lipid cocktails for 6 h to induce insulin resistance in an acute fashion indicates that a reduction in ceramide accumulation in skeletal muscle can prevent defects in glucose disposal (Holland et al. 2007). In genetically obese Zucker diabetic fatty (ZDF) rats and fat-fed mice, long-term administration of the inhibitor results in improvement of glucose tolerance. In contrast, administration of unsaturated fatty acids neither effects on Akt-dependent ceramide synthesis nor produces insulin sensitivity (Storlien et al. 1987). This may be due to increased polyunsaturated fatty acid-mediated utilization of glucose (Opara et al. 1994). In addition, the presence of polyunsaturated fatty acids induces a certain degree of unsaturation for maintaining plasticity, which facilitates movement of the glucose receptor to the cell's surface. In the human brain and skeletal muscles, insulin sensitivity is correlated with the amount of DHA present in the neural and skeletal muscle cell phospholipid (Salmeron et al. 2001). Although little is known about the molecular mechanism involved in beneficial effects of n-3 fatty acids, it is suggested that n-3 fatty acids, like insulin, not only stimulate leptin gene expression and secretion but also increase basal glucose uptake and regulate lipogenesis by modulating transcription factors (Fedor and Kelley 2009). Similar to n-3 fatty acids, leptin has multiple beneficial effects (both systemic and local) on energy homeostasis and glucose, so it is possible that some effects of n-3 fatty acids may limit obesity and improve insulin sensitivity via increased leptin production (Fedor and Kelley 2009; Pérez-Matute et al. 2005; Kalupahana et al. 2010).

Like n-3 fatty acids, oleic acid also produces beneficial effects on palmitic acid-mediated impairment of the insulin-dependent activation of Akt, a process, which is linked strongly to the hormonal regulation of glucose transport and glycogen synthesis. Oleic acid acts by preventing palmitate-induced ceramide synthesis at the level of dihydroceramide desaturase (DES1). It prevents palmitate-induced increase in mRNA for DES1, an enzyme that plays a crucial role in de novo synthesis of ceramide through addition of the 4,5-*trans* double bond to its immediate metabolic precursor dihydroceramide. Thus, DES1 may significantly impact ceramide signaling processes since the enzyme essentially modulates the dihydroceramide/ceramide ratio in cells (Kraveka et al. 2007). Oleic acid prevents the increase in ceramide through attenuation of the increase in message and activity of DES1. Knockdown of DES1 also protects from palmitate-induced insulin resistance, and overexpression of this enzyme ameliorated the protective effect of oleic acid. Together, these findings not only provide insight into the mechanisms of oleic acid-mediated protection against metabolic syndrome but also provide novel evidence for fatty acid-mediated regulation of a key enzyme of ceramide biosynthesis (Hu et al. 2011). Another monounsaturated fatty acid, palmitoleic acid ((*Z*)-9-hexadecenoic acid, 16:1Δ9), which differs from palmitate by the presence of one double bond, also increases insulin sensitivity and prevents inflammation (Mozaffarian et al. 2010).

Marked alterations in levels of serum prostaglandin F1α (6-keto-PGF1α), thromboxane $B_2$, and leukotriene $B_4$ ($LTB_4$) have been reported to occur in serum of patients with MetS (Karaman et al. 2010). Levels of anti-inflammatory oxilipines ($LTB_4$) are markedly increased in serum of MetS patients regardless of the presence or absence of insulin resistance. MetS patients with insulin resistance not only show elevated $LTB_4$ and 6-keto-PGF1α levels but also display enhanced synthesis of thromboxane $A_2$, which is responsible for vasoconstriction, platelet formation, and development of endothelial dysfunction supporting the view that disturbed synthesis of eicosanoids in MetS may contribute to the pathogenesis of this condition. Animals consuming a high-fat diet exhibit increase in oxidative stress, elevation in inflammatory cytokines, and decrease in levels of adiponectin along with reduction in vascular and adipose tissue levels of EETs and HO-1 compared to control rats (11 % dietary fat). Treatment with 11-(nonyloxy)undec-8(*Z*)-enoic acid (NUDA), an EET agonist, not only reverses serum adiponectin and vascular and adipose tissue levels of EETs and HO-1 but also reduces blood pressure, subcutaneous and visceral fat content, and serum TNF-α and IL-6 levels in rats on high-fat diet (Sodhi et al. 2012). These findings suggest that the EETs stimulate HO-1 expression via suppression of Bach 1, and interplay of these two systems affords vascular and metabolic protection in diet-induced obesity.

Like MetS, neurological disorders are also accompanied by increase in levels of prostaglandins, leukotrienes, and thromboxanes (Farooqui 2011). These mediators interact with their receptors and modulate onset of inflammation, oxidative stress, immune responses, and gene expression. In addition, these mediators contribute to vascular function by regulating cerebral blood flow (Farooqui 2011).

## 4.4 Diacylglycerol in Metabolic Syndrome and Neurological Disorders

Diacylglycerol (DAG) is a key lipid intermediate linking nutrient excess to the antagonism of insulin signaling (Shmueli et al. 1993). This molecule has been shown to accumulate in muscles obtained from insulin-resistant rodents and humans (Itani et al. 2002; Montell et al. 2001). In neural and nonneural cells (muscles and liver), insulin-induced stimulation of phospholipase C (PLC) results in degradation of membrane phosphatidylinositol 4,5-bisphosphate (PtdIns-4,5-$P_2$) leading to the formation of DAG and inositol 1,4,5-trisphosphate (Ins-1,4,5-$P_3$) (Fig. 4.8). DAG stimulates various isoforms of protein kinase C (PKC) and Ins-1,4,5-$P_3$ mobilizes calcium from intracellular stores. It should be noted that DAG-induced insulin resistance depends on activation of protein kinase CβII (PKCε), and DAG-containing polyunsaturated fatty acids are better protein kinase C activators than saturated diacylglycerol species (Schmitz-Peiffer and Biden 2008). Increase in DAG levels in the

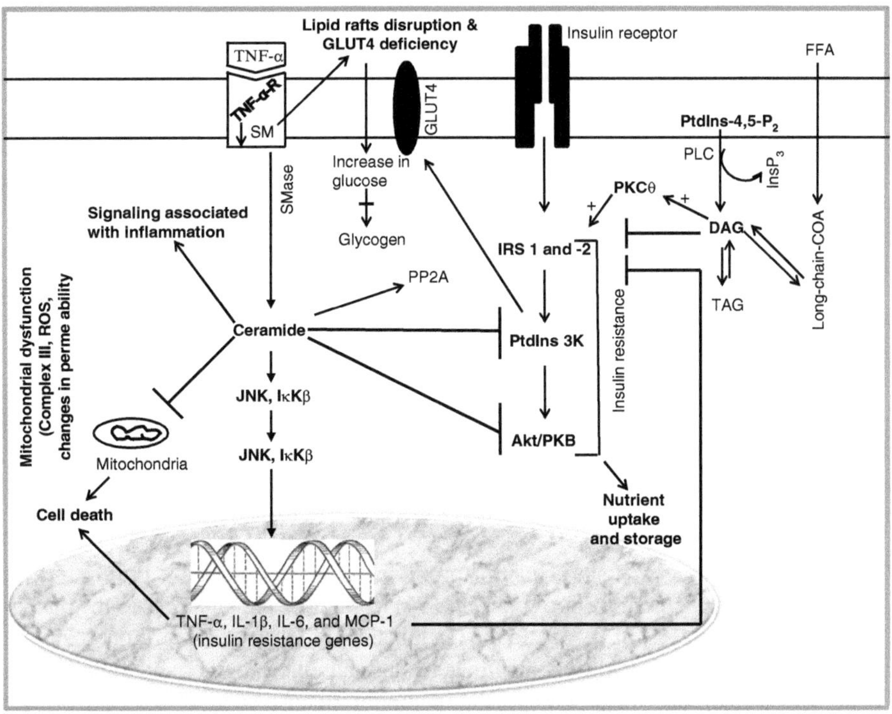


**Fig. 4.8** Modulation of insulin signaling by diacylglycerol and ceramide. Sphingomyelin (SM); sphingomyelinase (SMase); phosphatidylinositol 4,5-bis phosphate (PtdIns 4,5-bis$P_2$); inositol 1,4,5-trisphosphate ($InsP_3$); diacylglycerol (DAG); phospholipase C (PLC); protein kinase θ (PKCθ); tumor necrosis factor-α (TNF-α); tumor necrosis factor-α (TNF-α) receptor; protein phosphatase 2A (PP2A); c-Jun N-terminal kinase (JNK); insulin receptor substrate (IRS); phosphatidylinositol 3-kinase (PtdIns 3K); protein kinase B (Akt/PKB); and reactive oxygen species (ROS)

hypothalamus induces elevation in expression of leptin and produces insulin resistance in the brain (Holland et al. 2011; Benoit et al. 2009). Part of this effect is mediated by saturated fatty acids, which activate neuronal JNK and NF-κB signaling pathways with direct effects on leptin and insulin signaling (Zhang et al. 2008a). Disruption of signaling through TLR4/MyD88, IKKβ/NF-κB, and ER stress pathways in neurons protects mice from diet-induced obesity and its downstream metabolic effects (Zhang et al. 2008b; Kleinridders et al. 2009; Ozcan et al. 2009). No information is available on the effect of DAG ingestion on signal transduction processes in the brain. Similarly, levels of DAG in the brains of MetS patients have not been determined. However, ischemic injury, AD, and depression are accompanied by breakdown of neural membrane phospholipids and increase in levels of DAG levels. Accumulation of DAG in the brain is followed by its rapid hydrolysis by DAG lipase into fatty acids and monoacylglycerol (MAG) (Farooqui et al. 2000; Farooqui 2011).

The body's insulin sensitivity is negatively related to intramyocellular TAG content (Perseghin et al. 1999), and physical exercise not only increases muscle TAG levels but improves insulin sensitivity (Hoppeler et al. 1985). It is also reported that TAG accumulation protects against FFA-induced lipotoxicity (Listenberger et al. 2003). Increase in diacylglycerol acyltransferase 1 (DGAT1) and acyl-CoA:cholesterol acyltransferase (ACAT), enzymes that catalyze the synthesis of TAG and cholesteryl ester synthesis from DAG, respectively, retards the fat-mediated insulin resistance (Liu et al. 2007; Schenk and Horowitz 2007). Therefore, it is likely that muscle TAG does not exert unfavorable metabolic actions and might even be protective against accumulation of other, more toxic, lipid classes in adipocytes and arteries (Liu et al. 2007; Schenk and Horowitz 2007). Therefore, DGAT and ACAT have been considered as potential therapeutic targets for MetS. Inhibitors of DGAT have been recently synthesized. In in vitro systems these inhibitors can act as therapeutic agent for the treatment of MetS (Yanai et al. 2007; Nakada et al. 2010).

## 4.5 Ceramide in Metabolic Syndrome and Neurological Disorders

In muscle and adipose tissues, insulin promotes glucose uptake by promoting the translocation of glucose transporter 4 (GLUT4) from intracellular stores to the plasma membrane (Summers et al. 1998). In the liver, insulin blocks glucose efflux by inhibiting gluconeogenesis. In addition, insulin activates anabolic enzymes and inhibits catabolic ones to facilitate the storage of the incoming glucose as glycogen (Chabowski et al. 2004). As stated above, ceramide is a second messenger of sphingolipid metabolism. Ceramide impairs mitochondrial function and cell viability through several mechanisms, including $Ca^{2+}$ mobilization (Pinton et al. 2001), modulation of mitochondrial respiratory complex I and III activities (France-Lanord et al. 1997), and the formation of a channel in the outer mitochondrial membrane (Siskind et al. 2002).

Ceramide also induces insulin resistance by activating proinflammatory cytokines and inhibiting insulin-stimulated signaling through PtdIns 3K/Akt (Fig. 4.8) (Holland and Summers 2008; Kitatani et al. 2008; Farooqui 2011). Ceramide inhibits insulin stimulation of PtdIns 3K/Akt by two independent mechanisms: (a) translocation of Akt to the plasma membrane and (b) stimulation of Akt dephosphorylation by protein phosphatase 2A (PP2A) (Stratford et al. 2004) that in turn dephosphorylates and deactivates Akt. Ceramide also activates JNK and IκKβ, which inhibits insulin signal transduction by serine phosphorylation of IRS-1. In addition, degradation of sphingomyelin by sphingomyelinases decreases sphingomyelin and disrupts membrane lipid rafts. This process induces abnormalities in the translocation of GLUT4 resulting in increase in the rate of glucose influx into the tissues.

Accumulation of ceramide is crucial for the development of obesity-induced insulin resistance (Holland et al. 2007; Holland and Scherer 2009). The systemic administration of recombinant adiponectin to mice fed with a high-fat diet or to leptin-deficient ob/ob mice reduces hepatic ceramide content. Elevation or reduction in adiponectin levels decreases or increases hepatic ceramide content, respectively (Holland and Scherer 2009; Holland et al. 2011). In the liver overexpression of adiponectin receptors increases hepatic ceramidase activity and concomitantly reduces hepatic ceramide content, indicating an association among adiponectin levels, hepatic ceramide content, and insulin sensitivity. This suggestion is supported by the observation that adiponectin-mediated AMPK activation can be blocked by a broad-spectrum ceramidase inhibitor implying that sphingosine-1-phosphate (S1P) generation may be required for the activation of AMPK by adiponectin (Holland et al. 2011; Lancaster and Febbraio 2011). S1P, a potent inducer of proliferation, is known to inhibit apoptosis (Fig. 4.4) (Farooqui 2011). Collective evidence suggests that adiponectin not only modulates glucose and lipid metabolism, insulin sensitivity, blood pressure, and angiogenesis but also reduces inflammation and inhibits apoptosis.

Biosynthesis of ceramide starts either with degradation of sphingomyelin by sphingomyelinases or with condensation of serine with acyl CoA such as palmitoyl CoA (de novo synthesis). The dependence of ceramide synthesis on palmitate, a saturated fatty acid, makes this sphingolipid an attractive candidate metabolite linking lipid oversupply to insulin signaling. Exposing muscle cells to palmitate not only elevates ceramide synthesis and inhibits insulin-mediated stimulation of Akt/protein kinase B but also negates the antagonistic effect of saturated free fatty acids toward Akt/PKB (Chavez et al. 2003). In obesity, the adipose tissue, skeletal muscle, and liver exhibit major abnormalities in sphingolipid metabolism that not only result in increased ceramide production, inflammation, and activation of proinflammatory cytokines but also produce impairments in glucose homeostasis and insulin responsiveness (Summers 2006). Thus, levels of ceramide are increased in the muscles or liver from insulin-resistant rodents (Gorska et al. 2004) and humans (Adams et al. 2004). In addition, levels of circulating ceramide and sphingosine-1-phosphate (S1P) with atherogenic lipid profiles and the development of insulin resistance are also elevated in genetically obese (ob/ob) mice (Majumdar and Mastrandrea 2012). As stated above, increase in ceramide mass may either be due to the accelerated hydrolysis of sphingomyelin or by increase in de novo synthesis of ceramide produced largely by increase in serine palmitoyltransferase activity. Sphingomyelin deficiency may lead to alterations in lipid raft formation, thereby changing GLUT4 translocation and leading to glucose accumulation in the plasma (Silva et al. 2009; Han et al. 2011). It is realized that this model is speculative and has many unanswered questions that need further testing in subsequent studies.

Another mechanism that may account for increased inflammation associated with MetS is activation of the innate immune signaling pathway triggered by TLR4 recognition of saturated fatty acids, an event that is essential for lipid-induced insulin resistance (Holland et al. 2011). In vitro and in vivo in rodents indicate that TLR4 is an upstream signaling component required for saturated fatty acid-induced ceramide biosynthesis (Holland et al. 2011). This increase in ceramide production is associated with the upregulation of genes driving ceramide biosynthesis, an event dependent of the activity of the proinflammatory kinase IKKβ. Importantly, increased ceramide production is not required for TLR4-dependent induction of inflammatory cytokines, but it is essential for TLR4-dependent insulin resistance supporting the view that ceramide may be a key component of the signaling networks that link lipid-induced inflammatory pathways to the antagonism of insulin action that contributes to diabetes in MetS (Holland et al. 2011). Collectively, these studies indicate that the treatment of cultured cells with ceramide and its analogs retards insulin-stimulated glucose uptake, GLUT4 translocation, and/or glycogen synthesis (Han et al. 2011). As stated earlier, ceramide blocks Akt signaling either via PP2A-induced dephosphorylation of Akt or through ceramide-mediated inhibition of PKCε translocat ion (Resjo et al. 2002; Schmitz-Peiffer 2010; Powell et al. 2003).

In another system, chronic treatment of genetically obese (ob/ob) and high-fat diet-induced obese mice for 8 weeks with myriocin (an inhibitor of de novo ceramide synthesis) reduces levels of circulating ceramides (Yang et al. 2009). Inhibition of de novo ceramide biosynthesis not only reduces adipose expression of suppressor of cytokine signaling 3 (SOCS3) but also induces adipose uncoupling protein 3 (UCP3). In addition, ceramide directly mediates the induction SOCS3 and inhibits UCP3 mRNA in cultured adipocytes supporting the view that ceramide plays a direct role in the regulation of metabolism and energy expenditure. Inhibition of de novo ceramide synthesis has no effect on adipose tumor necrosis factor-α (TNF-α) expression. However, decrease in ceramide dramatically reduces levels of plasminogen activator inhibitor-1 (PAI-1) and monocyte chemoattactant protein-1 (MCP-1) in adipose tissues. These observations support the role of ceramide biosynthesis in the regulation of body weight, energy expenditure, and the metabolic syndrome (Yang et al. 2009).

Like ceramide, $GM_3$ ganglioside also contributes to the development of insulin resistance. This sphingolipid not only inhibits tyrosine phosphorylation of insulin receptors and IRS-1 but also retards the activity of phospholipase $A_2$ (Kabayama et al. 2005; Yang et al. 1994). It is proposed that $GM_3$ induces TNF-α-mediated insulin resistance causes enhancement of tyrosine phosphorylation of insulin receptors (Tagami et al. 2002; Kabayama et al. 2005; Yamashita et al. 2003). Improved phosphorylation of insulin receptors is also observed in $GM_3$ synthase knockout mice (Zhao et al. 2007). These studies indicate that complex sphingolipids play an important role in insulin resistance.

Studies on plasma sphingoid base compositions in healthy individuals and individuals with nondiabetic metabolic syndrome and diabetic patients indicate that deoxysphingolipids (dSLs) are significantly elevated in the plasma of patients with either nondiabetic metabolic syndrome or type II diabetes mellitus compared with controls supporting the view that dSLs are relevant biomarkers for both the metabolic syndrome and type II diabetes mellitus (Othman et al. 2012).

Alterations in ceramide levels and metabolism in brain tissues have been reported to occur in patients with neurological disorders. Thus, levels of ceramide are significantly increased in ischemic injury, AD, and depression (Farooqui 2011). This increase in ceramide levels is accompanied by elevation in acid sphingomyelinase (ASMase) and acid ceramidase expression in AD (Farooqui 2011). Moreover, significant correlations are observed between the brain ASMase activity and the levels of Aβ peptide and hyperphosphorylated tau protein in AD patients. Treatment of neural cell cultures with Aβ oligomers not only activates ASMase but also increases ceramide and induces apoptotic cell death. In addition, there is evidence that ceramide induces oxidative stress, insulin resistance, impairment in energy metabolism, and apoptotic cell death in Alzheimer's disease (Farooqui 2011). Elevation in ceramide levels is known to promote apoptosis through the activation of caspase-3, the protease, which is responsible for the cleavage of polyADP-ribose polymerase. Activation of caspase-3 results in degradation of a number of enzymes (protein kinase C, $cPLA_2$, PLC, phosphatases, cyclooxygenases), cytoskeletal proteins (α-spectrin, β-spectrin, actin, and vimentin), and members of the Bcl-2 family of apoptosis-related proteins as well as DNA-modulating enzymes. The degradation of the above mentioned proteins results in abnormal signal transduction and key morphological changes associated with apoptotic cell death (Farooqui 2011).

## 4.6 Endocannabinoids in Metabolic Syndrome and Neurological Disorders

Endocannabinoids are ARA-containing phospholipid-derived lipid mediators capable of binding to G protein-coupled receptors called CB1 and CB2 receptors. The two best characterized endocannabinoids include N-arachidonoylethanolamine (AEA; anandamide) and 2-arachidonoylglycerol (2-AG) (Farooqui 2011). Endocannabinoids are not stored in cells, but are rapidly synthesized, released, and degraded by intracellular enzymes. AEA and 2-AG are hydrolyzed by fatty acid amide hydrolase (FAAH) and monoacylglycerol lipase (MAGL), respectively (Di Marzo and Petrosino 2007). In addition to CB1 and CB2 receptors, endocannabinoids also act as retrograde signaling messengers for GABAergic and glutamatergic synapses (Farooqui 2011). CB1 receptors are distributed in brain areas associated with motor control, emotional responses, motivated behavior, and energy homeostasis. In the periphery, the same receptor is expressed in the adipose tissue, pancreas, liver, GI tract, skeletal muscles, heart, and the reproductive system. The CB2 receptor is mainly expressed in the immune system regulating its functions. Endocannabinoid signaling involves inhibition of cAMP synthesis and $Ca^{2+}$ mobilization, as well as activation of $K^+$ efflux in neuronal cells. CB1 is highly expressed in several organs and tissues, with an outstanding abundance in the central nervous system, whereas lower expression levels are found in peripheral tissues. CB2 receptor is predominantly located on immune peripheral cells (Di Marzo and Petrosino 2007). The actual expression level of CB2 receptor in normal conditions is still under intense debate, but several studies have revealed that CB2 receptors are upregulated after chronic proinflammatory stimuli (Zhang et al. 2008b). CB2 receptors seem to be involved in a protective cellular cascade activation ameliorating endothelium function (Zhao et al. 2010).

The endocannabinoid system and in particular the CB1 receptors are profoundly involved in the regulation of energy balance, body weight, and food intake at both the central and peripheral levels (Di Marzo et al. 2001). The stimulation of CB1 receptors in the brain is believed to control and mediate the effects on appetite, whereas in visceral tissues CB1 receptor activation is responsible for energy homeostasis and governs emotions and behaviors such as anxiety, fear, appetite, food, and water intake. Thus, constant stimulation of endocannabinoid system (ECS) is linked with metabolic dysfunctions and inflammation in the hypothalamus and peripheral tissues of obese subjects. In contrast, inhibition of ECS results in weight loss in animal and human subjects. In mice with high-fat diet-induced obesity, the upregulation of CB1 receptors and/or endocannabinoid concentrations have been reported to contribute to insulin resistance, dyslipoproteinemia, abdominal obesity, type II diabetes, and nonalcoholic fatty liver diseases (Di Marzo 2008). Thus, it is becoming increasingly evident that abnormalities in ECS contribute to visceral fat accumulation and disruption of energy homeostasis, which are characteristics of the metabolic syndrome. Constant activation of ECS has been linked to metabolic processes that are associated with the hypothalamus and peripheral tissues of obese patients. In contrast, inhibition of ECS results in weight loss in animal and human subjects (Kim et al. 2011). Collective evidence suggests that overactivation of ECS is associated with various metabolic conditions, such as dyslipidemia, insulin resistance, lipogenesis, excessive weight gain, and increasing intra-abdominal obesity.

In fact, elevation in AEA and 2-AG levels in the skeletal muscle, heart, and kidney has been reported to occur in overweight and obese subjects (Matias et al. 2008). Studies in animal models of obesity indicate that mutations in either the leptin or the leptin receptor gene, for example, ob/ob mice, db/db mice, and fa/fa rats, may play a pathophysiological role in obesity. In normal rats, leptin injection reduces hypothalamic levels of AEA and 2-AG, and endocannabinoid levels are elevated in the above animal models of leptin deficiency (ob/ob mice) or defective leptin receptors signaling (db/db mice, fa/fa rats).

It is well known that both inhibitory and stimulatory factors from the gastrointestinal tract regulate food intake. The effects of satiety hormone, cholecystokinin (CCK), are mediated via vagal afferent neurons, which not only express CCK-1 receptor on vagal afferent neurons but also express cannabinoid CB1 receptors. Retrograde tracing studies indicate that these neurons project to the stomach and duodenum. RT-PCR, immunohistochemistry, and in situ hybridization studies on the expression of CB1 receptors in rat nodose ganglia indicate that these receptors are increased by withdrawal of food for ≥12 h. However, after re-feeding of fasted rats results in a rapid loss of CB1 receptor expression. These effects can be blocked by the CCK-1 receptor antagonist (lorglumide) and mimicked by administration of CCK to fasted rats. Because CCK is a satiety factor that acts via the vagus nerve, CB1 agonists stimulate food intake. These data suggest a new mechanism modulating the effect on food intake of satiety signals from the gastrointestinal tract (Zhang et al. 2008a, b; Zhao et al. 2010; Di Marzo 2008; Nogueiras et al. 2008).

Blockade of the CB1 receptor has been recently used for the treatment of obesity. The first selective CB1 receptor antagonist (rimonabant) has already successfully completed in phase III clinical trials. It is reported that rimonabant treatment results into sustained weight loss and a reduction in waist circumference (Nogueiras et al. 2008). Rimonabant-treated patients not only show significant improvement in high-density lipoprotein cholesterol profiles, triglyceride levels, and insulin resistance but also show significant reduction in the prevalence of metabolic syndrome. Currently, one of the most discussed aspects of endocannabinoid system function is to what extent the endocannabinoid system might affect metabolism independently of its control over body weight and food intake. Specifically, a food intake- and body weight-independent role in the regulation of glucose homeostasis and insulin sensitivity could have major impact on the potential of drug candidates targeting the endocannabinoid system for the prevention and treatment of metabolic syndrome.

Alterations in endocannabinoids have been reported to occur in mood disorders and neurodegenerative diseases. Accumulating evidence suggests that an unbalance in the endocannabinoid system (decrease in neuronal cannabinoid CB1 receptors, increase of glial cannabinoid CB2 receptors, and overexpression of FAAH in astrocytes in experimental models of AD as well as in postmortem brain tissue of AD patients) may be involved not only in alterations in neurotransmitter systems but also in upregulation in inflammatory processes.

Endocannabinoids (AEA and 2-AG) are substrates for FAAH and both can be converted into ARA. The overexpression of FAAH in astrocytes surrounding neuritic plaques suggests that astrocytes, via FAAH, may be a significant source of ARA and related proinflammatory eicosanoids near these plaques in AD brain (Micale et al. 2007). Endocannabinoids produce beneficial effects in neurodegenerative disorders such as AD and Huntington disease (HD) (Mechoulam et al. 2002). These beneficial effects are due to endocannabinoid receptor-dependent and endocannabinoid receptor-independent mechanisms, which include the antioxidant activity of cannabinoids, activation of cytoprotective signaling pathways as neurotrophic factors (e.g., brain-derived neurotrophic factor, BDNF) or protein kinase A and B, and modulation of immune responses (Marsicano et al. 2002).

Endocannabinoid-based drugs produce beneficial effects on neurological disorders that involve the basal ganglia Parkinson's disease (PD) and HD (Fernández-Ruiz 2009). Beneficial effects include the alleviation of specific motor symptoms, namely, choreic movements with CB1/transient receptor potential vanilloid type 1 agonists in HD and bradykinesia with CB1 antagonists and tremor with CB1 agonists in PD. In AD, β-amyloid (Aβ) accumulation is accompanied by endocannabinoids released from neurons and glial cells resulting in elevation in ERK activity and induction of BDNF and activation of CB1-induced neuroprotective pathways (Marsicano et al. 2003). In addition, endocannabinoids modulate the release of inflammatory mediators in microglia through cannabinoid CB2 receptors. This supports the view that endocannabinoids may trigger $CB_2$-dependent proinflammatory cytokine release and neuroinflammation, a process closely associated with the pathogenesis of AD (Maresz et al. 2005).

## 4.7 4-Hydroxynonenal in Metabolic Syndrome and Neurological Disorders

Nonenzymic peroxidation of ARA generates 4-HNE, a nine-carbon α,β-unsaturated aldehyde. This aldehyde is one of the major end products of nonenzymic oxidation of ARA and an important mediator of neural and nonneural cell damage because of its ability to react with lysine, cysteine, and histidine residues in proteins but also with free amino acids, deoxyguanosine, and aminoglycerophospholipids (Esterbauer et al. 1991). The C3 position of 4-HNE is a highly reactive site that undergoes a Michael addition reaction with cellular thiols and hence readily forms adducts with glutathione or protein-containing thiol groups. 4-HNE may cause a number of deleterious effects in cells including inhibition of DNA and RNA synthesis, disturbance in calcium homeostasis, and inhibition of mitochondrial respiration. These events may play a substantial role in the disruption of the energy-producing capacity of mitochondria in MetS. It is likely that in the brains of MetS subjects, by inhibiting glucose transporter, glutamate transporter, and sodium–potassium ATPases, 4-HNE depolarizes neuronal membranes leading to the opening of NMDA receptor channels and influx of calcium ions into the cell (Mark et al. 1997; Keller and Mattson 1998; Kadoya et al. 2003). As a second messenger, $Ca^{2+}$ mediates physiological responses of neurons to neurotransmitters and neurotrophic factors. The proapoptotic effects of $Ca^{2+}$ are mediated by a diverse range of $Ca^{2+}$-sensitive factors that are compartmentalized in various intracellular organelles including the endoplasmic reticulum, cytoplasm, and mitochondria in MetS. The $Ca^{2+}$ dynamics of these organelles appear to be modulated by the apoptosis-regulating Bcl-2 family proteins (McConkey 1996). Moreover, elevated cytoplasmic $Ca^{2+}$ levels activate the mitogen-activated protein kinase (MAPK) cascade, and the PtdIns 3K-Akt pathway, being also critical for amino acid-mediated activation of mammalian target of rapamycin (mTOR) (Xu et al. 2011). Activation of MAPK and/or mTOR pathways, in turn, may promote cell survival or cell death, depending on stimuli.

4-HNE also impairs glucose transport and decreases cellular ATP levels by depressing mitochondrial function (Keller et al. 1997). 4-HNE contributes to the pathogenesis of insulin resistance in MetS. Levels of 4-HNE are increased in the blood and muscle tissue of obese subjects compared to normal weight subjects (Vincent et al. 2001; Russell et al. 2003). Improved insulin sensitivity in muscle cells following exercise or dietary energy restriction reduces levels of 4-HNE (Johnson et al. 2007). Levels of 4-HNE are also elevated in adipocytes in obesity where it may impair the function of proteins that play important roles not only in biosynthesis of complex lipids (Grimsrud et al. 2007) but also in dampening of inflammation (Zarrouki et al. 2007). 4-HNE not only damages pancreatic β cells but impairs the ability of muscle and liver cells to respond to insulin. These processes may contribute to insulin resistance (Mattson 2009). 4-HNE promotes atherosclerosis by modifying lipoproteins and may induce cardiac cell damage by impairing metabolic enzymes. Acute and repeated exposure of adipocytes to physiological 4-HNE not only promotes oxidative stress, impairs adipogenesis, and alters the expression of adipokines but also upregulates genes for expression of lipolytic enzymes resulting in increase in FFA release (Dasuri et al. 2013). It is also reported that long-term high-fat diet-mediated obesity in mouse is accompanied by decrease in plasma adiponectin and increase in adipose tissue 4-HNE content. Exposure of adipocytes to exogenous 4-HNE reduces adiponectin secretion in a dose-dependent manner (Wang et al. 2012). Collective evidence suggests that 4-HNE can not only differentially regulate adiponectin gene expression but also produce diverse effects in adipocyte homeostasis and adipocyte differentiation, which may be closely associated with the pathogenesis of obesity and metabolic syndrome (Wang et al. 2012; Dasuri et al. 2013).

4-HNE produces neurodegeneration by modifying membrane-associated glucose and glutamate transporters, ion-motive ATPases, enzymes associated with β-amyloid metabolism, and cytoskeletal proteins (Mark et al. 1997; Keller and Mattson 1998). Exercise and dietary energy restriction may not only reduce 4-HNE generation but also increase cellular systems for 4-HNE detoxification, such as glutathione and oxidoreductases. Accumulating evidence suggests that excessive generation and accumulation of 4-HNE may contribute to obesity, insulin resistance, and MetS in rodents and humans (Singh et al. 2008; Dasuri et al. 2013; Wang et al. 2012). Thus, consumption of high in ω-6 fat and high-carbohydrate diet in human subjects increases the level of 4-HNE in the plasma rapidly (within minutes to hours) (Devaraj et al. 2008). Now it is becoming increasingly evident that high-fat and high-carbohydrate diet promotes the MetS and obesity when consumed regularly, suggesting a role for 4-HNE very early in the development of MetS and obesity. In contrast, when moderately obese subjects are maintained on an alternate-day calorie restriction diet, their levels of circulating 4-HNE and other markers of inflammation (eicosanoids, platelet-activating factor, CRP) are significantly reduced, supporting the view that membrane lipid peroxidation can be reversed by dietary restriction. Levels of 4-HNE are markedly increased in brain tissue from patients with stroke, AD, and depression, and it is suggested that protein, lipid, and nucleic acid modifications by 4-HNE may be involved in neurodegenerative processes associated with cell death in neurological diseases (Farooqui 2011) (Fig. 4.9).

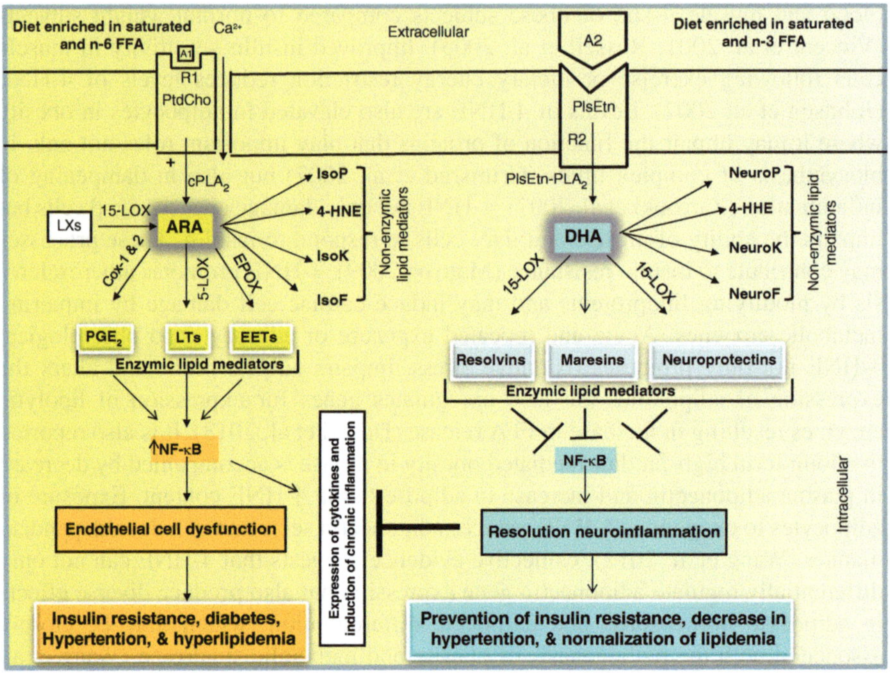


**Fig. 4.9** Generation of enzymic and nonenzymic lipid mediators from arachidonic and docosahexaenoic acids. Agonists (A1 and A2); receptors (R1 and R2); phosphatidylcholine (PtdCho); ethanolamine plasmalogen (PlsEtn); arachidonic acid (ARA); docosahexaenoic acid (DHA); phospholipase $A_2$ ($PLA_2$); plasmalogen-selective phospholipase $A_2$ (PlsEtn-$PLA_2$); 5-lipoxygenase (5-LOX); 15-lipoxygenase (15-LOX); epoxygenase (EPOX); cyclooxygenase-1 and cyclooxygenase-2 (COX-1 and COX-2); prostaglandin $E_2$ ($PGE_2$); leukotrienes (LTs); epoxyeicosatrienoic acids (EETs); isoprostane (IsoP); 4-hydroxynonenal (4-HNE); isoketal (IsoK); isofuran (IsoF); neuroprostane (NeuroP); 4-hydroxyhexenal (4-HHE); neuroketal (NeuroK); and neurofuran (NeuroF)

## 4.8 Oxy/Hydroxycholesterol in Metabolic Syndrome and Neurological Disorders

It is well known that oxidative stress is involved in insulin resistance, hypertension, endothelial dysfunction, and dyslipidemia. These factors contribute to type II diabetes, the MetS, and cardiovascular disease (CVD). Elevated levels of oxy/hydroxysterols have been observed in patients with diabetes, hypercholesterolemia, advanced carotid atherosclerosis, and individuals at increased risk for CVD (Murakami et al. 2001). These metabolites not only play an important role in the regulation of cholesterol metabolism (Krook 2006) but also act as important modulators of insulin by acting as agonists of liver X receptors (LXRs) (Murakami et al. 2001). Oxy/hydroxysterols have been shown to mediate proinflammatory, proapoptotic, and pro-fibrogenic effects in various types of cell cultures (Sottero et al. 2009). Increase in levels of oxy/hydroxysterols in obesity and the metabolic syndrome are linked with enhanced inflammatory stress, as indicated by the increase in C-reactive protein, a pentameric acute-phase protein that has emerged as the golden marker for inflammation (Knopp and Paramsothy 2006). C-reactive protein is produced dominantly by the liver. A major function of C-reactive protein is its ability to bind phosphocholine and recognize some foreign pathogens as well as phospholipids of damaged cells. It activates the complement system not only by acting as a ligand but also through its interactions with phagocytic cells. These observations support the view that C-reactive protein can initiate the elimination of targeted cells by its interaction with both humoral and cellular effector systems of inflammation. Other proinflammatory effects of C-reactive protein include induction of inflammatory cytokines and tissue factor in monocytes. C-reactive protein also assists in complement binding to foreign and damaged cells and affects the humoral response to disease (Gabay and Kusher 1999).

Brain cholesterol is mostly synthesized by de novo synthesis in astrocytes and plasma lipoproteins cannot cross the BBB. The BBB is permeable to oxy/hydroxysterols. Thus, excess cholesterol in the brain may, under normal physiological conditions, be excreted in the blood. The major oxysterol involved in this excretion mechanism appears to be 24-hydroxycholesterol, almost exclusively synthesized in the brain by the enzyme cholesterol 24-hydroxylase (cytochrome P450-46A1) (Bjorkhem and Meaney 2004; Farooqui 2011). Another oxy/hydroxysterol of major interest in brain pathophysiology is 27-hydroxycholesterol. This mediator is produced in the brain by the cytochrome P450-27A1. In contrast to cholesterol, 27-hydroxycholesterol can enter into the brain from the blood circulation (Bjorkhem 2006). Levels of hydroxyl and ketocholesterol are also elevated neurological disorders. It is reported that hydroxyl and ketocholesterols promote neurodegeneration in various types of neuronal cultures (Farooqui 2011). Thus, in SH-SY5Y human neuroblastoma cells, 27-hydroxycholesterol enhances Aβ1–42 production by upregulating APP and BACE1 protein levels (Prasanthi et al. 2009; Famer et al. 2007). It is also shown that high cholesterol levels reduce insulin-degrading enzyme (IDE) and low-density lipoprotein receptor-related protein (LRP-1) levels in rabbit hippocampus (Prasanthi et al. 2008). Although these proteins are involved in Aβ clearance from the brain, the molecular mechanism and functional link among cholesterol metabolism, IDE, LRP-1, and Aβ regulation remain unknown (Tanzi et al. 2004). Recent studies suggest that leptin (16 kDa adipocytokine) regulates Aβ production and tau phosphorylation in vivo and in vitro (Tezapsidis et al. 2009). Studies on the effect of high cholesterol or 27-hydroxycholesterol on the expression of leptin on AD-like pathology indicate that feeding rabbits a 2 % cholesterol-enriched diet for 12 weeks reduces the levels of leptin by about 80 % and treatment of organotypic slices from adult rabbit hippocampus with 27-hydroxycholesterol reduces leptin levels by approximately 30 % (Marwarha et al. 2010). 27-Hydroxycholesterol produces a 1.5-fold increase in Aβ40 and a threefold increase in Aβ42 and in phosphorylated tau. Treatment with leptin reverses the 27-hydroxycholesterol-mediated increase in Aβ and phosphorylated tau by decreasing the levels of BACE1 and GSK3β, respectively. Similarly, treatment of human neuroblastoma SH-SY5Y cells and organotypic slices from rabbit hippocampus with 27-hydroxycholesterol produces AD-like pathology by increasing Aβ production and triggers apoptotic cell death (Prasanthi et al. 2009). Collective evidence suggests that cholesterol-enriched diets and cholesterol metabolites induce AD-like pathology by altering leptin signaling (Marwarha et al. 2010).

## 4.9 Biochemical Events Associated with Lipid Mediator-Mediated Insulin Resistance

Insulin resistance is caused by defects in intracellular insulin signaling. The most important stage in the development of insulin resistance is impairment of insulin-stimulated skeletal muscle glucose uptake. There is evidence that intramyocellular lipid-derived mediators may contribute to the inhibition of insulin signaling (Erion and Shulman 2010). Induction of low-grade inflammation, activation of endoplasmic reticulum (ER) stress pathways, and accumulation of lipid mediators have been suggested to contribute to insulin resistance in MetS (Shulman 2000; Samuel et al. 2010). Thus, increase in intracellular DAG levels inhibits insulin signaling through the activation of novel PKCε (Samuel et al. 2010), which in turn blocks insulin receptor kinase phosphorylation of insulin receptor substrates 1 and 2. Fatty acid-mediated insulin resistance in the muscle is associated not only with reduction in insulin-stimulated phosphatidylinositol 3-kinase (PtdIns 3K) (Dresner et al. 1999) but also with elevation in plasma fatty acid concentrations. In addition, the accumulation of intracellular ceramides prevents the activation of Akt2/PKB, which promotes the uptake of glucose through the translocation of an insulin responsive glucose transporter, termed glucose transporter 4 (GLUT4). This process is also facilitated by decrease in sphingomyelin and disruption of lipid rafts resulting in increase in the rate of glucose influx into the tissue. Involvement of inflammation in the onset of insulin resistance is also supported by increase in adipocytokines (e.g., TNF-α, IL-1β, and IL-6), monocyte chemoattractant protein-1 (MCP-1), and plasminogen activator inhibitor-1 (PAI-1). Elevation in these parameters interferes with insulin signaling not only through activation of the JNK but also through the inhibition of IκB kinase-β pathways (Tilg and Moschen 2008). The administration of TNF-α to healthy rats (Lang et al. 1992) and humans antagonizes insulin-stimulated glucose disposal and genetic deletion of TNF-α or TNF-α receptors partially protecting mice from obesity-induced insulin resistance (Uysal et al. 1997). In isolated cells or tissues, TNF-α inhibits insulin signaling by promoting the serine phosphorylation of IRS-1 (Paz et al. 1997). Finally, the unfolded protein response or ER stress pathways are also implicated in the pathogenesis of insulin resistance. Emerging evidence suggests that excess dietary fat intake for several weeks in animals and humans causes muscle insulin resistance along with cytosolic lipid accumulation. Insulin-resistance is accompanied not only by increased availability of free fatty acids along with elevation in triacylglycerol storage but also by increase in intracellular concentrations of lipid mediators (DAG, ceramides, and long-chain acyl CoA).

DAG and ceramides mediate impairment of the insulin signaling pathway via activation of inflammatory messengers (e.g., $PKC_{\varepsilon}$), which leads to inhibitory serine phosphorylation of IRS.

## 4.10 Interactions Among Lipid Mediator in Metabolic Syndrome and Neurological Disorders

It is obvious from the above description that like MetS, levels of lipid mediators are markedly increased in neurological disorders including stroke, AD, and depression (Table 4.1). Under normal conditions, homeostasis among phospholipid-, sphingolipid-, and cholesterol-derived lipid mediators is based not only on activities of lipid mediator-synthesizing enzymes and organization of signaling network but also on the complexity and interconnectedness of their metabolism (Farooqui 2011). Thus, phospholipid-, sphingolipid-, and cholesterol-derived lipid mediators and their downstream targets constitute a complex lipid signaling network with multiple nodes of interaction and cross-regulation. In MetS, increase in plasma concentrations of proinflammatory cytokines, adipokines, and lipids such as non-esterified fatty acids in obese individuals may not only reduce insulin action but also initiate insulin resistance, hyperglycemia, and metabolic dysfunction (van Gaal et al. 2006). It should be noted that insulin mediates anti-inflammatory effects in contrast to the glucose, which produces proinflammatory and oxidative effects (Dandona et al. 2009). In neurological disorders increased activities of lipid mediator-synthesizing enzymes, which are modulated by insulin in the brain may result in marked elevation in levels of lipid mediators. Enhancement in levels of lipid mediators results in oxidative stress and neuroinflammation (Farooqui 2011). These abnormalities disturb the signaling networks leading to the loss of communication among various receptors including TLR and G protein-coupled receptors. This process not only threatens the integrity of neural cell lipid homeostasis but also facilitates neural cell death (Farooqui 2011). In addition to the above processes, the organization and compartmentalization of phospholipids, sphingolipids, and cholesterol in neural membranes of various subcellular fractions provides structural and functional integrity that facilitates the appropriate interactions with integral membrane proteins. An organized compartmentalization of phospholipids, sphingolipids, and cholesterol is needed for modulating regular cellular functions such as cell proliferation, differentiation, communication, and controlled adaptive responses through interactions among phospholipid-, sphingolipid-, and cholesterol-derived lipid mediators (Ivanova et al. 2004). However, under pathological conditions, high levels and intense interactions among phospholipid-, sphingolipid-, and cholesterol-derived lipid mediators produce harmful effects that facilitate oxidative stress, neuroinflammation, and apoptotic cell death (Farooqui 2011). At the molecular level, close interactions between ARA and ARA-derived metabolites and ceramide and its metabolites (ceramide-1-phosphate) occur in various types of neuronal cultures (Farooqui 2011). These biochemical events are closely associated with oxidative stress, inflammation, and apoptosis.

**Table 4.1** Levels of lipid mediators in MetS and neurological disorders

| Lipid mediator | MetS | Stroke | AD | Depression | Reference |
|---|---|---|---|---|---|
| Prostaglandins | Increased | Increased | Increased | Increased | Farooqui (2010, 2011); Novgorodtseva et al. (2011); Henkel et al. (2011) |
| Leukotrienes | Increased | Increased | Increased | Increased | Farooqui (2010, 2011); Novgorodtseva et al. (2011); Henkel et al. (2011) |
| Thromboxanes | Increased | Increased | Increased | Increased | Farooqui (2010, 2011); Novgorodtseva et al. (2011); Henkel et al. (2011) |
| Platelet-activating factor | Decreased | Increased | Increased | – | Farooqui (2010, 2011); McLarnon et al. (2005); Derbent et al. (2011) |
| Diacylglycerol | – | Increased | – | – | Farooqui (2011) |
| Isoprostanes | No effect | Increased | Increased | – | Seet et al. (2010); Tsai et al. (2009) |
| Endocannabinoids | Increased | Increased | Increased | Increased | Mechoulam et al. (2002) |
| Ceramide | Increased | Increased | Increased | Increased | Farooqui (2011); Holland et al. (2007) |
| Sphingosine-1-phosphate | Increased | Increased | Decreased | Increased | Majumdar and Mastrandrea (2012); Farooqui (2011); Yung et al. (2012); Zhou et al. (2010); Kornhuber et al. (2010) |
| Deoxysphingolipids | Increased | – | – | – | Othman et al. (2012) |
| Oxy/hydroxysterols | Increased | Increased | Increased | Increased | Farooqui (2011); Alkazemi et al. (2008) |

## 4.11 Modulation of Lipid Mediators by Diet

Dietary fat is an important energy-producing macronutrient. The fatty acid component of dietary fat is essential for growth and development but is a precursor of lipid mediators. Levels of lipid mediators in the brain and visceral tissues are partly regulated by diet (Farooqui 2009, 2011). The high intake of food enriched in ARA-containing fats (vegetable oils) elevates levels of eicosanoids and PAF. These lipid mediators by acting through their receptors (Fig. 4.9) not only modulate vasodilation and vasoconstriction in the cardiovascular and cerebrovascular systems but also regulate the expression of proinflammatory cytokines and chemokines. ARA-containing diet also increases levels of lipid peroxides and isoprostanes, which mediate vasoconstrictive effects in cardiovascular and cerebrovascular systems (Montuschi et al. 2007; Farooqui 2011). Accumulating evidence suggests that ARA-derived lipid mediators produce prothrombotic, proaggregatory, and proinflammatory effects in the body. In contrast, diet enriched in EPA and DHA generates docosanoids, which induce anti-inflammatory, antithrombotic, antiarrhythmic, hypolipidemic, vasodilatory, and anti-excitotoxic effects (Farooqui 2012a). The molecular mechanisms of EPA and DHA action are only partially understood. These fatty acids not only induce changes in membrane structures, but their direct interactions with ion channels and transcription factors produce downregulation of eicosanoid biosynthesis (Calder 2006; Jump 2004; Farooqui 2012a). EPA- and DHA-derived lipid mediators downregulate expression of proinflammatory cytokines (Farooqui 2011). EPA- and DHA-derived lipid mediators act as agonists at specific receptors (CMKLR1, BLT1, ALX/FPR2, and GPR32) (Farooqui 2009) found on immune cells, neutrophils, and macrophages, but information on their occurrence in the brain is not available.

ARA is an essential fatty acid and its presence in diet is required for normal health. However, now a days diets contain very high amounts of saturated fat and with ARA to DHA ratio of about 20:1. The Paleolithic diet on which human beings have evolved, and lived for most of their existence, had a ratio of 1:1 and was high in fiber and rich in fruits, vegetables, lean meat, and fish (Simopoulos 2006, 2008; Farooqui 2009). Changes in eating habits, natural versus processed food, which is enriched in corn-based livestock, and increase in the consumption of vegetable oil in past 20 years along with decrease in consumption of seafood have altered the ARA to DHA ratio in favor of ARA (Simopoulos 2006; Kang and Weylandt 2008). In contrast, consumption of DHA-enriched diet has anti-inflammatory effects that are partly mediated by repression of genes that code for proinflammatory cytokines. DHA-enriched diet also elevates the levels of docosanoids that have antioxidant and antiapoptotic effects in neural and nonneural tissues (Farooqui 2011). ARA and DHA also interact with several transcription factors (PPAR, LXR, NF-κB, and SREBP), which differently to modulate lipid metabolism for maintaining cellular homeostasis (Jump 2004). Fatty acids or their metabolites bind directly to specific transcription factors to regulate gene transcription. These fatty acids also modulate gene expression indirectly through their interactions with COX, LOX, PKC, or sphingomyelinase signal transduction pathways as well as pathways that involve changes in membrane lipid/lipid raft composition that affects G-protein receptor or tyrosine kinase-linked receptor signaling (Farooqui 2009). Collectively, these studies support the view that levels of ARA- and DHA-derived lipid mediators are partly modulated by the diet (Farooqui 2009, 2012a) and the ratio between ARA and DHA must be properly maintained in the human diet. The primary goal of future research on neurochemistry of lipid mediators should be focused on: how to increase the levels of DHA-derived lipid mediators in the brain tissue through diet and how to avoid neurotraumatic, neurodegenerative, and neuropsychiatric diseases.

## 4.12 Conclusion

Lipid mediators are important endogenous regulators derived from enzymic degradation of phospholipids, sphingolipids, and cholesterol by phospholipases, sphingomyelinases, and cytochrome P450 hydroxylases, respectively. In phospholipids polyunsaturated fatty acids (DHA and ARA) are located at the sn-2 position of glycerol moiety and released by the action of $PLA_2$s. Oxygenation of ARA generates proinflammatory mediators (prostaglandins, leukotrienes, and thromboxanes), whereas DHA generates anti-inflammatory lipid mediators, such as resolvins, protectins, neuroprotectins, and maresins, which directly or indirectly suppress the activity of NF-κB. Diet and lifestyle play an important role in the development of MetS, and nutritional modulation may alter the development of MetS. The fatty acid composition in patients with MetS is typically characterized by high levels of saturated and n-6 fatty acids and low levels of n-3 fatty acids. This is because diet enriched in saturated fatty acids induces insulin resistance, whereas diet enriched in polyunsaturated fatty acids retards insulin resistance. Degradation of PtdIns 4,5-bisphosphate by PLC generates DAG and hydrolysis of sphingomyelin by sphingomyelinase generates ceramide. Insulin resistance is a key feature of MetS that is caused by the accumulation of saturated fatty acids, DAG, and ceramide. These metabolites interfere with leptin/insulin signaling and glucose homeostasis in visceral tissues. Neurological disorders are also accompanied by increase in levels of fatty acids, ceramide, and DAG in the brain. Accumulation of these lipid mediators in the brain is responsible for insulin resistance and neurodegeneration in the brain.

## References


Adams JM II, Pratipanawatr T, Berria R, Wang E, DeFronzo RA, Sullards MC, Mandarino LJ (2004) Ceramide content is increased in skeletal muscle from obese insulin-resistant humans. Diabetes 53:25–31

Alkazemi D, Egeland G, Vaya J, Meltzer S, Kubow S (2008) Oxysterol as a marker of atherogenic dyslipidemia in adolescence. J Clin Endocrinol Metab 93:4282–4289

Alkazemi D, Egeland GM, Jackson Roberts L, Kubow S (2012) Isoprostanes and isofurans as non-traditional risk factors for cardiovascular disease among Canadian Inuit. Free Radic Res 46:1258–1266

Angelico F, Loffredo L, Pignatelli P, Augelletti T, Carnevale R, Pacella A, Albanese F, Mancini I, Di Santo S, Del Ben M, Violi F (2012) Weight loss is associated with improved endothelial dysfunction via NOX2-generated oxidative stress down-regulation in patients with the metabolic syndrome. Intern Emerg Med 7:219–227

Belfort R, Mandarino L, Kashyap S, Wirfel K, Pratipanawatr T, Berria R, DeFronzo RA, Cusi K (2005) Dose-response effect of elevated plasma free fatty acid on insulin signaling. Diabetes 54:1640–1648

Benoit SC, Kemp CJ, Elias CF, Abplanalp W, Herman JP, Migrenne S, Lefevre AL, Cruciani-Guglielmacci C, Magnan C, Yu F, Niswender K, Irani BG, Holland WL, Clegg DJ (2009) Palmitic acid mediates hypothalamic insulin resistance by altering PKC-theta subcellular localization in rodents. J Clin Invest 119:2577–2589

Bjorkhem I (2006) Crossing the barrier: oxysterols as cholesterol transporters and metabolic modulators in the brain. J Intern Med 260:493–508
Bjorkhem I, Meaney S (2004) Brain cholesterol: long secret life behind a barrier. Arterioscler Thromb Vasc Biol 24:806–815
Björkhem I, Lütjohann D, Diczfalusy U, Stahle L, Ahlborg G, Wahren J (1998) Cholesterol homeostasis in human brain: turnover of 24S-hydroxycholesterol and evidence for a cerebral origin of most of this oxysterol in the circulation. J Lipid Res 39:1594–1600
Boden G, Laakso M (2004) Lipids and glucose in type 2 diabetes. Diabetes Care 27:2253–2259
Boden G, Cheung P, Stein TP, Kresge K, Mozzoli M (2002) FFA cause hepatic insulin resistance by inhibiting insulin suppression of glycogenolysis. Am J Physiol Endocrinol Metab 283:12–19
Börgeson E, Mcgillicuddy FC, Harford KA, Corrigan N, Higgins DF, Maderna P, Roche HM, Godson C (2012) Lipoxin A4 attenuates adipose inflammation. FASEB J 26:4287–4294
Bruneau C, Staedel-Flaig C, Crémel G, Leray C, Beck JP, Hubert P (1987) Influence of lipid environment on insulin binding in cultured hepatoma cells. Biochim Biophys Acta 928:287–296
Buteau J, Accili D (2007) Regulation of pancreatic beta-cell function by the forkhead protein FoxO1. Diabetes Obes Metab 9(Suppl 2):140–146
Calder PC (2006) Polyunsaturated fatty acids and inflammation. Prostaglandins Leukot Essent Fatty Acids 75:197–202
Chabowski A, Coort SL, Calles-Escandon J, Tandon NN, Glatz JF, Luiken JJ, Bonen A (2004) Insulin stimulates fatty acid transport by regulating expression of FAT/CD36 but not FABPpm. Am J Physiol Endocrinol Metab 287:E781–E789
Chavez JA, Knotts TA, Wang LP, Li G, Dobrowsky RT, Florant GL, Summers SA (2003) A role for ceramide, but not diacylglycerol, in the antagonism of insulin signal transduction by saturated fatty acids. J Biol Chem 278:10297–10303
Dandona P, Chaudhuri A, Ghanim H, Mohanty P (2009) Insulin as an anti-inflammatory and anti-atherogenic modulator. J Am Coll Cardiol 53(Suppl 1):S14–S20
Das UN (2010) A defect in Δ6 and Δ5 desaturases may be a factor in the initiation and progression of insulin resistance, the metabolic syndrome and ischemic heart disease in South Asians. Lipids Health Dis 9:130
Dasuri K, Ebenezer P, Fernandez-Kim SO, Zhang L, Gao Z, Bruce-Keller AJ, Freeman LR, Keller JN (2013) Role of physiological levels of 4-hydroxynonenal on adipocyte biology: implications for obesity and metabolic syndrome. Free Radic Res 47:8–19
de la Monte SM, Wands JR (2005) Review of insulin and insulin-like growth factor expression, signaling, and malfunction in the central nervous system: relevance to Alzheimer's disease. J Alzheimers Dis 7:45–61
Derbent A, Kargili A, Koca C, Gümüş İİ, Sevgili S, Simavli S, Karakurt F, Turhan NÖ (2011) Serum platelet-activating factor acetylhydrolase activity: relationship with metabolic syndrome in women with history of gestational diabetes mellitus. Gynecol Endocrinol 27:128–133
D'Ercole AJ, Ye P, Calikoglu AS, Gutierrez-Ospina G (1996) The role of the insulin-like growth factors in the central nervous system. Mol Neurobiol 13:227–255
Devaraj S, Wang-Polagruto J, Polagruto J, Keen CL, Jialal I (2008) High-fat, energy-dense, fast-food-style breakfast results in an increase in oxidative stress in metabolic syndrome. Metabolism 57:867–870
Di Marzo V (2008) The endocannabinoid system in obesity and type 2 diabetes. Diabetologia 51:1356–1367
Di Marzo V, Petrosino S (2007) Endocannabinoids and the regulation of their levels in health and disease. Curr Opin Lipidol 18:129–140
Di Marzo V, Goparaju SK, Wang L, Liu J, Bàtkai S, Jàrai Z, Fezza F, Miura GI, Palmiter RD, Sugiura T, Kunos G (2001) Leptin-regulated endocannabinoids are involved in maintaining food intake. Nature 410:822–825
Dimopoulos N, Watson M, Sakamoto K, Hundal HS (2006) Differential effects of palmitate and palmitoleate on insulin action and glucose utilization in rat L6 skeletal muscle cells. Biochem J 399:473–481

Dresner A, Laurent D, Marcucci M, Griffin ME, Dufour S, Cline GW, Slezak LA, Andersen DK, Hundal RS, Rothman DL, Petersen KF, Shulman GI (1999) Effects of free fatty acids on glucose transport and IRS-1-associated phosphatidylinositol 3-kinase activity. J Clin Invest 103:253–259

Edidin M (2003) The state of lipid rafts: from model membranes to cells. Annu Rev Biophys Biomol Struct 32:257–283

El Akoum S, Lamontagne V, Cloutier I, Tanguay JF (2011) Nature of fatty acids in high fat diets differentially delineates obesity-linked metabolic syndrome components in male and female C57BL/6 J mice. Diabetol Metab Syndr 3:34

Erion DM, Shulman GI (2010) Diacylglycerol-mediated insulin resistance. Nat Med 16:400–402

Esterbauer H, Schaur RJ, Zollner H (1991) Chemistry and biochemistry of 4-hydroxynonenal, malonaldehyde and related aldehydes. Free Radic Biol Med 11:81–128

Evans JL, Goldfine ID, Maddux BA, Grodsky GM (2002) Oxidative stress and stress-activated signaling pathways: a unifying hypothesis of type 2 diabetes. Endocr Rev 23:599–622

Famer D, Meaney S, Mousavi M, Nordberg A, Björkhem I, Crisby M (2007) Regulation of alpha- and beta-secretase activity by oxysterols: cerebrosterol stimulates processing of APP via the alpha-secretase pathway. Biochem Biophys Res Commun 359:46–50

Fantini J, Yahi N (2010) Molecular insights into amyloid regulation by membrane cholesterol and sphingolipids: common mechanisms in neurodegenerative diseases. Expert Rev Mol Med. doi:10.1017/S1462399410001602

Farooqui AA (2009) Effect of fish oil on human brain. Springer, New York

Farooqui AA (2010) Neurochemical aspects of neurotraumatic and neurodegenerative diseases. Springer, New York

Farooqui AA (2011) Lipid mediators and their metabolism in the brain. Springer, New York

Farooqui AA (2012a) n-3 Fatty acid-derived lipid mediators in the brain: new weapons against oxidative stress and inflammation. Curr Med Chem 19:532–543

Farooqui AA, Farooqui T, Panza F, and Frisardi V (2012) Metabolic syndrome as a risk factor for neurological disorders. Cell Mol Life Sci 69:741–762

Farooqui AA (2012b) Phytochemicals, signal transduction, and neurological disorders. Springer, New York

Farooqui AA, Horrocks LA, Farooqui T (2000) Glycerophospholipids in brain: their metabolism, incorporation into membrane, functions and involvement in neurological disorders. Chem Phys Lipids 106:1–29

Farooqui AA, Farooqui T, Panza F, Frisardi V (2012) Metabolic syndrome as a risk factor for neurological disorders. Cell Mol Life Sci 69:741–762

Fedor D, Kelley DS (2009) Prevention of insulin resistance by n-3 polyunsaturated fatty acids. Curr Opin Clin Nutr Metab Care 12:138–146

Fernández-Ruiz J (2009) The endocannabinoid system as a target for the treatment of motor dysfunction. Br J Pharmacol 156:1029–1040

Flachs P, Horakova O, Brauner P, Rossmeisl M, Pecina P, Franssen-van Hal N, Ruzickova J, Sponarova J, Drahota Z, Vlcek C, Keijer J, Houstek J, Kopecky J (2005) Polyunsaturated fatty acids of marine origin upregulate mitochondrial biogenesis and induce beta-oxidation in white fat. Diabetologia 48:2365–2375

France-Lanord V, Brugg B, Michel PP, Agid Y, Ruberg M (1997) Mitochondrial free radical signal in ceramide-dependent apoptosis: a putative mechanism for neuronal death in Parkinson's disease. J Neurochem 69:1612–1621

Frisardi V, Panza F, Seripa D, Imbimbo BP, Vendemiale G, Pilotto A, Solfrizzi V (2010) Nutraceutical properties of Mediterranean diet and cognitive decline: possible underlying mechanisms. J Alzheimers Dis 22(3):715–740

Fu X, Menke JG, Chen Y, Zhou G, MacNaul KL, Wright SD, Sparrow CP, Lund EG (2001) 27-Hydroxycholesterol is an endogenous ligand for liver X receptor in cholesterol-loaded cells. J Biol Chem 276:38378–38387

Gabay C, Kusher I (1999) Acute-phase proteins and other systemic responses to inflammation. N Engl J Med 340:448–454

Garg A, Bantel JP, Henry RR, Coulston AM, Griver KA, Raatz SK, Brinkley L, Chen YD, Grundy SM, Huet BA (1999) Effects of varying carbohydrate content of diet in patients with noninsulin dependent diabetes mellitus. JAMA 27:1421–1428

Ginsberg BH, Jabour J, Spector AA (1982) Effect of alterations in membrane lipid unsaturation on the properties of the insulin receptor of ehrlich ascites cells. Biochim Biophys Acta 690:157–164

Ginsberg BH, Chatterjee P, Yorek MA (1991) Insulin sensitivity is increased in friend erythroleukemia cells enriched in polyunsaturated fatty acid. Receptor 1:155–166

Gonzalez-Periz A, Claria J (2010) Resolution of adipose tissue inflammation. Sci World J 10:832–856

González-Périz A, Horrillo R, Ferré N, Gronert K, Dong B, Morán-Salvador E, Titos E, Martínez-Clemente M, López-Parra M, Arroyo V, Clària J (2009) Obesity-induced insulin resistance and hepatic steatosis are alleviated by omega-3 fatty acids: a role for resolvins and protectins. FASEB J 23:1946–1957

Gorska M, Dobrzyn A, Zendzian-Piotrowska M, Gorski J (2004) Effect of streptozotocin-diabetes on the functioning of the sphingomyelin-signalling pathway in skeletal muscles of the rat. Horm Metab Res 36:14–21

Grimsrud PA, Picklo MJ Sr, Griffin TJ, Bernlohr DA (2007) Carbonylation of adipose proteins in obesity and insulin resistance: identification of adipocyte fatty acid-binding protein as a cellular target of 4-hydroxynonenal. Mol Cell Proteomics 6:624–637

Gross LS, Li L, Ford ES, Liu S (2004) Increased consumption of refined carbohydrate and the epidemic of type 2 diabetes in the United States: an ecologic assessment. Am J Clin Nutr 79:774–779

Haffner SM (2006) The metabolic syndrome: inflammation, diabetes mellitus, and cardiovascular disease. Am J Cardiol 97:3A–11A

Hajduch E, Litherland GJ, Hundal HS (2001) Protein kinase B: a key regulator of glucose transport? FEBS Lett 492:199–203

Han X, Rozen S, Boyle SH, Hellegers C, Cheng H, Burke JR, Welsh-Bohmer KA, Doraiswamy PM, Kaddurah-Daouk R (2011) Metabolomics in early Alzheimer's disease: identification of altered plasma sphingolipidome using shotgun lipidomics. PLoS One 6:e21643

Helmersson J, Vessby B, Larsson A, Basu S (2004) Association of type 2 diabetes with cyclooxygenase-mediated inflammation and oxidative stress in an elderly population. Circulation 109:1729–1734

Henkel J, Gärtner D, Dorn C, Hellerbrand C, Schanze N, Elz SR, Püschel GP (2011) Oncostatin M produced in Kupffer cells in response to PGE2: possible contributor to hepatic insulin resistance and steatosis. Lab Invest 91:1107–1117

Hering H, Lin CC, Sheng M (2003) Lipid rafts in the maintenance of synapses, dendritic spines, and surface AMPA receptor stability. J Neurosci 23:3262–3271

Holland WL, Scherer PE (2009) PAQRs: a counteracting force to ceramides? Mol Pharmacol 75:740–743

Holland WL, Summers SA (2008) Sphingolipids, insulin resistance, and metabolic disease: new insights from *in vivo* manipulation of sphingolipid metabolism. Endocr Rev 29:381–402

Holland WL, Brozinick JT, Wang LP, Hawkins ED, Sargent KM, Liu Y, Narra K, Hoehn KL, Knotts TA, Siesky A, Nelson DH, Karathanasis SK, Fontenot GK, Birnbaum MJ, Summers SA (2007) Inhibition of ceramide synthesis ameliorates glucocorticoid-, saturated-fat-, and obesity-induced insulin resistance. Cell Metab 593:167–179

Holland WL, Bikman BT, Wang LP, Yuguang G, Sargent KM, Bulchand S, Knotts TA, Shui G, Clegg DJ, Wenk MR, Pagliassotti MJ, Scherer PE, Summers SA (2011) Lipid-induced insulin resistance mediated by the proinflammatory receptor TLR4 requires saturated fatty acid-induced ceramide biosynthesis in mice. J Clin Invest 121:1858–1870

Hoppeler H, Howald H, Conley K, Lindstedt SL, Claassen H, Vock P, Weibel ER (1985) Endurance training in humans: aerobic capacity and structure of skeletal muscle. J Appl Physiol 59:320–327

Hou Q, Huang Y, Amato S, Snyder SH, Huganir RL, Man HY (2008) Regulation of AMPA receptor localization in lipid rafts. Mol Cell Neurosci 38:213–223

Hu W, Ross JS, Geng T, Brice SE, Cowart LA (2011) Differential regulation of dihydroceramide desaturase by palmitate vs. monounsaturated fatty acids: implications to insulin resistance. J Biol Chem 285:35792–35802

Ikeda M, Kihara A, Igarashi Y (2006) Lipid asymmetry of the eukaryotic plasma membrane: functions and related enzymes. Biol Pharm Bull 29:1542–1546

Itani SI, Ruderman NB, Schmieder F, Boden G (2002) Lipid-induced insulin resistance in human muscle is associated with changes in diacylglycerol, protein kinase C, and IkappaB-alpha. Diabetes 51:2005–2011

Ivanova PT, Milne SB, Forrester JS, Brown HA (2004) LIPID arrays: new tools in the understanding of membrane dynamics and lipid signaling. Mol Interv 4:86–96

Johnson JB, Summer W, Cutler RG, Martin B, Hyun DH, Dixit VD, Pearson M, Nassar M, Telljohann R, Maudsley S, Carlson O, John S, Laub DR, Mattson MP (2007) Alternate day calorie restriction improves clinical findings and reduces markers of oxidative stress and inflammation in overweight adults with moderate asthma. Free Radic Biol Med 42:665–674

Jump DB (2004) Fatty acid regulation of gene transcription. Crit Rev Clin Lab Sci 41:41–78

Kabayama K, Sato T, Kitamura F, Uemura S, Kang BW, Igarashi Y, Inokuchi J (2005) TNFalpha-induced insulin resistance in adipocytes as a membrane microdomain disorder: involvement of ganglioside GM3. Glycobiology 15:21–29

Kadoya A, Miyake H, Ohyashiki T (2003) Contribution of lipid dynamics on the inhibition of bovine brain synaptosomal $Na^+$-$K^+$-ATPase activity induced by 4-hydroxy-2-nonenal. Biol Pharm Bull 26:787–793

Kalupahana NS, Claycombe K, Newman SJ, Stewart T, Siriwardhana N, Matthan N, Lichtenstein AH, Moustaid-Moussa N (2010) Eicosapentaenoic acid prevents and reverses insulin resistance in high-fat diet-induced obese mice via modulation of adipose tissue inflammation. J Nutr 140:1915–1922

Kang JX, Weylandt KH (2008) Modulation of inflammatory cytokines by omega-3 fatty acids. Subcell Biochem 49:133–143

Karaman IK, Lobanova EG, Iubitskaia NS (2010) Disturbed synthesis of eicosanoids in patients with metabolic syndrome. Klin Med (Mosk) 88:46–49

Keller JN, Mattson MP (1998) Roles of lipid peroxidation in modulation of cellular signaling pathways, cell dysfunction, and death in the nervous system. Rev Neurosci 9:105–116

Keller JN, Mark RJ, Bruce AJ, Blanc E, Rothstein JD, Uchida K, Wäg G, Mattson MP (1997) 4-Hydroxynonenal, an aldehydic product of membrane lipid peroxidation, impairs glutamate transport and mitochondrial function in synaptosomes. Neuroscience 80:685–696

Kim J, Li Y, Watkins BA (2011) Endocannabinoid signaling and energy metabolism: a target for dietary intervention. Nutrition 27:624–632

Kitatani K, Idkowiak-Baldys J, Hannun YA (2008) The sphingolipid salvage pathway in ceramide metabolism and signaling. Cell Signal 20:1010–1018

Kleinridders A, Schenten D, Könner AC, Belgardt BF, Mauer J, Okamura T, Wunderlich FT, Medzhitov R, Brüning JC (2009) MyD88 signaling in the CNS is required for development of fatty acid-induced leptin resistance and diet-induced obesity. Cell Metab 10:249–259

Knopp RH, Paramsothy P (2006) Oxidized LDL and abdominal obesity: a key to understanding the metabolic syndrome. Am J Clin Nutr 83:1–2

Kornhuber J, Tripal P, Reichel M, Mühle C, Rhein C, Muehlbacher M, Groemer TW, Gulbins E (2010) Functional inhibitors of acid sphingomyelinase (FIASMAs): a novel pharmacological group of drugs with broad clinical applications. Cell Physiol Biochem 26:9–20

Kraveka JM, Li L, Szulc ZM, Bielawski J, Ogretmen B, Hannun YA, Obeid LM, Bielawska A (2007) Involvement of dihydroceramide desaturase in cell cycle progression in human neuroblastoma cells. J Biol Chem 282:16718–16728

Krook A (2006) Can the liver X receptor work its magic in skeletal muscle too? Diabetologia 49(95):819–821

Lancaster GI, Febbraio MA (2011) Adiponectin sphings into action. Nat Med 17:37–38

Lang CH, Dobrescu C, Bagby GJ (1992) Tumor necrosis factor impairs insulin action on peripheral glucose disposal and hepatic glucose output. Endocrinology 130:43–52

Listenberger LL, Han X, Lewis SE, Cases S, Farese RV Jr, Ory DS, Schaffer JE (2003) Triglyceride accumulation protects against fatty acid-induced lipotoxicity. Proc Natl Acad Sci U S A 100:3077–3082

Liu L, Zhang Y, Chen N, Shi X, Tsang B, Yu YH (2007) Upregulation of myocellular DGAT1 augments triglyceride synthesis in skeletal muscle and protects against fat-induced insulin resistance. J Clin Invest 117:1679–1689

Lucero HA, Robbins PW (2004) Lipid rafts-protein association and the regulation of protein activity. Arch Biochem Biophys 426:208–224

Ludwig DS, Majzoub JA, Al-Zahrani A, Dallal GE, Blanco I, Roberts SB (1999) High glycemic index foods, overeating, and obesity. Pediatrics 103:E26

Lumeng CN, Bodzin JL, Saltiel AR (2007) Obesity induces a phenotypic switch in adipose tissue macrophage polarization. J Clin Invest 117:175–184

Majumdar I, Mastrandrea LD (2012) Serum sphingolipids and inflammatory mediators in adolescents at risk for metabolic syndrome. Endocrine 41:442–449

Manco M, Calvani M, Mingrone G (2004) Effects of dietary fatty acids on insulin sensitivity and secretion. Diabetes Obes Metab 6:402–413

Maresz K, Carrier EJ, Ponomarev ED, Hillard CJ, Dittel BN (2005) Modulation of the cannabinoid CB2 receptor in microglial cells in response to inflammatory stimuli. J Neurochem 95:437–445

Mark RJ, Lovell MA, Markesbery WR, Uchida K, Mattson MP (1997) A role for 4-hydroxynonenal, an aldehydic product of lipid peroxidation, in disruption of ion homeostasis and neuronal death induced by amyloid β-peptide. J Neurochem 68:255–264

Marsicano G, Moosmann B, Hermann H, Lutz B, Behl C (2002) Neuroprotective properties of cannabinoids against oxidative stress: role of the cannabinoid receptor CB1. J Neurochem 80:448–456

Marsicano G, Goodenough S, Monory K, Hermann H, Eder M, Cannich A, Azad SC, Cascio MG, Gutiérrez SO, van der Stelt M, López-Rodriguez ML, Casanova E, Schütz G, Zieglgänsberger W, Di Marzo V, Behl C, Lutz B (2003) CB1 cannabinoid receptors and on-demand defense against excitotoxicity. Science 302:84–88

Marwarha G, Dasari B, Prasanthi JR, Schommer J, Ghribi O (2010) Leptin reduces the accumulation of Abeta and phosphorylated tau induced by 27-hydroxycholesterol in rabbit organotypic slices. J Alzheimers Dis 19:1007–1019

Matias I, Petrosino S, Racioppi A, Capasso R, Izzo AA, Di Marzo V (2008) Dysregulation of peripheral endocannabinoid levels in hyperglycemia and obesity: effect of high fat diets. Mol Cell Endocrinol 286:S66–S78

Mattson MP (2009) Roles of the lipid peroxidation product 4-hydroxynonenal in obesity, the metabolic syndrome, and associated vascular and neurodegenerative disorders. Exp Gerontol 44:625–633

McConkey DJ (1996) The role of calcium in the regulation of apoptosis. Scanning Microsc 10:777–793

McLarnon JG, Choi HB, Lue LF, Walker DG, Kim SU (2005) Perturbations in calcium-mediated signal transduction in microglia from Alzheimer's disease patients. J Neurosci Res 81:426–435

Mechoulam R, Spatz M, Shohami E (2002) Endocannabinoids and neuroprotection. Sci STKE 129:RE5

Merrill AH Jr (2002) De novo sphingolipid biosynthesis: a necessary, but dangerous, pathway. J Biol Chem 277(29):25843–25846

Micale V, Mazzola C, Drago F (2007) Endocannabinoids and neurodegenerative diseases. Pharmacol Res 56:382–392

Montell E, Turini M, Marotta M, Roberts M, Noé V, Ciudad CJ, Macé K, Gómez-Foix AM (2001) DAG accumulation from saturated fatty acids desensitizes insulin stimulation of glucose uptake in muscle cells. Am J Physiol Endocrinol Metab 280:E229–E237

Montuschi P, Barnes P, Roberts LJ II (2007) Insights into oxidative stress: the isoprostanes. Curr Med Chem 14:703–717

Morrison CD, Morton GJ, Niswender KD, Gelling RW, Schwartz MW (2005) Leptin inhibits hypothalamic Npy and Agrp gene expression via a mechanism that requires phosphatidylinositol 3-OH-kinase signaling. Am J Physiol Endocrinol Metab 289:E1051–E1057

Mozaffarian D, Cao H, King IB, Lemaitre RN, Song X, Siscovick DS, Hotamisligil GS (2010) Circulating palmitoleic acid and risk of metabolic abnormalities and new-onset diabetes. Am J Clin Nutr 92:1350–1358

Muoio DM, Newgard CB (2006) Obesity-related derangements in metabolic regulation. Annu Rev Biochem 75:367–401

Murakami M (2011) Lipid mediators in life sciences. Exp Anim 60:7–20

Murakami H, Tamasawa N, Matsui J, Yamato K, JingZhi G, Suda T (2001) Plasma levels of soluble vascular adhesion molecule-1 and cholesterol oxidation product in type 2 diabetic patients with nephropathy. J Atheroscler Thromb 8:21–24

Nahle Z, Hsieh M, Pietka T, Coburn CT, Grimaldi PA, Zhang MQ, Das D, Abumrad NA (2008) CD36-dependent regulation of muscle FoxO1 and PDK4 in the PPAR delta/beta-mediated adaptation to metabolic stress. J Biol Chem 283:14317–14326

Nakada Y, Ogino M, Asano K, Aoki K, Miki H, Yamamoto T, Kato K, Masago M, Tamura N, Shimada M (2010) Novel acyl coenzyme A: diacylglycerol acyltransferase 1 inhibitors-synthesis and biological activities of N-(substituted heteroaryl)-4-(substituted phenyl)-4-oxobutanamides. Chem Pharm Bull(Tokyo) 58:673–679

Neschen S, Morino K, Rossbacher JC, Pongratz RL, Cline GW, Sono S, Gillum M, Shulman GI (2006) Fish oil regulates adiponectin secretion by a peroxisome proliferator-activated receptor-gamma-dependent mechanism in mice. Diabetes 55:924–928

Neschen S, Morino K, Dong J, Wang-Fischer Y, Cline GW, Romanelli AJ, Rossbacher JC, Moore IK, Regittnig W, Munoz DS, Kim JH, Shulman GI (2007) n-3 Fatty acids preserve insulin sensitivity in vivo in a peroxisome proliferator-activated receptor-alpha-dependent manner. Diabetes 56:1034–1041

Nogueiras R, Rohner-Jeanrenaud F, Woods SC, Tschöp MH (2008) The endocannabinoid system and the control of glucose homeostasis. J Neuroendocrinol 20(Suppl 1):147–151

Novgorodtseva TP, Karaman YK, Zhukova NV, Lobanova EG, Antonyuk MV, Kantur TA (2011) Composition of fatty acids in plasma and erythrocytes and eicosanoids level in patients with metabolic syndrome. Lipids Health Dis 10:82

Opara EC, Garfinkel M, Hubbard VS, Burch WM, Akwari OE (1994) Effect of fatty acids on insulin release: role of chain length and degree of unsaturation. Am J Physiol 266(4 Pt 1): E635–E639

Othman A, Rütti MF, Ernst D, Saely CH, Rein P, Drexel H, Porretta-Serapiglia C, Lauria G, Bianchi R, von Eckardstein A, Hornemann T (2012) Plasma deoxysphingolipids: a novel class of biomarkers for the metabolic syndrome? Diabetologia 55:421–431

Ozcan L, Ergin AS, Lu A, Chung J, Sarkar S, Nie D, Myers MG Jr, Ozcan U (2009) Endoplasmic reticulum stress plays a central role in development of leptin resistance. Cell Metab 9:35–51

Paz K, Hemi R, LeRoith D, Karasik A, Elhanany E, Kanety H, Zick Y (1997) A molecular basis for insulin resistance. Elevated serine/threonine phosphorylation of IRS-1 and IRS-2 inhibits their binding to the juxtamembrane region of the insulin receptor and impairs their ability to undergo insulin-induced tyrosine phosphorylation. J Biol Chem 272:29911–29918

Perseghin G, Scifo P, De Cobelli F, Pagliato E, Battezzati A, Arcelloni C, Vanzulli A, Testolin G, Pozza G, Del Maschio A, Luzi L. (1999) Intramyocellular triglyceride content is a determinant of in vivo insulin resistance in humans: a $^{1}H$-$^{13}C$ nuclear magnetic resonance spectroscopy assessment in offspring of type 2 diabetic parents. Diabetes 48:1600–1606

Pérez-Matute P, Marti A, Martínez JA, Fernández-Otero MP, Stanhope KL, Havel PJ, Moreno-Aliaga MJ (2005) Eicosapentaenoic fatty acid increases leptin secretion from primary cultured rat adipocytes: role of glucose metabolism. Am J Physiol Regul Integr Comp Physiol 288:R1682–R1688

Phillis JW, Horrocks LA, Farooqui AA (2006) Cyclooxygenases, lipoxygenases, and epoxygenases in CNS: their role and involvement in neurological disorders. Brain Res Rev 52:201–243

Pinton P, Ferrari D, Rapizzi E, Di Virgilio F, Pozzan T, Rizzuto R (2001) The Ca2+ concentration of the endoplasmic reticulum is a key determinant of ceramide-induced apoptosis: significance for the molecular mechanism of Bcl-2 action. EMBO J 20:2690–2701
Plata-Salaman CR (1991) Insulin in the cerebrospinal fluid. Neurosci Biobehav Rev 15(2): 243–258
Portois L, Hacquebard M, Malaisse WJ, Carpentier YA (2009) The metabolic syndrome of omega3-depleted rats. III. Brain phospholipids. Int J Mol Med 24:269–278
Powell DJ, Hajduch E, Kular G, Hundal HS (2003) Ceramide disables 3-phosphoinositide binding to the pleckstrin homology domain of protein kinase B (PKB)/Akt by a PKCzeta-dependent mechanism. Mol Cell Biol 23:7794–7808
Prasanthi JRP, Schommer E, Thomasson S, Thompson A, Ghribi O (2008) Regulation of beta-amyloid levels in the brain of cholesterol-fed rabbit, a model system for sporadic Alzheimer's disease. Mech Ageing Dev 129:649–655
Prasanthi JR, Huls A, Thomasson S, Thompson A, Schommer E, Ghribi O (2009) Differential effects of 24-hydroxycholesterol and 27-hydroxycholesterol on beta-amyloid precursor protein levels and processing in human neuroblastoma SH-SY5Y cells. Mol Neurodegener 4:1
Prentki M, Murthy Madiraju SR (2008) Glycerolipid metabolism and signaling in health and disease. Endocr Rev 29:647–676
Resjo S, Goransson O, Harndahl L, Zolnierowicz S, Manganiello V, Degerman E (2002) Protein phosphatase 2A is the main phosphatase involved in the regulation of protein kinase B in rat adipocytes. Cell Signal 14:231–238
Riccardi G, Rivellese AA (2000) Dietary treatment of the metabolic syndrome – the optimal diet. Br J Clin Nutr 83:S143–S148
Riccardi G, Giacco R, Rivellese AA (2004) Dietary fat, insulin sensitivity and the metabolic syndrome. Clin Nutr 23:447–456
Russell DW (2000) Oxysterol biosynthetic enzymes. Biochim Biophys Acta 1529:126–135
Russell AP, Gastaldi G, Bobbioni-Harsch E, Arboit P, Gobelet C, Dériaz O, Golay A, Witztum JL, Giacobino JP (2003) Lipid peroxidation in skeletal muscle of obese as compared to endurance-trained humans: a case of good vs. bad lipids? FEBS Lett 551:104–106
Salmeron J, Hu FB, Manson JE, Stampfer MJ, Colditz GA, Rimm EB, Willett WC (2001) Dietary fat intake and risk of type 2 diabetes in women. Am J Clin Nutr 73:1019–1026
Samuel VT, Petersen KF, Shulman GI (2010) Lipid-induced insulin resistance: unravelling the mechanism. Lancet 375:2267–2277
Schenk S, Horowitz JF (2007) Acute exercise increases triglyceride synthesis in skeletal muscle and prevents fatty acid-induced insulin resistance. J Clin Invest 117:1690–1698
Schmitz-Peiffer C (2010) Targeting ceramide synthesis to reverse insulin resistance. Diabetes 59:2351–2353
Schmitz-Peiffer C, Biden TJ (2008) Protein kinase C function in muscle, liver, and beta-cells and its therapeutic implications for type 2 diabetes. Diabetes 57:1774–1783
Schroeder F, Petrescu AD, Huang H, Atshaves BP, McIntosh AL, Martin GG, Hostetler HA, Vespa A, Landrock D, Landrock KK, Payne HR, Kier AB (2008) Role of fatty acid binding proteins and long chain fatty acids in modulating nuclear receptors and gene transcription. Lipids 43:1–17
Seet RC, Lee CY, Lim EC, Quek AM, Huang SH, Khoo CM, Halliwell B (2010) Markers of oxidative damage are not elevated in otherwise healthy individuals with the metabolic syndrome. Diabetes Care 33:1140–1142
Shmueli E, Alberti KG, Record CO (1993) Diacylglycerol/protein kinase C signalling: a mechanism for insulin resistance? J Intern Med 234:397–400
Shulman GI (2000) Cellular mechanisms of insulin resistance. J Clin Invest 106:171–176
Silva LC, Futerman AH, Prieto M (2009) Lipid raft composition modulates sphingomyelinase activity and ceramide-induced membrane physical alterations. Biophys J 96:3210–3222
Simons K, Ikonen E (2000) How cells handle cholesterol. Science 290:1721–1726
Simons K, Toomre D (2000) Lipid rafts and signal transduction. Nat Rev Mol Cell Biol 1:31–39
Simopoulos AP (2006) Evolutionary aspects of diet, the omega-6/omega-3 ratio and genetic variation: nutritional implications for chronic diseases. Biomed Pharmacother 60:502–507

Simopoulos AP (2008) The importance of the omega-6/omega-3 fatty acid ratio in cardiovascular disease and other chronic diseases. Exp Biol Med (Maywood) 233:674–688

Singh SP, Niemczyk M, Saini D, Awasthi YC, Zimniak L, Zimniak P (2008) Role of the electrophilic lipid peroxidation product 4-hydroxynonenal in the development and maintenance of obesity in mice. Biochemistry 47:3900–3911

Siskind LJ, Kolesnick RN, Colombini M (2002) Ceramide channels increase the permeability of the mitochondrial outer membrane to small proteins. J Biol Chem 277:26796–26803

Sodhi K, Puri N, Inoue K, Falck JR, Schwartzman ML, Abraham NG (2012) EET agonist prevents adiposity and vascular dysfunction in rats fed a high fat diet via a decrease in Bach 1 and an increase in HO-1 levels. Prostaglandins Other Lipid Mediat 98:133–142

Soto C, Estrada LD (2008) Protein misfolding and neurodegeneration. Arch Neurol 65:184–189

Sottero B, Gamba P, Gargiulo S, Leonarduzzi G, Poli G (2009) Cholesterol oxidation products and disease: an emerging topic of interest in medicinal chemistry. Curr Med Chem 16:685–705

Storlien LH, Kraegen EW, Chisholm DJ, Ford GL, Bruce DG, Pascoe WS (1987) Fish oil prevents insulin resistance induced by high-fat feeding in rats. Science 237:885–888

Stratford S, Hoehn KL, Liu F, Summers SA (2004) Regulation of insulin action by ceramide: dual mechanisms linking ceramide accumulation to the inhibition of Akt/protein kinase B. J Biol Chem 279:36608–36615

Su X, Abumrad NA (2009) Cellular fatty acid uptake: a pathway under construction. Trends Endocrinol Metab 20:72–77

Summers SA (2006) Ceramides in insulin resistance and lipotoxicity. Prog Lipid Res 45:42–72

Summers SA, Garza LA, Zhou H, Birnbaum MJ (1998) Regulation of insulin-stimulated glucose transporter GLUT4 translocation and Akt kinase activity by ceramide. Mol Cell Biol 18:5457–5464

Tagami S, Inokuchi JJ, Kabayama K, Yoshimura H, Kitamura F, Uemura S, Ogawa C, Ishii A, Saito M, Ohtsuka Y, Sakaue S, Igarashi Y (2002) Ganglioside GM3 participates in the pathological conditions of insulin resistance. J Biol Chem 277:3085–3092

Tanzi RE, Moir RD, Wagner SL (2004) Clearance of Alzheimer's Abeta peptide: the many roads to perdition. Neuron 43:605–608

Teran-Garcia M, Adamson AW, Yu G, Rufo C, Suchankova G, Dreesen TD, Tekle M, Clarke SD, Gettys TW (2007) Polyunsaturated fatty acid suppression of fatty acid synthase (FASN): evidence for dietary modulation of NF-Y binding to the Fasn promoter by SREBP-1c. Biochem J 402:591–600

Tezapsidis N, Johnston JM, Smith MA, Ashford JW, Casadesus G, Robakis NK, Wolozin B, Perry G, Zhu X, Greco SJ, Sarkar S (2009) Leptin: a novel therapeutic strategy for Alzheimer's disease. J Alzheimers Dis 16:731–740

Tilg H, Moschen AR (2008) Inflammatory mechanisms in the regulation of insulin resistance. Mol Med 14:222–231

Tillman TS, Cascio M (2003) Effects of membrane lipids on ion channel structure and function. Cell Biochem Biophys 38:161–190

Tsai IJ, Croft KD, Mori TA, Falck JR, Beilin LJ, Puddey IB, Barden AE (2009) 20-HETE and F2-isoprostanes in the metabolic syndrome: the effect of weight reduction. Free Radic Biol Med 46:263–270

Unger J, McNeill TH, Moxley RT, White M, Moss A, Livingston JN (1989) Distribution of insulin receptor-like immunoreactivity in the rat forebrain. Neuroscience 31:143–157

Urakawa H, Katsuki A, Sumida Y, Gabazza EC, Murashima S, Morioka K, Maruyama N, Kitagawa N, Tanaka T, Hori Y, Nakatani K, Yano Y, Adachi Y (2003) Oxidative stress is associated with adiposity and insulin resistance in men. J Clin Endocrinol Metab 88:4673–4676

Uysal KT, Wiesbrock SM, Marino WM, Hotamisligil GS (1997) Protection from obesity-induced insulin resistance in mice lacking TNF-α function. Nature 389:610–614

van Gaal LF, Mertens IL, De Block CE (2006) Mechanisms linking obesity with cardiovascular disease. Nature 444:875–880 van Schothorst EM, Flachs P, Franssen-van Hal NL, Kuda O, Bunschoten A, Molthoff J, Vink C, Hooiveld GJ, Kopecky J, Keijer J (2009) Induction of lipid oxidation by polyunsaturated fatty acids of marine origin in small intestine of mice fed a high-fat diet. BMC Genomics 10:110

Vincent HK, Powers SK, Dirks AJ, Scarpace PJ (2001) Mechanism for obesity-induced increase in myocardial lipid peroxidation. Int J Obes Relat Metab Disord 25:378–388

Wang Z, Dou X, Gu D, Shen C, Yao T, Nguyen V, Braunschweig C, Song Z (2012) 4-Hydroxynonenal differentially regulates adiponectin gene expression and secretion via activating PPARγ and accelerating ubiquitin-proteasome degradation. Mol Cell Endocrinol 349:222–231

Xu B, Chen S, Luo Y, Chen Z, Liu L, Zhou H, Chen W, Shen T, Han X, Chen L, Huang S (2011) Calcium signaling is involved in cadmium-induced neuronal apoptosis via induction of reactive oxygen species and activation of MAPK/mTOR network. PLoS One 6(4):e19052

Yamaji-Hasegawa A, Tsujimoto M (2006) Asymmetric distribution of phospholipids in biomembranes. Biol Pharm Bull 29:1547–1553

Yamashita T, Hashiramoto A, Haluzik M, Mizukami H, Beck S, Norton A, Kono M, Tsuji S, Daniotti JL, Werth N, Sandhoff R, Sandhoff K, Proia RL (2003) Enhanced insulin sensitivity in mice lacking ganglioside GM3. Proc Natl Acad Sci U S A 100:3445–3449

Yanai H, Tomono Y, Ito K, Furutani N, Yoshida H, Tada N (2007) Diacylglycerol oil for the metabolic syndrome. Nutr J 6:43–45

Yang H-C, Farooqui AA, Horrocks LA (1994) Effects of sialic acid, sialoglycoconjugates on cytosolic phospholipases $A_2$ from bovine brain. Biochem Biophys Res Commun 199:1158–1166

Yang G, Badeanlou L, Bielawski J, Robert AJ, Hannun YA, Samad F (2009) Central role of ceramide biosynthesis in body weight regulation, energy metabolism, and the metabolic syndrome. Am J Physiol Endocrionol Metab 297:E211–E214

Yaqoob P, Shaikh SR (2010) The nutritional and clinical significance of lipid rafts. Curr Opin Clin Nutr Metab Care 13:156–166

Yung LM, Wei Y, Qin T, Wang Y, Smith CD, Waeber C (2012) Sphingosine kinase 2 mediates cerebral preconditioning and protects the mouse brain against ischemic injury. Stroke 43:199–204

Zajchowski LD, Robbins SM (2002) Lipid rafts and little caves – compartmentalized signalling in membrane microdomains. Eur J Biochem 269:737–752

Zarrouki B, Soares AF, Guichardant M, Lagarde M, Géloën A (2007) The lipid peroxidation end-product 4-HNE induces COX-2 expression through p38MAPK activation in 3T3-L1 adipose cell. FEBS Lett 581:2394–2400

Zhang X, Zhang G, Zhang H, Karin M, Bai H, Cai D (2008a) Hypothalamic IKKbeta/NF-kappaB and ER stress link overnutrition to energy imbalance and obesity. Cell 135:61–73

Zhang M, Martin BR, Adler MW, Razdan RK, Ganea D, Tuma RF (2008b) Modulation of the balance between cannabinoid CB(1) and CB(2) receptor activation during cerebral ischemic/reperfusion injury. Neuroscience 152:753–760

Zhao H, Przybylska M, Wu IH, Zhang J, Siegel C, Komarnitsky S, Yew NS, Cheng SH (2007) Inhibiting glycosphingolipid synthesis improves glycemic control and insulin sensitivity in animal models of type 2 diabetes. Diabetes 56:1210–1218

Zhao Y, Yuan Z, Liu Y, Xue J, Tian Y, Liu W, Zhang W, Shen Y, Xu W, Liang X, Chen T (2010) Activation of cannabinoid CB2 receptor ameliorates atherosclerosis associated with suppression of adhesion molecules. J Cardiovasc Pharmacol 55:292–298

Zhou Y, Lekic T, Fathali N, Ostrowski RP, Martin RD, Tang J, Zhang JH (2010) Isoflurane posttreatment reduces neonatal hypoxic-ischemic brain injury in rats by the sphingosine-1-phosphate/phosphatidylinositol-3-kinase/Akt pathway. Stroke 41:1521–1527

# Chapter 5
# Molecular Aspects of Obesity and Insulin Resistance in Metabolic Syndrome and Neurological Disorders

## 5.1 Introduction

Obesity is a chronic and multifactorial (interaction of social, behavioral, psychological, metabolic, cellular, and molecular factors) condition caused by an imbalance in energy intake and energy expenditure leading to the pathological growth of adipocytes. Factors related to the cause of obesity include lack of exercise, sedentary lifestyle, accessible energy-dense foods, consumption of fructose-sweetened beverages and refined carbohydrates, and endocrine dysfunction such as hypothyroidism. Other determinants of obesity include increase in television watching, decrease in sleep time (Taheri et al. 2004), consumption of trans fat (Kavanagh et al. 2007), and differences in macronutrient quality (e.g., lower vs. higher glycemic-load of carbohydrates) that might alter metabolism or appetite (Thomas et al. 2007). At the cellular level, obesity is caused not only by the hypertrophy of adipocytes and increase in numbers of adipocytes (hyperplasia) but also through the generation of new adipocytes from precursor cells (Sikaris 2004). Body mass index (BMI) is a simple index of weight for height that is commonly used to classify underweight, overweight, and obesity in adult humans (ages of 18 and 65 years). BMI of 18.5–24.9 is considered as normal. Individuals below BMI of 18.5 are considered as underweight whereas individuals above BMI of 25 are considered as overweight (Fig. 5.1). Obesity (body mass index (BMI) >30) not only increases the risk of many visceral and neurological diseases but also enhances all causes of mortality and reduces life expectancy (Fontaine et al. 2003). Caucasian individuals who have a BMI >40 between the ages of 20 and 29 years are expected to decrease their life expectancy by approximately 6 years and 12 years, respectively (Fontaine et al. 2003; Huffman and Barzilai 2009). BMI is a commonly used indicator of obesity. BMI is associated with an unfavorable lipid profile consisting of elevated triglycerides, total cholesterol, and low-density lipoprotein (LDL) cholesterol and low high-density lipoprotein (HDL) cholesterol in men and women as young as 20 years of age (Denke et al. 1994). In addition, plasma total cysteine (tCys) levels are consistently correlated with BMI, fat mass, and odds of obesity in large human studies

A.A. Farooqui, *Metabolic Syndrome: An Important Risk Factor for Stroke, Alzheimer Disease, and Depression*, DOI 10.1007/978-1-4614-7318-3_5,

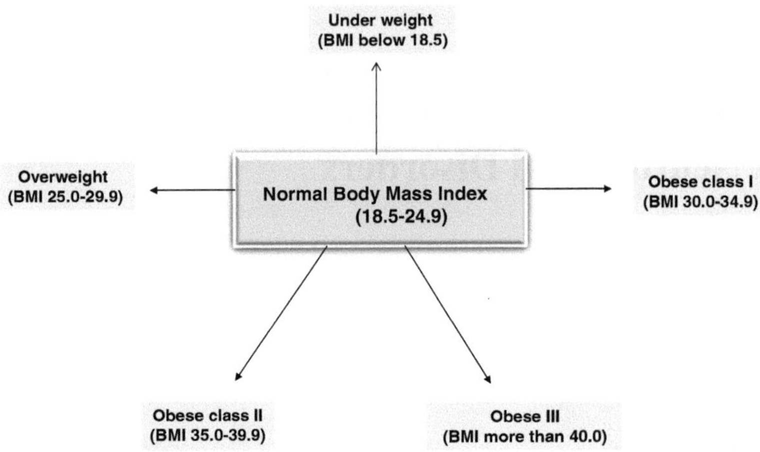


**Fig. 5.1** Relationship between body mass index and weight

(Elshorbagy et al. 2009), and higher consumption of the cysteine precursor methionine is associated with increased BMI and prevalence of diabetes, metabolic syndrome (MetS), and cardiovascular disease (El-Khairy et al. 2001; Virtanen et al. 2006; Giral et al. 2008). High dietary cystine inhibits metabolic rate, lowers insulin sensitivity, and increases visceral fat deposition in conjunction with changes in the expression of several genes involved in lipid and glucose metabolism. Cystine supplementation is known to enhance total fat mass, lean mass, and bone mineral content (BMC), with no effect on bone mineral density (BMD). Fasting plasma tCys and total glutathione (tGSH) are not increased, but levels of plasma taurine are elevated in the cystine-supplemented animals. The mechanism involved in sulfur amino acid-mediated adiposity in human is not fully understood. However, in vitro studies have indicated that in rat adipocytes cysteine not only inhibits lipolysis in a concentration-dependent manner, but high intake of cystine lowers energy expenditure and decreases glucose tolerance, and up-regulates lipogenic and diabetogenic enzymes (Elshorbagy et al. 2012).

Although in the clinical setting BMI is usually used to estimate body fat and to assess risk among adults, the use of BMI has limitations because it does not account for body size and body fat distribution such as abdominal obesity (Michels et al. 1998). There are two types of fats in the human body, namely, subcutaneous fat and visceral (abdominal) fat. Subcutaneous fat is located just underneath the skin, whereas visceral fat is located in the abdomen and surrounding vital organs. Visceral fat can infiltrate the liver and other organs, streak through muscles, and even strangle the heart. Visceral fat is linked with bad cholesterol and hypertension to diabetes, heart disease, and stroke. Subcutaneous fat has low degree of vascularization and is located farther away from the portal vein. The molecular characteristics of visceral fat are different from subcutaneous fat. Thus, visceral fat not only releases free fatty acids, inflammatory cytokines, and other adipose tissue metabolites directly into the portal circulation (Bergman et al. 2006) but also displays increased glucocorticoid receptor activity and inflammatory cytokine secretion and shows reduced secretion of the insulin-sensitizing adipokine, adiponectin (Wajchenberg et al. 2002; Gesta et al. 2007). In contrast, subcutaneous fat releases very little free fatty acids and cytokines. Gene expression profiling studies indicate that there are intrinsic molecular differences between visceral and subcutaneous adipose tissue depots in both humans and mice (Gesta et al. 2006). These studies support the view that there are metabolic and genetic differences between visceral and subcutaneous fats in mice.

A rapid elevation in obesity rate throughout the world has resulted in marked increase in chronic diseases throughout the world. There has been considerable increase in research activities on this growing epidemic. This has led to an extensive analysis of external factors influencing weight gain, such as the density of fast-food restaurants and the prevalence of food-related marketing (WHO 2011a, b). It is proposed that increase in the accessibility of processed energy-dense foods may be responsible for the obesity epidemic. Advanced age is also associated with increase in adiposity, decrease in lean mass, and increase in risk for obesity-related diseases (Sakamoto et al. 1993). In overweight subjects chronic inflammatory and stress responses are a central feature of obesity, insulin resistance, and type II diabetes and contribute to the metabolic imbalance (Hossain et al. 2007). In addition, obesity is commonly associated with vasodilatory dysfunctions (Shankar and Steinberg 2005).

## 5.2 Obesity and Insulin Resistance

Insulin resistance is defined by reduction of its capacity to stimulate glucose utilization, either by insulin deficiency or by impairment in its secretion and/or utilization. Insulin regulates glucose uptake and free fatty acid (FFA) levels in the blood. In adipose tissue, insulin decreases lipolysis, thereby reducing FFA efflux from adipocytes. In liver, insulin inhibits gluconeogenesis by reducing key enzyme activities, and in skeletal muscle insulin predominantly induces glucose uptake by stimulating the translocation of the GLUT4 glucose transporter to the plasma membrane. Insulin resistance is mediated by both genetic and acquired factors. Although very little is known about the genetic causes or predispositions of insulin resistance in prediabetic populations, it is proposed that defects in oxidative metabolism and inherited defects in the basic insulin signaling cascade are closely associated with genetic causes of insulin resistance (Morino et al. 2006, 2008; Thaler and Schwartz 2010; Thaler et al. 2012). It is also stated that genetic component may interact with environmental factors to promote a pronounced pathophysiological abnormality during insulin resistance (Thaler and Schwartz 2010; Thaler et al. 2012). In Western countries, the most common acquired factors that produce insulin resistance are obesity, sedentary lifestyle, and aging. All these processes are interrelated (Mokdad et al. 2003; Hamilton et al. 2007; Thaler and Schwartz 2010). The molecular mechanism associated with insulin resistance is not fully understood. However, accumulation of lipids in the liver is considered to be one of the primary mechanisms involved in

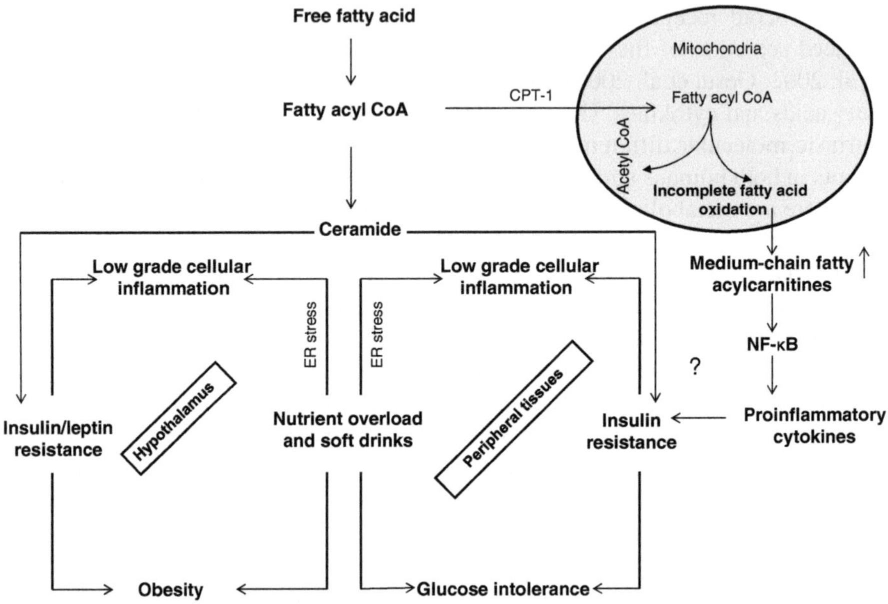


**Fig. 5.2** Interactions between obesity and diabetes. Induction of nutrient-mediated obesity results in cellular inflammation and insulin and leptin resistance in hypothalamus and peripheral tissues leading to type II diabetes and MetS. Carnitine palmitoyltransferase-I (CPT-1)

insulin resistance and type II diabetes. Thus, elevation in triacylglycerol (TAG), diacylglycerol (DAG), acylcarnitines, and ceramide may be closely associated with the molecular mechanism of insulin resistance (Itani et al. 2002; Adams et al. 2004, 2009; Holland et al. 2007) (Fig. 5.2). DAG modulates isoforms of protein kinase C (PKCθ and PKCε), which can regulate insulin-mediated signal transduction via serine phosphorylation of insulin receptor substrate (IRS)-1 (Yu et al. 2002). Ceramides may promote insulin resistance by preventing insulin-stimulated Akt serine phosphorylation, activation, and translocation of Akt to its substrate (Summers et al. 1998). In addition, ceramide initiates inflammatory signaling pathways, leading to the activation of both c-Jun $NH_2$-terminal kinase (JNK) and nuclear factor-κB/inducer of κ kinase (Ruvolo 2003). All these processes have reported to promote the development of insulin resistance (Cai et al. 2005; Chung et al. 2008; Ikonen and Vainio 2005). In the presence of a robust compensatory insulin secretory response to insulin resistance, glucose levels can remain relatively normal. However, the inability of insulin-secreting pancreatic β cells to produce insulin results in type I diabetes, whereas insulin resistance, disturbance in glucose homeostasis, and impairment in glucose tolerance caused by nutrient overload lead to the development of type II diabetes (Stumvoll et al. 2005; Thaler and Schwartz 2010). Insulin resistance may also contribute to the activation of the immune system in essential hypertension due to the formation of methylglyoxal (MG) and uric acid. MG reacts with proteins to form advanced glycation end products (AGEs) and contributes to oxidative stress.

The generation of.methylglyoxal may not only cause alterations in the functioning of the renin–angiotensin system but trigger the production of proinflammatory cytokines which result in the release of more reactive oxygen species (ROS). Generation of uric acid inhibits nitric oxide synthase and reduces the bioavailability of NO resulting in impairment of endothelium-dependent vasodilation (Cheng et al. 2005). Collective evidence suggests that insulin resistance not only contributes to defects in insulin receptor function, abnormalities in insulin signaling, alterations in glucose metabolism, induction of hyperinsulinemia, hyperglycemia, and inflammation but also increases blood pressure (Wang and Jin 2009).

Endoplasmic reticulum (ER) stress may also contribute to the development of insulin resistance in over-nourished or obese rodents. The ER is a well-organized protein-folding machine composed of protein chaperones, proteins that catalyze protein folding, and sensors that detect the presence of misfolded or unfolded proteins (Malhotra and Kaufman 2007). Furthermore, the ER also contains a sensitive surveillance mechanism that prevents misfolded proteins from transiting the secretory pathway. The efficiency of protein-folding reactions depends not only on appropriate environmental and genetic factors but also on metabolic conditions. Conditions that disrupt protein folding threaten cells with decrease in viability and longevity. Accumulation of unfolded proteins in the ER lumen initiates activation of an adaptive signaling cascade known as the unfolded protein response (UPR) (Malhotra and Kaufman 2007). Appropriate adaptation to misfolded protein accumulation in the ER lumen requires regulation at all levels of gene expression including transcription, translation, translocation into the ER lumen, and ER-associated degradation leading to the activation of protective, apoptotic, and inflammatory responses (Malhotra and Kaufman 2007). Several major transducers of the UPR have been identified. These include PKR-like ER kinase (PERK), inositol-requiring enzyme 1 (IRE1), and activating transcription factor 6 (ATF6). The activation of these factors transmits signals from the ER to the cytoplasm or nucleus and activates three pathways: (a) suppression of protein translation to avoid the generation of more unfolded proteins (Harding et al. 2000), (b) induction of genes encoding ER molecular chaperones to facilitate protein folding (Li et al. 2000), and (c) activation of ER-associated degradation (ERAD) to reduce unfolded protein accumulation in the ER (Ng et al. 2000). If these strategies fail, the cells are unable to maintain ER homeostasis and undergo apoptosis due to increase in ER stress (Urano et al. 2000), which activates metabolic pathways that trigger insulin resistance, release of macrophage chemoattractant proteins, and initiate chronic inflammation. The infiltrated macrophages in turn release inflammatory proteins causing further recruitment of macrophages to adipose tissue and the release of inflammatory cytokines. IRE1 also induces an inflammatory signaling cascade by activating IKK, the MAPKs p38 and JNK, and finally the major inflammatory transcription factor NF-κB. Consequently, obesity-induced ER stress leads to insulin receptor substrate (IRS) serine phosphorylation and inhibits insulin signaling (Ozcan et al. 2004, 2006; Xu et al. 2003; Lumeng et al. 2007; Thaler and Schwartz 2010). Furthermore, increase in mitochondrial oxidative phosphorylation due to increase in influx of nutrients increases mitochondrial superoxide ($O_2^{\bullet -}$) production contributing to the pathogenesis of insulin resistance in animal and cellular models (Hoehn et al. 2009). A decrease in number of mitochondria without a concomitant reduction in nutrient uptake can also increase net substrate flux through the remaining mitochondria resulting in elevation in $O_2^{\bullet-}$ production per energy unit. Based on these studies, it is suggested that mitochondrial $O_2^{\bullet-}$ production may represent a link between mitochondrial function and insulin resistance (Hoehn et al. 2009). It is also shown that overexpression of muscle-specific antioxidant enzymes (Anderson et al. 2009) or treatment with the mitochondrial superoxide dismutase (SOD) mimetics (Hoehn et al. 2009) and mitochondria-specific free radical scavengers (Anderson et al. 2009) protects rodents from developing insulin resistance following high-fat overfeeding. Increase in inflammatory response in hypothalamus produces insulin and leptin resistance contributing to the defective food intake both in genetic or dietary fat-induced obesity (De Souza et al. 2005). Collective evidence suggests that mitochondrial oxidative stress, ER stress, intracellular ceramide accumulation, and the induction of JNK, IKK, or PKCθ may contribute to the development of insulin resistance in over-nourished or obese rodents (Savage et al. 2007). It should also be noted that inflammatory response to dietary fat is mediated and supported by ER stress and TLR (Toll-like receptor) signaling, which results in the activation of NF-κB and production of inflammatory cytokines, such as IL-1β, IL-6, and TNF-α (Akira et al. 2006; Hayden and Ghosh 2008).

Most overweight/obese individuals are insulin resistant and they become insulin sensitive with their weight loss supporting the view that there may be a link between insulin resistance and obesity (Reaven 2005). Increase in renin–angiotensin–aldosterone system activities may produce insulin resistance through the stimulation of Ang II type 1 receptors, which trigger increased production of reactive oxygen species (ROS) and reactive nitrogen species (RNS) in adipocytes, skeletal muscle, and cardiovascular tissue of obese individuals (Sharma and Engeli 2006). Insulin resistance is not only associated with intramuscular lipid accumulation but also involves genetic modifications such as muscle-specific overexpression of lipoprotein lipase and fatty acid translocase/CD36 (Ibrahimi et al. 1999; Kim et al. 2001). Insulin resistance is the pathogenetic link among different metabolic abnormalities that occur in hypertension, type II diabetes, and MetS. The pathogenesis of obesity and associated insulin resistance involves overactivation of the endoplasmic reticulum-stress (ER-stress) response signaling (Ozcan et al. 2004, 2006). Whole-body and tissue-specific knockout studies in mice have indicated that protein tyrosine phosphatase 1B (PTP1B) is a major regulator of insulin sensitivity and adiposity (Bence et al. 2006). This process involves the regulation of insulin and leptin signaling pathways in muscle, liver, adipose tissue, and hypothalamus (Delibegovic et al. 2009). In these tissues, PTP1B negatively regulates insulin and leptin signaling contributing to insulin and leptin resistance. The ER-stress response and PTP1B expression are interlinked and this may be one of the mechanisms behind improvement in insulin sensitivity and lower accumulation of lipid in L-PTP1B$^{-/-}$ mice (Agouni et al. 2011).

Insulin resistance can be induced by different environmental factors, including dietary habits. Thus, consumption of energy-dense/high-fat diets is strongly and positively associated with obesity, which negatively effects insulin sensitivity, particularly when the excess of body fat is located in abdominal region. Nevertheless, the link between consumption of high-fat diet and obesity is not limited to the high-energy content of fatty foods. In some individuals, the ability to oxidize and metabolize dietary fat is also impaired, genetically predisposing subjects to obesity. Epidemiological and intervention studies have also indicated that in humans saturated fat significantly worsen insulin-resistance. In contrast, the consumption of monounsaturated and polyunsaturated fatty acids improves insulin resistance through the modification of cell membranes composition. These observations support the view that obesity can be controlled by reducing the intake of energy dense diet with saturated fatty acid (Riccardi et al. 2004). Due to the consumption of energy-dense food, the rate of prevalence of many obesity-related diseases, such as cardiovascular diseases, type II diabetes, MetS, hypertension, dyslipidemia, myocardial infarction, stroke, certain cancers, sleep apnea, osteoarthritis, and neurological disorders, is increasing with an alarming rate (Raffaitin et al. 2011; Flegal et al. 2007; Farooqui et al. 2012). Collective evidence suggests that insulin resistance is not only associated with impaired insulin signaling, impaired fibrinolysis, and inflammation but also with elevated activity of plasminogen activator inhibitor (PAI-1), impaired fibrinolysis, and hypercoagulability. Insulin resistance may result from abnormalities in key molecules of the insulin signaling pathways, including overexpression of phosphatases and downregulation and/or activation of protein kinase cascades (Avramoglu et al. 2006), leading to abnormalities in the expression and action of various cytokines, growth factors, and peptides and overproduction of VLDL (Fonseca et al. 2004).

The prevalence of obesity will rise to approximately 700 million people worldwide in 2015 (WHO 2011a, b). Obesity has become a big public health problem in the United States, with a prevalence among adults increasing to 32 % from 13 % between 1960s and 2004 (Wang and Beydom 2007). Currently, 66 % of adults and 16 % of children and adolescents are overweight or obese (Wang and Beydom 2007). Although, as stated above, many factors (genetic, environmental, socioeconomical, behavioral, lack of physical activity, and psychological factors) modulate the onset of obesity (Fig. 5.3), the main cause of the development of obesity is a positive energy balance, which consists in imbalance between energy intake and expenditure, lasting for several years (Businaro et al. 2012). Such a balance is regulated by a complex network of signal transduction processes that connect the endocrine system with the brain signaling (Straub et al. 2011).

Overnutrition and lack of physical activity, which are known to cause obesity, impair systemic metabolic homeostasis and are metabolic stressors for intracellular organelles, such as the ER. The hypothalamus, a key processing area within the brain through regulatory neuropeptides such as leptin, cholecystokinin, ghrelin, orexin, insulin, neuropeptide Y (NPY) signaling, is recognized as the main brain region that regulates food intake as it relates to caloric and nutrition requirements (Dietrich and Horvath 2009; Blouet and Schwartz 2010). Thus, it is becoming increasingly evident that food intake is regulated by a complex signaling network involving hypothalamic as well as extra-hypothalamic structures in the brain

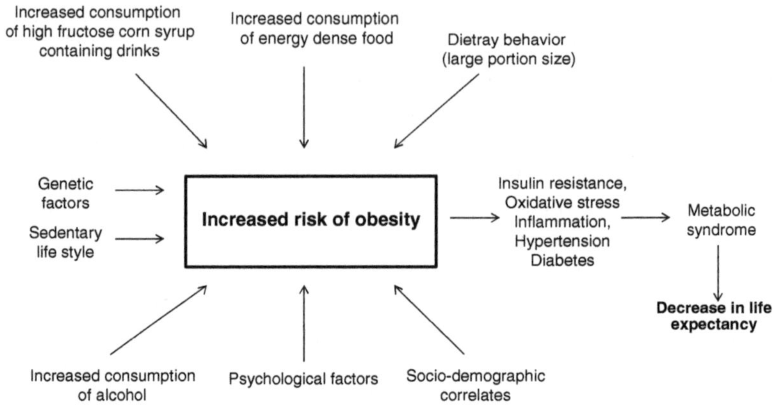


**Fig. 5.3** Risk factors for increased risk of obesity

(Williams et al. 2011). Peripheral signals from adipose stores, the gastrointestinal tract, and endocrine system communicate and modulate neurons within the arcuate nucleus of the hypothalamus. When fat stores are reduced and energy levels are low, hunger signals are induced through elevation in the ghrelin and decrease in levels of insulin, glucose, leptin, and cholecystokinin (CCK). These changes increase the activity of NPY and agouti-related protein (AgRP) neurons, which in turn produce reduction in the melanocortin system, producing an increase in melanin-concentrating hormone (MCH) and orexin (ORX) signaling causing a marked orexigenic effect. High levels of glucose, insulin, and CCK and reduction in levels of ghrelin result in an increase in pro-opiomelanocortin (POMC). POMC increases α-melanocyte-stimulating hormone (α-MSH) leading to termination of feeding signals (Coll et al. 2007; Valassi et al. 2008; Abizaid and Horvath 2008; Schwartz et al. 2000). Thus, insulin and leptin control food intake via regulating POMC and AgRP expression. FOXO1 is a downstream effector of insulin signaling and Sirt1 is an $NAD^+$-dependent deacetylase; both are closely associated with the regulation of metabolism not only in the liver, pancreas, muscle, and adipose tissue but also in the hypothalamus (Sasaki and Kitamura 2010). Both FOXO1 and SIRT1 are expressed in AgRP and POMC neurons (Fig. 5.4). In fasted rats, FOXO1 is localized in the nucleus, whereas in refed rats, it is localized in the cytoplasm. Unlike FOXO1, hypothalamic SIRT1 protein is decreased during fasting due to increase in ubiquitination of SIRT1. In rodents, overexpression of FOXO1 in the hypothalamus through the microinjection of adenovirus produces hyperphagia and body weight gain, and simultaneous overexpression of SIRT1 suppresses these processes (Sasaki and Kitamura 2010). Collective evidence suggests that FOXO1 and the transcription factor STAT3 exert opposing effects on the expression of AgRP and POMC through transcriptional squelching, and SIRT1 suppresses AgRP expression (Sasaki and Kitamura 2010). The macronutrient components of high-fat diet, which promote obesity, may also impact neurotransmitter signaling in the brain. Thus, high-fat obesogenic diet reduces dopamine levels in the nucleus accumbens, as well as shifts the

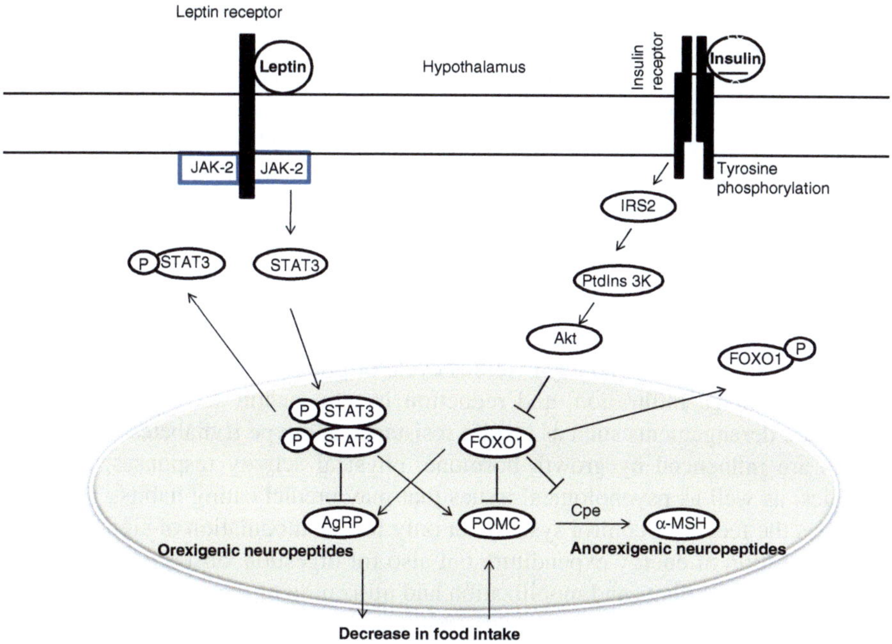


**Fig. 5.4** Modulation of appetite by insulin and leptin signaling in the hypothalamus. α-Melanocyte-stimulating hormone (α-MSH); insulin receptor substrate-2 (IRS2); Janus kinase 2 (JAK2); phosphoinositide 3-kinase (PtdIns 3K); pro-opiomelanocortin (POMC); signal transducer and activator of transcription 3 (STAT3); and carboxypeptidase (Cpe)

reactivity of the mesocorticolimbic system in such a manner that more palatable diet is required to achieve similar food-induced increases in extracellular dopamine as seen in chow-fed controls (Archer and Mercer 2007; Geiger et al. 2008; Volkow et al. 2011). Certain foods, particularly those rich in sugar and fat, are potent reward inducers that promote eating (even in the absence of an energetic requirement) and trigger learned associations between the stimulus and the reward (conditioning) (Lenoir et al. 2007). The stimulation of orosensory system with either sucrose or fat is sufficient to stimulate dopamine release in the nucleus accumbens (Hajnal et al. 2004; Liang et al. 2006). It should be noted that fat and sugar seem to modulate reward systems differently. Sugar produces addictive-like behaviors, whereas fat does not (Avena et al. 2009). Autoradiographic studies reveal an increase in $D_1$ receptor binding in the nucleus accumbens and decrease in $D_2$ receptor binding in the striatum relative to nonpurified diet-fed rats (Colantuoni et al. 2001). Rats with intermittent sugar and nonpurified diet access also show a decrease in $D_2$ receptor mRNA in the nucleus accumbens and increased $D_3$ receptor mRNA in the nucleus accumbens and dorsal striatum compared with nonpurified diet-fed controls (Spangler et al. 2004). Similarly, sugar-bingeing rats show a significant decrease in enkephalin mRNA (Spangler et al. 2004), whereas μ-opioid receptor binding is significantly enhanced in the nucleus accumbens shell, cingulate, hippocampus, and locus coeruleus (Colantuoni et al. 2001).

A major function of adipose tissue is to remove and store circulating lipid, which protects other cells and tissues in the body from the cytotoxic effects of FFA in the circulation. Failure of adipose tissue to effectively remove FFA from the circulation has been reported to contribute to hypertension and atherosclerosis and ultimately the onset and development of diabetes and MetS (Moller and Kaufman 2005). Adipose tissue is dynamically linked with the immune system through specific signaling pathways, with Jun N-terminal kinase and I kappa B kinase β (IKKβ)/nuclear factor-κ-light-chain-enhancer of activated B cells (Olefsky and Glass 2010; Odegaard and Chawla 2011; Zeyda and Stulnig 2007). Chronic overnutrition and lack of physical activity induce a vicious cycle of proinflammatory cytokine secretion, macrophage infiltration, and reduction in adiponectin secretion, leading to metabolic derangements such as insulin resistance and type II diabetes. These processes are influenced by growth hormone, physical activity responses, age, and genetics, as well as psychological issues that may predict eating habits. The brain provides the feedback control system not only for the modulation of signal process and integration of energy expenditure but also for digestion, absorption, transport, and storage of nutrients and mobilization and utilization of fuels. Signals regarding alterations in fuel utilization are tightly regulated and come primarily from adipocytes and from the gastrointestinal tract.

Obesity-mediated inflammatory response involves a number of well-known components associated with classical events involved in inflammatory response to pathogens. These components include (a) systemic enhancement in circulating inflammatory cytokines, adipokines, and acute-phase proteins (C-reactive protein, haptoglobin, and Aβ), (b) recruitment of white blood cells (WBC) to inflamed tissues, (c) activation of tissue WBC, (d) induction of reparative tissue responses, and (e) accumulation of macrophages in adipose tissue. However, the nature of obesity-mediated inflammation is unique in comparison to other inflammatory paradigms of infections and autoimmune diseases (Alam et al. 2012). Several mechanisms may be associated with obesity-mediated inflammation in peripheral tissues and hypothalamus. These mechanisms include the activation of TLR4, induction of endoplasmic reticulum stress, and activation of serine/threonine kinases, such as IKKβ (Thaler and Schwartz 2010; Thaler et al. 2012). Although the relative contribution made by these mechanisms remains unknown, early onset of inflammation in hypothalamus relative to that in peripheral tissues suggests that different processes may cause inflammation in peripheral tissues and hypothalamus. Moreover, the nature of the hypothalamic inflammation occurring during the first days of high-fat diet (HFD) feeding may differ fundamentally from that involved with chronic HFD exposure. Based on several studies, it is proposed that the rapid onset of mediobasal hypothalamus (MBH) inflammation is a manifestation of neuron injury, and associated neuroprotective responses is consistent with previous evidence of apoptosis and glial ensheathment of hypothalamic arcuate nucleus (ARC) neurons in animals exposed to rendered obese b chronic HFD feeding (Horvath et al. 2010; Moraes et al. 2009).

Obesity is closely associated with decrease in life expectancy due to the prevalence of cardiovascular diseases, diabetes, colon cancer, and other chronic neurological diseases. In addition, obesity also has negative impact on cognitive function due to vascular defects, impaired insulin metabolism, and defect in glucose transport mechanisms in the brain. Molecular mechanisms underlying cognitive dysfunction and early death in obesity are not fully understood. However, it is proposed that exaggerated metabolic demands trigger stress responses in adipocytes that may contribute to oxidative stress, chronic inflammation, cell injury, and death (Ozcan et al. 2004). It is also indicated that the disequilibrium between oxygen demand and supply in expanding adipose tissues in obesity can lead to local hypoxia, which may cause adipocyte dysfunction by inducing alterations in adipokine expression and promoting chronic inflammation in adipose tissue (Hosogai et al. 2007). Secretion of adipokines by adipose tissues promotes the production of reactive oxygen species (ROS). Thus, adipose tissues are considered independent factors for the generation of systemic oxidative stress. There are several mechanisms by which obesity produces oxidative stress. One mechanism is the oxidation of fatty acids by mitochondria and peroxisomes, which produces ROS. Another mechanism for ROS production is coupled with over-consumption of oxygen through oxidative phosphorylation in mitochondria (Esposito et al. 2006). Lipid- and carbohydrate-rich diets may also contribute to the generation of ROS because they can alter oxygen metabolism.

Finally, increase of adipose tissue is accompanied by diminished activities of antioxidant enzymes such as superoxide dismutase (SOD), catalase (CAT), and glutathione peroxidase (GPx) (Chrysohoou et al. 2007). In obesity, the immune system produces both superoxide and nitric oxide, which may react together to produce significant amounts of peroxynitrite anion ($ONOO^-$). Peroxynitrite is a potent oxidizing agent that can cause DNA fragmentation and lipid peroxidation (Pacher et al. 2007). Many studies have demonstrated the formation of peroxynitrite in diabetic vasculature, both in experimental models and in humans (Pacher and Szabo 2008; Molnar et al. 2006). Peroxynitrite induces diabetic complications and vascular alterations through activation of the nuclear enzyme poly(ADP-ribose) polymerase (PARP) enzymes. Activated PARP-1 cleaves $NAD^+$ into nicotinamide and ADP-ribose and polymerizes the latter on nuclear acceptor proteins. Peroxynitrite-mediated overactivation of PARP utilizes $NAD^+$ and consequently ATP culminating in cell dysfunction, apoptosis, or necrosis (Virag and Szabo 2002). It is also suggested that peroxynitrite produces its toxic effect through the cytotoxic effects of high glucose (Du et al. 2001).

Studies on oxidative markers in obesity subjects indicate that oxidative damage is associated with increased BMI and percentage of body fat (Vincent and Taylor 2006). Conversely, parameters of antioxidant capacity are inversely related to the amount of body fat and central obesity (Hartwich et al. 2007). Thus, mounting evidence suggests that possible mechanisms of obesity-related oxidative stress include increase in oxygen consumption and subsequent production of ROS derived from the increase in mitochondrial respiration, diminished antioxidant capacity, fatty acid oxidation, lipid oxidizability, and cell injury producing increase in rates of ROS production (Vincent and Taylor 2006; Brown et al. 2009). The increase in obesity-mediated ROS production is due to the induction of proinflammatory cytokines (TNF-α, IL-1, and IL-6), which can lead to further increase in ROS production. In addition, obesity also stimulates nicotinamide adenine dinucleotide phosphate (NADPH) oxidase activity, which contributes to ROS production (Bastard et al. 2006). High concentrations of ROS modulate platelet function through different mechanisms, including reduction in NO bioavailability, calcium mobilization abnormalities, and overexpression of membrane glycoproteins (Krötz et al. 2004). Based on these observations, it is proposed that obesity can lead to stressful proinflammatory state that increases the risk of diabetes along with cardiovascular and neurological diseases decreasing life expectancy (Yudkin et al. 2000; Altintas et al. 2011; Farooqui et al. 2012). These suggestions are also supported by the presence of shorter telomeres (240 bp) in obese women compared to lean women of similar age (Valdes et al. 2005), supporting the view that shorter telomerases may result in shorter age.

Adipose tissue is an active endocrine organ, which secretes numerous bioactive molecules, collectively known as adipocytokines (leptin, adiponectin, and resistin), cytokines (TNF-α, IL-1β, and IL-6), and chemokines (MCP1). These molecules not only regulate carbohydrate and lipid metabolism but also modulate immune function and blood coagulability and serve as blood biomarkers of cardiometabolic risk. Adipocytokines also regulate metabolic and inflammatory processes in an autocrine and paracrine manner (Ahima et al. 2000). Adipocytokines, cytokines, and chemokines contribute to low-grade inflammation and initial adipose macrophage infiltration, which promote impairment in adipocyte insulin signaling and induction of insulin resistance through the interference of insulin/insulin-like growth factor I receptor (IGF-IR) signaling pathways (Jiao et al. 2009). Normally, insulin/IGF-IR signaling is accompanied by Tyr phosphorylation of insulin receptor substrates (IRS) followed by activation of two major downstream pathways: The phosphatidylinositol 3-kinase (PtdIns 3K)–Akt pathway is largely responsible for insulin-mediated glucose uptake and suppression of gluconeogenesis. The other major pathway involves MAPK signaling. This pathway regulates gene expression and is closely associated with cell growth and differentiation. In this pathway, the phosphorylation of IRS on critical Ser residue, as opposed to tyrosine, renders IRS inactive and prone to degradation resulting in a blockade of insulin signaling (Boura-Halfon and Zick 2009; Sun and Liu 2009). Serine phosphorylation is modulated by proinflammatory signaling, such as IKKβ, p38 mitogen-activated protein kinase (MAPK), C-Jun N-terminal kinase (JNK) that are activated by cytokines/chemokines such as TNF-α, interleukin-1β, monocyte chemotactic protein 1, and C-reactive protein produced by activated macrophages. Another important mechanism that contributes to insulin resistance is inactivation of IRS by lipid metabolite-induced endoplasmic reticulum stress and JNK activation (Hotamisligil 2008; Zeyda and Stulnig 2009). Lipid rafts, which are present in striated muscles, adipose tissue, liver, and pancreatic β cells, have also implicated in the development of insulin resistance in obesity (Ikonen and Vainio 2005). This suggestion is supported by two key insulin receptor-mediated signaling pathways, one involving with the small guanosine triphosphatase TC10 and the other utilizing PtdIns 3K. Lipid rafts are clearly important for insulin signaling and may thereby contribute to the insulin resistance during obesity or diabetes. Another line of evidence corroborating the involvement of lipid rafts in obesity is related to findings in Obese Zucker fa/fa rats and ob/ob mice with increased levels of $GM_3$ ganglioside synthase mRNA in their adipose tissues. The addition of $GM_3$ to 3T3-LI adipocytes suppresses insulin-mediated phosphorylation of the insulin receptor, supporting the view that lipid rafts containing $GM_3$ ganglioside are associated with insulin receptor signaling (Inokuchi 2006; Yaqoob 2009). At a later stage, obesity involves massive infiltration and activation of macrophages, which become important source for adipose tissue chemokines and cytokines. Obese people have higher circulating concentrations of many inflammatory mediators (C-reactive protein, tumor necrosis factor-α (TNF-α), interleukin-6 (IL-6), and macrophage migration inhibitory factor (MIF)) than lean human subjects contributing to the development of systemic metabolic dysfunction associated with obesity-related disorders (Yudkin et al. 2000; Egger and Dixon 2011). These inflammatory mediators contribute to insulin resistance and other metabolic disturbances. Blood concentrations of inflammatory mediators are lowered following weight loss. As stated above, excessive energy intake combined with low energy expenditure induces lipid accumulation not only in the adipose tissue but also in the liver, muscle, and other internal organs, predisposing to the development of insulin resistance and metabolic disturbances (Hotamisligil 2003). Although, obesity has long been recognized as an independent risk factor for atherosclerosis, cardiovascular diseases, and diabetes mellitus, recent studies indicate obesity as an important risk factor for MetS and neurological disorders (Farooqui et al. 2012).

## 5.3 Obesity, Adipocytokines, and Development of Insulin Resistance

Gender is an important genetic factor that determines adiposity. Although there is no evidence for significant site- and sex-related differences in early development, a striking difference in total fat mass and regional differences in adipose tissue distribution develops in the childhood. Adult males have approximately 1.3–1.7 times lean body mass than females. Females generally have a characteristic gynecoid body fat distribution, with adipose tissue prominently developing in the subcutaneous depots around the hips and thighs; males in contrast have an android body fat distribution, with fat distributed prominently in the abdominal area (Bottner et al. 2004). This difference is largely regulated by endocrine factors, with critical roles played by sex hormones (Shi and Clegg 2009). Thus, despite having lower fat mass, the prevalence of diabetes and early abnormalities of glucose metabolism is higher in men than in women. Obesity is characterized not only by disturbances in the control of glucose and fat metabolism which is associated with an atherogenic lipid profile in blood but also by deleterious fat accumulation in non-adipose tissues like liver, heart, and muscle. As a result, individuals with obesity suffer from high risk of developing low-grade chronic inflammation and cerebrovascular and cardiovascular complications, myocardial infarction, and kidney failure (Kassi et al. 2011). The fundamental mechanism responsible for insulin resistance in obesity is poorly understood. However, increasing evidence supports the view that a complex proinflammatory state, together with increased lipid accumulation in extra adipose tissues such as the liver and skeletal muscle, may be closely associated with the pathogenesis of insulin resistance. As stated above, cytokines (TNF-α, IL-6), MCP-1, and plasminogen activator inhibitor-1 (PAI-1) play a central role in the development of systemic proinflammatory state observed in obesity (Juhan-Vague et al. 2003; Loskutoff and Samad 1998). PAI-1 is the primary inhibitor of plasminogen activation in vivo and an established risk factor not only for cardiovascular and cerebrovascular diseases but also for obesity-induced insulin resistance, diabetes, and weight gain (De et al. 2005). Increase in activation of proinflammatory pathways such as inhibitor of NF-κB (IκB) kinase/NF-κB (IKK/NF-κB) (Itani et al. 2002) and JNK (Watt et al. 2006) causes an aberrant cascade of cellular events ultimately leading to impaired insulin signaling and skeletal muscle insulin resistance. Activation of both JNK/NF-κB pathways is elevated in skeletal muscle from obese and type II diabetic subjects and is strongly correlated with the severity of insulin resistance (Hirosumi et al. 2002). In fact, the activation of the NF-κB system appears to represent a link between obesity, inflammation of adipose tissue, and insulin resistance (Shoelson et al. 2003). In contrast, reducing intracellular fatty acid metabolite accumulation by enhancing intramuscular triacylglycerol synthesis may improve insulin sensitivity via a reversal in the activation of these proinflammatory pathways.

Among various adipocytokines, rutin, TNF-α, IL-6, and retinol-binding protein 4 induce prohyperglycemic effects, whereas leptin, adiponectin, visfatin, and omertin produce antihyperglycemic effects (Fig. 5.5). Dysregulated secretion of adipocytokines from the adipose tissues in obesity contributes to the insulin resistance in peripheral tissues (muscle, liver) and increased risk for cardiovascular disease. Another mechanism by which obesity leads to insulin resistance in the liver and muscle is via increased intracellular lipid accumulation in these tissues (Kraegen et al. 2001; Lelliott and Vidal-Puig 2004). High intake of *n*-6 PUFAs is associated with obesity. It is shown that feeding mice with a diet containing the *n*-6 PUFA (linoleic acid) during the pregnancy–lactation period leads to obesity in the offspring (Massiera et al. 2003). This effect, however, can be retarded by inclusion of the *n*-3 PUFA (α-linolenic acid) in the diet (Massiera et al. 2003). Based on several studies on different types of fatty acid consumption, it is proposed that a complex interplay among nutrient overload, systemic fatty acids excess, inflammation, endoplasmic reticulum stress, and oxidative stress occurs in obesity and this interplay is closely associated with the pathogenesis of insulin resistance (Hotamisligil 2006). Other important events in obesity are lipolysis in adipocyte and deposition of free fatty acids in muscle and liver ("lipid dumping"), which promote lipotoxicity and induction of fatty liver (hepatosteatosis) (Mittra et al. 2008; Bays et al. 2004).

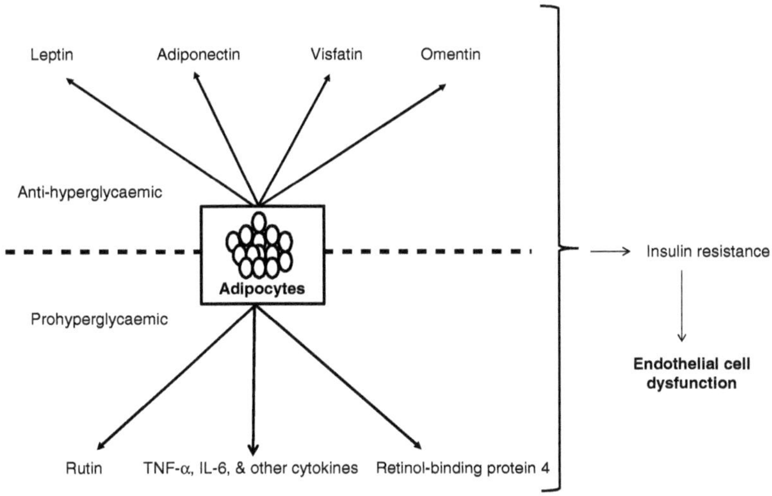


**Fig. 5.5** Release of adipocytokines and cytokines from adipocytes

These fatty acids not only interfere with glucose transporter translocation through alterations in PKCθ activity, but they also downregulate insulin receptor substrate-2 and increase the expression of the lipogenic transcription factor SREBP-1c (Unger 2003). In addition, a number of unsaturated fatty acids bind with the peroxisome proliferator-activated receptor gamma (PPARγ) leading to the translocation of the receptor to the nucleus, where it regulates gene expression. However, not all FFAs contribute to insulin resistance, lipotoxicity, and induction of fatty liver (Ziegler et al. 2001; Bays et al. 2008a, b). For example, *n*-3 fatty acids do not contribute to lipotoxicity and fatty liver development. It is not known also if increase in hepatic TAG content, which is associated with obesity, is simply a marker for increased free fatty acid delivery to the liver or if hepatic TAGs themselves actively participate in creating derangements leading to MetS. Saturated fatty acids produce proinflammatory effects through the activation of Toll-like receptors (TLR) and/or G protein-coupled receptors (GPCR) (Olefsky and Glass 2010). It is hypothesized that saturated fatty acids and metabolites derived from mitochondrial oxidation may be biomarkers that predict inflammatory response and insulin resistance in diet-induced obesity. In addition, elevation in circulating saturated free fatty acids increases the flux of FFAs in liver inducing lipotoxicity, causing impaired β cell function with relative insulinopenia in pancreas and muscle accumulation of intramyocellular lipids (diacylglycerol, fatty acyl CoA, and ceramides) in the muscles contributing to insulin resistance (Bays et al. 2004, 2008a). Insulin resistance in obesity is manifested by decrease in insulin-stimulated glucose transport and metabolism in adipocytes and skeletal muscle and by impaired suppression of hepatic glucose output (Reaven 1995). These functional defects may result from impaired insulin signaling not only in the above-mentioned target tissues but also in adipocytes

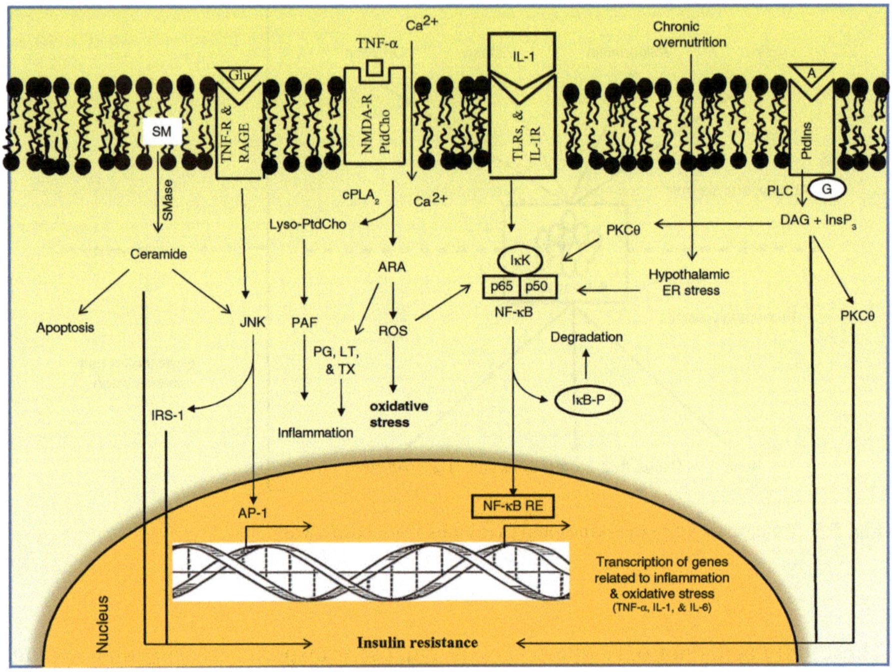


**Fig. 5.6** Degradation of phospholipids and sphingolipids and contribution of transcription factors in the pathogenesis of insulin resistance. Phosphatidylcholine (PtdCho); phosphatidylinositol (PtdIns); sphingomyelin (SM); cytosolic phospholipase $A_2$ (cPLA$_2$); sphingomyelinase (SMase); arachidonic acid (ARA); diacylglycerol (DAG); protein kinase C (PKC); reactive oxygen species (ROS); cyclooxygenase-2 (COX-2); and plasma membrane (PM). Platelet-activating factor (PAF); insulin receptor substrate-1 (IRS1); proinflammatory genes include TNF-α, IL-1β, nitric oxide synthase, cyclooxygenase-2 (COX-2), sPLA$_2$. Ceramide induces apoptosis. Serine/threonine kinase c-Jun N-terminal kinase (JNK) and inhibitor of κB-kinase β (IκKβ) to activate the transcription factors Jun and NF-κB

through downregulation of the major insulin-responsive glucose transporter, GLUT4. In both muscle and adipocytes, insulin binding to its receptor, receptor phosphorylation and tyrosine kinase activity, and phosphorylation of IRSs are also reduced. In addition, infusion of FFA also impairs the ability of insulin to suppress hepatic glucose production and to stimulate glucose uptake into skeletal muscle (Lam et al. 2003; Boden et al. 2005), which in turn may cause insulin resistance. Potential intracellular mechanisms by which FFAs mediate insulin resistance are not fully understood. However, it is reported that PKC-θ, IKKα-, and JNK-knockout mice are substantially protected from FFA-induced insulin resistance (Fig. 5.6) (Kim et al. 2004; Hirosumi et al. 2002). These observations support the view that insulin resistance may be associated with signaling networks that involve suppression of PKCθ, NF-κB, and JNK activities. Detailed investigations are needed on signaling pathways involved in fatty acid–induced insulin resistance.

### 5.3.1 Involvement of Leptin in Obesity

As stated above, adipose tissue is not only a site for the storage of triacylglycerols (TAG) but also acts as endocrine organ and is also associated with production and secretion of adipokines, such as leptin, adiponectin, and resistin. Insulin and leptin are secreted in direct proportion, and adiponectin in negative proportion, to the size of the adipose mass. These three hormones are key molecules in the regulation of lipid metabolism. During states of positive energy balance, as it occurs in obesity, the adipose tissue expands and more leptin and insulin (Baile et al. 2000) but less adiponectin are secreted (Matsubara et al. 2002). Some adipokines induce inflammatory effects, while other adipokines (adiponectin) produce anti-inflammatory effects and reduce obesity by regulating food intake and therefore exerting a direct effect on energy balance and weight control. Adipokines also modulate lipid and glucose/insulin metabolism, oxidative stress, and vascular integrity (Lowell 1999; Berg et al. 2001, 2002). The molecular mechanism by which adipokines regulate glucose metabolism and enhancement in insulin sensitivity may be linked with increase in fatty acid oxidation and glucose uptake. These processes involve the activation of AMP-activated protein kinase (AMPK) and PPARα (Yamauchi and Kamon 2002). Hypoxia (Fukuyama et al. 2007), hypoglycemia (McCrimmon et al. 2004), and adipocytokines, such as leptin and adiponectin (Zhou et al. 2009; Minokoshi et al. 2002), activate AMPK. The activation of AMPK is associated with numerous energy-conserving cellular processes, such as promotion of glucose uptake and glucose transportation (Xiao et al. 2010); stimulation of glycogen, cholesterol, fatty acid, triacylglycerol synthesis and glucose and fatty acid oxidation (Hardie and Sakamoto 2006); and acceleration of mitochondrial biogenesis (Kukidome et al. 2006). In addition to its roles in energy homeostasis, AMPK is also involved in a protective role in atherosclerosis (Zang et al. 2006) and inflammation (Jeong et al. 2009) as well as some beneficial angiogenic effects (Li et al. 2008).

AMPK is a heterotrimeric protein that consists of a catalytic α subunit and the regulatory β and γ subunits. It is associated with the response to energetic stress (Hardie 2007), because of its sensitivity to increased cellular AMP levels (Corton et al. 1994). Pharmacological activation of AMPK (primarily via the AMP analog ZMP) enhances catabolic processes such as GLUT4 translocation (Koistinen et al. 2003), glucose uptake (Bergeron et al. 1999), long-chain fatty acid uptake (Shearer et al. 2004), and substrate oxidation (Bergeron et al. 1999; Atkinson et al. 2002) Activation of AMPK in muscles involves leptin (a 167 amino acid polypeptide hormone), which is a product of the *ob* gene (Zhang et al. 1994). Leptin is almost exclusively expressed and produced by white adipose tissue, specifically by differentiated adipocytes (Ahima et al. 2000). Levels of leptin in circulation correlate with both the BMI and total amount of body fat (Frederich et al. 1995). Increase in circulating leptin is a marker for leptin resistance and is closely associated with obesity. Leptin also regulates immune function by modulating starvation-induced immunosuppression (Tanaka et al. 2010, 2011). Deficient leptin signaling impairs cellular responses, whereas immune- and malnutrition-related diseases are associated with increased synthesis of leptin and of inflammatory cytokines (Kiguchi et al. 2009). Consumption of high-fat diet and fructose, which does not initiate insulin secretion, lowers levels of circulating leptin resulting in overeating and weight gain. Leptin regulates appetite, body weight, energy balance, and insulin sensitivity by interacting with transmembrane leptin receptors (Ob-Ra and Ob-Rb). These receptors are highly expressed in the hippocampus, neocortex, hypothalamic and brain stem autonomic nuclei, choroid plexus, and microvessels. In choroid plexus, Ob-Ra and Ob-Rb play a role in leptin uptake or efflux from the cerebrospinal fluid and in receptor-mediated transport of leptin across the blood–brain barrier into the brain (Tartaglia et al. 1995; Bjørbæk et al. 1998; Oral et al. 2002). The binding of leptin to its receptors in the brain stimulates the production of pro-opiomelanocortin (POMC). The two products of POMC are α-melanocyte-stimulating hormone (α-MSH) and adrenocorticotropin (ACTH). α-MSH binds to melanocortin-4 receptors in the hypothalamic paraventricular nucleus that cause a decrease in food intake.

Leptin receptors contain three tyrosine residues within its cytosolic domain. Interactions of leptin with its receptors cross-phosphorylate and activate the Janus kinases (JAK1, JAK2, JAK3, and TYK2), which in turn phosphorylate tyrosine residues in cytosolic domain (Ghilardi and Skoda 1997). This JAK2 activation mediates tyrosine phosphorylation at 985 and 1,138 amino acid residue of ObRb (Bjorbaek and Kahn 2004; Banks et al. 2000). This phosphorylation provides binding motifs for Src homology 2 (SH2) domain-containing proteins such as signal transducer and activator of transcription 3 (STAT3) and SH2 domain–phosphotyrosine phosphatase (SHP2) (Bjorbaek et al. 1999). STAT3 proteins interact with Y1138, become tyrosine phosphorylated by JAK2, then dissociate and form dimers in the cytoplasm, finally migrating to the nucleus to regulate gene transcription (Li 2008). STAT3 activation is closely associated with the regulation of body weight by leptin, as specific knockout (KO) of the Y1138 residue of ObRb in mice produces severe obesity (Bates et al. 2003, 2005). Tyrosine 985 of Ob-Rb binds SHP-2 and mediate the activation of extracellular signal-regulated kinase (ERK)1/2 pathway and induction of c-Fos expression in transfected cells (Banks et al. 2000; Bjorbaek et al. 1999). Although the cellular and molecular mechanism by which the JAK2/STAT3 pathway is involved in neuronal function remains unknown. It is proposed that JAK can regulate the expression or function of several neurotransmitter receptors, including γ-aminobutyric acid (GABA) (Lund et al. 2008), muscarinic acetylcholine (Chiba et al. 2009), *N*-methyl-D-aspartate (NMDA), and α-amino-3-hydroxy-5-methyl-4-isoxazole propionic acid (AMPA) receptors (Mahmoud and Grover 2006; Xu et al. 2008).

Leptin also activates phosphoinositol 3-kinase (PtdIns 3K) in neural and non-neuronal tissues (Harvey et al. 2000; Zhao et al. 2002). The activation of PtdIns 3K is involved in both insulin receptor substrate (IRS)-1 (Niswender et al. 2004) and IRS-2 (Zhao et al. 2002). Evidence also indicates that JAK/PtdIns 3K signaling is important in neurons (Yadav et al. 2005), and leptin-mediated effects in neurons have been shown to be inhibited by JAK inhibitors (Hsu et al. 2006) and PtdIns 3K inhibitors (O'Malley et al. 2005). Leptin also produces serine phosphorylation of

Akt and glycogen synthase kinase 3 but to a lesser extent than insulin, and the combination of these hormones does not produce additive effect (Szanto and Kahn 2000), supporting the view that complex interactions between the leptin and insulin signaling pathways can potentially lead to differential modification of the metabolic and mitotic effects of insulin exerted through IRS-1 and IRS-2 and the downstream kinases that they activate. Furthermore, the ability of leptin to reduce food intake when administered into the ventral tegmental area depends on JAK–STAT3 signaling (Morton et al. 2009), which may represent a level of molecular specificity in relation to the regulation of food intake and food reward, suggesting that manipulation of effector proteins linked to insulin and leptin receptor activation may be capable of altering food intake (Davis et al. 2010).

Leptin levels are modulated by the age. Thus, aged rats show a lower response to leptin-induced Ob-Ra/Ob-Rb receptor signaling than young rats. This is because of decrease in downstream signaling that involves the activation of STAT3 by leptin (Scarpace et al. 2000). In addition, there is also an age-linked decrease in the uptake of leptin by hypothalamic nuclei, which correlates to a decrease in Ob-Ra/Ob-Rb expression (Fernández-Galaz et al. 2001). Another mechanism that may modulate the age-related decline in leptin responsiveness is an increase in the levels of the suppressor of cytokine signaling 3 (SOCS3), which blocks the transduction of the signals induced by interactions of leptin with Ob-R (Bjorbaek et al. 1999; Wang et al. 2000). Thus, elevation in SOCS3 expression has been reported to occur in the aged hypothalamus (Peralta et al. 2002). Similarly, increase in protein tyrosine phosphatase 1B expression may be associated with age-related leptin resistance in the hypothalamus (Morrison et al. 2007). Collective evidence suggests that aged subjects respond to alterations in leptin signaling less than young individuals.

### 5.3.2 Involvement of Adiponectin in Obesity

Adiponectin is produced mainly by adipocytes. It acts through two different types of cellular receptors. The activation of these receptors results in a reduction of insulin resistance as well as in the regulation of many biologic processes including inflammation and immunity (Tilg and Moschen 2006). In obese animals, treatment with adiponectin decreases hyperglycemia and improves insulin sensitivity (Tilg and Moschen 2006). Thus, unlike other adipokines that impair insulin sensitivity and promote vascular diseases, adiponectin possess antidiabetic, anti-atherogenic and anti-inflammatory properties (Kadowaki et al. 2006) not only due to its ability to block the adhesion of monocytes with endothelial cells and transform macrophages into foam cells, but also to inhibit TNF-α expression (Sikaris 2004), decrease CRP levels, and increase nitric oxide (NO) production in endothelial cells (Ouedraogo et al. 2007). Adiponectin, an adipocyte-derived 30 kDa secretory protein, consists of a signal sequence followed by a nonconserved N-terminal domain, 22 collagen repeats, and a C-terminal globular domain (gAd) which is structurally related to TNF-α (Hug and Lodish 2005). In circulation, adiponectin occurs in three forms: a trimer (low-molecular-weight form), a hexamer of medium-molecular-weight form, and a larger multimeric high-molecular-weight form (Pajvani et al. 2003). The high-molecular-weight form is the active form, which has a beneficial role in human and rodents with regard to prevention against atherosclerosis. Adiponectin is decreased in obesity. This decrease in adiponectin levels is modulated by interactions between genetic factors and environmental factors causing obesity and leading to the development of insulin resistance, type II diabetes, MetS, and atherosclerosis (Kadowaki et al. 2008). Consumption of DHA increases the expression of adiponectin in adipose tissues contributing to the insulin-sensitizing and antisteatotic effects of DHA. The increase in adiponectin is also accompanied by DHA-mediated phosphorylation of AMP-activated protein kinase (AMPK), a fuel-sensing enzyme downstream the adiponectin receptor that acts as a gatekeeper of the systemic energy balance by modulating glucose and lipid homeostasis in the adipose, liver, and muscle tissues (Long and Zierath 2006). AMPK responds to changes in the cellular energy state, so when the AMP-to-ATP ratio is increased, this enzyme is phosphorylated and becomes active to restore the energy levels by inhibiting ATP-consuming pathways and activating ATP-producing pathways (Long and Zierath 2006). In addition, adiponectin contributes to insulin sensitivity through the involvement of AMPK-dependent PPARγ activation (Nawrocki et al. 2006). Adiponectin also increases fat oxidation resulting in reduced circulating fatty acid levels and reduction in intramyocellular or liver triacylglycerol content. In addition to its insulin-sensitizing actions, adiponectin has central actions in the regulation of energy homeostasis. In addition, adiponectin inhibits endothelial cell proliferation and migration and possesses antiangiogenesis and antitumor activity (Bråkenhielm et al. 2004). Adiponectin also decreases gluconeogenesis, increases glucose uptake, and stimulates β-oxidation and triacylglycerol clearance (Yamauchi et al. 2001, 2003). Many of these cellular effects of adiponectin are exerted through interactions with two receptor isoforms with seven putative transmembrane domains. These adiponectin receptor 1 (AdipoR1) and adiponectin receptor 2 (AdipoR2) isoforms have distinct distribution patterns in various tissues (Yamauchi et al. 2007; Kadowaki et al. 2008). Overexpression of adiponectin in adipose tissue improves the systemic insulin sensitivity (Kim et al. 2007), whereas loss of function of adiponectin or its receptors produces decrease in insulin sensitivity. Administration of recombinant adiponectin to rodents increases glucose uptake and fat oxidation in muscle, reduces hepatic glucose production, and improves whole-body insulin sensitivity (Yamauchi et al. 2001). Adiponectin transgenic mice show partial amelioration of insulin resistance and diabetes (Yamauchi et al. 2003) and suppression of endogenous glucose production (Combs et al. 2004). In contrast, adiponectin-deficient mice exhibited insulin resistance and glucose intolerance (Nawrocki et al. 2006). Cell culture studies indicate that adiponectin increases both basal and insulin-stimulated glucose uptake by promoting GLUT4 translocation to the cell membrane (Ceddia et al. 2005; Fang et al. 2005; Mao et al. 2006) and increase fatty acid uptake and oxidation (Yoon et al. 2006) through the activation of AMP-activated protein kinase (AMPK), p38-MAPK, and PPARα pathways (Yamauchi et al. 2007; Yoon et al. 2006). Among these pathways activation of the pleiotropic enzyme AMPK is integral to

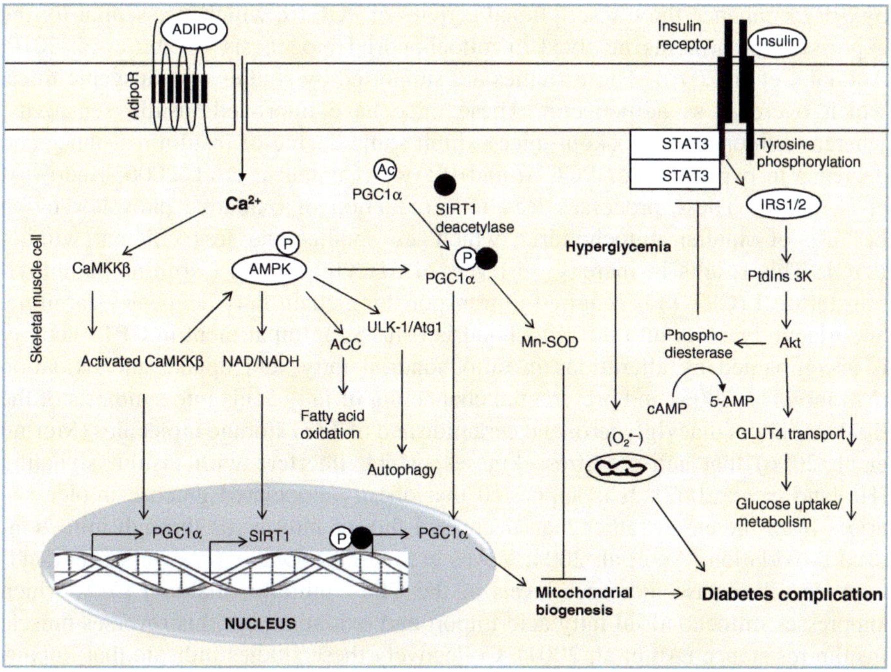


**Fig. 5.7** Modulation of PGC1α by adiponectin and its contribution to biogenesis of mitochondria. Adiponectin (ADIPO); adiponectin receptor (AdipoR); AMP-activated kinase (AMPK); acetyl-CoA carboxylase (ACC); peroxisome proliferator-activated response-gamma coactivator-1α (PGC1α); sirtuin1 (SIRT1); insulin receptor substrate-1/2 (IRS1/2); phosphatidylinositol 3-kinase (PtdIns 3K); protein kinase B (akt); glucose transporter4 (GLUT4); and phosphodiesterase 3B (PDE3B). *Downward arrow* indicates decrease and *upward arrow* indicates increase

the signaling effects of adiponectin (Yamauchi et al. 2007; Yoon et al. 2006). AMPK also activates eNOS in endothelial cells. This provides a link between adiponectin signaling and NO generation (Morrow et al. 2003). Pharmacological AMPK activation also ameliorates the increase in apoptotic cell death in endothelial cells caused by exposure to high glucose (Ido et al. 2002), supporting the view that AMPK may mediate cellular growth and differentiation responses produced by adiponectin in endothelial cells. Collective evidence suggests that adiponectin produces beneficial metabolic effects in skeletal muscle, but the precise underlying molecular mechanisms are not fully understood. It is suggested that adiponectin increases skeletal muscle mitochondrial mass and oxidative capacity by promoting extracellular $Ca^{2+}$ influx and subsequently activating $Ca^{2+}$/calmodulin-dependent protein kinase kinase β (CaMKKβ)–AMPK–SIRT1–peroxisome proliferator-activated receptor gamma coactivator-1α (PGC1α) pathway (Iwabu et al. 2010). In this pathway, the activation of CaMKKβ not only upregulates the expression of PGC1α but also phosphorylates and activates AMPK (Fig. 5.7). Activated AMPK increases the cellular $NAD^+$/NADH ratio, leading to activation of the SIRT1 deacetylase (Iwabu et al. 2010; Canto et al. 2010). Both phosphorylation of PGC1α by AMPK and its deacetylation by SIRT1 enhance the transcriptional activity of PGC1α, which is essential for the expression of proteins involved in mitochondrial biogenesis (Iwabu et al. 2010; Aquilano et al. 2010). These studies are supported by studies on transgenic mice, which overexpress adiponectin. These mice have improved insulin sensitivity, whereas adiponectin knockout mice exhibit some degree of insulin resistance and decrease in expression of PGC1α and PPARγ (Civitarese et al. 2006; Kadowaki et al. 2006). These processes lead to a reduction of oxidative phosphorylation because of smaller mitochondria, which are smaller and less efficient without PGC1. This results in increase in levels of triacylglycerols. Carnitine palmitoyltransferase-I (CPT-1) is required to transport long-chain fatty acids, acylcarnitine, and triacylglycerols into the mitochondria (Fig. 5.2). Impairment in CPT1 activity is accompanied by alterations in mitochondrial fatty acid uptake and oxidation (Morino et al. 2006) and preferential channeling of fatty acids into synthesis of the lipid mediators diacylglycerol and ceramide and of TAG storage molecules (Morino et al. 2006) that activate stress kinases, which interfere with insulin signaling (Holland et al. 2007). It is suggested that obesity-associated glucose intolerance arises from increased rather than decreased flux of fatty acids through mitochondrial β-oxidation (An et al. 2004; Koves et al. 2008). Overexpression of malonyl-CoA decarboxylase increases levels of the CPT1 inhibitor malonyl CoA, which suppresses mitochondrial fatty acid import and oxidation, but this reverses muscle insulin resistance (An et al. 2004). Collectively these studies indicate that impairment in the CPT1 activity may contribute to insulin resistance. Although the molecular mechanism of this process is not fully established, it is proposed that in T2DM tissues decrease in TCA cycle activity causes inefficient long-chain fatty acid β-oxidation that promotes the accumulation of acylcarnitine, which not only activates NFκB-associated pathways to inhibit insulin action but also correlates with impaired muscle insulin sensitivity and glucose metabolism (Noland et al. 2009). Acylcarnitine accumulation, or mitochondrial dysfunction, can be improved with exercise or in rodent models of improved glucose tolerance (Noland et al. 2009; Koves et al. 2005; Adams et al. 2009). This causes an increase of triacylglycerols and acylcarnitines in the cytoplasm. Increases of triacylglycerol in specific tissues such as cardiomyocytes, hepatocytes, and pancreatic β cells may cause lipotoxicity and contribute to the pathogenesis of type II diabetes (Herrara and Lindgren 2010; Kelley et al. 2002; Koves et al. 2008).

Beneficial effects of adiponectins are also supported by increased catabolism of ceramide in skeletal muscle (Holland et al. 2011). Emerging evidence suggests that adiponectin is protective against vascular dysfunction caused by obesity, through its multiple favorable effects on glucose and lipid metabolism as well as on vascular function. Adiponectin not only improves insulin sensitivity but also induces metabolic profiles. These processes decrease risk of heart disease. In addition, adiponectin protects the vasculature through its pleiotropic actions on endothelial cells, endothelial progenitor cells, smooth muscle cells, and macrophages (Li et al. 2011). Mounting evidence suggests that adiponectin is an important component of the adipo-vascular axis that mediates the cross talk between adipose tissue and vasculature. Human cross-sectional studies have indicated that plasma adiponectin levels are negatively correlated with obesity (Hotta et al. 2000), waist-to-hip ratio (Nakamura et al. 2004), insulin resistance (Hotta et al. 2000), dyslipidemia (Rothenbacher et al. 2005), diabetes (Xydakis et al. 2004), and cardiovascular disease (Bjorbaek and Kahn 2004). A low plasma adiponectin level is observed in the MetS (Considine et al. 1996) and is considered as an independent risk factor for future development of type II diabetes (Sader et al. 2003). Although molecular mechanisms involved in above processes are not clearly understood, several mechanisms have been proposed. These mechanisms include (a) sensitization of insulin, which in turn regulates glucose metabolism through stimulation of AMPK, a stress kinase (Yamauchi and Kamon 2002); (b) enhancement in oxidation of muscle fat and glucose transport, which has been shown to cause AMPK activation and acetyl-CoA carboxylase inhibition (Tomas et al. 2002); (c) inhibition of hepatic gluconeogenesis through decrease in the expression of phosphoenolpyruvate carboxylase and glucose-6-phosphatase (Berg et al. 2001; Yamauchi and Kamon 2002); and (d) PPARα activation and reduction in triacylglycerol content of skeletal muscles and liver (Yamauchi et al. 2003, 2007).

## 5.4 Obesity and Insulin Resistance

A metabolic consequence of obesity, especially central adiposity, is the development of insulin resistance, which produces an increase in the secretion of insulin from the pancreas. Increased insulin production leads to compensatory hyperinsulinemia, which may contribute to insulin resistance (Johansen et al. 2010). The development of insulin resistance is one of the earliest negative effects of obesity and is associated with the early alterations in glucose metabolism, chronic inflammation, oxidative stress, and decreased levels of adipose hormone adiponectin and PPARγ, key regulators for adipogenesis (Jazet et al. 2003). Accumulating evidence thus supports the view that a major cause of obesity is insulin resistance. The release of free fatty acids is a critical mechanistic link between obesity and insulin resistance (Cornier et al. 2008). Under normal metabolism, insulin-mediated inhibition of adipocyte hormone-sensitive lipase activity reduces the release of free fatty acids from adipose tissue. This process is disrupted in obese and insulin-resistant individuals, leading to persistent free fatty acid elevations (Craft 2009). Normalizing free fatty acid levels causes 50 % increase in insulin sensitivity in obese adults (Kivipelto and Solomon 2008). Free fatty acids also produce defect in insulin-stimulated glucose transport and/or phosphorylation that is caused by a defect in insulin signaling (Boden and Chen 1995). Plasma free fatty acids can easily enter cells, where they are oxidized to generate energy in the form of ATP or are re-esterified for storage as triglycerides (Fig. 5.8). Furthermore, free fatty acids may also interfere with insulin-mediated stimulation of glucose transport by modulating glucose transporter gene transcription and mRNA stability (Armoni et al. 2005). The metabolic relationship between free fatty acid elevations and development of type

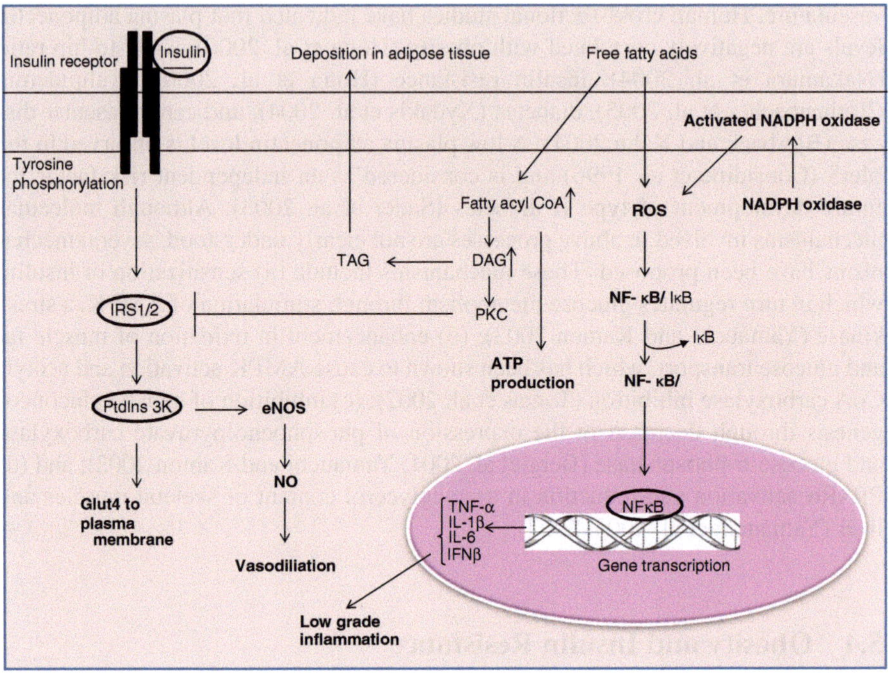


**Fig. 5.8** Potential mechanisms that contribute to free fatty acid-mediated insulin resistance. Insulin receptor substrate 1/2 (IRS1/2); phosphatidylinositol 3-kinase (PtdIns 3K); glucose transporter4 (GLUT4); reactive oxygen species (ROS); diacylglycerol (DAG); triacylglycerol (TAG); endothelial nitric oxide synthase (eNOS); nitric oxide (NO); tumor necrosis factor-α (TNF-α); interleukin-1β (IL-1β); interleukin-6 (IL-6); and nuclear factor-κB (NF-κB)

II diabetes is supported by the observation that normoglycemic individuals with a family history of diabetes show high fasting free fatty acid levels (Whitmer et al. 2005) that increase plasma free fatty acid levels indicating the progression of diabetes (Reaven 2008).

As stated above, obesity is also accompanied by increase in systemic and local levels of inflammatory cytokines (TNF-α, IL-6, IL-10, and MCP-1). These cytokines are released from cells of innate immune system (macrophages, mast cells, and neutrophils). Cytokines not only accumulate in the adipose tissue but also stimulate the release of fatty acids through the stimulation of lipases and phospholipases in obesity. In addition, cells of adaptive immunity (regulatory T cells, $CD8^+$ T cells, and natural killer T cells) also contribute to adipose tissue inflammation in obesity. Studies on the role of adiponectin and leptin provide an important link between obesity and insulin resistance (Tilg and Moschen 2006). The onset of the insulin resistance is accompanied by not only weight gain but also by alterations in levels of adiponectin, leptin, and other proinflammatory cytokines (Muoio and Newguard 2005). As stated earlier, mice lacking adiponectin expression show decrease in insulin sensitivity. They are more likely to suffer from insulin resistance and mice (ob/ob) overexpressing adiponectin exhibit improvement in insulin sensitivity and metabolic derangements (Nawrocki et al. 2006; Kim et al. 2007). In addition, levels of adiponectin are explicitly correlated with fat cell size and are found to be negatively related to BMI (Johnson et al. 2001). Prospective studies on Pima Indian (Arizonian ethnicity having the highest prevalence of obesity associated with insulin resistance and type II diabetes) children (Lillioja et al. 1991) indicate that adiponectin deficiency in obesity causes insulin resistance (Shimada et al. 2004), and maintenance of adiponectin levels improves and restores insulin sensitivity.

## 5.5 Contribution of Oxidative Stress and Inflammation in Obesity

Oxidative stress is a redox-sensitive process with pathogenic outcome caused by an imbalanced ratio between reactive oxygen species (ROS) production and cell's antioxidant capacity. This imbalance causes damage to important biomolecules, such as lipids, proteins, and nucleic acids, with potential impact on the whole body (Halliwell 2006; Duracková 2010). Thus, oxidative stress is a threshold phenomenon characterized by a major increase in the amount of oxidized cellular components. ROS are atoms or molecules possessing one or more unpaired electrons in the outer orbit and therefore are prone to react chemically (Halliwell 2006). ROS include superoxide anions ($O_2^{\bullet-}$), hydroxyl ($^{\bullet}OH$), alkoxyl ($RO^{\bullet-}$), and peroxyl radicals ($ROO^{\bullet}$), and hydrogen peroxide ($H_2O_2$). ROS originate from several sources, including enzymic and non-enzymic oxidation of fatty acids, activation of NADPH oxidase, oxidation of excess of glucose through polyol pathway, and generation of methylglyoxal (Robertson 2004). ROS modulate the transcription factor, NF-κB through the activation of kinases that phosphorylate the inhibitory subunit of NF-κB, I-κB, causing its ubiquitination and release of NF-κB from the NF-κB complex. Free NF-κB migrates from cytosol to the nucleus, where it binds to the κB domain of the target gene promoter, leading to transcriptional activation of many proinflammatory enzymes, cytokines (TNF-α, IL-1β, and IL-10), chemokines, immune receptors, and cell surface adhesion molecules (Li and Stark 2002), which are involved in different cellular processes such as cell proliferation, survival, stress responses, cellular immunity, and inflammation.

As stated earlier, adipose tissue secretes adipokines, which in turn promote the generation of ROS supporting the view that adipose tissue may be an independent factor for the generation of systemic oxidative stress (Vincent and Taylor 2006). There are several mechanisms by which obesity produces oxidative stress. These mechanisms include the mitochondrial and peroxisomal oxidation of fatty acids. Another mechanism of obesity-mediated oxidative stress involves the overconsumption of oxygen, which generates ROS in the mitochondrial respiratory chain that is found coupled with oxidative phosphorylation in mitochondria (Fernandez-Sanchez et al. 2011). High-fat diet and consumption of fructose corn syrup containing soft drinks are also capable of producing ROS because they can stimulate oxygen consumption. Moreover, increase in adipose tissue in obesity significantly diminishes the activity of antioxidant enzymes such as superoxide dismutase, catalase, and glutathione peroxidase (Fernandez-Sanchez et al. 2011). Finally, production of high levels of ROS, induction of proinflammatory cytokines, and the decrease in antioxidant capacity may cause many metabolic abnormalities, such as endothelial dysfunction, which is characterized by a reduction in the bioavailability of vasodilators, particularly nitric oxide (NO), and an increase in endothelium-derived contractile factors, favoring atherosclerotic disease (Fernandez-Sanchez et al. 2011).

It is well known that visceral adipose tissues from experimental models of obesity and MetS produce TNF-α (Borst 2004), which inhibits the synthesis of several other adipocyte-specific proteins including adiponectin and enhances the release of FFA from adipose tissue (Ruan and Lodish 2003). TNF-α interferes with insulin signaling in a variety of non-insulin-producing cells, essentially inducing a state of insulin resistance (del Aguila et al. 1999). Neutralization of TNF-α improves insulin sensitivity by increasing the activity of the insulin receptor tyrosine kinase, specifically in muscle and fat tissues. In addition to its effect on the insulin signaling cascade, TNF-α is also capable of reducing GLUT4 gene expression (Saltiel 2001). At the cellular level, TNF-α is a potent inhibitor of the insulin-stimulated tyrosine phosphorylation on the β-chain of the insulin receptor and insulin receptor substrate-1, supporting the view that there is a defect at or near the tyrosine kinase activity of the insulin receptor. Based on the relationship among obesity, insulin resistance, and diabetes, it is suggested that TNF-α may play a crucial role in the systemic insulin resistance. Furthermore, TNF-α not only stimulates IL-6, which, in turn, promotes the production of C-reactive protein, but also activates plasminogen activator inhibitor-1 and fibrinogen in the hepatocytes (Castro et al. 2003), supporting the view that TNF-α affects various biochemical and physiological processes which may be linked to the etiology of non-insulin-dependent diabetes mellitus with adipocytokines and insulin resistance.

## 5.6 Obesity and Metabolic Syndrome

Obesity may play an important role in the pathogenesis of MetS, a metabolic disorder, which encompasses insulin resistance, hypertension, dyslipidemia, high LDL cholesterol and high triacylglycerol levels, and central obesity (Fig. 5.9). In fact, obesity and insulin resistance are the cornerstones of MetS. Oxidative stress in obesity suppresses mRNA expression and secretion of adiponectin (Furukawa et al. 2004), resulting in reduction in plasma adiponectin levels. Decrease in adiponectin levels is found in cardiovascular diseases and hypertension which are two of the major components of MetS (Kadowaki et al. 2006). Low-grade chronic inflammation in obesity contributes to the development of insulin resistance, glucose intolerance, atherogenesis, and hypertension, parameters that are closely involved in the

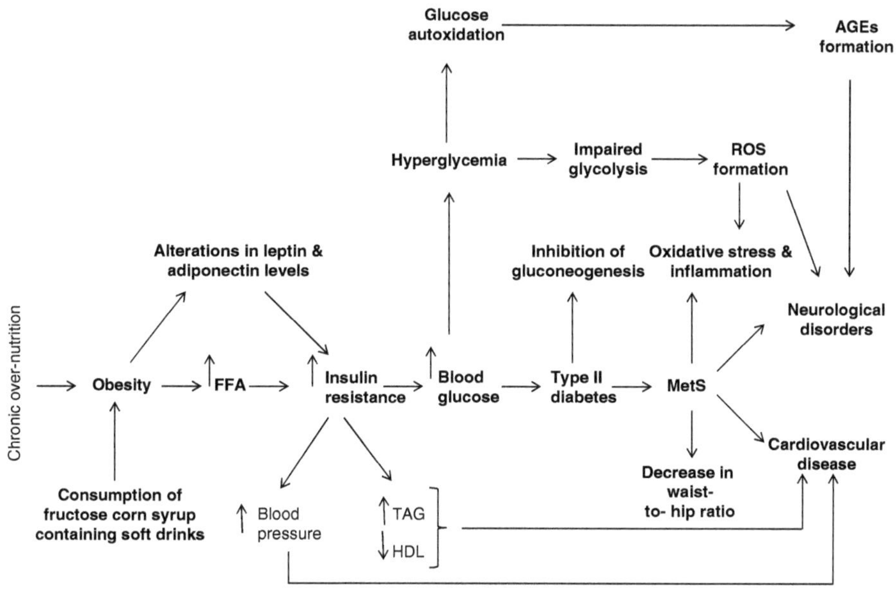


**Fig. 5.9** Effect of chronic overnutrition on pathogenesis of metabolic syndrome. Free fatty acid (FFA); triacylglycerol (TAG); high-density lipoprotein (HDL). *Downward arrow* indicates decrease and *upward arrow* indicates increase

pathogenesis of MetS (Grundy 2003; Moller and Kaufman 2005). Furthermore, inflammatory response (synthesis and release of TNF-α, IL-1β, and IL-6) that is observed in MetS is directly linked to the presence of obesity and seems to stem from the adipose tissue itself (Hotamisligil 2006), supporting the view that obesity contributes to the pathogenesis of MetS. Based on the above information, it is proposed that MetS subjects have significantly increased risk of developing not only metabolic diseases (coronary heart disease, diabetes mellitus, and peripheral vascular diseases) but also neurological disorders (stroke, Alzheimer disease, and depression). Although the molecular mechanisms and signal transduction processes involved in the pathogenesis of metabolic and neurological diseases remain elusive, oxidative stress and chronic inflammation have been implicated as possible causes of cell death in MetS and neurological disorders.

## 5.7 Obesity and Neurological Disorders

Molecular mechanisms associated with the pathogenesis of neurological disorders remain elusive. Causes of neuronal death in neurological disorders include decrease in cellular antioxidant defenses (activities of superoxide dismutase, glutathione peroxidase, catalase, and glutathione reductase), generation of ROS, increase in expression of proinflammatory cytokines, and accumulation of peroxidized lipids, proteins, and DNA oxidative products along with genetic and environmental factors supporting the view that neural cell death in neurological disorders is a multifactorial process involving genetic, environmental, and endogenous factors (Farooqui 2010). Endogenous factors that contribute to neurological disorders include excitotoxicity; oxidative stress; neuroinflammation; abnormal protein dynamics with defective protein degradation and aggregation related to the ubiquitin-proteasomal system resulting in generation and accumulation of misfolded proteins, autoimmunity, and mitochondrial dysfunction resulting in increase in $Ca^{2+}$ levels; and impairment in energy metabolism (Farooqui and Horrocks 2007; Farooqui 2010; Jellinger 2009).

Although, no information is available on the direct link between obesity and neurological disorders, considerable information is available on the effect of obesity on dementia, a major cause of disability, which is clinically defined not only by memory deficits and disturbances of other higher cortical functions along with deterioration in emotional control and social behavior (Sonnen et al. 2009). Two major types of dementia have been identified. Generalized atrophy in the cortical area of the brain results in dementia associated with Alzheimer disease and that due to vascular disorders mainly due to stroke (Farooqui 2010). According to diagnostic and statistical manual of mental disorders, the essential feature of dementia is the development of multiple cognitive deficits that include memory impairment and at least one of the following cognitive disturbances such as aphasia, apraxia, agnosia, or a disturbance in executive function. Worldwide, there is a new case of dementia every 7 s and more than 35 million people are currently estimated to have dementia with more than 4.6 million new cases diagnosed each year (Alzheimer's disease, Senile dementia 2010). The number of people suffering from dementia is expected to be 107 million by the year 2050. Dementia affects more than four million Americans and results in a total health care cost of more than $100 billion annually (Kasper et al. 2005; Chaudhuri et al. 2011).

It is difficult to relate obesity with impairment in cognitive function at the molecular level; however, it is becoming increasingly evident that obesity is closely associated not only with a variety of cardiovascular and cerebrovascular risk factors but also with long-term cognitive performance. In addition, lower cognitive abilities are a risk factor for obesity. Accumulating evidence suggests that cognitive performance may influence the pathogenesis of obesity and being overweight may induce the development of cognitive impairment. Obesity and diabetes increase the risk of dementia (Fig. 5.10), a condition of acquired cognitive defects sufficient to interfere with social or occupational functioning. It affects memory, judgment, speech, comprehension, execution, orientation, and learning (Xu et al. 2004; Qaseem et al. 2008). A major risk factor for dementia is advancing age. After the age of 65, the prevalence and onset of dementia double every 5 years (Alzheimer's Disease International 2010). Although environmental factors are known to contribute to the pathogenesis and development of the dementia syndrome, little is understood about the underlying mechanisms (World Health Organization 2006). Other risk factors such as (a) cardiovascular problems, (b) excessive alcohol consumption, (c) social isolation, (d) head injury, and (e) having one or two copies of

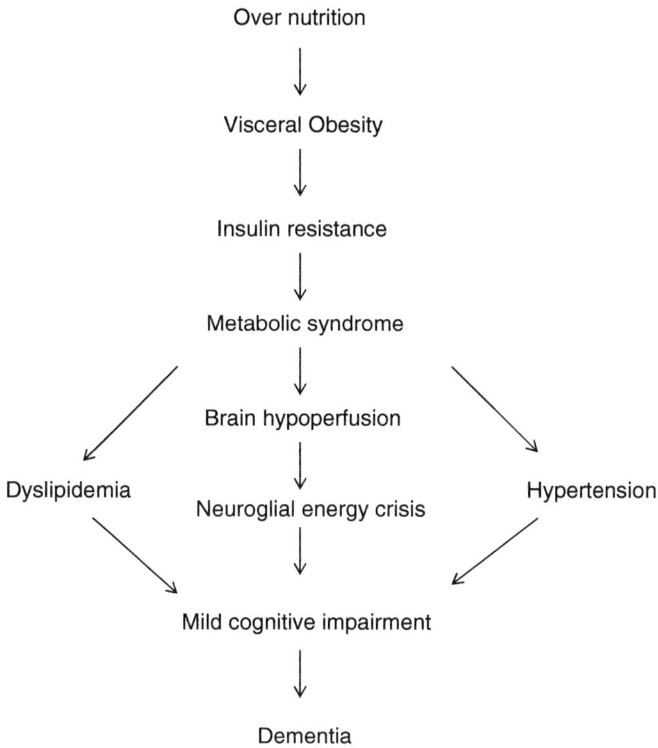


**Fig. 5.10** Obesity increases risk for dementia

the APOEϵ4 genetic variant also contribute to the pathogenesis of dementia syndrome. It is interesting to note that moderate alcohol consumption and estrogen reduce the risk of developing dementia syndrome (Alzheimer Scotland report 2011).

Insulin resistance is present in most patients with diabetes and Alzheimer disease (AD) and is associated with compensatory hyperinsulinemia, which is one of the suggested mechanisms to explain the increased risk of AD in diabetic subjects (Qiu and Folstein 2006). In addition, impairments in the leptin system may also contribute to the link among type II diabetes, obesity, and AD (Morley and Banks 2010), as obesity and diabetes are triggered by development of leptin resistance due to elevated leptin levels. Furthermore, clinical studies have indicated that the circulating leptin levels are significantly lower in AD patients than normal control subjects (Power et al. 2001). There is evidence from prospective studies that there is a lower risk of AD development in individuals with high leptin levels (Holden et al. 2009; Lieb et al. 2009). Moreover, leptin levels are also significantly reduced in murine models (APPSwe; PSIM146V) of AD, supporting the view that impairment in leptin system may be closely associated with the pathogenesis of AD (Fewlass et al. 2004).

Alterations in APP expression or function are involved in AD, which are closely associated with pathophysiology of obesity. It is well known that adipose tissues and adipocyte cell lines (Lee et al. 2008; Sommer et al. 2009) express APP. Furthermore, obese individuals have increased plasma levels of Aβ1–40 (Lee et al. 2008). Increased plasma levels of Aβ1–42 correlate with increased body fat in humans (Balakrishnan et al. 2005; Leahey et al. 2007). It is also demonstrated that mutated APP produces severe hypoleptinemia, decrease in adiposity, and reduction in adipocyte size. Although these effects can be reversed under obesogenic conditions, this reversal is accompanied by mutant APP-associated increases in insulin resistance, elevations in circulating triacylglycerol, and increases in macrophage infiltration of visceral adipose tissue (Freeman et al. 2012). Collective evidence suggests that changes in APP expression or function occur not only in the brain but also in the adipose tissues supporting the view that adipose tissue dysfunction and global changes take place in endocrine and metabolic function in obesity and AD (Puig et al. 2012; Freeman et al. 2012).

Significant reduction in components of the insulin signaling pathway, including IRS-1, PtdIns 3K, and phospho-Akt, is observed in the frontal cortex and hippocampus of autopsied AD brains, but not in control brains (Steen et al. 2005). Patients with diabetes and AD show impaired insulin signaling and accumulation of Aβ along with synaptic disconnection, tau hyperphosphorylation, Aβ oligomerization and accumulation, and impaired energy metabolism (Takeda et al. 2011). Insulin is also associated with reduction of Aβ levels through the involvement of insulin-degrading enzyme (IDE). Since IDE is much more selective for insulin than for Aβ, brain hyperinsulinism may deprive Aβ of its main clearance mechanism (Roriz-Filho et al. 2009). Based on the above information, AD is now considered as a syndrome with brain insulin resistance and insulin deficiency, with significant abnormalities in the expression of genes and activation of kinases that are regulated by insulin and insulin-like growth factor (IGF) signaling (Hoyer 2004; Rivera et al. 2005; Steen et al. 2005). Thus, AD is now considered as "type III diabetes" (de la Monte and Wands 2008; de la Monte 2012).

In diabetes chronic hyperglycemia induces biologic toxicity via glycation of proteins, altered redox potential, altered signal transduction, and generation of reactive oxygen species (McCall 2005). The resulting oxidative stress contributes to vascular/endothelial dysfunction and this microvascular damage may be an important determinant of the observed cognitive deficits. High levels of advanced glycation end product-modified proteins (AGE-proteins) and ROS are present in diabetic and AD patients. RAGEs (receptors for AGE) play a prominent role in pathogenesis of diabetes and neurological disorders (Arab et al. 2011). Thus, diabetic mice (32 % HbA1c vs. 12 % in control mice) with impaired cognitive function show increased expression of RAGEs in neurons and glial cells and damage to white matter and myelin (Toth et al. 2006), supporting the view that RAGEs play an important role in the development of cerebral dysfunction (Yan et al. 1997). In humans, patients with diabetes and Alzheimer disease show greater *N*-carboxymethyllysine (a type of AGE) staining on brain slices obtained postmortem than patients with Alzheimer disease alone (Girones et al. 2004). The hypothesis that AGE-induced brain injury may be a mechanism through which hyperglycemia and diabetes alter cerebral function is also supported by studies in a rat model of focal cerebral ischemia, where infusion of AGEs increases cerebral infarct size and the coadministration of aminoguanidine, an inhibitor of AGE cross-linking, attenuates the infarct volume (Zimmerman et al. 1995).

RAGE also acts as a putative receptor for Aβ (Deane et al. 2003; Nogueira-Machado and Chaves 2008) and transduces an inflammatory response in the brain. High-affinity binding properties between RAGE and Aβ and its long-term presence as immunogenic complex antigen in blood circulation may produce adverse consequences potentially relevant to AD pathogenesis including potentiation of Aβ aggregation, perturbation of neuronal function, elevation of oxidative stress, amyloidosis, augmentation of microglial inflammatory responses, vascular dysfunction, and stimulation of autoimmunity to both Aβ and RAGE (Mruthinti et al. 2006; Fang et al. 2010), supporting the view that there may be a pathological relationship between diabetes and AD.

## 5.8 Conclusion

Obesity is the combined result of dysfunction of feeding center (hypothalamus) in the brain, imbalance in energy intake and expenditure, and genetic variations resulting in increased adipose tissue mass. Abdominal obesity is considered by the International Diabetes Federation as an essential component of the MetS, a cluster of cardiovascular risk factors also including glucose and lipid abnormalities and arterial hypertension. An important event in adipose tissue with chronic low-grade inflammation is insulin resistance, which causes hyperinsulinemia to maintain normal glucose levels. When the pancreas cannot sustain hyperinsulinemia to overcome insulin resistance, prediabetes or type II diabetes and MetS ensue. In fact, these processes are accompanied by the activation of both JNK and IKK pathways through classical receptor-mediated mechanisms. JNK and IKK activation cause insulin resistance by disrupting serine phosphorylation of IRS-1, a protein that connects the insulin receptor to the PtdIns 3K signaling cascade. In parallel the activation of the above mentioned kinases is accompanied by increase in expression and levels of adipokines (TNF-α, IL-6, and MCP-1). These adipokines initiate and maintain the induction of low grade inflammation in visceral tissues.

Obesity involves a variety of metabolic signals feeding back to the brain, for example, leptin, insulin, or different cytokines. Among adipocytokines, leptin acts on key neuronal circuits within the brain to reduce food intake, limit body adiposity, and enhance peripheral glucose metabolism. The absence of leptin leads to obesity, indicating that leptin plays an essential role in the maintenance of energy homeostasis. Leptin resistance is closely associated with obesity and diabetes. In contrast, adiponectin produces anti-inflammatory effects and reduces obesity by regulating food intake, therefore exerting a direct effect on energy balance and weight control. Other cytokines and chemokines contribute to low-grade inflammation and initial adipose macrophage infiltration, which promote impairment in adipocyte insulin signaling and induction of insulin resistance through the interference of insulin/insulin-like growth factor I receptor (IGF-IR) signaling pathways. Accumulating evidence suggests that obesity is closely associated with changes in the serum levels of leptin, insulin, glucose, and corticosterone/cortisol. The interplay among these molecules is complex. In normal individuals, insulin is secreted in response to elevated glucose levels and stimulates glucose uptake by muscle and liver cells, corticosterone induces gluconeogenesis, leptin production and secretion from adipocytes can be stimulated by insulin, and circulating leptin enters the brain and interacts with neurons in the hypothalamus to suppress appetite. Oxidative stress and low-grade inflammation are triggered by an imbalance between increased production of ROS and/or reduced antioxidant activity and secretion of adipocytokines and cytokines, respectively. Both inflammation and oxidative stress along with insulin resistance induce damage to visceral and neural cells in type II diabetes, MetS, and AD.

## References

Abizaid A, Horvath TL (2008) Brain circuits regulating energy homeostasis. Regul Pept 149:3–10

Adams JM II, Pratipanawatr T, Berria R, Wang E, DeFronzo RA, Sullards MC, Mandarino LJ (2004) Ceramide content is increased in skeletal muscle from obese insulin-resistant humans. Diabetes 53:25–31

Adams SH, Hoppel CL, Lok KH, Zhao L, Wong SW, Minkler PE, Hwang DH, Newman JW, Garvey WT (2009) Plasma acylcarnitine profiles suggest incomplete long-chain fatty acid beta-oxidation and altered tricarboxylic acid cycle activity in type 2 diabetic African-American women. J Nutr 139:1073–1081

Agouni A, Mody N, Owen C, Czopek A, Zimmer D, Bentires-Alj M, Bence KK, Delibegović M (2011) Liver-specific deletion of protein tyrosine phosphatase (PTP) 1B improves obesity- and pharmacologically induced endoplasmic reticulum stress. Biochem J 438:369–378

Ahima RS, Prabakaran D, Flier JS (2000) Leptin. Annu Rev Physiol 62:413–437

Akira S, Uematsu S, Takeuchi O (2006) Pathogen recognition and innate immunity. Cell 124:783–801

Alam I, Ng TP, anf Larbi NA (2012) Does inflammation determine whether obesity is metabolically healthy or unhealthy? The aging perspective. Mediators Inflamm 2012:456456

Altintas MM, Rossetti MA, Nayer B, Puig A, Zagallo P, Ortega LM, Johnson KB, McNamara G, Reiser J, Mendez AJ, Nayer A (2011) Apoptosis, mastocytosis, and diminished adipocytokine gene expression accompany reduced epididymal fat mass in long-standing diet-induced obese mice. Lipids Health Dis 10:198

Alzheimer Scotland (2011) Risk factors in dementia. Alzheimer Scotland, Edinburgh

Alzheimer's Disease International (2010) World Alzheimer Report 2010. ADI, Illinois

Alzheimer's disease, Senile dementia [Last cited on 2010 Jan 8]. http://www.Dementia/pages.prodigy.net/naturedoctor/alzheimers.html

An J, Muoio DM, Shiota M, Fujimoto Y, Cline GW, Shulman GI, Koves TR, Stevens R, Millington D, Newgard CB (2004) Hepatic expression of malonyl-CoA decarboxylase reverses muscle, liver and whole-animal insulin resistance. Nat Med 10:268–274

Anderson EJ, Lustig ME, Boyle KE, Woodlief TL, Kane DA, Lin CT, Price JW III, Kang L, Kabinovitch PS, Szeto HH, Houmard JA, Cortright RN, Wasserman DH, Nufer PD (2009)

Mitochondrial $H_2O_2$ emission and cellular redox state link excess fat intake to insulin resistance in both rodents and humans. J Clin Invest 119:573–581

Aquilano K, Vigilanza P, Baldelli S, Pagliei B, Rotilio G, Ciriolo MR (2010) Peroxisome proliferator-activated receptor gamma co-activator 1alpha (PGC-1alpha) and sirtuin 1 (SIRT1) reside in mitochondria: possible direct function in mitochondrial biogenesis. J Biol Chem 285:21590–21599

Arab L, Sadeghi R, Walker DG, Lue LF, Sabbagh MN (2011) Consequences of Aberrant Insulin Regulation in the Brain: Can Treating Diabetes be Effective for Alzheimer's Disease. Curr Neuropharmacol 9:693–705

Archer ZA, Mercer JG (2007) Brain responses to obesogenic diets and diet-induced obesity. Proc Nutr Soc 66:124–130

Armoni M, Harel C, Bar-Yoseph F, Milo S, Karnieli E (2005) Free fatty acids repress the Glut4 gene expression in cardiac muscle via novel response elements. J Biol Chem 280:34786–34795

Atkinson LL, Fischer MA, Lopaschuk GD (2002) Leptin activates cardiac fatty acid oxidation independent of changes in the AMP-activated protein kinase-acetyl-CoA carboxylase-malonyl-CoA axis. J Biol Chem 277:29424–29430

Avena NM, Rada P, Hoebel BG (2009) Sugar and fat bingeing have notable differences in addictive-like behavior. J Nutr 139:623–628

Avramoglu RK, Basciano H, Adeli K (2006) Lipid and lipoprotein dysregulation in insulin resistant states. Clin Chim Acta 368:1–19

Baile CA, Della-Fera MA, Martin RJ (2000) Regulation of metabolism and body fat mass by leptin. Annu Rev Nutr 20:105–127

Balakrishnan K, Verdile G, Mehta PD, Beilby J, Nolan D, Galvão DA, Newton R, Gandy SE, Martins RN (2005) Plasma Abeta42 correlates positively with increased body fat in healthy individuals. J Alzheimers Dis 8:269–282

Banks AS, Davis SM, Bates SH, Myers MG Jr (2000) Activation of downstream signals by the long form of the leptin receptor. J Biol Chem 275:14563–14572

Bastard JP, Maachi M, Lagathu C, Kim MJ, Caron M, Vidal H, Capeau J, Feve B (2006) Recent advances in the relationship between obesity, inflammation, and insulin resistance. Eur Cytokine Netw 17:4–12

Bates SH, Stearns WH, Dundon TA, Schubert M, Tso AW, Wang Y, Banks AS, Lavery HJ, Haq AK, Maratos-Flier E, Neel BG, Schwartz MW, Myers MG Jr (2003) STAT3 signalling is required for leptin regulation of energy balance but not reproduction. Nature 421:856–859

Bates SH, Kulkarni RN, Seifert M, Myers MG Jr (2005) Roles for leptin receptor/STAT3-dependent and -independent signals in the regulation of glucose homeostasis. Cell Metab 1:169–178

Bays H, Mandarino L, DeFronzo RA (2004) Role of the adipocyte, free fatty acids, and ectopic fat in pathogenesis of type 2 diabetes mellitus: peroxisomal proliferator-activated receptor agonists provide a rational therapeutic approach. J Clin Endocrinol Metab 89:463–478

Bays HE, Tighe AP, Sadovsky R, Davidson MH (2008a) Prescription omega-3 fatty acids and their lipid effects: physiologic mechanisms of action and clinical implications. Expert Rev Cardiovasc Ther 6:391–409

Bays HE, Gonzalez-Campoy JM, Bray GA, Kitabchi AE, Bergman DA, Schorr AB, Rodbard HW, Henry RR (2008b) The pathogenic potential of adipose tissue and the metabolic consequences of adipocyte hypertrophy and increased visceral adiposity. Expert Rev Cardiovasc Ther 6:343–368

Bence KK, Delibegovic M, Xue B, Gorgun CZ, Hotamisligil GS, Neel BG, Kahn BB (2006) Neuronal PTP1B regulates body weight, adiposity and leptin action. Nat Med 12:917–924

Berg AH, Combs TP, Du X, Brownlee M, Scherer PE (2001) The adipocyte-secreted protein Acrp30 enhances hepatic insulin action. Nat Med 7:947–953

Berg AH, Combs TP, Du X, Brownlee M, Scherer PE (2002) ACRP30/adiponectin; adipokine regulating glucose and lipid metabolism. Trends Endocrinol Metab 13:84–89

Bergeron R, Russell RR III, Young LH, Ren JM, Marcucci M, Lee A, Shulman GI (1999) Effect of AMPK activation on muscle glucose metabolism in conscious rats. Am J Physiol 276:E938–E944

Bergman RN, Kim SP, Catalano KJ, Hsu IR, Chiu JD, Kabir M, Hucking K, Ader M (2006) Why visceral fat is bad: mechanisms of the metabolic syndrome. Obesity (Silver Spring) 14(Suppl 1):16S–19S

Bjorbaek C, Kahn BB (2004) Leptin signaling in the central nervous system and the periphery. Recent Prog Horm Res 59:305–331

Bjorbaek C, El-Haschimi K, Frantz JD, Flier JS (1999) The role of SOCS-3 in leptin signaling and leptin resistance. J Biol Chem 274:30059–30065

Bjørbæk C, Elmquist JK, Michl P, Ahima RS, van Bueren A, McCall AL, Flier JS (1998) Expression of leptin receptor isoforms in rat brain microvessels. Endocrinology 139:3485–3491

Blouet C, Schwartz GJ (2010) Hypothalamic nutrient sensing in the control of energy homeostasis. Behav Brain Res 209:1–12

Boden G, Chen X (1995) Effects of fat on glucose uptake and utilization in patients with noninsulin-dependent diabetes. J Clin Invest 96:1261–1268

Boden G, She P, Mozzoli M, Cheung P, Gumireddy K, Reddy P, Xiang X, Luo Z, Ruderman N (2005) Free fatty acids produce insulin resistance and activate the proinflammatory nuclear factor-κB pathway in rat liver. Diabetes 54:3458–3465

Borst SE (2004) The role of TNF-alpha in insulin resistance. Endocrine 23:177–182

Bottner A, Kratzsch J, Muller G, Kapellen TM, Bluher S, Keller F, Bluher M, Kiess W (2004) Gender differences of adiponectin levels develop during the progression of puberty and are related to serum androgen levels. J Clin Endocrinol Metab 89:4053–4061

Boura-Halfon S, Zick Y (2009) Phosphorylation of IRS proteins, insulin action, and insulin resistance. Am J Physiol Endocrinol Metab 296:E581–E591

Bråkenhielm E, Veitonmäki N, Cao R, Kihara S, Matsuzawa Y, Zhivotovsky B, Funahashi T, Cao Y (2004) Adiponectin-induced antiangiogenesis and antitumor activity involve caspase-mediated endothelial cell apoptosis. Proc Natl Acad Sci USA 101:2476–2481

Brown LA, Kerr CJ, Whiting P (2009) Oxidant stress in healthy normal-weight, overweight, and obese individuals. Obesity (Silver Spring) 17:460–466

Businaro R, Ippoliti F, Ricci S, Canitano N, Fuso A (2012) Alzheimer's disease promotion by obesity: induced mechanisms-molecular links and perspectives. Curr Gerontol Geriatr Res 2012:986823

Cai D, Yuan M, Frantz DF, Melendez PA, Hansen L, Lee J, Shoelson SE (2005) Local and systemic insulin resistance resulting from hepatic activation of IKK-beta and NF-kappaB. Nat Med 11:183–190

Canto C, Jiang LQ, Deshmukh AS, Mataki C, Coste A, Lagouge M, Zierath JR, Auwerx J (2010) Interdependence of AMPK and SIRT1 for metabolic adaptation to fasting and exercise in skeletal muscle. Cell Metab 11:213–219

Castro JP, El-Atat FA, McFarlane SI, Aneja A, Sowers JR (2003) Cardiometabolic syndrome: pathophysiology and treatment. Curr Hypertens Rep 5:393–401

Ceddia RB, Somwar R, Maida A, Fang X, Bikopoulos G, Sweeney G (2005) Globular adiponectin increases GLUT4 translocation and glucose uptake but reduces glycogen synthesis in rat skeletal muscle cells. Diabetologia 48:132–139

Chaudhuri K, Samarakoon SM, Chandola HM, Kumar R, Ravishankar B (2011) Evaluation of diet and life style in etiopathogenesis of senile dementia: a survey study. Ayu 32:171–176

Cheng ZJ, Vapaatalo H, Mervaala E (2005) Angiotensin II and vascular inflammation. Med Sci Monit 11:RA194–RA205

Chiba T, Yamada M, Aiso S (2009) Targeting the JAK2/STAT3 axis in Alzheimer's disease. Expert Opin Ther Targets 13:1155–1167

Chrysohoou C, Panagiotakos DB, Pitsavos C, Skoumas I, Papademetriou L, Economou M, Stefanadis C (2007) The implication of obesity on total antioxidant capacity apparently healthy men and women: the ATTICA study. Nutr Metab Cardiovasc Dis 17:590–597

Chung J, Nguyen AK, Henstridge DC, Holmes AG, Chan MH, Mesa JL, Lancaster GI, Southgate RJ, Bruce CR, Duffy SJ, Horvath I, Mestril R, Watt MJ, Hooper PL, Kingwell BA, Vigh L, Hevener A, Febbraio MA (2008) HSP72 protects against obesity-induced insulin resistance. Proc Natl Acad Sci USA 105:1739–1744

Civitarese AE, Ukropcova B, Carling S, Hulver M, DeFronzo RA, Mandarino L, Ravussin E, Smith SR (2006) Role of adiponectin in human skeletal muscle bioenergetics. Cell Metab 4:75–87

Colantuoni C, Schwenker J, McCarthy J, Rada P, Ladenheim B, Cadet JL, Schwartz GJ, Moran TH, Hoebel BG (2001) Excessive sugar intake alters binding to dopamine and mu-opioid receptors in the brain. Neuroreport 12:3549–3552

Coll AP, Farooqi IS, O'Rahilly S (2007) The hormonal control of food intake. Cell 129:251–262

Combs TP, Pajvani UB, Berg AH, Lin Y, Jelicks LA, Laplante M, Nawrocki AR, Rajala MW, Parlow AF, Cheeseboro L, Ding YY, Russell RG, Lindemann D, Hartley A, Baker GR, Obici S, Deshaies Y, Ludgate M, Rossetti L, Scherer PE (2004) A transgenic mouse with a deletion in the collagenous domain of adiponectin displays elevated circulating adiponectin and improved insulin sensitivity. Endocrinology 145:367–383

Considine RV, Sinha MK, Heiman ML, Kriauciunas A, Stephens TW, Nyce MR, Ohannesian JP, Marco CC, McKee LJ, Bauer TL (1996) Serum immunoreactive-leptin concentrations in normal-weight and obese humans. N Engl J Med 334:292–295

Cornier MA, Dabelea D, Hernandez TL, Lindstrom RC, Steig AJ, Stob NR, Van Pelt RE, Wang H, Eckel RH (2008) The metabolic syndrome. Endocr Rev 29:777–822

Corton JM, Gillespie JG, Hardie DG (1994) Role of the AMP-activated protein kinase in the cellular stress response. Curr Biol 4:315–324

Craft S (2009) The role of metabolic disorders in Alzheimer disease and vascular dementia: two roads converged. Arch Neurol 66:300–305

Davis JF, Choi DL, Benoit SC (2010) Insulin, leptin and reward. Trends Endocrinol Metab 21:68–74

de la Monte SM (2012) Brain insulin resistance and deficiency as therapeutic targets in Alzheimer's disease. Curr Alzheimer Res 9:35–66

de la Monte SM, Wands JR Jr (2008) Alzheimer's disease is type 3 diabetes-evidence reviewed. J Diabetes Sci Technol 2:1101–1113

De Souza CT, Araujo EP, Bordin S, Ashimine R, Zollner RL, Boschero AC, Saad MJ, Velloso LA (2005) Consumption of a fat-rich diet activates a proinflammatory response and induces insulin resistance in the hypothalamus. Endocrinology 146:4192–4199

De TB, Smith LH, Vaughan DE (2005) Plasminogen activator inhibitor-1: a common denominator in obesity, diabetes and cardiovascular disease. Curr Opin Pharmacol 5:149–154

Deane R, Du Yan S, Submamaryan RK, LaRue B, Jovanovic S, Hogg E, Welch D, Manness L, Lin C, Yu J, Zhu H, Ghiso J, Frangione B, Stern A, Schmidt AM, Armstrong DL, Arnold B, Liliensiek B, Nawroth P, Hofman F, Kindy M, Stern D, Zlokovic B (2003) RAGE mediates amyloid-beta peptide transport across the blood-brain barrier and accumulation in brain. Nat Med 9:907–913

del Aguila LF, Claffey KP, Kirwan JP (1999) TNF-alpha impairs insulin signaling and insulin stimulation of glucose uptake in C2C12 muscle cells. Am J Physiol 276:E849–E855

Delibegovic M, Zimmer D, Kauffman C, Rak K, Hong EG, Cho YR, Kim JK, Kahn BB, Neel BG, Bence KK (2009) Liver-specific deletion of protein-tyrosine phosphatase 1B (PTP1B) improves metabolic syndrome and attenuates diet-induced endoplasmic reticulum stress. Diabetes 58:590–599

Denke MA, Sempos CT, Grundy SM (1994) Excess body weight. An under-recognized contributor to dyslipidemia in white American women. Arch Intern Med 154:401–410

Dietrich M, Horvath T (2009) Feeding signals and brain circuitry. Eur J Neurosci 30:1688–1696

Du XL, Edelstein D, Dimmeler S, Ju Q, Sui C, Brownlee M (2001) Hyperglycemia inhibits endothelial nitric oxide synthase activity by posttranslational modification at the Akt site. J Clin Invest 108:1341–1348

Duracková Z (2010) Some current insights into oxidative stress. Physiol Res 59:459–469
Egger G, Dixon J (2011) Non-nutrient causes of low-grade, systemic inflammation: support for a "canary in the mineshaft" view of obesity in chronic disease. Obes Rev 12:339–345
El-Khairy L, Ueland PM, Refsum H, Graham IM, Vollset SE (2001) Plasma total cysteine as a risk factor for vascular disease: the European Concerted Action Project. Circulation 103:2544–2549
Elshorbagy AK, Refsum H, Smith AD, Graham IM (2009) The association of plasma cysteine and gamma-glutamyltransferase with BMI and obesity. Obesity (Silver Spring) 17:1435–1440
Elshorbagy AK, Church C, Valdivia-Garcia M, Smith AD, Refsum H, Cox R (2012) Dietary cystine level affects metabolic rate and glycaemic control in adult mice. J Nutr Biochem 23:332–340
Esposito K, Ciotola M, Giugliano D (2006) Oxidative stress in the metabolic syndrome. J Endocrinol Invest 29:791–795
Fang X, Palanivel R, Zhou X, Liu Y, Xu A, Wang Y, Sweeney G (2005) Hyperglycemia- and hyperinsulinemia-induced alteration of adiponectin receptor expression and adiponectin effects in L6 myoblasts. J Mol Endocrinol 35:465–476
Fang F, Lue LF, Yan S, Xu H, Luddy JS, Chen D, Walker DG, Stern DM, Yan S, Schmidt AM, Chen JX, Yan S (2010) RAGE-dependent signaling in microglia contributes to neuroinflammation, Abeta accumulation, and impaired learning/memory in a mouse model of Alzheimer's disease. FASEB J 24:1043–1055
Farooqui AA (2010) Neurochemical aspects of neurotraumatic and neurodegenerative diseases. Springer, New York
Farooqui AA, Horrocks LA (2007) Glycerophospholipids in the brain. Springer, New York
Farooqui AA, Farooqui T, Panza F, Frisardi V (2012) Metabolic syndrome as a risk factor for neurological disorders. Cell Mol Life Sci 69:741–762
Fernández-Galaz C, Fernández-Agulló T, Campoy F, Arribas C, Gallardo N, Andres A, Ros M, Carrascosa M (2001) Decreased leptin uptake in hypothalamic nuclei with ageing in Wistar rats. J Endocrinol 171:23–32
Fernandez-Sanchez A, Madrigal-Santillan E, Bautista M, Esquivel-Soto J, Morales-Gonzalez A, Esquivel-Chirino C, Durante-Montiel I, Sanchez-Rivera G, Valadez-Vega C, Morales-Gonzalez JA (2011) Inflammation, oxidative stress, and obesity. Int J Mol Sci 12:3117–3132
Fewlass DC, Noboa K, Pi-Sunyer FX, Johnston JM, Yan SD, Tezapsidis N (2004) Obesity-related leptin regulates Alzheimer's Abeta. FASEB J 18:1870–1878
Flegal KM, Graubard BI, Williamson DF, Gail MH (2007) Cause-specific excess deaths associated with underweight, overweight, and obesity. JAMA 298:2028–2037
Fonseca V, Desouza C, Asnani S, Jialal I (2004) Nontraditional risk factors for cardiovascular disease in diabetes. Endocr Rev 25:153–175
Fontaine KR, Redden DT, Wang C, Westfall AO, Allison DB (2003) Years of life lost due to obesity. JAMA 289:187–193
Frederich RC, Hamann A, Anderson S, Lollmann B, Lowell BB, Flier JS (1995) Leptin levels reflect body lipid content in mice: evidence for diet-induced resistance to leptin action. Nat Med 1:1311–1314
Freeman LR, Zhang L, Dasuri K, Fernandez-Kim SO, Bruce-Keller AJ, Keller JN (2012) Mutant amyloid precursor protein differentially alters adipose biology under obesogenic and non-obesogenic conditions. PLoS One 7:e43193
Fukuyama Y, Ohta K, Okoshi R, Suehara M, Kizaki H, Nakagawa K (2007) Hypoxia induces expression and activation of AMPK in rat dental pulp cells. J Dent Res 86:903–907
Furukawa S, Fujita T, Shimabukuro M, Iwaki M, Yamada Y, Nakajima Y, Nakayama O, Makishima M, Matsuda M, Shimomurai I (2004) Increased oxidative stress in obesity and its impact on metabolic syndrome. J Clin Invest 114:1752–1761
Geiger BM, Behr GG, Frank LE, Caldera-Siu AD, Beinfeld MC, Kokkotou EG, Pothos EN (2008) Evidence for defective mesolimbic dopamine exocytosis in obesity-prone rats. FASEB J 22:2740–2746

Gesta S, Blüher M, Yamamoto Y, Norris AW, Berndt J, Kralisch S, Boucher J, Lewis C, Kahn CR (2006) Evidence for a role of developmental genes in the origin of obesity and body fat distribution. Proc Natl Acad Sci USA 103:6676–6681

Gesta S, Tseng YH, Kahn CR (2007) Developmental origin of fat: tracking obesity to its source. Cell 131:242–256

Ghilardi N, Skoda RC (1997) The leptin receptor activates Janus kinase 2 and signals for proliferation in a factor-dependent cell line. Mol Endocrinol 11:393–399

Giral P, Jacob N, Dourmap C, Hansel B, Carrie A, Bruckert E (2008) Elevated gamma-glutamyltransferase activity and perturbed thiol profile are associated with features of metabolic syndrome. Arterioscler Thromb Vasc Biol 28:587–593

Girones X, Guimera A, Cruz-Sanchez CZ, Ortega A, Sasaki N, Makita Z, Lafuente JV, Kalaria R, Cruz-Sanchez FF (2004) N ε-carboxymethyllysine in brain aging, diabetes mellitus, and Alzheimer's disease. Free Radic Biol Med 36:1241–1247

Grundy SM (2003) Inflammation, hypertension and the metabolic syndrome. JAMA 290:3000–3002

Hajnal A, Smith GP, Norgren R (2004) Oral sucrose stimulation increases accumbens dopamine in the rat. Am J Physiol Regul Integr Comp Physiol 286:R31–R37

Halliwell B (2006) Oxidative stress and neurodegeneration: where are we now? J Neurochem 97:1634–1658

Hamilton MT, Hamilton DG, Zderic TW (2007) The role of low energy expenditure and sitting on obesity, metabolic syndrome, type 2 diabetes, and cardiovascular disease. Diabetes 56:2655–2667

Hardie DG (2007) AMP-activated/SNF1 protein kinases: conserved guardians of cellular energy. Nat Rev Mol Cell Biol 8:774–785

Hardie DG, Sakamoto K (2006) AMPK: a key sensor of fuel and energy status in skeletal muscle. Physiology (Bethesda) 21:48–60

Harding HP, Novoa I, Zhang H, Wek R, Schapira M, Ron D (2000) Regulated translation initiation controls stress-induced gene expression in mammalian cells. Mol Cell 6:1099–1108

Hartwich J, Góralska J, Siedlecka D, Gruca A, Trzos M, Dembinska-Kiec A (2007) Effect of supplementation with vitamin E and C on plasma hsCRP level and cobalt-albumin binding score as markers of plasma oxidative stress in obesity. Genes Nutr 2:151–154

Harvey J, McKay NG, Walker KS, Van der Kaay J, Downes CP, Ashford ML (2000) Essential role of phosphoinositide 3-kinase in leptin-induced K(ATP) channel activation in the rat CRI-G1 insulinoma cell line. J Biol Chem 275:4660–4669

Hayden MS, Ghosh S (2008) Shared principles in NF-κB signaling. Cell 132:344–362

Herrara BM, Lindgren CM (2010) The genetics of obesity. Curr Diab Rep 10:498–505

Hirosumi J, Tuneman G, Chang L, Gorgun CZ, Uysal KT, Karin M, Hotamisligil S (2002) A central role for JNK in obesity and insulin resistance. Nature 420:333–336

Hoehn KL, Salmon AB, Hohnen-Behrens C, Turner N, Hoy AJ, Maghzal GJ, Stocker R, Van Remmen H, Kraegen EW, Cooney GJ, Richardson AR, James DE (2009) Insulin resistance is a cellular antioxidant defense mechanism. Proc Natl Acad Sci USA 106:17787–17792

Holden KF, Lindquist K, Tylavsky FA, Rosano C, Harris TB, Yaffe K (2009) Serum leptin level and cognition in the elderly: findings from the Health ABC Study. Neurobiol Aging 30:1483–1489

Holland WL, Brozinick JT, Wang LP, Hawkins ED, Sargent KM, Liu Y, Narra K, Hoehn KL, Knotts TA, Siesky A, Nelson DH, Karathanasis SK, Fontenot GK, Birnbaum MJ, Summers SA (2007) Inhibition of ceramide synthesis ameliorates glucocorticoid-, saturated-fat- and obesity-induced insulin resistance. Cell Metab 5:167–179

Holland WL, Miller RA, Wang ZV, Sun K, Barth BM, Bui HH, Davis KE, Bikman BT, Halberg N, Rutkowski JM, Wade MR, Tenorio VM, Kuo MS, Brozinick JT, Zhang BB, Birnbaum MJ, Summers SA, Scherer PE (2011) Receptor-mediated activation of ceramidase activity initiates the pleiotropic actions of adiponectin. Nat Med 17:55–63

Horvath TL, Sarman B, García-Cáceres C, Enriori PJ, Sotonyi P, Shanabrough M, Borok E, Argente J, Chowen JA, Perez-Tilve D, Pfluger PT, Brönneke HS, Levin BE, Diano S, Cowley MA, Tschöp MH (2010) Synaptic input organization of the melanocortin system predicts diet-induced hypothalamic reactive gliosis and obesity. Proc Natl Acad Sci USA 107:14875–14880

Hosogai N, Fukuhara A, Oshima K, Miyata Y, Tanaka S, Segawa K, Furukawa S, Tochino Y, Komuro R, Matsuda M, Shimomura I (2007) Adipose tissue hypoxia in obesity and its impact on adipocytokine dysregulation. Diabetes 56:901–911

Hossain P, Kawar B, El Nahas M (2007) Obesity and diabetes in the developing world: a growing challenge. N Engl J Med 356:213–215

Hotamisligil GS (2003) Inflammatory pathways and insulin action. Int J Obes Relat Metab Disord 27(Suppl 3):S53–S55

Hotamisligil GS (2006) Inflammation and metabolic disorders. Nature 444:860–867

Hotamisligil GS (2008) Inflammation and endoplasmic reticulum stress in obesity and diabetes. Int J Obes (Lond) 32(Suppl 7):S52–S54

Hotta K, Funahashi T, Arita Y, Takahashi M, Matsuda M, Okamoto Y, Iwahashi H, Kuriyama H, Ouchi N, Maeda K, Nishida M, Kihara S, Sakai N, Nakajima T, Hasegawa K, Muraguchi M, Ohmoto Y, Nakamura T, Yamashita S, Hanafusa T, Matsuzawa Y (2000) Plasma concentrations of a novel adipose-specific protein, adiponectin, in type 2 diabetic patients. Arterioscler Thromb Vasc Biol 20:1595–1599

Hoyer S (2004) Glucose metabolism and insulin receptor signal transduction in Alzheimer disease. Eur J Pharmacol 490:115–125

Hsu HT, Chang YC, Chiu YN, Liu CL, Chang KJ, Guo IC (2006) Leptin interferes with adrenocorticotropin/3′,5′-cyclic adenosine monophosphate (cAMP) signaling, possibly through a Janus kinase 2-phosphatidylinositol 3-kinase/Akt-phosphodiesterase 3-cAMP pathway, to down-regulate cholesterol side-chain cleavage cytochrome P450 enzyme in human adrenocortical NCI-H295 cell line. J Clin Endocrinol Metab 91:2761–2769

Huffman DM, Barzilai N (2009) Role of visceral adipose tissue in aging. Biochim Biophys Acta 1790:1117–1123

Hug C, Lodish HF (2005) The role of the adipocyte hormone adiponectin in cardiovascular disease. Curr Opin Pharmacol 5:129–134

Ibrahimi A, Bonen A, Blinn WD, Hajri T, Li X, Zhong K, Cameron R, Abumrad NA (1999) Muscle-specific overexpression of FAT/CD36 enhances fatty acid oxidation by contracting muscle, reduces plasma triglycerides and fatty acids, and increases plasma glucose and insulin. J Biol Chem 274:26761–26766

Ido Y, Carling D, Ruderman N (2002) Hyperglycemia-induced apoptosis in human umbilical vein endothelial cells: inhibition by the AMP-activated protein kinase activation. Diabetes 51:159–167

Ikonen E, Vainio S (2005) Lipid microdomains and insulin resistance: is there a connection? Sci STKE 2005:pe3

Inokuchi J (2006) Insulin resistance as a membrane microdomain disorder. Biol Pharm Bull 29:1532–1537

Itani SI, Ruderman NB, Schmieder F, Boden G (2002) Lipid-induced insulin resistance in human muscle is associated with changes in diacylglycerol, protein kinase C, and IkappaB-alpha. Diabetes 51:2005–2011

Iwabu M, Yamauchi T, Okada-Iwabu M, Sato K, Nakagawa T, Funata M, Yamaguchi M, Namiki S, Nakayama R, Tabata M, Ogata H, Kubota N, Takamoto I, Hayashi YK, Yamauchi N, Waki H, Fukayama M, Nishino I, Tokuyama K, Ueki K, Oike Y, Ishii S, Hirose K, Shimizu T, Touhara K, Kadowaki T (2010) Adiponectin and AdipoR1 regulate PGC-1alpha and mitochondria by $Ca^{2+}$ and AMPK/SIRT1. Nature 464:1313–1319

Jazet IM, Pijl H, Meinders AE (2003) Adipose tissue as an endocrine organ: impact on insulin resistance. Neth J Med 61:194–212

Jellinger KA (2009) Recent advances in our understanding of neurodegeneration. J Neural Transm 116:1111–1162

Jeong HW, Hsu KC, Lee JW, Ham M, Huh JY, Shin HJ, Kim WS, Kim JB (2009) Berberine suppresses proinflammatory responses through AMPK activation in macrophages. Am J Physiol Endocrinol Metab 296:E955–E964

Jiao P, Chen Q, Shah S, Tao B, Tzameli I, Yan W, Xu H (2009) Obesity-related upregulation of monocyte chemotactic factors in adipocytes: involvement of nuclear factor-κB and c-Jun NH2-terminal kinase pathways. Diabetes 58:104–115

Johansen D, Stocks T, Jonsson H, Lindkvist B, Bjorge T, Concin H, Almquist M, Haggstrom C, Engeland A, Ulmer H, Hallmans G, Selmer R, Nagel G, Tretli S, Stattin P, Manjer J (2010) Metabolic factors and the risk of pancreatic cancer: a prospective analysis of almost 580,000 men and women in the Metabolic Syndrome and Cancer Project. Cancer Epidemiol Biomarkers Prev 19:2307–2317

Johnson JA, Ynada Y, Flancbaum L, Albu J (2001) Cytokine release in deep subcutaneous, abdominal and visceral tissue depots. Diabetes 50:A88

Juhan-Vague I, Alessi MC, Mavri A, Morange PE (2003) Plasminogen activator inhibitor-1, inflammation, obesity, insulin resistance and vascular risk. J Thromb Haemost 1:1575–1579

Kadowaki T, Yamauchi T, Kubota N, Hara K, Ueki K, Tobe K (2006) Adiponectin and adiponectin receptors in insulin resistance, diabetes, and the metabolic syndrome. J Clin Invest 116:1784–1792

Kadowaki T, Yamauchi T, Kubota N (2008) The physiological and pathophysiological role of adiponectin and adiponectin receptors in the peripheral tissues and CNS. FEBS Lett 582:74–80

Kasper DL, Braunwald E, Fauci AS, Hauser SL, Longo DL, Jameson JL (2005) Harrison's principles of Internal medicine, 16th edn. McGraw-Hill Medical publishing division, New York, pp 2393–2399

Kassi E, Pervanidou P, Kaltsas G, Chrousos G (2011) Metabolic syndrome: definitions and controversies. BMC Med 9:48

Kavanagh K, Jones KL, Sawyer J, Kelley K, Carr JJ, Wagner JD, Rudel LL (2007) Trans fat diet induces abdominal obesity and changes in insulin sensitivity in monkeys. Obesity (Silver Spring) 15:1675–1684

Kelley DE, He J, Menshikova EV, Ritov VB (2002) Dysfunction of mitochondria in human skeletal muscle in type 2 diabetes. Diabetes 51:2944–2950

Kiguchi N, Maeda T, Kobayashi Y, Fukazawa Y, Kishioka S (2009) Leptin enhances CC-chemokine ligand expression in cultured murine macrophage. Biochem Biophys Res Commun 384:311–315

Kim JK, Fillmore JJ, Chen Y, Yu C, Moore IK, Pypaert M, Lutz EP, Kako Y, Velez-Carrasco W, Goldberg IJ, Breslow JL, Shulman GI (2001) Tissue-specific overexpression of lipoprotein lipase causes tissue specific insulin resistance. Proc Natl Acad Sci USA 98:7522–7527

Kim JK, Fillman JJ, Sunshine MJ, Albrecht B, Higashimori T, Kim DW, Soos TJ, Cline GW, O'Brien WR, Littman DR, Shulman GI (2004) PKC-θ knockout mice are protected from fat-induced insulin resistance. J Clin Invest 114:823–827

Kim JY, van de Wall E, Laplante M, Azzara A, Trujillo ME, Hofmann SM et al (2007) Obesity-associated improvements in metabolic profile through expansion of adipose tissue. J Clin Invest 117:2621–2637

Kivipelto M, Solomon A (2008) Alzheimer's disease—the ways of prevention. J Nutr Health Aging 12:89S–94S

Koistinen HA, Galuska D, Chibalin AV, Yang J, Zierath JR, Holman GD, Wallberg-Henriksson H (2003) 5-Amino-imidazole carboxamide riboside increases glucose transport and cell-surface GLUT4 content in skeletal muscle from subjects with type 2 diabetes. Diabetes 52:1066–1072

Koves TR, Li P, An J, Akimoto T, Slentz D, Ilkayeva O, Dohm GL, Yan Z, Newgard CB, Muoio DM (2005) Peroxisome proliferator-activated receptor-gamma co-activator 1alpha-mediated metabolic remodeling of skeletal myocytes mimics exercise training and reverses lipid-induced mitochondrial inefficiency. J Biol Chem 280:33588–33598

Koves TR, Ussher JR, Noland RC, Sientz D, Mosedale M, Ilkayeva O, Bain J, Stevens R, Dyck JR, Newgard CB, Lopaschuk GD, Muoio DM (2008) Mitochondrial overload and incomplete fatty acid oxidation contribute to skeletal muscle insulin resistance. Cell Metab 7:45–56
Kraegen EW, Cooney GJ, Ye JM, Thompson AL, Furler SM (2001) The role of lipids in the pathogenesis of muscle insulin resistance and beta cell failure in type II diabetes and obesity. Exp Clin Endocrinol Diabetes 109(Suppl 2):S189–S201
Krötz F, Sohn H-Y, Pohl U (2004) Reactive oxygen species: players in the platelet game. Arterioscler Thromb Vasc Biol 24:1988–1996
Kukidome D, Nishikawa T, Sonoda K, Imoto K, Fujisawa K, Yano M, Motoshima H, Taguchi T, Matsumura T, Araki E (2006) Activation of AMP-activated protein kinase reduces hyperglycemia-induced mitochondrial reactive oxygen species production and promotes mitochondrial biogenesis in human umbilical vein endothelial cells. Diabetes 55:120–127
Lam TT, Van De Werve G, Giacca A (2003) Free fatty acids increase basal hepatic glucose production and induce hepatic insulin resistance at different sites. Am J Physiol Endocrinol Metab 284:E281–E290
Leahey TM, Myers TA, Gunstad J, Glickman E, Spitznagel MB, Alexander T, Juvancic-Heltzel J (2007) Abeta40 is associated with cognitive function, body fat and physical fitness in healthy older adults. Nutr Neurosci 10:205–209
Lee YH, Tharp WG, Maple RL, Nair S, Permana PA et al (2008) Amyloid precursor protein expression is upregulated in adipocytes in obesity. Obesity (Silver Spring) 16:1493–1500
Lelliott C, Vidal-Puig AJ (2004) Lipotoxicity, an imbalance between lipogenesis de novo and fatty acid oxidation. Int J Obes Relat Metab Disord 28(Suppl 4):S22–S28
Lenoir M, Serre F, Cantin L, Ahmed SH (2007) Intense sweetness surpasses cocaine reward. PLoS One 2:e698
Li WX (2008) Canonical and non-canonical JAK-STAT signaling. Trends Cell Biol 18:545–551
Li X, Stark GR (2002) NFkappaB-dependent signaling pathways. Exp Hematol 30:285–296
Li M, Baumeister P, Roy B, Phan T, Foti D, Luo S, Lee AS (2000) ATF6 as a transcription activator of the endoplasmic reticulum stress element. Thapsigargin stress-induced changes and synergistic interactions with NF-Y and YY1. Mol Cell Biol 20:5096–5106
Li X, Han Y, Pang W, Li C, Xie X et al (2008) AMP-activated protein kinase promotes the differentiation of endothelial progenitor cells. Arterioscler Thromb Vasc Biol 28:1789–1795
Li FY, Cheng KK, Lam KS, Vanhoutte PM, Xu A (2011) Cross-talk between adipose tissue and vasculature: role of adiponectin. Acta Physiol (Oxf) 203:167–180
Liang NC, Hajnal A, Norgren R (2006) Sham feeding corn oil increases accumbens dopamine in the rat. Am J Physiol Regul Integr Comp Physiol 291:R1236–R1239
Lieb W, Beiser AS, Vasan RS, Tan ZS, Au R, Harris TB, Roubenoff R, Auerbach S, DeCarli C, Wolf PA, Seshadri S (2009) Association of plasma leptin levels with incident Alzheimer disease and MRI measures of brain aging. JAMA 302:2565–2572
Lillioja S, Nyomba BL, Saad MF, Ferraro R, Castillo C, Bennett PH (1991) Exaggerated early insulin release and insulin resistance in a diabetes-prone population: a metabolic comparison of Pima Indians and Caucasians. J Clin Endocrinol Metab 73:866–876
Long YC, Zierath JR (2006) AMP-activated protein kinase signaling in metabolic regulation. J Clin Invest 116:1776–1783
Loskutoff DJ, Samad F (1998) The adipocyte and hemostatic balance in obesity: studies of PAI-1. Arterioscler Thromb Vasc Biol 18:1–6
Lowell BB (1999) PPARγ: an essential regulator of adipogenesis and modulator of fat cell function. Cell 99:239–242
Lumeng CN, Deyoung SM, Bodzin JL, Saltiel AR (2007) Increased inflammatory properties of adipose tissue macrophages recruited during diet-induced obesity. Diabetes 56:16–23
Lund IV, Hu Y, Raol YH, Benham RS, Faris R, Russek SJ, Brooks-Kayal AR (2008) BDNF selectively regulates GABAA receptor transcription by activation of the JAK/STAT pathway. Sci Signal 1:ra9
Mahmoud GS, Grover LM (2006) Growth hormone enhances excitatory synaptic transmission in area CA1 of rat hippocampus. J Neurophysiol 95:2962–2974

Malhotra JD, Kaufman RJ (2007) The endoplasmic reticulum and the unfolded protein response. Semin Cell Dev Biol 18:716–731
Mao X, Hong JY, Dong LQ (2006) The adiponectin signaling pathway as a novel pharmacological target. Mini Rev Med Chem 6:1331–1340
Massiera F, Saint-Marc P, Seydoux J, Murata T, Kobayashi T, Narumiya S, Guesnet P, Amri EZ, Negrel R, Ailhaud G (2003) Arachidonic acid and prostacyclin signaling promote adipose tissue development: a human health concern? J Lipid Res 44:271–279
Matsubara M, Maruoka S, Katayose S (2002) Inverse relationship between plasma adiponectin and leptin concentrations in normal-weight and obese women. Eur J Endocrinol 147:173–180
McCall AL (2005) Altered glycemia and brain-update and potential relevance to the aging brain. Neurobiol Aging 26:70–75
McCrimmon RJ, Fan X, Ding Y, Zhu W, Jacob RJ, Sherwin RS (2004) Potential role for AMP-activated protein kinase in hypoglycemia sensing in the ventromedial hypothalamus. Diabetes 53:1953–1958
Michels KB, Greenland S, Rosner BA (1998) Does body mass index adequately capture the relation of body composition and body size to health outcomes? Am J Epidemiol 147:167–172
Minokoshi Y, Kim YB, Peroni OD, Fryer LG, Muller C, Carling D, Kahn BB (2002) Leptin stimulates fatty-acid oxidation by activating AMP-activated protein kinase. Nature 415:339–343
Mittra S, Bansal VS, Bhatnagar PK (2008) From a glucocentric to a lipocentric approach towards metabolic syndrome. Drug Discov Today 13:211–218
Mokdad AH, Ford ES, Bowman LA, Dietz WH, Vinicor E, Bales VS, Mark JS (2003) Prevalence of obesity, diabetes, and obesity-related health risk factors, 2001. JAMA 289:76–79
Moller DE, Kaufman KD (2005) Metabolic syndrome: a clinical and molecular perspective. Annu Rev Med 56:45–62
Molnar A, Toth A, Bagi Z, Papp Z, Edes I, Vaszily M, Galajda Z, Papp JG, Varro A, Szuts V, Lacza Z, Gero D, Szabo C (2006) Activation of the poly(ADP-ribose) polymerase pathway in human heart failure. Mol Med 12:143–152
Moraes JC, Coope A, Morari J, Cintra DE, Roman EA, Pauli JR, Romanatto T, Carvalheira JB, Oliveira AL, Saad MJ, Velloso LA (2009) High-fat diet induces apoptosis of hypothalamic neurons. PLoS One 4:e5045
Morino K, Petersen KF, Shulman GI (2006) Molecular mechanisms of insulin resistance in humans and their potential links with mitochondrial dysfunction. Diabetes 55(Suppl 2):S9–S15
Morino K, Neschen S, Bilz S, Sono S, Tsirigotis D, Reznick RM, Moore I, Nagai Y, Samuel V, Sebastian D, White M, Philbrick W, Shulman GI (2008) Muscle-specific IRS-1 Ser→Ala transgenic mice are protected from fat-induced insulin resistance in skeletal muscle. Diabetes 57:2644–2651
Morley JE, Banks WA (2010) Lipids and cognition. J Alzheimers Dis 20:737–747
Morrison CD, White CL, Wang Z, Lee SY, Lawence DS, Cefalu WT, Zhang ZY, Gettys TW (2007) Increased hypothalamic protein tyrosine phosphatase 1B contributes to leptin resistance with age. Endocrinology 148:433–440
Morrow VA, Foufelle F, Connell JMC, Petrie JR, Gould GW, Salt IP (2003) Direct activation of AMP-activated protein kinase stimulates nitric-oxide synthesis in human aortic endothelial cells. J Biol Chem 278:31629–31639
Morton GJ, Blevins JE, Kim F, Matsen M, Figlewicz DP (2009) The action of leptin in the ventral tegmental area to decrease food intake is dependent on Jak-2 signaling. Am J Physiol Endocrinol Metab 297:E202–E210
Mruthinti S, Schade RF, Harrell DU, Gulati NK, Swamy-Mruthinti S, Lee GP, Buccafusco JJ (2006) Autoimmunity in Alzheimer's disease as evidenced by plasma immunoreactivity against RAGE and Abeta42: complication of diabetes. Curr Alzheimer Res 3:229–235
Muoio DM, Newguard CB (2005) Metabolism: A is for adipokine. Nature 436:337–338
Nakamura Y, Shimada K, Fukuda D, Shimada Y, Ehara S, Hirose M, Kataoka T, Kamimori K, Shimodozono S, Kobayashi Y, Yoshiyama M, Takeuchi K, Yoshikawa J (2004) Implications of plasma concentrations of adiponectin in patients with coronary artery disease. Heart 90:528–533

Nawrocki AR, Rajala MW, Tomas E, Pajvani UB, Saha AK, Trumbauer ME, Pang Z, Chen AS, Ruderman NB, Chen H, Rossetti L, Scherer PE (2006) Mice lacking adiponectin show decreased hepatic insulin sensitivity and reduced responsiveness to peroxisome proliferator-activated receptor gamma agonists. J Biol Chem 281:2654–2660

Ng DT, Spear ED, Walter P (2000) The unfolded protein response regulates multiple aspects of secretory and membrane protein biogenesis and endoplasmic reticulum quality control. J Cell Biol 150:77–88

Niswender KD, Baskin DG, Schwartz MW (2004) Insulin and its evolving partnership with leptin in the hypothalamic control of energy homeostasis. Trends Endocrinol Metab 15:362–369

Nogueira-Machado JA, Chaves MM (2008) From hyperglycemia to AGE-RAGE interaction on the cell surface: a dangerous metabolic route for diabetic patients. Expert Opin Ther Targets 12:871–882

Noland RC, Koves TR, Seiler SE, Lum H, Lust RM, Ilkayeva O, Stevens RD, Hegardt FG, Muoio DM (2009) Carnitine insufficiency caused by aging and overnutrition compromises mitochondrial performance and metabolic control. J Biol Chem 284:22840–22852

O'Malley D, Irving AJ, Harvey J (2005) Leptin-induced dynamic changes in the actin cytoskeleton mediate the activation and synaptic clustering of BK channels. FASEB J 19:1917–1919

Odegaard JI, Chawla A (2011) Alternative macrophage activation and metabolism. Annu Rev Pathol 6:275–297

Olefsky JM, Glass CK (2010) Macrophages, inflammation, and insulin resistance. Annu Rev Physiol 72:219–246

Oral EA, Simha V, Ruiz E, Andewelt A, Premkumar A, Snell P, Wagner AJ, De Paoli AM, Reitman ML, Taylor SI, Gorden P, Garg A (2002) Leptin—replacement therapy for lipodystrophy. N Engl J Med 346:570–578

Ouedraogo R, Gong Y, Berzins B, Wu X, Mahadev K, Hough K, Chan L, Goldstein BJ, Scalia R (2007) Adiponectin deficiency increases leukocyte-endothelium interactions via upregulation of endothelial cell adhesion molecules *in vivo*. J Clin Invest 117:1718–1761

Ozcan U, Cao Q, Yilmaz E, Lee AH, Iwakoshi NN, Ozdelen E, Tuncman G, Görgün C, Glimcher LH, Hotamisligil GS (2004) Endoplasmic reticulum stress links obesity, insulin action, and type 2 diabetes. Science 306:457–461

Ozcan U, Yilmaz E, Ozcan L, Furuhashi M, Vaillancourt E, Smith RO, Gorgun CZ, Hotamisligil GS (2006) Chemical chaperones reduce ER stress and restore glucose homeostasis in a mouse model of type 2 diabetes. Science 313:1137–1140

Pacher P, Szabo C (2008) Role of the peroxynitrite-poly(ADP-ribose) polymerase pathway in human disease. Am J Pathol 173:2–13

Pacher P, Beckman JS, Liaudet L (2007) Nitric oxide and peroxynitrite in health and disease. Physiol Rev 87:315–424

Pajvani UB, Du X, Combs TP, Berg AH, Rajala MW, Schulthess T, Engel J, Brownlee M, Scherer PE (2003) Structure–function studies of the adipocyte-secreted hormonec Acrp30/adiponectin. Implications for metabolic regulation and bioactivity. J Biol Chem 278:9073–9085

Peralta S, Carrascosa JM, Gallardo N, Ros M, Arribas C (2002) Ageing increases SOCS-3 expression in rat hypothalamus: effects of food restriction. Biochem Biophys Res Commun 296:425–428

Power DA, Noel J, Collins R, O'Neill D (2001) Circulating leptin levels and weight loss in Alzheimer's disease patients. Dement Geriatr Cogn Disord 12:167–170

Puig KL, Floden AM, Adhikari R, Golovko MY, Combs CK (2012) Amyloid precursor protein and proinflammatory changes are regulated in brain and adipose tissue in a murine model of high fat diet-induced obesity. PLoS One 7:e30378

Qaseem A, Snow V, Cross J Jr, Forciea MA, Hopkins R Jr, Shekelle P, Adelman A, Mehr D, Schellhase K, Campos-Outcalt D, Santaguida P, Owens DK, American College of Physicians/ American Academy of Family Physicians Panel on Dementia (2008) Current pharmacologic treatment of dementia: a clinical practice guideline from the American College of Physicians and the American Academy of Family Physicians. Ann Intern Med 148:370–378

Qiu WQ, Folstein MF (2006) Insulin, insulin-degrading enzyme and amyloid-beta peptide in Alzheimer's disease: review and hypothesis. Neurobiol Aging 27:190–198

Raffaitin C, Feart C, Le Goff M, Amieva H, Helmer C, Akbaraly TN, Tzourio C, Gin H, Barberger-Gateau P (2011) Metabolic syndrome and cognitive decline in French elders: the Three-City Study. Neurology 76:518–525

Reaven GM (1995) Pathophysiology of insulin resistance in human disease. Physiol Rev 75:473–486

Reaven G (2005) All obese individuals are not created equal: insulin resistance is the major determinant of cardiovascular disease in overweight/obese individuals. Diab Vasc Dis Res 2:105–112

Reaven G (2008) Why a cluster is truly a cluster: insulin resistance and cardiovascular disease. Clin Chem 54:785–787

Riccardi G, Giacco R, Rivellese AA (2004) Dietary fat, insulin sensitivity and the metabolic syndrome. Clin Nutr 23:447–456

Rivera EJ, Goldin A, Fulmer N, Tavares R, Wands JR, de la Monte SM (2005) Insulin and insulin-like growth factor expression and function deteriorate with progression of Alzheimer's disease: link to brain reductions in acetylcholine. J Alzheimers Dis 8:247–268

Robertson RP (2004) Chronic oxidative stress as a central mechanism for glucose toxicity in pancreatic islet β cells in diabetes. J Biol Chem 279:42351–42354

Roriz-Filho J, Sá-Roriz TM, Rosset I, Camozzato AL, Santos AC, Chaves ML, Moriguti JC, Roriz-Cruz M (2009) (Pre)diabetes, brain aging, and cognition. Biochim Biophys Acta 1792:432–443

Rothenbacher D, Brenner H, Marz W, Koenig W (2005) Adiponectin, risk of coronary heart disease and correlations with cardiovascular risk markers. Eur Heart J 26:1640–1646

Ruan H, Lodish HF (2003) Insulin resistance in adipose tissue: direct and indirect effects of tumor necrosis factor-alpha. Cytokine Growth Factor Rev 14:447–455

Ruvolo PP (2003) Intracellular signal transduction pathways activated by ceramide and its metabolites. Pharmacol Res 47:383–392

Sader S, Nian M, Liu P (2003) Leptin: a novel link between obesity, diabetes, cardiovascular risk, and ventricular hypertrophy. Circulation 108:644–646

Sakamoto S, Ishikawa K, Senda S, Nakajima S, Matsuo H (1993) The effect of obesity on ventilator and anaerobic threshold during exercise. J Med Syst 17:227–231

Saltiel AR (2001) You are what you secrete. Nat Med 7:887–888

Sasaki T, Kitamura T (2010) Roles of FoxO1 and Sirt1 in the central regulation of food intake. Endocr J 57:939–946

Savage DB, Petersen KF, Shulman GI (2007) Disordered lipid metabolism and the pathogenesis of insulin resistance. Physiol Rev 87:507–520

Scarpace PJ, Matheny M, Moore RL, Tümer N (2000) Impaired leptin responsiveness in aged rats. Diabetes 49:431–435

Schwartz MW, Woods SC, Porte D, Seeley RJ, Baskin DG (2000) Central nervous system control of food intake. Nature 404:661–671

Shankar SS, Steinberg HO (2005) Obesity and endothelial dysfunction. Semin Vasc Med 5:56–64

Sharma AM, Engeli S (2006) The role of renin-angiotensin system blockade in the management of hypertension associated with the cardiometabolic syndrome. J Cardiometab Syndr 1:29–35

Shearer J, Fueger PT, Vorndick B, Bracy DP, Rottman JN, Clanton JA, Wasserman DH (2004) AMP kinase-induced skeletal muscle glucose but not long-chain fatty acid uptake is dependent on nitric oxide. Diabetes 53:1429–1435

Shi H, Clegg DJ (2009) Sex differences in the regulation of body weight. Physiol Behav 97:199–204

Shimada K, Miyazaki T, Daida H (2004) Adiponectin and atherosclerotic disease. Clin Chim Acta 344:1–12

Shoelson SE, Lee J, Yuan M (2003) Inflammation and the IKKβ/IκB/NF-κB axis in obesity- and diet-induced insulin resistance. Int J Obes Relat Metab Disord 27:S49–S52

Sikaris K (2004) The clinical biochemistry of obesity. Clin Biochem Rev 25:165–181

Sommer G, Kralisch S, Lipfert J, Weise S, Krause K, Jessnitzer B, Lössner U, Blüher M, Stumvoll M, Fasshauer M (2009) Amyloid precursor protein expression is induced by tumor necrosis factor alpha in 3 T3-L1 adipocytes. J Cell Biochem 108:1418–1422

Sonnen JA, Larson EB, Haneuse S, Woltjer R, Li G, Crane PK, Craft S, Montine TJ (2009) Neuropathology in the adult changes in thought study: a review. J Alzheimers Dis 18:703–711

Spangler R, Wittkowski KM, Goddard NL, Avena NM, Hoebel BG, Leibowitz SF (2004) Opiate-like effects of sugar on gene expression in reward areas of the rat brain. Brain Res Mol Brain Res 124:134–142

Steen E, Terry BM, Rivera EJ, Cannon JL, Neely TR, Tavares R, Xu XJ, Wands JR, de la Monte SM (2005) Impaired insulin and insulin-like growth factor expression and signaling mechanisms in Alzheimer's disease—is this type 3 diabetes? J Alzheimers Dis 7:63–80

Straub RH, Buttgereit F, Cutolo M (2011) Alterations of the hypothalamic-pituitary-adrenal axis in systemic immune diseases—a role for misguided energy regulation. Clin Exp Rheumatol 29(5 Suppl 68):S23–S31

Stumvoll M, Goldstein BJ, van Haeften TW (2005) Type 2 diabetes: principles of pathogenesis and therapy. Lancet 365:1333–1346

Summers SA, Garza L, Zhou H, Birnbaum MJ (1998) Regulation of insulin-stimulated glucose transporter GLUT4 translocation and Akt kinase activity by ceramide. Mol Cell Biol 18:5457–5464

Sun XJ, Liu F (2009) Phosphorylation of IRS proteins Yin-Yang regulation of insulin signaling. Vitam Horm 80:351–387

Szanto I, Kahn CR (2000) Selective interaction between leptin and insulin signaling pathways in a hepatic cell line. Proc Natl Acad Sci USA 97:2355–2360

Taheri S, Lin L, Austin D, Young T, Mignot E (2004) Short sleep duration is associated with reduced leptin, elevated ghrelin, and increased body mass index. PLoS Med 1:e62

Takeda S, Sato N, Rakugi H, Morishita R (2011) Molecular mechanisms linking diabetes mellitus and Alzheimer disease: beta-amyloid peptide, insulin signaling, and neuronal function. Mol Biosyst 7:1822–1827

Tanaka M, Suganami T, Sugita S, Shimoda Y, Kasahara M, Aoee SS, Takeya MM, Takeda SS, Kamei YY, Ogawa YY (2010) Role of central leptin signaling in renal macrophage infiltration. Endocr J 57:61–72

Tanaka M, Suganami T, Kim-Saijo M, Toda C, Tsuiji M, Ochi K, Kamei Y, Minokoshi Y, Ogawa Y (2011) Role of central leptin signaling in the starvation-induced alteration of B-cell development. J Neurosci 31:8373–8380

Tartaglia LA, Dembski M, Weng X, Deng N, Culpepper J, Devos R, Richards GJ, Campfield LA, Clark FT, Deeds J, Muir C, Sanker S, Moriatry A, Moore KJ, Smutko JS, Mays GG, Wool EA, Monroe CA, Tepper RI (1995) Identification and expression cloning of a leptin receptor OB-R. Cell 83:1263–1271

Thaler JP, Schwartz MW (2010) Minireview: inflammation and obesity pathogenesis: the hypothalamus heats up. Endocrinology 151:4109–4115

Thaler JP, Yi C-X, Schur EA, Guyenet SJ, Hwang BH, Dietrich MO, Zhao X, Sarruf DA, Izgur V, Maravilla KR, Nguyen HT, Fischer JD, Matsen ME, Wisse BE, Morton GJ, Horvath TL, Baskin DG, Tschop MH, Schwartz MW (2012) Obesity is associated with hypothalamic injury in rodents and humans. J Clin Invest 122:153–162

Thomas DE, Elliott EJ, Baur L (2007) Low glycaemic index or low glycaemic load diets for overweight and obesity. Cochrane Database Syst Rev 2007, CD005105

Tilg H, Moschen AR (2006) Adipocytokines: mediators linking adipose tissue, inflammation and immunity. Nat Rev Immunol 6:772–783

Tomas E, Tsao TS, Saha AK, Murray HE, Zhang CC, Itani SI, Lodish HF, Ruderman NB (2002) Enhanced muscle fat oxidation and glucose transport by ACRP30 globular domain: acetyl-CoA carboxylase inhibition and AMP-activated protein kinase activation. Proc Natl Acad Sci USA 99:16309–16313

Toth C, Schmidt AM, Tuor UI, Francis G, Foniok T, Brussee V, Kaur J, Yan SF, Martinez JA, Barber PA, Buchan A, Zochodne DW (2006) Diabetes, leukoencephalopathy and rage. Neurobiol Dis 23:445–461

Unger RH (2003) Lipid overload and overflow: metabolic trauma and the metabolic syndrome. Trends Endocrinol Metab 14:398–403

Urano F, Wang X, Bertolotti A, Zhang Y, Chung P, Harding HP, Ron D (2000) Coupling of stress in the ER to activation of JNK protein kinases by transmembrane protein kinase IRE1. Science 287:664–666

Valassi E, Scacchi M, Cavagnini F (2008) Neuroendocrine control of food intake. Nutr Metab Cardiovasc Dis 18:158–168

Valdes AM, Andrew T, Gardner JP, Kimura M, Oelsner E, Cherkas LF, Aviv A, Spector TD (2005) Obesity, cigarette smoking, and telomere length in women. Lancet 366:662–664

Vincent HK, Taylor AG (2006) Biomarkers and potential mechanisms of obesity-induced oxidant stress in humans. Int J Obes (Lond) 30:400–418

Virag L, Szabo C (2002) The therapeutic potential of poly(ADP-ribose) polymerase inhibitors. Pharmacol Rev 54:375–429

Virtanen JK, Voutilainen S, Rissanen TH, Happonen P, Mursu J, Laukkanen JA, Poulsen H, Lakka TA, Salonen JT (2006) High dietary methionine intake increases the risk of acute coronary events in middle-aged men. Nutr Metab Cardiovasc Dis 16:113–120

Volkow ND, Wang GJ, Baler RD (2011) Reward, dopamine and the control of food intake: implications for obesity. Trends Cogn Sci 15:37–46

Wajchenberg BL, Giannella-Neto D, da Silva ME, Santos RF (2002) Depot-specific hormonal characteristics of subcutaneous and visceral adipose tissue and their relation to the metabolic syndrome. Horm Metab Res 34:616–621

Wang Y, Beydom MA (2007) The obesity epidemic in the United States—gender, age, socioeconomic, racial/ethnic, and geographic characteristics: a systematic review and meta-regression analysis. Epidemiol Rev 29:6–28

Wang Q, Jin T (2009) The role of insulin signaling in the development of beta-cell dysfunction and diabetes. Islets 1:95–101

Wang Z, Zhou YT, Kakuma T, Lee Y, Kalra SP, Kalra PS, Pan W, Unger RH (2000) Leptin resistance of adipocytes in obesity: role of suppressors of cytokine signaling. Biochem Biophys Res Commun 277:20–26

Watt MJ, Hevener A, Lancaster GI, Febbraio MA (2006) Ciliary neurotrophic factor prevents acute lipid-induced insulin resistance by attenuating ceramide accumulation and phosphorylation of JNK in peripheral tissues. Endocrinology 147:2077–2085

Whitmer RA, Sidney S, Selby J, Johnston SC, Yaffe K (2005) Midlife cardiovascular risk factors and risk of dementia in late life. Neurology 64:277–281

WHO (2011a) WHO calls for action to restrict marketing of unhealthy foods and drinks to children. BMJ 342:d503

WHO (World Health Organization) (2011b) Obesity and overweight. 2010. http://www.who.int/dietphysicalactivity/media/en/gsfs_obesity.pdf

Williams KW, Scott MM, Elmquist JK (2011) Modulation of the central melanocortin system by leptin, insulin, and serotonin: co-ordinated actions in a dispersed neuronal network. Eur J Pharmacol 660:2–12

World Health Organization (2006) Neurological disorders: public health challenges. WHO, Geneva

Xiao X, Su G, Brown SN, Chen L, Ren J, Zhao P (2010) Peroxisome proliferator-activated receptors gamma and alpha agonists stimulate cardiac glucose uptake via activation of AMP-activated protein kinase. J Nutr Biochem 21:621–626

Xu H, Barnes GT, Yang Q, Tan G, Yang D, Chou CJ, Sole J, Nichols A, Ross JS, Tartaglia LA, Chen H (2003) Chronic inflammation in fat plays a crucial role in the development of obesity-related insulin resistance. J Clin Invest 112:1821–1830

Xu WL, Qiu CX, Wahlin A, Winblad B, Fratiglioni L (2004) Diabetes mellitus and risk of dementia in the Kungsholmen project: a 6-year follow-up study. Neurology 63:1181–1186

Xu L, Rensing N, Yang XF, Zhang HX, Thio LL, Rothman SM, Weisenfeld AE, Wong M, Yamada KA (2008) Leptin inhibits 4-aminopyridine- and pentylenetetrazole-induced seizures and AMPAR-mediated synaptic transmission in rodents. J Clin Invest 118:272–280

Xydakis AM, Case CC, Jones PH, Hoogeveen RC, Liu MY, Smith EO, Nelson KW, Ballantyne CM (2004) Adiponectin, inflammation and the expression of the metabolic syndrome in obese individuals: the impact of rapid weight loss through caloric restriction. J Clin Endocrinol Metab 89:2697–2703

Yadav A, Kalita A, Dhillon S, Banerjee K (2005) JAK/STAT3 pathway is involved in survival of neurons in response to insulin-like growth factor and negatively regulated by suppressor of cytokine signaling-3. J Biol Chem 280:31830–31840

Yamauchi T, Kamon J (2002) Adiponectin stimulates glucose utilization and fatty acid oxidation by activating AMP-kinase. Nat Med 8:1288–1295

Yamauchi T, Kamon J, Waki H, Terauchi Y, Kubota N, Hara K, Mori Y, Ide T, Murakami K, Tsuboyama-Kasaoka N, Ezaki O, Akanuma Y, Gavrilova O, Vinson C, Reitman ML, Kagechika H, Shudo K, Yoda M, Nakano Y, Tobe K, Nagai R, Kimura S, Tomita M, Froguel P, Kadowaki T (2001) The fat-derived hormone adiponectin reverses insulin resistance associated with both lipoatrophy and obesity. Nat Med 7:941–946

Yamauchi T, Kamon J, Waki H, Imai Y, Shimozawa N, Hioki K (2003) Globular adiponectin protected ob/ob mice from diabetes and ApoE-deficient mice from atherosclerosis. J Biol Chem 278:2461–2468

Yamauchi T, Nio Y, Maki T, Kobayashi M, Takazawa T, Iwabu M, Okada-Iwabu M, Kawamoto S, Kubota N, Kubota T, Ito Y, Kamon J, Tsuchida A, Kumagai K, Kozono H, Hada Y, Ogata H, Tokuyama K, Tsunoda M, Ide T, Murakami K, Awazawa M, Takamoto I, Froguel P, Hara K, Tobe K, Nagai R, Ueki K, Kadowaki T (2007) Targeted disruption of AdipoR1 and AdipoR2 causes abrogation of adiponectin binding and metabolic actions. Nat Med 13:332–339

Yan SD, Stern D, Schmidt AM (1997) What's the RAGE? The receptor for advanced glycation end products (RAGE) and the dark side of glucose. Eur J Clin Invest 27:179–181

Yaqoob P (2009) The nutritional significance of lipid rafts. Annu Rev Nutr 29:257–282

Yoon MJ, Lee GY, Chung JJ, Ahn YH, Hong SH, Kim JB (2006) Adiponectin increases fatty acid oxidation in skeletal muscle cells by sequential activation of AMP-activated protein kinase, p38 mitogen-activated protein kinase, and peroxisome proliferator-activated receptor alpha. Diabetes 55:2562–2570

Yu C, Chen Y, Cline GW, Zhang D, Zong H, Wang Y, Bergeron R, Kim JK, Cushman SW, Cooney GJ, Atcheson B, White MF, Kraegen EW, Shulman GI (2002) Mechanism by which fatty acids inhibit insulin activation of insulin receptor substrate-1 (IRS-1)-associated phosphatidylinositol 3-kinase activity in muscle. J Biol Chem 277:50230–50236

Yudkin JS, Kumari M, Humphries SE, Mohamed-Ali V (2000) Inflammation, obesity, stress and coronary heart disease: is interleukin-6 the link? Atherosclerosis 148:209–214

Zang M, Xu S, Maitland-Toolan KA, Zuccollo A, Hou X, Jiang B, Wierzbicki M, Verbeuren TJ, Cohen RA (2006) Polyphenols stimulate AMP-activated protein kinase, lower lipids, and inhibit accelerated atherosclerosis in diabetic LDL receptor-deficient mice. Diabetes 55:2180–2191

Zeyda M, Stulnig TM (2007) Adipose tissue macrophages. Immunol Lett 112:61–67

Zeyda M, Stulnig TM (2009) Obesity, inflammation, and insulin resistance—a mini-review. Gerontology 55:379–386

Zhang Y, Proenca R, Maffei M, Barone M, Leopold L, Friedman JM (1994) Positional cloning of the mouse obese gene and its human homologue. Nature 372:425–432

Zhao AZ, Huan JN, Gupta S, Pal R, Sahu A (2002) A phosphatidylinositol 3-kinase phosphodiesterase 3B-cyclic AMP pathway in hypothalamic action of leptin on feeding. Nat Neurosci 5:727–728

Zhou L, Deepa SS, Etzler JC, Ryu J, Mao X et al (2009) Adiponectin activates AMP-activated protein kinase in muscle cells via APPL1/LKB1-dependent and phospholipase C/$Ca^{2+}$/$Ca^{2+}$/calmodulin-dependent protein kinase kinase-dependent pathways. J Biol Chem 284:22426–22435

Ziegler O, Quilliot D, Guerci B, Drouin P (2001) Macronutrients, fat mass, fatty acid flux and insulin sensitivity. Diabetes Metab 27:261–270

Zimmerman GA, Meistrell M III, Bloom O, Cockroft KM, Bianchi M, Risucci D, Broome J, Farmer P, Cerami A, Vlassara H, Tracey KJ (1995) Neurotoxicity of advanced glycation end-products during focal stroke and neuroprotective effects of aminoguanidine. Proc Natl Acad Sci USA 92:3744–3748

# Chapter 6
# Effect of Dietary Phytochemicals on Metabolic Syndrome and Neurological Disorders

## 6.1 Introduction

Metabolic syndrome (MetS) and neurological disorders are conditions that affect a large number of people. Both conditions are on increase due to the consumption of Western diet and sedentary lifestyle (Taubes 2008). Consumption of Western diet (high-fat, high-protein diet, which is also enriched in sugar and refined flour) along with lack of physical activity and stress are culprits that produce unhealthy effects in the Western countries. Thus, high-fat Western diet exerts a powerful adverse effect on health and well-being (Garrel et al. 1995; Baran et al. 2005) (Fig. 6.1). Consumption of Western diet results in the growing incidence of visceral obesity, dyslipidemia, hypertension, endothelial dysfunction, and impairment in insulin receptor (IR) signaling due to hyperglycemic condition. These processes initiate and result in MetS, a pathological condition characterized by the above-mentioned pathological parameters. People with MetS are twice as likely to die from heart attack or stroke compared with people without the syndrome and have a fivefold greater risk of developing type II diabetes (Farooqui et al. 2012). Impairment in insulin signaling (insulin resistance) is the main and common defect in MetS and neurological disorders. Insulin signaling affects blood glucose levels and also acts as a growth factor for all cells including neurons in the brain. Consumption of Western diet induces metabolic dysfunction capable of disrupting brain homeostasis and is likely to contribute the development of brain dysfunction associated with pathogenesis of neurological disorders. The common pathological processes in Alzheimer disease (AD) and type II diabetes, a major component of MetS, have recently been reviewed by several investigators (Farooqui et al. 2012; Aggarwal 2010). It is suggested that generation of free radicals (oxidative stress) plays a vital role in aging, MetS, and neurological disorders. The brain is especially vulnerable to oxidative damage because of its high oxygen consumption rate, its high levels of unsaturated fatty acids, and relative low activity of antioxidant enzymes as compared to other tissues. Consumption of high-fat diet containing n-6 fatty acids under aerobic metabolism produces not only reactive oxygen species (ROS) but high

A.A. Farooqui, *Metabolic Syndrome: An Important Risk Factor for Stroke, Alzheimer Disease, and Depression*, DOI 10.1007/978-1-4614-7318-3_6,

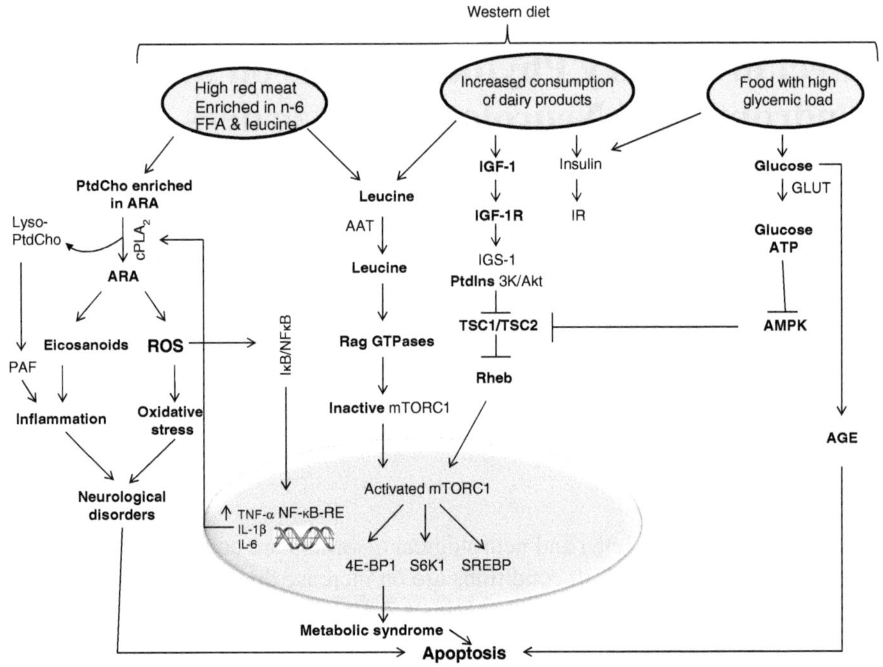


**Fig. 6.1** Effect of Western diet components on insulin signaling and metabolic syndrome. Free fatty acids (FFA); arachidonic acid (ARA); reactive oxygen species (ROS); cytosolic phospholipase $A_2$ ($cPLA_2$); insulin/insulin-like growth factor-1 (IGF-1); mammalian target of rapamycin complex 1 (mTORC1); AMP-activated protein kinase (AMPK); phosphoinositol 3-kinases (PtdIns 3Ks); insulin receptor (IR); glucose transporter (GLUT); tuberous sclerosis proteins TSC1 (hamartin); TSC2 (tuberin); nuclear factor-κB (NF-κB); nuclear factor-κB response element (NF-κB-RE); inhibitory subunit of NF-κB (IκB); tumor necrosis factor-α (TNF-α); interleukin-1β (IL-1β); interleukin-6 (IL-6). The Western diet with increased glucose-, insulin/IGF-1-, and leucine-induced stimulation of mTORC1 may be associated with the pathogenesis of type II diabetes. Pro-proliferative signals of the Western diet and along with the persistent increase in levels of leucine may modulate the functional capacity of the endoplasmic reticulum and thus β-cell homeostasis

levels of inflammatory eicosanoids and platelet-activating factor (PAF) (Fig. 6.1). Neural and visceral cells and tissues possess defense mechanisms against the removal of reactive oxygen species (ROS). The chemical reactivity of ROS varies from the very toxic hydroxyl [$^{\bullet}OH$] to the less reactive superoxide radical ($O_2^{\bullet-}$). The initial product, $O_2^{\bullet-}$, results from the addition of a single electron to molecular oxygen. $O_2^{\bullet-}$ is rapidly dismutated by superoxide dismutase (SOD), yielding $H_2O_2$ and $O_2$, which can be reused to generate superoxide radical (Farooqui 2012a). In the presence of reduced transition metals, $H_2O_2$, although less reactive than $O_2^{\bullet-}$ and highly diffusible, can be converted into the highly reactive hydroxyl radical $^{\bullet}HO$. The tight regulation of ROS generation and removal makes fluctuations in their levels transient, a feature which is a characteristic of second messengers. Under normal physiological conditions, a balance exists between the production of ROS and antioxidant mechanism. ROS act as second messengers and play important physiological roles not only in maintaining cell redox status, but also in regulating normal cellular functions, including cell proliferation, aggregation, chemotaxis and apoptosis. ROS also regulate intracellular signaling pathways by modulating activities of transcription factors, such as nuclear factor (NF-κB), activator protein 1, and hypoxia-inducible factor-1α. However, under pathological conditions, consumption of the Western diet results in overproduction of ROS and advanced glycation products (AGEs), both of which have been implicated in the pathogenesis of MetS and neurodegenerative diseases through increased oxidation of carbohydrates, lipids, and proteins (Vicente Miranda and Outeiro 2010; Farooqui 2010).

## 6.2 Effect of Diet on Metabolic Syndrome and Neurological Disorders

Epidemiological, population-based, and randomized clinical trials indicate that diet affects human health. Thus, diet rich in some beneficial food groups such as fruit, vegetables, whole grains, and fish can reduce the incidence of MetS and cardiovascular, cerebrovascular, neurological, and neoplastic diseases (Willett 2006). Studies with dietary approaches to stop hypertension (DASH) intervention also indicate that DASH diet not only reduces weight but also reduces hypertension in both men and women compared to age-matched control subjects (Azadbakht et al. 2005). The DASH diet is rich in calcium, magnesium, and potassium, which may also lower the risk of stroke. Fiber and other phytonutrients in fruit and vegetables may be protective by lowering cholesterol or decrease markers of inflammation. In contrast, Western diet (American diet) is enriched in "red meat (high amount of n-6 fatty acids and leucine), refined carbohydrate (sugar, white flour, and white rice), and soft drinks containing fructose corn syrup" that is associated with a higher risk for developing chronic visceral diseases (MetS, type II diabetes, and arthritis) as well as neurological and neuropsychiatric disorders (stroke, AD and PD, depression, and schizophrenia) (Fig. 6.1). Western diet with high contents of simple sugars and food with a high glycemic index is associated which hyperglycemia, insulin resistance, type II diabetes, hypertriglyceridemia, and low HDL cholesterol. Prolonged hyperglycemia causes tissue damage through several mechanisms including increase in flux of glucose and other sugars through the polyol pathway, increase in intracellular formation of AGEs, increase in expression of the receptor for advanced glycation end products (RAGE) and its activating ligands, activation of protein kinase C (PKC) isoforms, and overactivation of the hexosamine pathway. These mechanisms are activated by a single upstream event (Brownlee 2005), namely, mitochondrial overproduction of ROS. Western diet also induces oxidative stress and inflammation in cardiovascular and cerebrovascular systems. Elevations in FFA promote metabolic disturbances in part by impairing the ability of cells to store glucose and to respond to insulin leading to the development of type II diabetes. These processes are aided by increased expressions of cytokines and adipokines. The increase in plasma fatty acids and high cholesterol contents may cause activated redox cycling of the copper–albumin complex, excessive lipid peroxidation, increased expression of adipokines and cytokines, and accumulation of AGEs in body tissues. Formation of AGEs leads to protein cross-linking and contributes to the development and progression of several type II diabetes and MetS complications, such as cataract formation, atherosclerosis, neuropathy, and peripheral neuropathy.

Activation of RAGE also represents important mediators of cardiac and neural injuries following exposure to a high-fat diet (Ghosh et al. 2007). Circulation of AGE-modified proteins in the body can affect a range of cells and tissues. RAGEs have been shown to mediate their action through the activation of NFκB, and p21 ras dependent mitogen-activated protein kinase pathway (Lander et al. 1997). It is demonstrated that the RAGE–NF-κB axis operates in diabetic neuropathy, mediating functional sensory deficits (Bierhaus et al. 2004). In endothelial cells, interactions of AGE with RAGE alter the expression of several genes, including thrombomodulin, tissue factor, and vascular cell adhesion molecule 1 (VCAM1) (Vlassara et al. 1995). These effects not only mediate procoagulatory changes on the endothelial cell surface but also increase the adhesion of inflammatory cells to the endothelium. In addition, endothelial RAGE receptors induce an increase in vascular permeability caused by type II diabetes through the induction of vascular endothelial growth factor (VEGF) (Wautier et al. 1996; Yao and Brownlee 2010). In brain, the interaction of Aβ with RAGE produces neuronal stress and the activation of different signaling pathways, which mediate the effects of Aβ on microglial cells and blood–brain barrier. RAGE is an important player in the pathogenesis of AD. RAGE is expressed in astrocytes and recruits monocytes across the blood–brain barrier activating NF-κB and, consequently, increasing oxidative stress (Takeuchi et al. 2000; Yan et al. 1998). In fact, AGEs, Aβ, and RAGE colocalize in astrocytes, suggesting that Aβ may be taken via RAGE to be degraded by the lysosomal pathway in these cells (Sasaki et al. 2001; Yan et al. 1998).

High-fat diet not only generates elevated levels of proinflammatory eicosanoids, platelet-activating factor (PAF), and proinflammatory cytokines (TNF-α and IL-1β) but also upregulates the gp91(phox) subunit of NADPH oxidase and downregulates superoxide dismutase (SOD) isoforms, glutathione peroxidase (GPX), and heme oxygenase-2 (HO-2) in various body tissues (Roberts et al. 2006). This is accompanied by an increase in plasma malondialdehyde levels and impaired vasodilatory response to acetylcholine. It is also shown that these processes reduce levels of BDNF in the hippocampus and suggest that "Western" diet impairs synaptic function, induces metabolic perturbation, reduces BDNF, and produces atrophy of dendritic spines (Farooqui 2012a, b) (Fig. 6.2). These observations strongly support the view that American diet produces oxidative stress and inflammation (Roberts et al. 2006; Farooqui 2012b). Increased levels of oxidative stress in neural cells damage major classes of cellular molecules, including proteins, lipids, and nucleic acids. Some oxidative modifications of proteins have been reported in neuronal cells during aging include carbonyl formation (Butterfield et al. 1997; Farooqui 2012b); covalent modifications of cysteine, lysine, and histidine residues by the lipid

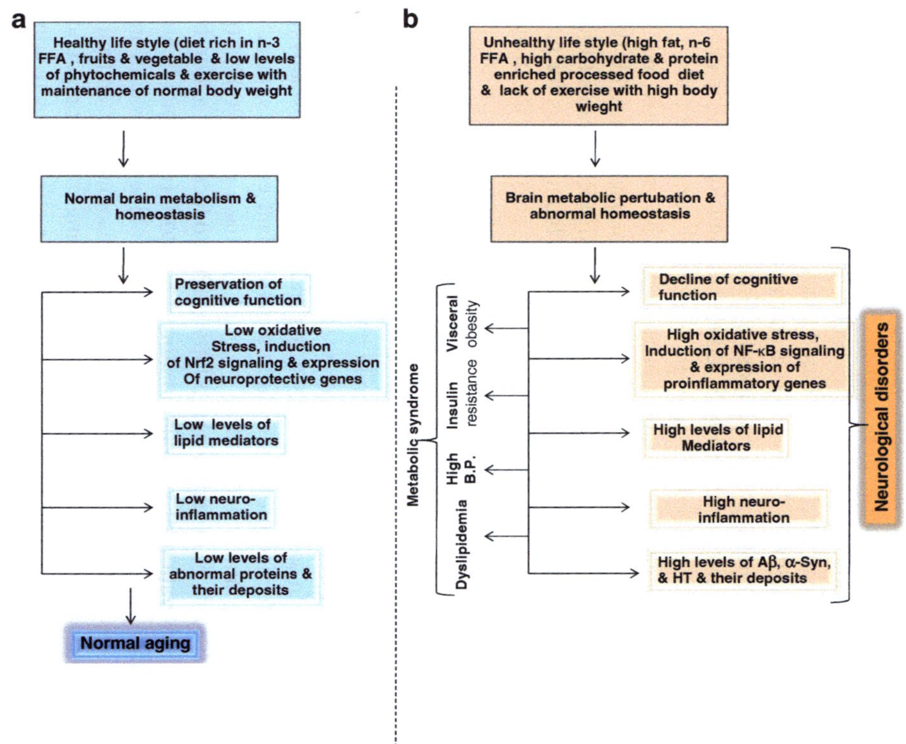


**Fig. 6.2** Effect of healthy and unhealthy lifestyle on parameter of neural cell death in metabolic syndrome and neurological disorders. Consumption of Western diet is associated not only with insulin resistance, adiposity, and low-grade inflammation (MetS) but also with induction of oxidative stress and neuroinflammation leading to neurological disorders. In contrast, consumption of diet enriched in fruits (phytochemicals), vegetables (phytochemicals), legumes, and cereals, with olive oil and n-3 fatty acids as the only sources of fat, and moderate consumption of red wine as a source of flavonoids especially during meals may delay the onset of MetS as well as neurological disorders

peroxidation product 4-hydroxynonenal (Papaioannou et al. 2001; Farooqui 2012a); nitration of proteins on tyrosine residues (Sloane et al. 1999); and generation of AGEs by glycation (Munch et al. 2000). Additionally, a common oxidative modification of DNA that is observed during brain aging is the formation of 8-hydroxydeoxyguanosine (Sohal et al. 1994).

n-3 fatty acids include α-linolenic acid (ALA), eicosapentaenoic acid (EPA), and docosahexaenoic acid (DHA). ALA is enriched in flaxseeds, whereas EPA and DHA are mostly derived from fish, which consumes golden marine algae. Diet rich in n-3 fatty acids, phytochemicals, and unrefined grain fiber inhibits oxidative stress and inflammation by blocking generation of eicosanoids, platelet-activating factor, and expression of proinflammatory cytokines (Farooqui et al. 2007a, b; Farooqui 2012c) (Fig. 6.2). n-3 fatty acids also decrease blood triacylglycerol concentrations and are weakly associated with HDL cholesterol concentrations, but there is no association with non-HDL cholesterol or adiposity (Belalcazar et al. 2009). In murine models of MetS, n-3 fatty acids strongly reduce insulin resistance, improve glucose tolerance, and decrease blood pressure. In neural and nonneural cells, n-3 fatty acids and phytochemicals block transcription factor NF-κB. This redox-sensitive transcription factor resides in the cytosol through complexation with IkB proteins. Upon phosphorylation, the IkBa is degraded and NF-κB migrates to the nucleus, where it initiates the transcription of many genes including genes for pro-inflammatory cytokines (TNF-α and IL-1β), proinflammatory enzymes (iNOS and COX-2), other genes (p53 and c-Myc) (Farooqui 2010). Alterations in brain energy metabolism due to n-3 deficiency are evidenced by a significant decrease in AMP-activated serine/threonine protein kinase (AMPK) phosphorylation and its upstream modulator tumor suppressor protein (LKB1) as well as a decrease in Sir2 levels. n-3 fatty acid deficiency also results in the reduction in phosphorylation of CREB, synapsin I, and synaptophysin levels leading to decrease in synaptic plasticity (Agrawal and Gomez-Pinilla (2012). In addition, n-3 fatty acids and phytochemicals activate genes that encode cytoprotective proteins including antioxidant enzymes, protein chaperones, growth factors, and mitochondrial proteins (Mattson et al. 2007; Son et al. 2008; Farooqui 2012c). For examples, n-3 fatty acids produce their action by (a) the inhibiting the generation of proinflammatory mediators (eicosanoids and platelet activating factor), suppression of proinflammatory cytokine expression, and generation of docosanoids with antioxidant, antithrombotic, vasodialatory, and anti-apoptotic properties; (b) sulforaphane (an anticancer and antimicrobial compound) and curcumin (principal component of tumeric) induce their effect by activating nuclear factor-E2-related factor-2 (Nrf2)-antioxidant response element (Nrf-2-ARE); (c) allicin (major component of garlic) and capsaicin (the active component of chili peppers) mediate their effect through the activation of transient receptor potential (TRP) ion channels; and (d) resveratrol (a component of red wine) produce its effect through the activation of sirtuin-1 (Mattson 2008; Son et al. 2008; Farooqui 2009, 2012b, c). Phytochemicals and n-3 fatty acids not only target oxidative stress and inflammation but also modulate angiogenesis, ionic homeostasis, and redox imbalance through cross talk signaling networks involved in controlling energy metabolism (Farooqui 2009, 2012b). Other effects of phytochemicals and n-3 fatty acids include alterations in immune function and changes in metabolism. At low doses, phytochemicals and n-3 fatty acids not only improve neuronal and cognitive functions, ocular health, and memory formation but also protect genomic DNA integrity, inducing neuronal regeneration, and help strengthen the immune system (Farooqui 2009). In addition, humans consuming Western diet also have higher prevalence of major depressive disorders. Their plasma levels of n-3 fatty acids are lower than normal subjects. DHA deficiency in depressive patients is associated with dysfunctions of neuronal membrane stability and transmission of serotonin, norepinephrine, and dopamine, which may be linked to the pathogenesis of mood and cognitive dysfunction in depression. On the other hand, EPA is important in balancing the immune function and physical healthy by reducing ARA level on cell membrane and prostaglandin $E_2$ ($PGE_2$) synthesis (Farooqui 2009).

Consumption of phytochemicals induces neurohormetic response that results in expression of adaptive stress resistance genes responsible for encoding antioxidant enzymes, protein chaperones, and neurotrophic factor (BDNF) (Mattson and Cheng 2006). Neurohormesis not only provides a framework for explaining the dose–response relationships between concentration of phytochemicals and beneficial effects but also explains a key insight for improving the accuracy of the therapeutic dose of phytochemicals within the highly heterogeneous human population (Calabrese et al. 2012). Based on the stimulation of signal transduction network and adaptive stress resistance genes, it is proposed that the use of phytochemicals from childhood to old age along with regular exercise 3–4 times per week is an inexpensive strategy for maintaining normal aging and delaying onset of age-related neurological disorders (stroke; Alzheimer disease, AD; and Parkinson disease, PD) (Joseph et al. 2009; Farooqui and Farooqui 2010, 2011; Farooqui 2012b). Risk factors for neurological disorders (stroke, AD, and PD) include old age, positive family history, unhealthy life style, consumption of high-fat diet, and exposure to toxic environment. These risk factors contribute to abnormal protein processing leading to oligomerization of unfolded proteins, generation of ROS, induction of neuroinflammation, and apoptotic cell death in neurological disorders (Farooqui 2010, 2012c; Farooqui and Farooqui 2011). Long-term use of diet containing n-3 fatty acids and phytochemicals may retard the effect of unhealthy lifestyle by delaying or slowing the onset of stroke, AD, and PD. These processes may lead to improvement in health status and quality of life in older age. Among various diets (Western, Indian, Chinese, and Mediterranean), the traditional Mediterranean diet contains fruits, vegetables, fish, nuts, olives (monounsaturated fatty acids), herbs, spices, whole grains, and moderate amount of red wine, which promote good heath by lowering incidence type II diabetes and reducing the risk of cardiovascular, cerebrovascular, and neurological disorders (Scholl 2012; Farooqui 2009, 2012c). In addition, the Mediterranean diet reduces cellular and circulating inflammatory mediators associated with atherosclerosis, insulin resistance, MetS, and neurological disorders (Reisin 2009; Farooqui et al. 2012). Results of Healthy Aging Longitudinal Study in Europe (HALE study) have indicated that consumption of the Mediterranean diet by 70–90-year-old individuals results in more than 50 % lower rate of all-cause mortality and reduction in both type II diabetes and cardiovascular

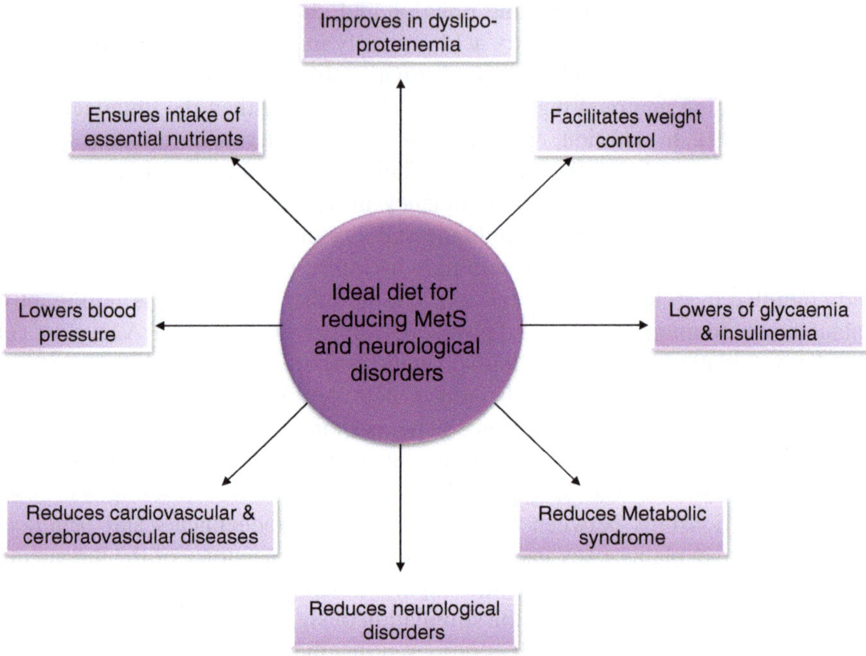


**Fig. 6.3** Characteristics of ideal healthy diet

diseases, compared to an age-matched control group of people maintaining their normal dietary and lifestyle patterns (Knoops et al. 2004). Diet is a major source of phytochemical intake, and diets rich in fruits and vegetables contain an abundance of various classes of phytochemicals and high levels of potassium, antioxidants, and dietary fiber. An ideal diet should be rich in fruits (phytochemicals), vegetables (phytochemicals), legumes, and cereals as sources of plant-derived proteins and vitamins, olive oil, and n-3 fatty acids as sources of fats, along with moderate consumption of red wine especially during meals, and very little red meat. This type of diet has been shown to be beneficial for cardiovascular and cerebrovascular diseases as well as endothelial dysfunction in obesity (Willett 2006). Endothelial dysfunction is a critical event in the pathogenesis of atherosclerosis and ischemic injury (Deanfield et al. 2007). It is characterized by a decrease in bioavailability of the endogenous vasodilator NO, which is generated from L-arginine (Huynh and Chin-Dusting 2006) along with upregulation of proinflammatory cytokines expression and cellular adhesion molecules. This type of diet would not only have the ability to improve dyslipoproteinemia, lower blood pressure, lower insulinemia, and promote weight control but also induce antioxidant and anti-inflammatory effects, provide all essential nutrients, and reduce the risk of type II diabetes, MetS, and cardiovascular and neurological disorders (Figs. 6.3 and 6.4).

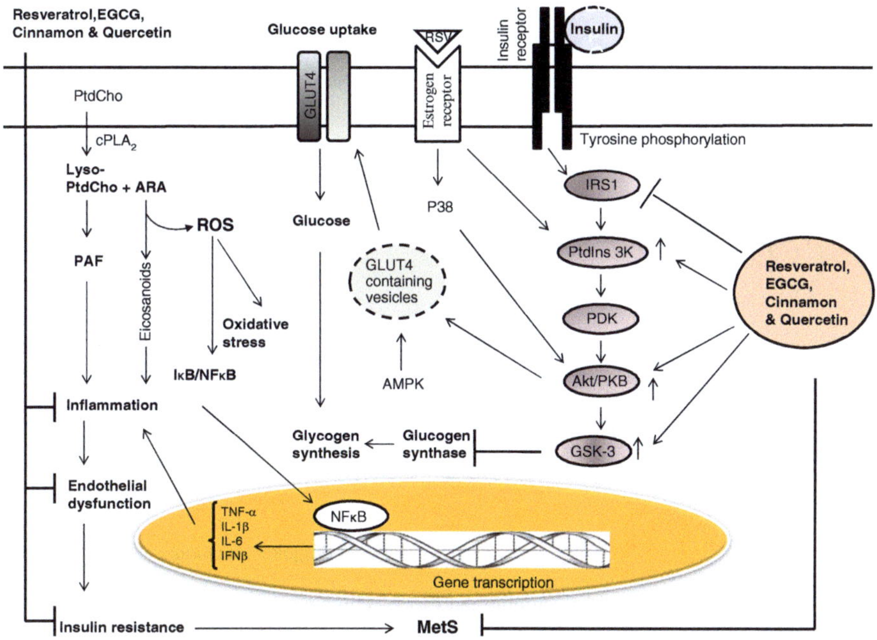


**Fig. 6.4** Effect of phytochemicals on insulin signaling and metabolic syndrome. Arachidonic acid (ARA); phosphatidylcholine (PtdCho); lyso-phosphatidylcholine (lyso-PtdCho); reactive oxygen species (ROS); cytosolic phospholipase $A_2$ ($cPLA_2$); AMP-activated protein kinase (AMPK); phosphatidylinositol 3-kinases (PtdIns 3K); insulin receptor (IR); glucose transporter 4 (GLUT4); nuclear factor-κB (NF-κB); nuclear factor-κB response element (NF-κB-RE); inhibitory subunit of NF-κB (IκB); resveratrol (RSV); tumor necrosis factor-α (TNF-α); interleukin-1β (IL-1β); interleukin-6 (IL-6); phosphatidylinositol 3-kinases (PtdIns 3K); serine/threonine protein kinase (Akt); glycogen synthase 3-kinase (GSK3); and green tea catechin (EGCG)

### 6.2.1 *Effect of Curcumin on Metabolic Syndrome*

Curcumin (diferuloylmethane), a hydrophobic polyphenol, is a principal active constituent of turmeric (orange-yellow curry powder) (Fig. 6.5), which is derived from the rhizome of the plant *Curcuma longa*. Curcumin has been used as an antioxidant, anti-inflammatory, anti-adipogenic, antitumor, antiviral, and antiarthritic agent (Anand et al. 2008; Aggarwal 2010; Farooqui 2012b). Underlying mechanisms of above mentioned effects are not fully understood. However, based on several investigations, it is proposed that molecular target for curcumin may include transcription factors (nuclear factor-κB, AP-1, STAT, HIF-1, Notch-1, Egr-1, β-catenin, PPAR-γ, and Nrf2), growth factors (vascular endothelial cell growth factor), proinflammatory cytokines (tumor necrosis factor-α, interleukin 1β and interleukin 6), and protein kinases (mammalian target of rapamycin, mitogen-activated protein kinases, and Akt). Curcumin also interacts and modulates cyclooxygenase-2 and

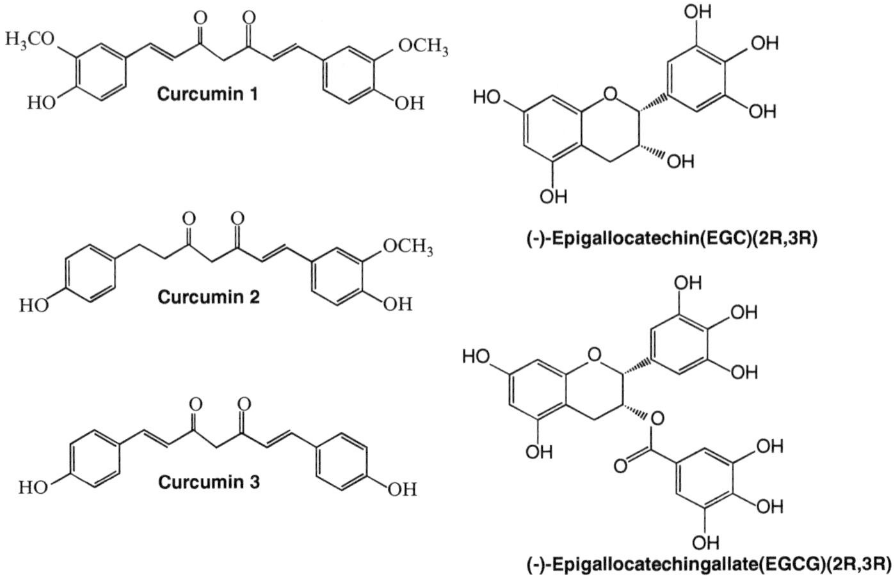


**Fig. 6.5** Chemical structures of curcumins and various components of green tea

5-lipoxygenase, DNA polymerase, and thioredoxin reductase, adhesion molecules, and apoptosis-related proteins (Anand et al. 2008; Aggarwal 2010; Farooqui 2012b). Curcumin exhibits poor bioavailability. There are three major reasons for the low bioavailability of curcumin: (a) its poor absorption, (b) its rapid metabolism, and (c) rapid systemic elimination and short biological half life (Anand et al. 2008). In rodents, curcumin undergoes rapid metabolism through conjugation and reduction reactions. It is partially absorbed by the digestive tract. Very little information is available on pharmacokinetic of curcumin in rodents and humans.

The bioavailability of curcumin can be improved by using adjuvants like piperine, preparing liposomes and nanoparticles, making curcumin–phospholipid conjugates, and synthesizing structural analogs (e.g., FLL32 and EF-24) (Shoba et al. 1998; Li et al. 2005; Liu et al. 2006; Gota et al. 2010). These procedures have resulted in longer circulation, increase in the cellular permeability, and resistance to metabolic processes. Among these approaches, the administration of curcumin as phospholipid conjugates, curcumin nanoparticles, and curcumin analogs has resulted in increased bioavailability of free curcumin (Begum et al. 2008; Gota et al. 2010).

Curcumin increases phosphorylation of 5′AMP-activated protein kinase, downregulates glycerol-3-phosphate acyl transferase-1, and upregulates the expression of carnitine palmitoyltransferase-1 leading to increase in oxidation and decrease in esterification of fatty acids (Ejaz et al. 2009). These results have been confirmed by real-time RT-PCR on subcutaneous adipose tissue. In addition, curcumin not only lowers serum cholesterol levels but also decreases the expression of PPARγ and CCAAT/enhancer-binding protein α, two key transcription factors during adipogenesis and lipogenesis. To explain the effect of curcumin on lipid and energy

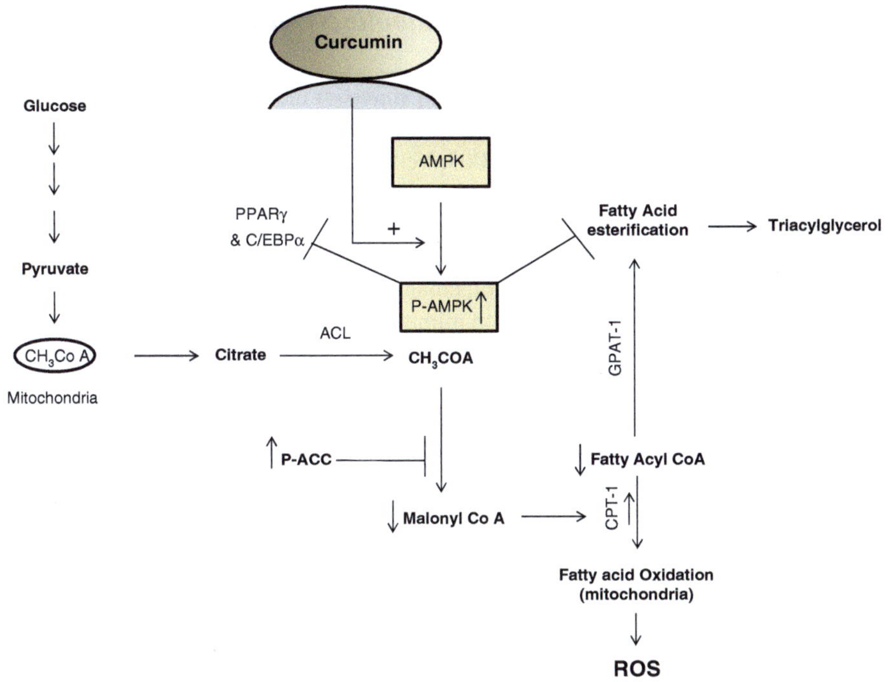


**Fig. 6.6** Effect of curcumin on AMP-activated protein kinase (AMPK); peroxisome proliferator-activated receptor γ (PPARγ); acetyl-CoA carboxylase (ACC); carnitine palmitoyltransferase-1 (CPT-1) and glycerol-3-phosphate acyl transferase-1 (GPAT-1); enhancer-binding protein α (C/EBP); and reactive oxygen species (ROS)

metabolism, a model has been proposed (Ejaz et al. 2009) (Fig. 6.6). According to this model, curcumin acts by increasing phosphorylation of 5′AMP-activated protein kinase (AMPK) and acetyl-CoA carboxylase (ACC) activities leading to the suppression of acetyl-CoA conversion to malonyl CoA. Low levels of malonyl CoA increase carnitine palmitoyltransferase-1 (CPT-1) expression leading to increase in fatty acid oxidation. The phosphorylation of AMPK also suppresses expression of glycerol-3-phosphate acyl transferase-1 (GPAT-1) resulting in reduction in fatty acid esterification. The phosphorylation of AMPK also inhibits PPARγ and C/EBPα transcription factors. In addition, curcumin also suppresses several transcription factors, including NF-κB, signal transducer and activators of transcription-3, and Wnt/β-catenin. Curcumin also activates Nrf2 cell signaling pathway, leading to the downregulation of various adipokines (TNF-α, interleukin-6, resistin, leptin, and monocyte chemotactic protein-1) and the upregulation of adiponectin (Aggarwal 2010). Collective evidence suggests that effect of curcumin on lipid metabolism in visceral cells (adipocytes, pancreatic cells, hepatic stellate cells, macrophages, and muscle cells) may contribute to lower body fat and body weight gain supporting the view that dietary curcumin may have a potential benefit in preventing obesity (Ejaz et al. 2009) through the reversal of insulin resistance, hyperglycemia, hyperlipidemia, inhibition of leptin expression, inhibition of cytokine expression, and other symptoms linked to MetS (Aggarwal 2010).

### 6.2.2 Effect of Green Tea on Metabolic Syndrome

The major catechins of green tea (*Camellia sinensis*) include (−)-epicatechin (EC), (−)-epicatechin-3-gallate(ECG),(−)-epigallocatechin(EGC),and(−)-epigallocatechin-3-gallate (EGCG) (Higdon and Frei 2003; Velayutham et al. 2008) (Fig. 6.5). EGCG is the major catechin (50–80 %) found in green tea. EGCG is quite stable in the stomach and small intestine. The bioavailability of green tea catechins depends upon their structural diversity. Catechin monomers can be easily absorbed through the gut barrier, whereas the large molecular weight catechins, such as EGCG, are poorly absorbed. EGCG acts as insulin mimetics in cultured adipocytes (Broadhurst et al. 2000; Anderson and Polansky 2002). EGCG has both anti-inflammatory and antioxidant properties. Because of these properties, it is thought that EGCG may produce therapeutic benefit in type II diabetes and MetS (Brown et al. 2008). Although the effectiveness of EGCG for treatment of type II diabetes in humans has not been established, in rodent models of MetS, dietary supplementation with EGCG improves glucose tolerance and insulin sensitivity by increasing GLUT1 and GLUT4 mRNA levels in the liver and increases GLUT2 and GLUT4 mRNA levels in the muscle (Potenza et al. 2007; Brown et al. 2008; Cao et al. 2007a). Green tea extract elevates GLUT1 and GLUT4 mRNA levels in the liver and increases GLUT2 and GLUT4 mRNA levels in the muscle. Green tea extract also increases glycogen synthase kinase 3β (GSK3β) and insulin receptor substrate-2 (IRS-2) mRNA levels in the liver (Cao et al. 2007a, b). Green tea not only increases GSK3β and phosphatidylinositol 3-kinase activities but also decreases Src homology 2 domain-containing transforming protein 1 (SHC1) mRNA levels in the liver (Cao et al. 2007a). EGCG also inhibits the activation of I-κ kinase (Chen et al. 2002), a serine/threonine kinase, implicated in the pathogenesis of insulin resistance (Tamura et al. 2007). RT-PCR studies on gene expression in liver tissue of EGCG-treated db/db mice have indicated that EGCG supplementation decreases blood glucose and FFA levels in a dose-dependent manner (Wolfram et al. 2006). EGCG treatment upregulates the expression of liver glucokinase (glycogenic enzyme) and carnitine palmitoyltransferase-1β but downregulates the expression of gluconeogenetic enzyme phosphoenolpyruvate carboxykinase. It is suggested that EGCG treatment not only reduces the endogenous liver glucose production but also increases glucose-induced insulin secretion (Wolfram et al. 2006). Treatment of H4IIE rat hepatoma cells or isolated hepatocytes with EGCG also results in upregulation of genes associated with glycogenesis, whereas genes involved in gluconeogenesis and in the synthesis of fatty acids, triacylglycerol, and cholesterol are downregulated (Shenouda and Vita 2007; Collins et al. 2007). These actions of EGCG are facilitated by the $Ca^{2+}$/calmodulin-dependent protein kinase and production of reactive oxygen species (Collins et al. 2007). EGCG not only mimics insulin action on the transcription factor forkhead box protein O1 (FOXO1a) but also inhibits β-2-aminobicyclo-(2.2.1)-heptane-2-carboxylic acid-mediated insulin secretion in pancreatic β-cells through the modulation of glutamate dehydrogenase (Anton et al. 2007; Li et al. 2006). EGCG increases both the expression and phosphorylation of

AMPK in 3T3-L1 cells and the phosphorylation of downstream target, acetyl-CoA carboxylase (ACC), which leads to suppression of esterification of fatty acids to triacylglycerol producing increased fatty acid oxidation (Murase et al. 2009). Thus, EGCG and other catechins contribute not only to decrease in adipogenesis but also in prevention of growth and expansion of adipose tissue.

EGCG also regulates endothelial cell function and production of vasoconstrictors and vasodilators, which subsequently impact on smooth muscle cell function and vascular tone leading to the modulation and control of blood pressure (Shenouda and Vita 2007). In cultured endothelial cells, EGCG activates endothelial nitric oxide synthase and stimulates the production of NO, which is a potent vasodilator (Shenouda and Vita 2007). Using pharmacological inhibitors, it is shown that EGCG stimulates NO production in the endothelial cells via the phosphatidylinositol 3-kinase/Akt pathway and may, in part, be the result of the inhibitory effect of EGCG on I-κ kinase activity. Accumulating information from cell culture and animal models suggests that regular consumption of EGCG may limit the development of insulin resistance and subsequent progression to type II diabetes and MetS during conditions of energy excess. In addition, EGCG has been reported to induce carbohydrate malabsorption, downregulate fatty acid synthase, suppress pancreatic and gastric lipase, induce thermogenesis, incite sympathetic nervous system activity and lipolysis, reduce adipocyte differentiation, and alter the satiety response (Cabrera et al. 2006; Moon et al. 2007; Zhong et al. 2006). EGCG also has anxiolytic, hypochromic, hepatoprotective, and kidney-protective properties. Large population, double-blind, and long-term studies are needed on the beneficial effects of EGCG in normal human subjects and patients with type II diabetes and MetS.

### 6.2.3 *Effect of Resveratrol on Metabolic Syndrome*

Resveratrol (3,4′,5-trihydroxystilbene) is a natural compound found in grapes, mulberries, pomegranates, peanuts, and other plants or food products, including raspberries, blueberries, Scots pine, eastern white pine, and knotweed. It belongs to a class of polyphenolic compounds called stilbenes (Saiko et al. 2008) (Fig. 6.7). Resveratrol promotes both antiaging, anticarcinogenic, cardioprotective, and cerebroprotective activities, which are attributed to its antioxidant, anti-inflammatory, and gene-modulating properties of stilbenes. It also acts as an analgesic. Resveratrol not only inhibits platelet aggregation and lipid peroxidation but also blocks eicosanoid synthesis, modulates lipoprotein metabolism, and exhibits vasorelaxing and anti-inflammatory activities (Das and Das 2007). Collective evidence suggests that resveratrol functions as an antioxidant, vasorelaxing, and anti-inflammatory agent by hampering free radical generation and scavenging free radicals, but also by inhibiting the release of proinflammatory cytokines.

In the absence of insulin, the stimulation of glucose uptake in the C2C12 skeletal muscle cells is mediated through the activation of AMP-activated protein kinase (AMPK) without the involvement of PtdIns 3K (Park et al. 2007). In the presence of

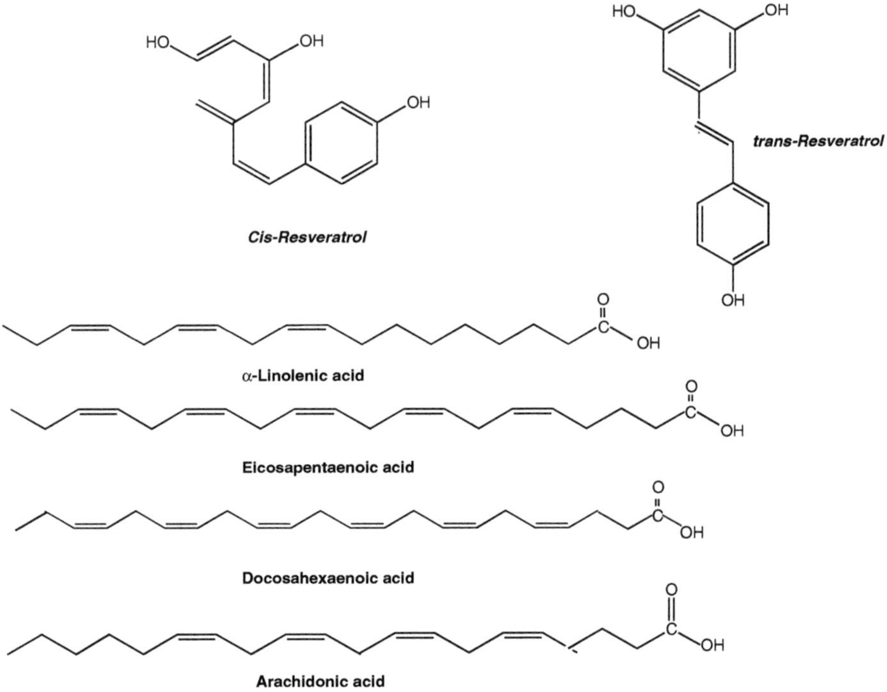


**Fig. 6.7** Chemical structures of resveratrol and n-3 fatty acids

insulin, resveratrol potentiates insulin-mediated glucose uptake via AMPK activation leading to activation of the PtdIns 3K-Akt signal pathway (Park et al. 2007) (Fig. 6.4). It is well known that mice that lack estrogen receptors (ERs) exhibit insulin resistance, impaired glucose tolerance, adipocyte hyperplasia, and hypertrophy. This supports the importance of the ER action in maintaining glucose homeostasis (Heine et al. 2000). Resveratrol has structural similarity with synthetic estrogen diethylstilbestrol. It binds with ER receptors and activates gene transcription in estrogen-sensitive tissues and cell lines (Henry and Witt 2002). In addition, ER-α is a positive regulator of GLUT4 expressions, whereas ER-β produces the suppression of GLUT4 (Barros et al. 2006a, b). In high-cholesterol fructose-fed rats, resveratrol increases insulin-stimulated and steady-state glucose uptake in the soleus muscle and liver. It not only stimulates the membrane trafficking activity of GLUT4 but also enhances phosphorylation of insulin receptor in insulin-resistant soleus muscles (Deng et al. 2008). The activation of ERs seems to be crucial for resveratrol-stimulating muscular glucose uptake via both insulin-dependent and insulin-independent pathways (Deng et al. 2008). In addition, resveratrol-mediated translocation of GLUT4 or GLUT1 does not involve the participation of Akt/protein kinases B (PKB). Although the molecular mechanism of these processes is not fully understood, it is suggested that sirtuin-dependent AMPK may be closely associated

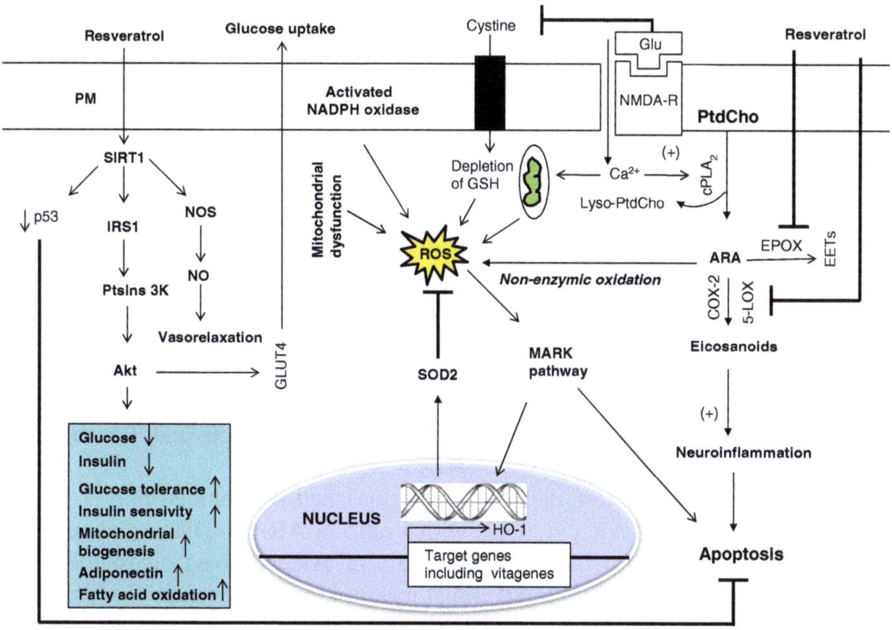


**Fig. 6.8** Effect of resveratrol on insulin signaling and neural membrane phospholipid degradation. *N*-methyl-D-aspartate receptor (NMDA-R); arachidonic acid (ARA); phosphatidylcholine (PtdCho); lyso-phosphatidylcholine (lyso-PtdCho); reactive oxygen species (ROS); cytosolic phospholipase $A_2$ ($cPLA_2$); phosphoinositol 3-kinases (PtdIns 3Ks); cyclooxygenase-2 (COX-2); 5-lipoxygenase (5-LOX); superoxide dismutase (SOD); serine/threonine protein kinase (Akt); sirtuin (SIRT1); nitric oxide synthase (NOS); nitric oxide (NO); and glucose transporter 4 (GLUT4)

with translocation and stimulation of intrinsic GLUT4 activity (Breen et al. 2008). Activation of SIRT1 (a member of sirtuin family), which catalyzes $NAD^+$-dependent deacetylation of histones, decreases glucose levels, improves insulin sensitivity, increases mitochondrial number and function, decreases adiposity, improves exercise tolerance, and potentially lowers body weight through the involvement of peroxisome proliferator-activated receptor coactivator-1α (PGC-1α) and FOXO protein (Lagouge et al. 2006; Milne et al. 2007) (Fig. 6.8).

## 6.2.4 *Effect of n-3 Fatty Acids on Metabolic Syndrome*

As stated above, n-3 family of essential fatty acids includes α-linolenic acid (ALA; 18:3n-3), docosahexaenoic acid (DHA; 22:6n-3), and eicosapentaenoic acid (EPA; 20:5n-3) (Fig. 6.7).

ALA is found in soybeans, rapeseed, walnuts, and dark green leafy vegetables (kale, spinach, broccoli, and Brussels sprouts). EPA and DHA are obtained from fish, which consumes golden marine algae. DHA and EPA are not synthesized by humans and therefore must be obtained from dietary sources or derived from the 18-carbon precursor, ALA, through a series of enzymic reactions involving chain elongation, desaturation, and β-oxidation in liver (Farooqui 2009). Neurons also lack the ability to synthesize arachidonic acid (ARA), another essential fatty acid belonging to n-6 family of fatty acids. The liver has the ability to transform linoleic acid (LA) into ARA. ALA is metabolized by $\Delta^6$-desaturation, elongation and $\Delta^5$-desaturation to yield EPA. DPA is formed by addition of C2 to EPA, which is converted subsequently to 24:5n-3 and 24:6n-3 by further chain elongation and $\Delta^6$-desaturation. These reactions occur in the endoplasmic reticulum. However, DHA is synthesized from 24:6n-3 by peroxisomal β-oxidation, which shortens the carbon chain by C2 (Fig. 3.3). This pathway has also been demonstrated in the pigs (Li et al. 2000) and baboons (Su et al. 1999a, b). Thus, DHA and ARA are obtained either directly from the diet or synthesized from ALA and LA in liver and transported to the brain through plasma lipoproteins (Scott and Bazan 1989; Farooqui 2009).

As stated earlier, Western diet is not only enriched in saturated fatty acids but high in cholesterol and ARA contents with a ratio of ARA to DHA of about 20:1. Consumption of Western diet elevates levels of ARA-derived eicosanoids and upregulates the expression of proinflammatory cytokines (TNF-α and IL-1β). ARA-derived eicosanoids have prothrombotic, proaggregatory, vasoconstrictive, and pro-inflammatory properties (Farooqui 2009). In addition, ARA is also a precursor of 4-hydroxynonal (4-HNE), a 9-carbon α, β-unsaturated aldehyde, which is generated by the lipid peroxidation of ARA. High concentrations of 4-HNE induce oxidative stress and apoptotic cell death (Farooqui 2011).

Rats fed with n-3 fatty acid (ALA, EPA, and DHA)-deficient diet show memory deficits in Barnes maze test. The memory deficit is further exacerbated by fructose intake. n-3 fatty acid-deficient diet and inclusion of fructose disrupt insulin receptor signaling in hippocampus as judged by a decrease in phosphorylation of insulin receptor and its downstream effector (Akt) (Agrawal and Gomez-Pinilla 2012). Consumption of high fructose along with n-3 fatty acid-deficient diet disrupts membrane homeostasis as judged by an increase in the ratio of n-6/n-3 fatty acids and elevation in 4-HNE levels. Disturbances in brain energy metabolism due to n-3 fatty acid deficiency and fructose treatments cause significant decrease in AMPK phosphorylation and its upstream modulator LKB1 as well as a decrease in Sir2 levels (Agrawal and Gomez-Pinilla 2012). n-3 fatty acid deficiency results in decreased phosphorylation of CREB, synapsin I, and synaptophysin (SYP) leading to metabolic dysfunction and alterations in synaptic plasticity in selected brain regions. Exposure to n-3 fatty acids results in enhancement in long-term potentiation and synaptic protein expression to increase the dendritic spine density, number of c-Fos-positive neurons, and neurogenesis in the hippocampus for learning memory processing (Farooqui 2009). All parameters of metabolic dysfunction related to the fructose treatment can be ameliorated by feeding n-3 fatty acid. These studies support the importance of n-3 fatty acid in diet against MetS (Agrawal and Gomez-Pinilla 2012).

In contrast, diet enriched in n-3 fatty acids generates docosanoids, which not only downregulate proinflammatory cytokines but also have anti-inflammatory, antithrombotic, antiarrhythmic, hypolipidemic, and vasodilatory effects. Lipid peroxidation of DHA results in the formation of 4-hydroxyhexanal (4-HHE). Like 4-HNE, 4-HHE also induces oxidative stress. Oxidative stress and inflammation are associated with the progression of insulin resistance and MetS (Urakawa et al. 2003). Some investigators have suggested that inflammation should be included as a component in the definition of MetS because it is an important part of the pathophysiology (Haffner 2006). Hyperglycemia, increase in plasma FFA levels, and hyperinsulinemia have been linked to increased production of reactive oxygen species (ROS), which activate nuclear factor-κB (NF-κB), a proinflammatory transcription factor, that triggers a signaling cascade leading to a continued synthesis of oxidative species, induction of proinflammatory cytokines, and low-grade chronic inflammation (Evans et al. 2002).

Another sensor of oxidative stress is forkhead homeobox type O (FOXO) transcription factors. FOXO is regulated by $H_2O_2$ (Essers et al. 2004). Recent studies have indicated that FOXO4 can be directly regulated by the cellular redox state through either monoubiquitination (van der Horst and Burgering 2007) or the induction of cysteine thiol–disulfide-dependent complexes of FOXO with the acetyltransferase p300/cAMP response element-binding protein (CREB)-binding protein (CBP) (Dansen et al. 2009). Interactions between p300/CBP and FOXO are required for FOXO-mediated transcription, since both CBP and p300 act as FOXO cofactors by weakening histone/DNA interactions (van der Heide and Smidt 2005). The modulation of biological activity of FOXO by p300/CBP-mediated acetylation depends on the formation of the redox-regulated complexes, supporting the view that FOXO transcription factors are bona fide sensors for the redox status of cells (Storz 2011) (Fig. 6.4). Collective evidence suggests that upregulation of ROS generation through inappropriate amount of dietary n-6 and n-3 fatty acids in Western diet may lead to increase not only in oxidative stress but also in abnormal inflammatory and immune responses in the brain and visceral tissues. An adequate ratio of n-6 and n-3 fatty acids (2–1:1) may promote a healthier balance between n-6 and n-3 PUFA-derived lipid mediators, which may maintain optimal visceral and neural membrane function (Farooqui 2009).

Thus, diet enriched in n-3 fatty acids (fish), whole grains, fruits and vegetables (Mediterranean diet) not only decreases levels of inflammatory lipid mediators, improves endothelial function, and significantly reduces waist circumference along with reduction in plasma glucose and decrease insulin resistance in animal models of MetS (Demarin et al. 2011). It is also shown that n-3 fatty acid-containing diet and Mediterranean diet improve plasma lipid profile: reduction of total and plasma LDL cholesterol levels, plasma triacylglycerol levels, and apoB and VLDL concentrations and an increase in plasma HDL cholesterol levels (Demarin et al. 2011; Lankinen et al. 2011). These effects are associated with increased plasma antioxidant capacity, improvement in endothelial function, reduction in insulin resistance, and reduction in incidence of MetS. The beneficial effects of n-3 fatty acid-containing diet also include reduction in cardiovascular and cerebrovascular diseases.

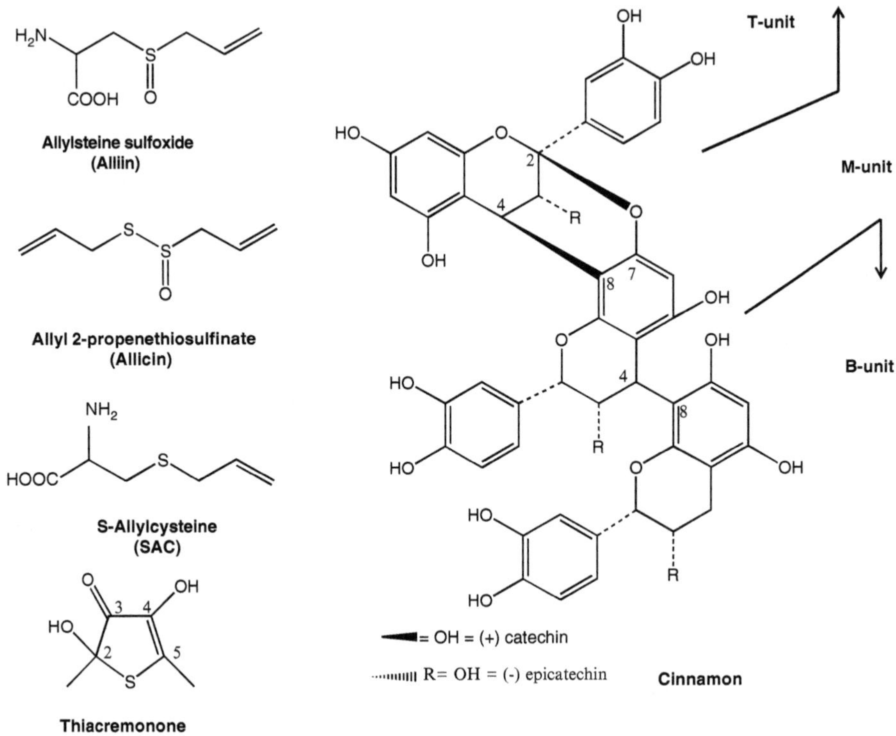


**Fig. 6.9** Chemical structures of garlic components and proanthocyanidins from cinnamon

## *6.2.5 Effect of Cinnamon on Metabolic Syndrome*

Cinnamon has been used as a spice and as traditional herbal medicine for centuries in Southeast Asia. In addition to containing polyphenolic compounds (proanthocyanidins and procyanidin), epicatechin, camphene, eugenol, phenol, salicylic acid, and tannins, cinnamon also contains manganese, dietary fiber, iron, and calcium. Cinnamon contains three major compounds—cinnamaldehyde, cinnamyl acetate, and cinnamyl alcohol. In rodents and humans, cinnamon polyphenol (Fig. 6.9) produces anti-inflammatory, antimicrobial, antioxidant, antitumor, cardiovascular, cholesterol-lowering, and immunomodulatory effects. In vitro studies have demonstrated that cinnamon may act as insulin mimetic to potentiate insulin activity or to stimulate cellular glucose metabolism. In patients with type II diabetes, cinnamon extracts have beneficial effects in reducing fasting plasma glucose, cholesterol, and triacylglycerols (Khan et al. 2003; Ziegenfuss et al. 2006). In overweight patients (Ziegenfuss et al. 2006; Roussel et al. 2009), nutritional intake of cinnamon also improves insulin sensitivity leading to beneficial antioxidant effects. In chow diet-fed rats, cinnamon extracts enhance insulin-stimulated insulin receptor β and insulin receptor substrate-1 (IRS-1) tyrosine phosphorylation levels and IRS-1/

phosphoinositide 3-kinase (PtdIns 3K) in skeletal muscle. Extensive in vitro studies have indicated that cinnamon improves insulin resistance by inhibiting and reversing impairments in insulin signaling in skeletal muscle. In adipose tissues, cinnamon acts by increasing the regulation of genes related to insulin signaling (CD36, LPL, FAS, GLUT4, and ACO) and blocking lipogenesis in adipose tissue (Sheng et al. 2008; Qin et al. 2009, 2010). Extensive in vitro studies have indicated that cinnamon improves insulin resistance by preventing and reversing impairments in insulin signaling in skeletal muscle. In adipose tissue, cinnamon increases the expression of peroxisome proliferator-activated receptors including, PPARγ (Sheng et al. 2008; Qin et al. 2009, 2010). In addition, in mouse 3T3-L1 adipocytes, cinnamon extracts also downregulate the expression of genes encoding insulin signaling pathway proteins including GSK3β, IGF-1R, IGF-2R, and PtdIns K, which may contribute to the potential health benefits of this phytochemical (Cao et al. 2010). Cinnamon also improves cholesterol levels, systolic blood pressure, and postprandial glucose levels in rodents. In T2DM patients, cinnamon lowers hemoglobin A1C (HbA1C) by 0.83 % compared with usual care alone lowering HbA1C by 0.37 % in a randomized, controlled trial (Crawford 2009).

It is well known that high-fructose feeding to rodents induces the symptoms of MetS including high blood pressure, insulin resistance, impaired glucose tolerance, and dyslipidemia (Lê and Tappy 2006). In addition, fructose feeding induces ventricular dilatation, ventricular hypertrophy, decreases ventricular contractile function, infiltration of inflammatory cells in heart, and hepatic steatosis (Patel et al. 2009). No information is available on the effect of cinnamon on body composition of humans in the MetS. Fructose decreases glucose infusion rates in rats fed with high-fat/high-fructose (HF/HF) diet (an animal model of MetS). Inclusion of cinnamon in the HF/HF diet increases the glucose infusion rates compared with control rats. The HF/HF diet induces a reduction in pancreas weight which can be prevented by the inclusion of cinnamon extract (Couturier et al. 2010). Mesenteric white fat accumulation is also observed in HF/HF rats compared with control rats. These deleterious effects of HF/HF diet can be alleviated by the inclusion of cinnamon in the diet (Couturier et al. 2010). Cinnamon extracts also improve glucose utilization in normal male rats fed with a high-fructose diet (HFD) (Qin et al. 2004). It is also reported that cinnamon extract prevents the development of insulin resistance, at least in part by enhancing insulin signaling and possibly via the NO pathway in skeletal muscle (Wang et al. 2007).

Recent studies have also indicated that cinnamic acid and its derivatives significantly inhibit the formation of advanced glycation end products (AGEs) by approximately 11.96–63.36 % in a concentration-dependent manner (Adisakwattana et al. 2012). Furthermore, cinnamic acid and its derivatives also reduce the level of fructosamine, the formation of *N*-(carboxymethyl)lysine, and the level of β-amyloid. Cinnamic acid and its derivatives also prevent oxidative protein damages, including effects on protein carbonyl formation and thiol oxidation of bovine serum albumin. Collectively these studies suggest that cinnamic acid and its derivatives prevent AGE-mediated diabetic complications (Adisakwattana et al. 2012).

### 6.2.6 Effect of Garlic on Metabolic Syndrome

It is well known that garlic (*Allium sativum*, family Lilliaceae) preparations possess cardioprotective, antioxidant, antineoplastic, and antimicrobial properties (Rahman and Lowe 2006; Tattelman 2005; Banerjee and Moulik 2000). Garlic preparations have been widely used for the prevention and treatment of cardiovascular and metabolic diseases, such as atherosclerosis, arrhythmia, hyperlipidemia, thrombosis, hypertension, and type II diabetes (Banerjee and Moulik 2000). Furthermore, garlic preparations also produce significant antiarrhythmic effect in both ventricular and supraventricular arrhythmias (Koch and Lawson 1996). Garlic preparations contain high concentrations of sulfur-containing compounds (Tattelman 2005). Garlic preparations contain several components, including allicin (a thiosulfinate), diallyl sulfide (DAS), diallyl disulfide (DADS), diallyl trisulfide (DATS), and dipropyl sulfide, along with ajoene, adenosine, vitamins E, C, minerals, polyphenols, inhibitors of adenosine deaminase, and cyclic AMP phosphodiesterase. Unlike antibiotics, garlic does not weaken the immune system but promotes it. It stimulates humoral and cell responses of the immune system. Allicin (Fig. 6.9) is the most active compound found in garlic. Allicin (thio-2-propene-1-sulfinic acid *S*-allyl ester) is formed when alliin (+*S-allyl*-L-cysteine sulfoxide), a sulfur-containing amino acid, comes into contact with the enzyme alliinase when raw garlic is chopped, crushed, or chewed. Allicin, unstable in aqueous solution, rapidly decomposes mainly to DAS, DADS, DATS, and ajoene (Amagase 2006). Alliinase is inactivated by heat during cooking, so heated garlic preparations have less medicinal value. Most medicinal effects of garlic preparations (antimicrobial, hypolipidemic, vasodilatory, antihypertensive, antioxidant, antineoplastic, and antithrombotic) are related to allicin and other breakdown products (Koch and Lawson 1996; Tattelman 2005).

In high-fructose corn syrup model of MetS, oral administration of raw garlic for a period of 8 weeks shows significant reduction of blood glucose and improvement of insulin sensitivity in rats. Other metabolic complications of MetS, such as increased serum triacylglycerol, insulin, and uric acid levels, are also normalized after garlic administration. It is suggested that lowering serum uric acid and triacylglycerol after garlic administration may be responsible for improving insulin resistance in fructose-fed rats (Nakagawa et al. 2006; Bruckdorfer et al. 1973). Fructose-fed rats show increase in serum levels of nitric oxide and decrease in $H_2S$ levels compared to normal subjects and control rats (Padiya et al. 2011). Elevated nitric oxide and low plasma levels of $H_2S$ are associated with increase in oxidative stress and marked endothelial dysfunction and insulin resistance. Chronic administration of garlic normalizes levels of both nitric oxide and $H_2S$ in fructose-fed rats. It is well known that oxidative stress plays an important role in the pathogenesis of type II diabetes. ROS are generated in type II diabetes by glucose oxidation. High levels of ROS and the simultaneous decline of endogenous antioxidants can lead to damage of cellular organelles and development of insulin resistance (Whiteman et al. 2010). In high-fructose-fed model of MetS, marked increase in TBARS levels is accompanied by reduction in glutathione levels in liver in comparison to control group. However, administration of raw garlic homogenate normalizes not only increases in TBARS but also decreases in glutathione levels in diabetic liver, supporting the view that oral administration of raw garlic homogenate increases insulin sensitivity and reduces metabolic complications along with oxidative stress in fructose-fed rat model of MetS (Padiya et al. 2011). Long-term, double-blind studies are needed on large human population to establish the role of garlic in controlling MetS and its complications.

## 6.3 Effect of Phytochemicals on Neurological Disorders

Neurological disorders include neurotraumatic, neurodegenerative, and neuropsychiatric disorders. Stroke, spinal cord injury, traumatic brain injury, and epilepsy fall under acute neural trauma or neurotraumatic diseases. Neurodegenerative diseases include Alzheimer disease (AD), Parkinson disease (PD), Huntington disease (HD), and amyotrophic lateral sclerosis (ALS). Neuropsychiatric disorders include both neurodevelopmental disorders and behavioral or psychological difficulties associated with some neurological disorders. Examples of neuropsychiatric disorders are schizophrenia, some forms of bipolar affective disorders, autism, mood disorders, depression, attention-deficit disorder, dementia, tardive dyskinesia, atypical spells, irritability, and organic mental disorder. Neuropsychiatric disorders are closely associated with the abnormalities in cerebral cortex and limbic system (thalamus, hypothalamus, hippocampus, and amygdala) (Farooqui 2010). The most important risk factors for stroke and neurodegenerative diseases are old age and a positive family history. Normal aging is accompanied by the ability of the brain to modify its own structural organization and functioning becomes weak and liable resulting in loss of some cognitive function (Farooqui and Farooqui 2010), but stroke and neurodegenerative diseases are accompanied by dramatic impairment in ability to modulate structural organization and functioning of the brain tissue. In addition, a common denominator among various neurodegenerative diseases and stroke is neuroinflammation (Farooqui et al. 2007a; Farooqui and Horrocks 2007; Farooqui 2009). Microglia, which are the resident macrophages in brain, secrete a barrage of factors (IL-1, TNF-$\alpha$, NO, $PGE_2$, and superoxide) that not only are toxic to neurons but also propagate neuronal injury.

Recent imaging studies have revealed that microglia dynamically interact with synapses where they not only serve as "synaptic sensors," responding to changes in neural activity and neurotransmitter release, but also play active roles at developing and mature synapses. Although molecular mechanisms and functional consequences of microglia–synapse interactions in the developing and mature brain are not fully understood, recent studies have indicated that microglia contribute to pathologies associated with abnormal and/or loss of neural circuits (Schafer et al. 2012).

Phytochemicals have been used for improving human health by various ancient medicinal systems (Indian, Chinese, Egyptian, Babylonian, and Greek). In recent years phytochemicals have gained considerable attention because it is becoming increasingly evident that these natural products are safer and produce lower side effects than synthetic drugs (Gershwin et al. 2010; Bent and Ko 2004). Although the precise molecular mechanisms through which phytochemicals exert their beneficial biological effects still remain the subject of intense investigations, it is suggested that health benefits of phytochemicals on visceral tissue and brain are due to their anti-inflammatory, antioxidant, anticarcinogenic, antiproliferative, hypocholesterolemic, and cellular repair properties (Farooqui 2012b). Phytochemicals mediate their effects through the modulation of signal transduction processes, which involve not only various transcription factors, growth factors, and inhibition of inflammatory cytokines expression but also regulation of enzymes, such phospholipases, cyclooxygenases, protein kinases, and protein phosphatases (Farooqui 2012a; Farooqui and Farooqui 2011). Phytochemicals also mediate their effects through the modulation of immune function.

The bioavailability of most phytochemicals to visceral organs is relatively higher than the brain not only because of the presence of blood–brain barrier (BBB) but also due to rapid metabolism and elimination of phytochemicals in the urine (Farooqui 2012b). Neurons are particularly sensitive to oxidative stress-mediated injury not only because of their large dependence on oxidative phosphorylation for energy as compared to glial cells but also due to the fact that many major antioxidant defense mechanisms are localized in astrocytes (Farooqui 2010, 2012b). The demand for oxygen consumption by neurons is extremely high and 1–2 % of the oxygen is converted into superoxide anion radicals ($O_2^{\bullet -}$) and hydrogen peroxide, leading to oxidative stress. ROS are known to trigger multiple cellular processes that include a transcriptional response. One of the redox-sensitive transcription factors is NF-κB. In the inactive state, NF-κB resides in the cytosol through complexation with IkB proteins. Upon phosphorylation IkBa is degraded, and NF-κB is released and translocated to the nucleus, where it initiates the transcription of many genes including genes for proinflammatory cytokines (TNF-α and IL-1β), proinflammatory enzymes (iNOS and COX-2), other genes (p53 and c-Myc) (Farooqui 2010, 2012a, b).

Aging, genetic disposition, and environmental factors induce oxidative stress, abnormal protein processing, induction of neuroinflammation, loss of synapse, abnormalities in neurotransmission, and ion homeostasis leading to symptoms of neurological disorders. Consumption of phytochemicals induces neurohormetic response that results in the expression of adaptive stress resistance genes responsible for encoding antioxidant enzymes, protein chaperones, and neurotrophic factor (BDNF) (Mattson and Cheng 2006). Neurohormesis not only provides a framework for explaining the dose–response relationships between concentration of phytochemicals and beneficial effects but also explains a key insight for improving the accuracy of the therapeutic dose of phytochemicals within the highly heterogeneous human population (Calabrese et al. 2012). Based on the stimulation of signal transduction network and adaptive stress resistance genes, it is proposed that the use of phytochemicals from childhood to old age along with regular exercise 3–4 times per week is an inexpensive strategy for maintaining normal aging and delaying onset of age-related neurological disorders (stroke; Alzheimer disease, AD; and Parkinson disease, PD) (Joseph et al. 2009; Farooqui and Farooqui 2010, 2011; Farooqui 2012b). Risk factors for neurological disorders (stroke, AD, and PD) include old age, positive family history, unhealthy life style, consumption of high-fat diet, and exposure to toxic environment. These risk factors contribute to abnormal protein processing leading to oligomerization of unfolded proteins, generation of ROS, induction of neuroinflammation, and apoptotic cell death in neurological disorders (Farooqui 2010, 2012b). Although phytochemicals have no effects on positive family history and gender, their long-term use (childhood to old age) may retard the effect of unhealthy lifestyle by delaying or slowing the onset of stroke, AD, PD, and other neurodegenerative diseases. These processes may lead to improvement in health status and quality of life in older age.

### 6.3.1 Effect of Curcumin on Neurological Disorders

Curcumin is a polyphenolic compound (Fig. 6.5), which produces antioxidant and anti-inflammatory effects not only by blocking oxidative stress and neuroinflammation in neurotraumatic and neurodegenerative diseases but also by restoring cellular homeostasis and rebalancing redox equilibrium (Farooqui 2012b). Curcumin also acts by inducing HO-1 expression and its activity in different brain cells via the activation of Nrf2/antioxidant responsive element (ARE) pathway (Scapagnini et al. 2011). Activation of Nrf2 target genes, and particularly HO-1, in astrocytes and neurons produces strongly protective effects against neuroinflammation, oxidative stress, and apoptotic cell death. The inhibition of oxidative stress and neuroinflammation by curcumin improves the outcome of ischemic injury after focal cerebral ischemia (Zhao et al. 2010). Collectively, these studies suggest that neuroprotective activity of curcumin in cerebral ischemia is mediated through its antioxidant activity (Zhao et al. 2010). In animal models of traumatic brain injury, curcumin produces beneficial effects by dramatically reducing oxidative damage and normalizing levels of BDNF and synapsin I, decreasing the levels of AMP-activated protein kinase (AMPK), and modulating CREB signaling (Wu et al. 2006; Sharma et al. 2009).

In an animal model of AD (3×Tg-AD mice), curcumin not only reduces senile plaque deposition but also inhibits JNK/IRS-1/tau signaling pathway that leads to AD-like p-IRS-1/insulin signaling defects (Ma et al. 2009a, b). Thus, treatment of the 3×Tg-AD mice on high-fat diet with curcumin for 4 months not only decreases levels of phosphorylated JNK, IRS-1, and tau but also retards the degradation of total IRS-1. Curcumin-mediated neurochemical changes in 3×Tg-AD mice are accompanied by improvement in Y-maze performance (Ma et al. 2009b). Mice consuming curcumin for 1 month show more significant effects on Y-maze performance, and the combination of curcumin and fish oil shows more significant inhibition of JNK, IRS-1, and tau phosphorylation. These results suggest that JNK-induced Aβ oligomer-mediated inactivation of IRS-1 and phospho-tau pathology along with dietary treatment with fish oil, curcumin, or a combination of both may produce improvement in insulin/trophic signaling and cognitive deficits in AD (Ma et al. 2009b). On neuronal surface, Aβ interacts with p75NTR and activates NF-κB-mediated proinflammatory signaling in a time and dose-dependent manner leading to neuronal cell death. Curcumin retards the activation of NF-κB and prevents Aβ-induced cell death in a human neuroblastoma cell line, suggesting a possibility of AD treatment with curcumin (Kuner et al. 1998; Ono et al. 2004). In addition, interactions between curcumin and Aβ cause the activation of early growth response-1 (Egr-1), a nuclear transcription factor, which upregulates the expression of cytokines (TNF-α and IL-1β) and chemokines (MIP-1β, MCP-1, and IL-8) in monocytes (Giri et al. 2003). Curcumin suppresses the activation of Egr-1 DNA-binding activity and abrogates Aβ-mediated expression of these cytokines and chemokines. In animal models of AD, mounting evidence suggests that curcumin or curcumin–phospholipid conjugate reduces Aβ-mediated increase in the ROS level, Aβ-mediated decrease in mitochondrial membrane potential, and Aβ-induced caspase activation.

Curcumin acts as antidepressant in animal models of depression. Thus, curcumin dose-dependently reduces despair immobility behavior in the tail suspension test and forced swimming test to a degree comparable to commonly prescribed tricyclic and selective serotonin reuptake antidepressants (Xu et al. 2005; Sanmukhani et al. 2011). Further in-depth behavioral comparisons combined with in vitro studies indicate that curcumin increases neural levels of serotonin, noradrenaline, and dopamine and inhibits monoamine oxidase activity with molecular mechanisms that appear to differ from typical tricyclic and selective serotonin reuptake antidepressants (Xu et al. 2005; Sanmukhani et al. 2011). Additional cellular mechanisms and antidepressant effects of curcumin indicate that in animal models of depression curcumin may also act by upregulating neurogenesis (Farooqui 2012b).

Exposure of rat cortical neurons with glutamate for 24 h not only produces a significant decrease in BDNF level but also decreases cell viability and enhances cell apoptosis. Pretreatment of neurons with curcumin reverses the BDNF expression and cell viability in a dose- and time-dependent manner (Wang et al. 2008). It is suggested that curcumin mediates its antidepressant activity not only by modulating the release of serotonin and dopamine but also by enhancing the level of neurotrophic factors such as BDNF (Wang et al. 2008). Collective evidence suggests that curcumin may also act by inhibiting the monoamine oxidase activity, modulating the release of serotonin and dopamine, and enhancing neurogenesis in the frontal cortex and hippocampus (Kulkarni et al. 2008, 2009).

### 6.3.2 Effect of Green Tea on Neurological Disorders

Green tea catechins (Fig. 6.5) interact with neurons and modulate neuronal signaling (Spencer 2009) and produce neuroprotective effects not only through the stimulation of (PtdIns 3K)/Akt-mediated antiapoptotic pathway but also through the inhibition of NF-κB and downregulation of the glycogen synthase kinase 3 (GSK3) activity (Koh et al. 2004). These enzymes and signal transduction pathways increase

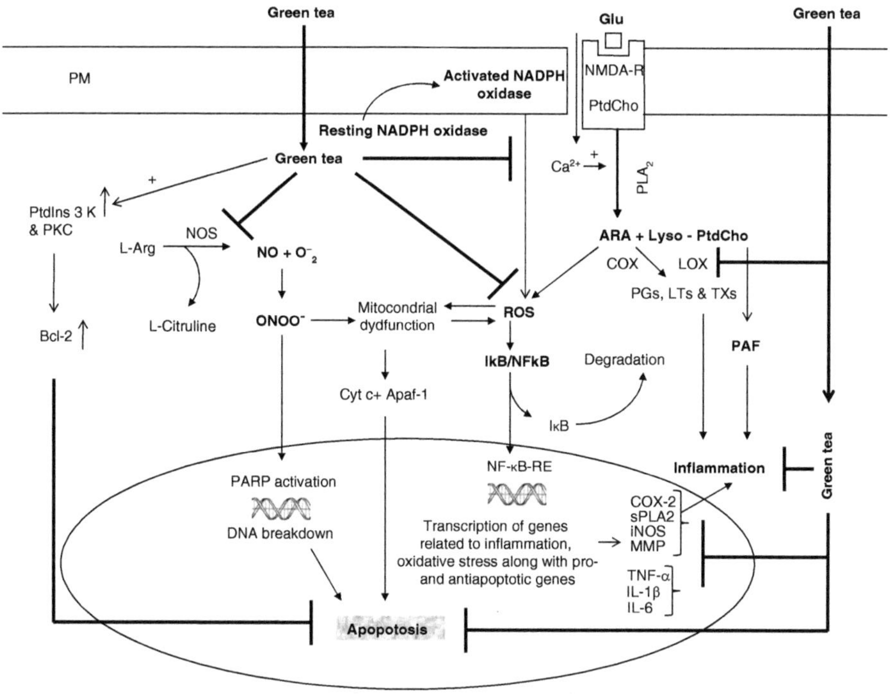


**Fig. 6.10** Effect of green tea catechins signal transduction processes. *N*-methyl-D-aspartate receptor (NMDA-R); arachidonic acid (ARA); phosphatidylcholine (PtdCho); lyso-phosphatidylcholine (lyso-PtdCho); reactive oxygen species (ROS); cytosolic phospholipase $A_2$ ($cPLA_2$); phosphoinositol 3-kinases (PtdIns 3K); cyclooxygenase-2 (COX-2); 5-lipoxygenase (5-LOX); nuclear factor-κB (NF-κB); nuclear factor-κB response element (NF-κB-RE); inhibitory subunit of NF-κB (IκB); resveratrol (RSV); tumor necrosis factor-α (TNF-α); interleukin-1β (IL-1β); interleukin-6 (IL-6); arginine (Arg); nitric oxide synthase (NOS); nitric oxide (NO); peroxynitrite ($ONOO^-$); protein kinase C (PKC); prostaglandins (PGs); leukotrienes (LTs); and thromboxanes (TXs)

the number and strength of connections between neurons, leading to an increase in BDNF, which supports and maintains cognitive function (Spencer 2009). Green tea catechins also inhibit NADPH oxidase, xanthine oxidase, cyclooxygenase, and lipoxygenase, suppress the activation of NF-κB, and activate adaptive cellular stress responses (Kim et al. 2010) (Fig. 6.10). In addition, catechins form complexes with iron and other transition metal ions. This binding of catechins with iron retards Fenton reaction, thereby inhibiting free radical generation (Mira et al. 2002). Collective evidence suggests that green tea catechins not only inhibit the PtdIns 3K/Akt pathway and block NF-κB- and AP-1-mediated transcription but retard aberrant arachidonic acid metabolism through cyclooxygenase and lipoxygenase. The oral administration of green tea extract to Wistar rats for 3 weeks before induction of ischemia not only causes reduction in eicosanoid levels, but also produces decrease in apoptotic cell death in the striatum, and cortical regions (Hong et al. 2001). Similarly, EGCG treatment immediately after transient global ischemia in gerbils produces significant reduction in neuronal cell damage in the hippocampal CA1 region (Lee et al. 2000) indicating that EGCG produces beneficial effects in ischemic–reperfusion injury.

Although consumption of green tea has no effect on symptoms of AD, several in vitro studies in cell culture and animal models of AD indicate that green tea extract protects neurons from the amyloid β-induced toxicity (Levites et al. 2003). In brain, amyloid precursor protein (APP) is processed either by a nonamyloidogenic pathways which involve cleavage of APP to soluble APP (sAPP) by the α-secretase activity or by the amyloidogenic β peptides by the β- and γ-secretases. EGCG enhances the nonamyloidogenic α-secretase pathway via PKC-dependent activation of α-secretase (Levites et al. 2003; Singh et al. 2008; Mandel et al. 2008), while (−)-epicatechin reduces the formation of amyloid β fibrils.

Collective evidence suggests that EGCG dose-dependently reduces the Aβ-induced toxicity by increasing brain α-secretase activity and decreasing brain β- and γ-secretase activities. In addition, EGCG inhibits the activation of extracellular signal-regulated kinase and nuclear transcription factor-κB in the Aβ-injected mouse brains. These studies strongly suggest that EGCG may contribute to the prevention of development or progression of AD in cell culture and animal models through its antioxidant and anti-inflammatory effects but also by reducing cerebral amyloidosis as well as cognitive deficits in a transgenic mouse model of AD (Rezai-Zadeh et al. 2008). These studies set the stage for long-term, double-blind clinical trials in large number of AD patients.

Nutritional studies indicate that the consumption of green tea may have beneficial effects in reducing the risk of PD (Checkoway et al. 2002). The molecular mechanisms underlying beneficial effects of EGCG have been investigated in animal model of PD and it is shown that pretreatment of mice with either green tea extract or EGCG prevents dopaminergic neuronal death mediated by *N*-methyl-4-phenyl-1,2,3,6-tetrahydropyridine (MPTP) (Levites et al. 2001). It is indicated that catechol-like structure in catechins may competitively inhibit the uptake by the presynaptic or vesicular transporters of the metabolite product of MPTP, 1-methyl-4-phenylpyridinium ion ($MPP^+$) (Pan et al. 2003), which has a structure similar to catechol. This competition may protect dopaminergic neuronal degeneration against the MPTP/$MPP^+$-mediated injury (Pan et al. 2003). In the substantia nigra, pretreatment of dopaminergic neurons with EGCGs not only produces neuroprotection against MPTP toxicity but also restores the depletion of striatal dopamine in mice. EGCG also attenuates the phosphorylation of c-Jun and modulates the phosphorylation of GSK3β (Ser9). These results suggest EGCGs protect dopaminergic neurons by inhibiting the JNK/c-Jun and GSK3β signal pathway (Ruan et al. 2009).

There is a growing body of data from animal and human studies supporting the role of green tea catechins in the modulation of behavior and mood through anxiolytic and antidepressant-like properties. The anxiolytic activity of EGCG is investigated after its acute administration in mice, using behavioral tests (elevated plus-maze and passive avoidance tests) as well as by electrophysiology on cultured hippocampal neurons. Patch-clamp studies show that EGCG (1–10 nM) has no effect on GABA currents. However, EGCG reverses GABA(A) receptor negative modulator methyl β-carboline-3-carboxylate (β-CCM) inhibition on GABA currents in a concentration-dependent manner supporting the view that EGCG can induce anxiolytic activity which may result from an interaction with GABA(A) receptors (Vignes et al. 2006). In addition green tea extracts have been shown to have antidepressant-like effects with multiple proposed mechanisms including antioxidant activity, regulation of adrenocorticotropic hormone (ACTH) serum levels, inhibition of prostaglandins and inflammatory cytokines, increase of anti-inflammatory cytokines, and inhibition of MAO activity (Dias et al. 2012).

*Ginkgo biloba* extract has been used for the symptomatic treatment of dementia (Kasper and Schubert 2009). The molecular basis of these effects is not yet fully understood, but as mentioned above that there is evidence of neuroprotective properties, including inhibition of synaptotoxic Aβ oligomer formation as well as antagonism of Aβ toxicity (Bastianetto et al. 2000). In addition, *Ginkgo biloba* extract not only improves mitochondrial function (Eckert et al. 2003; Abdel-Kader et al. 2007) but also decreases blood viscosity and enhances microperfusion (Költringer et al. 1995). It is also reported that *Ginkgo biloba* extract also modulates serotonin system (Ramassamy et al. 1992), enhances dopamine levels in prefrontal cortical areas (Yoshitake et al. 2010), inhibits the norepinephrine transporter (Fehske et al. 2009), attenuates hyperactivated hypothalamus–pituitary–adrenal (HPA) axis, and improves neuronal insulin sensitivity. All these processes may produce beneficial effects in dementia and behavioral disorders (Farooqui 2012b).

### *6.3.3 Effect of Resveratrol on Neurological Disorders*

Ischemic injury is accompanied by excitotoxicity, stimulation of $Ca^{2+}$-dependent enzymes, mitochondrial dysfunction, ATP depletion, stimulation in NF-κB activity, and increase expression and release of cytokines and chemokines (Farooqui 2010). Resveratrol (Fig. 6.7) treatment efficiently restores the disrupted mitochondrial integrity after ischemic damage in hippocampus (Yousuf et al. 2009). Resveratrol treatment not only inhibits the stimulation of NF-κB but also retards the activation of microglia, decreases the number of reactive astrocytes, and reduces the production of proinflammatory factors through cellular cascade signaling pathways (Wang et al. 2002). Resveratrol also confers vasculoprotection by downregulating neural and endothelial cells' VCAM and ICAM1 expression (Carluccio et al. 2003). In animal model of ischemic injury, pretreatment with resveratrol not only ameliorates neurological scores, reduces infarct volume and brain water content, and decreases MDA levels but also restores the SOD activity and upregulates the protein and mRNA expression of Nrf2 and HO-1 (Ren et al. 2011). These studies indicate that resveratrol produces neuroprotective effects in animal models of cerebral ischemia.

In animal models of traumatic brain injury (TBI) and spinal cord injury SCI), resveratrol administration not only restores neural morphology and increases the number of neurons but also suppresses the expression of proinflammatory cytokines including IL-1β, IL-10, TNF-α, and myeloperoxidase (MPO) after TBI and SCI, suggesting an anti-inflammatory effect of resveratrol. Finally, resveratrol treatment also inhibits TBI- and SCI-mediated apoptosis and expression of apoptosis-related gene Bax, Bcl-2, and caspase-3, which suggests an antiapoptotic role of resveratrol after the injury (Sönmez et al. 2007; Kaplan et al. 2005; Liu et al. 2011). Studies on the effect of red wine (resveratrol) in the Tg2576 mouse model indicate that moderate consumption of red wine promotes nonamyloidogenic, α-secretase-mediated APP processing and prevents the generation of amyloidogenic Aβ1–42-mediated cognitive deterioration (Wang et al. 2006). Similarly, in primary neuron cultures derived from Tg2576 embryos, resveratrol also lowers Aβ production through promotion of nonamyloidogenic processing of APP. Resveratrol treatment markedly inhibits polymerization of the β-amyloid peptide (Riviere et al. 2007). It is also suggested that resveratrol acts as an antioxidant by preventing the formation of toxic Aβ oligomers and protofibrillar intermediates (Jang and Surh 2003; Savaskan et al. 2003; Han et al. 2004). Recent studies indicate that intracerebroventricular injection of resveratrol reduces neurodegeneration in the hippocampus and prevents learning impairment in the p25 transgenic AD mouse model by a mechanism that may involve a decrease in the acetylation of known sirtuin 1 (SIRT1) substrates (Kim et al. 2007). For example, peroxisome proliferator-activated receptor gamma coactivator (PGC-1)α and p53.

Resveratrol also stimulates the deacetylation of FOXO3/4 transcription factors, thereby enhancing gene expression of antioxidative molecules and upregulates DNA repair (Brunet et al. 2004). Collectively, these studies suggest that beneficial effects of resveratrol may include antioxidation, free radical scavenger action, and modulation of neuronal energy homeostasis. In cell culture and animal model of Parkinson disease (PD), resveratrol not only prevents 6-hydroxydopamine (6-OHDA)-induced ROS generation in Neuroblastoma SH-SY5Y cells but also protects against rotenone-induced apoptosis and enhances degradation of α-synucleins in α-synuclein-expressing PC12 cell lines via autophagy induction (Chao et al. 2006; Khan et al. 2010; Wu et al. 2011). Accumulating evidence suggests that the beneficial effects of resveratrol are not only limited to its antioxidant and anti-inflammatory action but also include activation of sirtuin 1 (SIRT1) and vitagenes, which can prevent the deleterious effects triggered by oxidative stress (Wu et al. 2011). Collectively, these studies suggest that resveratrol may produce neuroprotective effects in cell culture and animal models of PD. Double-blind and long-term studies are needed in a large population of PD patients to determine the beneficial effects of resveratrol.

### 6.3.4 *Effect of n-3 Fatty Acids on Neurological Disorders*

As stated above, Western diet is rich in n-6 fatty acids (Fig. 6.7) due to the consumption of red meat and vegetable oils. This diet elevates levels of eicosanoids and platelet-activating factor and upregulates the expression of proinflammatory cytokines. These metabolites promote hyperneuroinflammation and oxidative stress

(Farooqui et al. 2007a). In contrast, consumption of n-3 fatty acids (Fig. 6.7) results in anti-inflammatory, antithrombotic, antiarrhythmic, hypolipidemic, vasodilatory, and immunosuppressive effects against stroke (Farooqui 2009; Okada et al. 1996; Cao et al. 2007a, b; Bas et al. 2007; Ghosh et al. 2007). Epidemiological studies on the effect n-3 fatty acids on stroke indicate that n-3 fatty acids show modest beneficial effects ischemic rather than hemorrhagic stroke (Iso et al. 2001; He et al. 2002). n-3 fatty acid consumption and administration promote cerebral blood flow; inhibit $PLA_2$, cyclooxygenase, and lipoxygenase activities; and reduce levels of brain post-ischemic eicosanoids. DHA not only decreases in blood–brain barrier disruption and reduces brain edema (Hossain et al. 1998) but also has antioxidant properties (Hossain et al. 1999). DHA inhibits production of inflammatory cytokines by downregulating NF-κB. In addition, the infusion of neuroprotectin $D_1$ (NPD1; 10R,17S-dihydroxy-docosa-4Z,7Z,11E,13E,15Z,19Z-hexaenoic acid), an endogenous lipid mediator derived from DHA, following ischemic–reperfusion injury downregulates neuroinflammation and oxidative stress and blocks neurodegeneration. $NPD_1$ also upregulates the antiapoptotic Bcl-2 proteins (Bcl-2 and bclxL) and decreases the expression of the proapoptotic proteins (Bax and Bad) (Bazan 2005; Belayev et al. 2011). $NPD_1$ blocks reperfusion-induced leukocyte infiltration, proinflammatory signaling, and infarct size. $NPD_1$ not only inhibits cytokine-mediated cyclooxygenase-2 expression but also promotes homeostatic regulation of the integrity of neural cells particularly during oxidative stress, and this protective signaling may be relevant to neural cell survival following ischemic injury (Bazan 2009; Belayev et al. 2011). These processes strengthen the survival mechanisms through ERK-mediated and/or Bcl-2-mediated prosurvival cascade.

Injections of n-3 fatty acids (ALA and DHA) 30 min after SCI in rats not only produce significant improvement in locomotor performance and neuroprotection but also reduce lesion size, inhibit apoptosis, and increase neuronal and oligodendrocyte survival (Lang-Lazdunski et al. 2003; King et al. 2006; Michael-Titus 2007). It is suggested that the neuroprotective effect of n-3 fatty acids may involve their antioxidant activity and generation of resolvins and neuroprotectins, which protect neuronal cells from apoptotic cell death (King et al. 2006; Michael-Titus 2007; Ward et al. 2010). In contrast, injections of ARA in rats produce a significantly worse outcome of injured animals following SCI than control rats supporting the view that consumption of n-3 fatty acids produces neuroprotective effects, whereas consumption of n-6 fatty acids mediates damaging effects in SCI (King et al. 2006; Michael-Titus 2007; Ward et al. 2010). Similarly, n-3 fatty acids produce beneficial effects by blocking oxidative stress and neuroinflammation (Wu et al. 2004). Although the molecular mechanism of n-3 fatty acid action is not fully understood, it is becoming evident that dietary n-3 fatty acids not only reduce oxidative stress but also normalize levels of BDNF, synapsin I, and CREB and restore learning and memory disability (Wu et al. 2003, 2004). In contrast, consumption of high-saturated-fat diet reduces levels of BDNF, compromises neuroplasticity, impairs cognitive function, and aggravates the outcome of TBI (Wu et al. 2005).

Fish consumption and supplementation of n-3 fatty acids in diet reduce the risk of having AD (Freund-Levi et al. 2006; Farooqui 2009) through multiple mechanisms. These mechanisms include antioxidant effects of n-3 fatty acids, reduction in levels of eicosanoids, generation of neuroprotective metabolites of DHA, and stimulation of trophic factors production (Cole et al. 2009). In rat model of AD, dietary intake of DHA significantly reduces the levels of Aβ40, cholesterol, and saturated fatty acids (Hashimoto et al. 2006, 2008). Chronic pre-administration of DHA also prevents β-amyloid-induced impairment of an avoidance ability-related memory function in a rat model of AD (Hashimoto et al. 2006) and protects mice from synaptic loss and dendritic pathology in another model of AD (Calon et al. 2004). DHA and its metabolite, neuroprotectin $D_1$ ($NPD_1$), not only decrease β-amyloid secretion from aging brain cells but also prevent apoptosis (Lukiw et al. 2005; Bazan 2005). DHA also inhibits c-Jun N-terminal kinase and the phosphorylation of adaptor protein insulin receptor substrate-1 (IRS-1) and tau in cultured hippocampal neurons (Ma et al. 2009b).

n-3 fatty acids also produce neuroprotective effects in animal models of PD. Thus, the consumption of n-3 fatty acids protects rodents from MPTP-induced neurotoxicity (Bousquet et al. 2008). The dietary consumption of n-3 fatty acid not only prevents the MPTP-mediated decrease in tyrosine hydroxylase-labeled nigral cells but also downregulates Nurr1 mRNA and dopamine transporter mRNA levels in the substantia nigra (Bousquet et al. 2008). Although n-3 fatty acids dietary treatment has no effect on striatal dopaminergic terminals, high levels of n-3 fatty acids in diet protect against the MPTP-mediated decrease in dopamine and its metabolite, dihydroxyphenylacetic acid, in the striatum. These observations suggest that the consumption of high-n-3-fatty-acid-containing diet produces neuroprotective effects in an animal model of Parkinsonism (Bousquet et al. 2008). In 6-hydroxydopamine (6-OHDA)-induced model of PD, n-3 fatty acids act by increasing dopamine turnover in the surviving neurons without modifying neuronal population (Delattre et al. 2010). Although the molecular mechanism associated with neuroprotective effects of n-3 fatty acids is not known, it is suggested that n-3 Fatty acids inhibit the synthesis and release of proinflammatory cytokines such as TNF-α and IL-1β and IL-2 that are released during induction and maintenance of inflammatory processes in the early course of PD (Farooqui 2009). In addition, n-3 fatty acids have antidepressant effect in PD patients and improve their quality of life.

### *6.3.5 Effect of Cinnamon on Neurological Disorders*

As stated above, cinnamon bark contains many components, including cinnamaldehyde, eugenol, cinnamyl acetate, cinnamyl alcohol, coumarin, cinnamic acid esters, and proanthocyanidins (polyphenolic bioflavonoid with known antioxidant activity) (Fig. 6.9). Many of these compounds mediate anti-inflammatory, antimicrobial, antioxidant, antitumor, cardiovascular, cholesterol-lowering, and immunomodulatory effects under in vitro conditions. Animal studies have demonstrated that cinnamon not only acts as insulin mimetic to potentiate insulin activity or to stimulate cellular glucose metabolism but also has strong hypoglycemic properties (Qin et al. 2010). Studies on the effect of procyanidin type A trimer (trimer 1) derived from cinnamon bark have indicated that in C6 glial cultures trimer 1 blocks oxygen–glucose deprivation (OGD)-mediated glial cell swelling, inhibits increase in free radical production, prevents increase in $Ca^{2+}$, and retards OGD-induced changes in inner mitochondrial membrane potential ($\Delta\Psi$(m) (Panickar et al. 2012). It is also shown that trimer 1 also significantly attenuates the OGD-induced decrease in glutamate uptake. Treatment with cyclosporin A, a blocker of the mitochondrial permeability pore (mPT), partially prevents the OGD-induced decline in glutamate uptake indicating the involvement of mitochondria in this process (Panickar et al. 2012).

Studies on the effect of cinnamon extract in cell culture model of AD indicate that cinnamon extract markedly inhibits the formation of toxic Aβ oligomers and prevents the toxicity of Aβ on neuronal PC12 cells (Frydman-Marom et al. 2011). When administered to an AD fly model system, cinnamon extract rectifies their reduced longevity, facilitates locomotion recovery, and totally abolishes the accumulation of tetrameric species of Aβ in their brain (Frydman-Marom et al. 2011). Furthermore, oral administration of cinnamon extract in an aggressive transgenic mice model of AD results not only in marked decrease in 56 kDa Aβ oligomers and reduction in number of plaques but also in improvement in cognitive behavior (Frydman-Marom et al. 2011). In addition, proanthocyanidin trimer and cinnamaldehyde which are present in cinnamon have been reported to inhibit tau aggregation and filament formation in vitro (Peterson et al. 2009). Collectively, these studies suggest that cinnamon extract contains components that can serve as an efficacious inhibitor of both the oligomerization and fibrillization of Aβ and tau protein aggregation. More studies are needed on the effect of cinnamon-derived compounds in cell culture and animal models of neurological disorders.

### 6.3.6 *Effect of Garlic on Neurological Disorders*

Studies on the effect of aqueous garlic extract in middle cerebral artery occlusion (MCAO) indicate that AGE significantly attenuates neurochemical and neurobehavioral changes produced by MCAO-mediated neuronal injury (Saleem et al. 2006). Similarly, treatment of animals which have been subjected to 1 h of ischemia plus 24 h of reperfusion with aqueous garlic extract at the onset of reperfusion indicates that AGE treatment diminishes the neurological alterations, the infarct area, and the histological damage caused by cerebral ischemic injury (Colín-González et al. 2011). It is also shown that aqueous garlic extract administration also attenuates the elevation in 8-OHdG levels, TNF-α levels, and COX-2 protein levels. Based on these results, it is suggested that the neuroprotective effect of aqueous garlic extract is not only due to its antioxidant and anti-inflammatory properties of aqueous garlic extract. In addition, aqueous garlic extract also lowers blood pressure (BP) due to the generation of hydrogen sulfide which relaxes vascular smooth muscle, induces vasodilation of blood vessels, and significantly reduces blood pressure (Kimura and Kimura 2004). In addition, studies on ischemic injury in the central nervous system indicate that exogenous $H_2S$ not only protects neurons from glutamate toxicity (Kimura and Kimura 2004), but also decreases the expression of apoptosis-inducing proteins (Minamishima et al. 2009). Collectively, these studies indicate that beneficial effects of garlic on ischemic/reperfusion injury are due to antioxidant properties of organosulfur compounds, generation of $H_2S$, and prevention of apoptotic cell death.

Studies on the effect of aqueous garlic extract in animal models of AD have indicated that *S*-allyl-L-cysteine (SAC), a component of aqueous garlic extract, lowers Aβ levels and toxicity (Ray et al. 2011a, b). Treatment of neuronal cultures with SAC results in protection from $H_2O_2$-mediated oxidative stress. Pretreatment with aqueous garlic extract alone also induces neuropreservation in 80 % neurons in cultures from ROS-mediated oxidative damage. In addition, aqueous garlic extract also preserves presynaptic protein called synaptosomal-associated protein of 25 kDa (SNAP25) from ROS-mediated insult. Thus, treatment with aqueous garlic extract and SAC independently enhances SNAP25 levels (~70 %) and synaptophysin in Alzheimer amyloid precursor protein transgenic mice, which is significantly decreased in AD (Ray et al. 2011a, b).

Pretreatment of C57BL/6J mice with SAC for 17 days, followed by $MPP^+$ injections, not only attenuates $MPP^+$-mediated neurotoxicity but also results in significant decrease in dopamine levels in the striatum (Rojas et al. 2011). In addition, neuroprotective effect of SAC against $MPP^+$ neurotoxicity also inhibits lipid peroxidation and improves the impairment in locomotion (Rojas et al. 2011). Collectively these studies suggest that in mice, SAC attenuates $MPP^+$-induced neurotoxicity in the striatum and that an antioxidant effect against oxidative stress may be partly responsible for its observed neuroprotective effects (Rojas et al. 2011). It is proposed that organosulfur compounds in garlic extracts produce their beneficial effects through antiaging, antioxidant, anti-inflammatory, and antiapoptotic mechanisms. In addition, the generation of $H_2S$ may also contribute to the beneficial physiological effects that garlic exerts on the cardiovascular and cerebrovascular systems. Based on above information, it can be proposed that large, long-term, fully blinded, and well-controlled clinical studies using a standardized preparation of garlic are necessary to evaluate the beneficial effects of garlic in patients with neurological disorders.

## 6.4 Conclusion

MetS has gained considerable importance because it is an important risk factor not only for type II diabetes and cardiovascular disease but also for neurological disorders, such as stroke, AD, and depression. At the present time, the MetS is increasing with alarming rate throughout the world. Patients with metabolic syndrome are twice as likely to die from cardiovascular disease, stroke, AD, and PD compared with people without the MetS and have a fivefold greater risk of developing type II diabetes. Consumption of Western diet, which is not only enriched in saturated and n-6 fatty acids and cholesterol but also high in animal proteins, produces oxidative stress, inflammation, and insulin resistance. Dietary excess energy is stored in adipocytes, which release proinflammatory cytokines and hormones that cause vascular injury due to endothelial dysfunction. Chronic hyperglycemia is accompanied by nonenzymic protein glycation by glucose and fructose. These sugars react with the free amino groups of protein to initiate a complex cascade of repeated condensations, rearrangements, and oxidative modifications, ultimately forming AGEs. Accumulation of AGEs in body tissues is a leading cause of atherosclerosis, diabetic complications (retinopathy, nephropathy, and neuropathy), and neurotraumatic, neurodegenerative, and neuropsychiatric diseases. The consumption of diet enriched in fruits (polyphenols), vegetables (fiber), dry fruits (n-3 fatty acids, proteins of plant origin), and whole grains decreases oxidative stress and inflammation caused by ROS, AGEs, lipid mediators, and cytokines. Growing evidence indicates that dietary phytochemicals influence carbohydrate and lipid metabolism at many levels. In animal models and in limited number of human studies carried out so far, phytochemicals have been shown to attenuate not only postprandial glycemic responses, fasting hyperglycemia, improved acute insulin secretion, and insulin sensitivity but also oxidative stress and inflammation. Greater understanding about the pathogenesis of MetS and neurological disorders indicates that nutritional supplementation with phytochemicals along with diet enriched in whole grain products, fruits, vegetables, and dry fruits (n-3 fatty acids, proteins of plant origin) may prevent or delay the onset of MetS, cardiovascular diseases, and neurological disorders.

## References


Abdel-Kader R, Hauptmann S, Keil U, Scherping I, Leuner K, Eckert A, Müller WE (2007) Stabilization of mitochondrial function by Ginkgo biloba extract (EGb 761®). Pharmacol Res 56:493–502

Adisakwattana S, Sompong W, Meeprom A, Ngamukote S, Yibchok-Anun S (2012) Cinnamic acid and its derivatives inhibit fructose-mediated protein glycation. Int J Mol Sci 13:1778–1789

Aggarwal BB (2010) Targeting inflammation-induced obesity and metabolic diseases by curcumin and other nutraceuticals. Annu Rev Nutr 30:173–199

Agrawal R, Gomez-Pinilla F (2012) "Metabolic syndrome" in the brain: Deficiency in omega-3-fatty acid exacerbates dysfunctions in insulin receptor signaling and cognition. J Physiol 590:2485–2499

Amagase H (2006) Clarifying the real bioactive constituents of garlic. J Nutr 136:716S–725S

Anand P, Thomas SG, Kunnumakkara AB, Sundaram C, Harikumar KB, Sung B, Tharakan ST, Misra K, Priyadarsini IK, Rajasekharan KN, Aggarwal BB (2008) Biological activities of curcumin and its analogues (Congeners) made by man and Mother Nature. Biochem Pharmacol 76:1590–1611

Anderson RA, Polansky MM (2002) Tea enhances insulin activity. J Agric Food Chem 50:7182–7186

Anton S, Melville L, Rena G (2007) Epigallocatechin gallate (EGCG) mimics insulin action on the transcription factor FOXO1a and elicits cellular responses in the presence and absence of insulin. Cell Signal 19:378–383

Azadbakht L, Mirmiran P, Esmaillzadeh A, Azizi T, Azizi F (2005) Beneficial effects of a dietary approaches to stop hypertension eating plan on features of the metabolic syndrome. Diabetes Care 28:2823–2831

Banerjee KS, Moulik SK (2000) Effect of garlic on cardiovascular disorders: a review. Nutri J 1:1–14

Baran SE, Campbell AM, Kleen JK, Foltz CH, Wright RL, Diamond DM, Conrad CD (2005) Combination of high fat diet and chronic stress retracts hippocampal dendrites. Neuroreport 16:39–43

Barros RP, Machado UF, Gustafsson JA (2006a) Estrogen receptors: new players in diabetes mellitus. Trends Mol Med 12:425–431

Barros RP, Machado UF, Warner M, Gustafsson JA (2006b) Muscle GLUT4 regulation by estrogen receptors ERbeta and ERalpha. Proc Natl Acad Sci USA 103:1605–1608

Bas O, Songur A, Sahin O, Mollaoglu H, Ozen OA, Yaman M, Eser O, Fidan H, Yagmurca M (2007) The protective effect of fish n-3 fatty acids on cerebral ischemia in rat hippocampus. Neurochem Int 50:548–554

Bastianetto S, Ramassamy C, Doré S, Christen Y, Poirier J, Quirion R (2000) The Ginkgo biloba extract (EGb 761®) protects hippocampal neurons against cell death induced by beta-amyloid. Eur J Neurosci 12:1882–1891

Bazan NG (2005) Neuroprotectin $D_1$ ($NPD_1$): a DHA-derived mediator that protects brain and retina against cell injury-induced oxidative stress. Brain Pathol 15:159–166

Bazan NG (2009) Neuroprotectin $D_1$-mediated anti-inflammatory and survival signaling in stroke, retinal degenerations and Alzheimer's disease. J Lipid Res 50(Suppl):S400–405

Begum AN, Jones MR, Lim GP, Morihara T, Kim P, Heath DD, Rock CL, Pruitt MA, Yang F, Hudspeth B, Hu S, Faull KF, Teter B, Cole GM, Frautschy SA (2008) Curcumin structure function, bioavailability, and efficacy in models of neuroinflammation and Alzheimer's disease. J Pharmacol Exp Ther 326:196–208

Belalcazar LM, Reboussin DM, Haffner SM, Reeves RS, Schwenke DC, Hoogeveen RC, Pi-Sunyer FX, Ballantyne CM, Look AHEAD (Action for Health in Diabetes) Obesity, Inflammation, and Thrombosis Research Group (2009) Marine omega-3 fatty acid intake: associations with cardiometabolic risk and response to weight loss intervention in the Look AHEAD (Action for Health in Diabetes) study. Diabetes Care 33:197–199

Belayev L, Khoutorova L, Atkins KD, Eady TN, Hong S, Lu Y, Obenaus A, Bazan NG (2011) Docosahexaenoic acid therapy of experimental ischemic stroke. Transl Stroke Res 2:33–41

Bent S, Ko R (2004) Commonly used herbal medicines in the United States: a review. Am J Med 116:478–485

Bierhaus A, Haslbeck KM, Humpert PM, Liliensiek B, Dehmer T, Morcos M, Sayed AA, Andrassy M, Schiekofer S, Schneider JG, Schulz JB, Heuss D, Neundörfer B, Dierl S, Huber J, Tritschler H, Schmidt AM, Schwaninger M, Haering HU, Schleicher E, Kasper M, Stern DM, Arnold B, Nawroth PP (2004) Loss of pain perception in diabetes is dependent on a receptor of the immunoglobulin superfamily. J Clin Invest 114:1741–51

Bousquet M, Saint-Pierre M, Julien C, Salem N Jr, Cicchetti F, Calon F (2008) Beneficial effects of dietary omega-3 polyunsaturated fatty acid on toxin-induced neuronal degeneration in an animal model of Parkinson's disease. FASEB J 22:1213–1225

Breen DM, Sanli T, Giacca A, Tsiani E (2008) Stimulation of muscle cell glucose uptake by resveratrol through sirtuins and AMPK. Biochem Biophys Res Commun 374:117–122

Broadhurst CL, Polansky MM, Anderson RA (2000) Insulin-like biological activity of culinary and medicinal plant aqueous extracts *in vitro*. J Agric Food Chem 48:849–852

Brown AL, Lane J, Coverly J, Stocks J, Jackson S, Stephen A, Bluck L, Coward A, Hendrickx H (2008) Effects of dietary supplementation with the green tea polyphenol epigallocatechin-3-gallate on insulin resistance and associated metabolic risk factors: randomized controlled trial. Br J Nutri 101:886–894

Brownlee M (2005) The pathobiology of diabetic complications: a unifying mechanism. Diabetes 54:1615–1625

Bruckdorfer KR, Kang SS, Yudkin J (1973) Plasma concentrations of insulin, corticosterone, lipids and sugars in rats fed on meals with glucose and fructose. Proc Nutr Soc 32:12–13

Brunet A, Sweeney LB, Sturgill JF, Chua KF, Greer PL, Lin Y, Tran H, Ross SE, Mostoslavsky R, Cohen HY, Hu LS, Cheng HL, Jedrychowski MP, Gygi SP, Sinclair DA, Alt FW, Greenberg ME (2004) Stress-dependent regulation of FOXO transcription factors by the SIRT1 deacetylase. Science 303:2011–2015

Butterfield DA, Howard BJ, Yatin S, Allen KL, Carney JM (1997) Free radical oxidation of brain proteins in accelerated senescence and its modulation by N-tert-butyl-alpha-phenylnitrone. Proc Natl Acad Sci USA 94:674–78

Cabrera C, Artacho R, Giménez R (2006) Beneficial effects of green tea – a review. J Am Coll Nutr 25:79–99

Calabrese V, Cornelius C, Dinkova-Kostova AT, Iavicoli I, Di Paola R, Koverech A, Cuzzocrea S, Rizzarelli E, Calabrese EJ (2012) Cellular stress responses, hormetic phytochemicals and vitagenes in aging and longevity. Biochim Biophys Acta 1822:753–783

Calon F, Lim GP, Yang FS, Morihara T, Teter B, Ubeda O, Rostaing P, Triller A, Salem NJ, Ashe KH, Frautschy SA, Cole GM (2004) Docosahexaenoic acid protects from dendritic pathology in an Alzheimer's disease mouse model. Neuron 43:633–645

Cao H, Hininger-Favier I, Kelly MA, Benaraba R, Dawson HD, Coves S, Roussel AM, Anderson RA (2007a) Green tea polyphenol extract regulates the expression of genes involved in glucose uptake and insulin signaling in rats fed a high fructose diet. J Agric Food Chem 55:6372–6378

Cao D, Yang B, Hou L, Xu J, Xue R, Sun L, Zhou C, Liu Z (2007b) Chronic daily administration of ethyl docosahexaenoate protects against gerbil brain ischemic damage through reduction of arachidonic acid liberation and accumulation. J Nutr Biochem 18:297–304

Cao H, Graves DJ, Anderson RA (2010) Cinnamon extract regulates glucose transporter and insulin-signaling gene expression in mouse adipocytes. Phytomedicine 17:1027–1032

Carluccio MA, Siculella L, Ancora MA, Massaro M, Scoditti E, Storelli C, Visioli F, Distante A, De Caterina R (2003) Olive oil and red wine antioxidant polyphenols inhibit endothelial activation: antiatherogenic properties of Mediterranean diet phytochemicals. Arterioscler Thromb Vasc Biol 23:622–629

Chao J, Yu MS, Ho HS, Wang M, Chang RC (2006) Dietary oxyresveratrol prevents parkinsonian mimetic 6-hydroxydopamine neurotoxicity. Free Rad Biol Med 45:1019–1026

Checkoway H, Powers K, Smith-Weller T, Franklin GM, Longstreth WT Jr, Swanson PD (2002) Parkinson's disease risks associated with cigarette smoking, alcohol consumption, and caffeine intake. Am J Epidemiol 155:732–738

Chen PC, Wheeler DS, Malhotra V, Odoms K, Denenberg AG, Wong HR (2002) A green tea-derived polyphenol, epigallocatechin-3-gallate, inhibits I kappa B kinase activation and IL-8 gene expression in respiratory epithelium. Inflammation 26:233–241

Cole GM, Ma QL, Frautschy SA (2009) Omega-3 fatty acids and dementia. Prostaglandins Leukot Essent Fatty Acids 81:213–221

Colín-González AL, Ortiz-Plata A, Villeda-Hernández J, Barrera D, Molina-Jijón E, Pedraza-Chaverrí J, Maldonado PD (2011) Aged Garlic Extract Attenuates Cerebral Damage and Cyclooxygenase-2 Induction after Ischemia and Reperfusion in Rats. Plant Foods Hum Nutr 66:348–354

Collins QF, Liu HY, Pi J, Liu Z, Quon MJ, Cao W (2007) Epigallocatechin-3-gallate (EGCG), a green tea polyphenol, suppresses hepatic gluconeogenesis through 5′-AMP-activated protein kinase. J Biol Chem 282:30143–30149

Couturier K, Batandier C, Awada M, Hininger-Favier I, Canini F, Anderson RA, Leverve X, Roussel AM (2010) Cinnamon improves insulin sensitivity and alters the body composition in an animal model of the metabolic syndrome. Arch Biochem Biophys 501:158–161

Crawford P (2009) Effectiveness of cinnamon for lowering hemoglobin A1C in patients with type 2 diabetes: a randomized, controlled trial. J Am Board Fam Med 22:507–512

Dansen TB, Smits LM, van Triest MH, de Keizer PL, van Leenen D, Koerkamp MG, Szypowska A, Meppelink A, Brenkman AB, Yodoi J, Holstege FC, Burgering BM (2009) Redox-sensitive

cysteines bridge p300/CBP-mediated acetylation and FoxO4 activity. Nat Chem Biol 5:664–672
Das S, Das DK (2007) Anti-inflammatory responses of resveratrol. Inflamm Allergy Drug Targets 6:168–173
Deanfield JE, Halcox JP, Rabelink TJ (2007) Endothelial function and dysfunction. Testing and clinical relevance Circulation 115:1285–1295
Delattre AM, Kiss A, Szawka RE, Anselmo-Franci JA, Bagatini PB, Xavier LL, Rigon P, Achaval M, Iagher F, de David C, Marroni NA, Ferraz AC (2010) Evaluation of chronic omega-3 fatty acids supplementation on behavioral and neurochemical alterations in 6-hydroxydopamine-lesion model of Parkinson's disease. Neurosci Res 66:256–264
Demarin V, Lisak M, Morovic S (2011) Mediterranean diet in healthy lifestyle and prevention of stroke. Acta Clin Croat 50:67–77
Deng J-Y, Hsieh P-S, Huang J-P, Lu L-S, Hung L-M (2008) Activation of estrogen receptor is crucial for resveratrol-stimulating muscular glucose uptake via both insulin-dependent and -independent pathways. Diabetes 57:1814–1823
Dias GP, Cavegn N, Nix A, do Nascimento Bevilaqua MC, Stangl D, Zainuddin MS, Nardi AE, Gardino PF, Thuret S (2012) The role of dietary polyphenols on adult hippocampal neurogenesis: molecular mechanisms and behavioural effects on depression and anxiety. Oxid Med Cell Longev 2012:541971
Eckert A, Keil U, Kressmann S, Schindowski K, Leutner S, Leutz S, Müller WE (2003) Effects of EGb 761® Ginkgo biloba extract on mitochondrial function and oxidative stress. Pharmacopsychiatry 36(Suppl 1):S15–S23
Ejaz A, Wu D, Kwan P, Meydani M (2009) Curcumin inhibits adipogenesis in 3 T3-L1 adipocytes and angiogenesis and obesity in C57/BL mice. J Nutr 139:919–925
Essers MA, Weijzen S, de Vries-Smits AM, Saarloos I, de Ruiter ND, Bos JL, Burgering BM (2004) FOXO transcription factor activation by oxidative stress mediated by the small GTPase Ral and JNK. EMBO J 23:4802–4812
Evans JL, Goldfine ID, Maddux BA, Grodsky GM (2002) Oxidative stress and stress-activated signaling pathways: a unifying hypothesis of type 2 diabetes. Endocr Rev 23:599–622
Farooqui AA (2009) Beneficial effects of fish oil human brain. Springer, New York
Farooqui AA (2010) Neurochemical aspects of neurotraumatic and neurodegenerative diseases. Springer, New York
Farooqui AA (2011) Lipid mediators and their metabolism in the brain. Springer, New York
Farooqui AA (2012a) Generation of reactive oxygen species in the brain: signaling for neural cell survival or suicide. In: Farooqui T, Farooqui AA (eds) Oxidative stress in vertebrates and invertebrates. Wiley-Blackwell, Hoboken, NJ, pp 3–15
Farooqui AA (2012b) Phytochemicals, signal transduction, and neurological disorders. Springer, New York
Farooqui AA (2012c) n-3 Fatty acid-derived lipid mediators in the brain: new weapons against oxidative stress and inflammation. Curr Med Chem 19:532–543
Farooqui T, Farooqui AA (2010) Aging: an important factor for the pathogenesis of neurodegenerative diseases. Mech Ageing Dev 130:203–215
Farooqui AA, Farooqui T (2011) Perspective and direction for future research on phytochemicals. In: Farooqui AA, Farooqui T (eds) Phytochemicals and human health: pharmacological and molecular aspects. Nova Science Publishers, Inc, New York, pp 501–513
Farooqui AA, Horrocks LA (2007) Glycerophospholipids in the brain: phospholipases $A_2$ in neurological disorders. Springer, New York
Farooqui AA, Horrocks LA, Farooqui T (2007a) Modulation of inflammation in brain: a matter of fat. J Neurochem 101:577–599
Farooqui AA, Ong WY, Horrocks LA, Chen P, Farooqui T (2007b) Comparison of biochemical effects of statins and fish oil in brain: the battle of the titans. Brain Res Rev 56:443–471
Farooqui AA, Farooqui T, Panza F, Frisardi V (2012) Metabolic syndrome as a risk factor for neurological disorders. Cell Mol Life Sci 69:741–762

Fehske CJ, Leuner K, Müller WE (2009) Ginkgo biloba extract (EGb 761®) influences monoaminergic neurotransmission via inhibition of NE uptake, but not MAO activity after chronic treatment. Pharmacol Res 60:68–73

Freund-Levi Y, Eriksdotter-Jonhagen M, Cederholm T, Basun H, Faxen-Irving G, Garlind A, Vedin I, Vessby B, Wahlund LO, Palmblad J (2006) Omega-3 fatty acid treatment in 174 patients with mild to moderate Alzheimer disease: OmegAD study: a randomized double-blind trial. Arch Neurol 63:1402–1408

Frydman-Marom A, Levin A, Farfara D, Benromano T, Scherzer-Attali R, Peled S, Vassar R, Segal D, Gazit E, Frenkel D, Ovadia M (2011) Orally administrated cinnamon extract reduces β-amyloid oligomerization and corrects cognitive impairment in Alzheimer's disease animal models. PLoS One 6:e16564

Garrel DR, Razi M, Lariviere F, Jobin N, Naman N, Emptoz-Bonneton A, Pugeat MM (1995) Improved clinical status and length of care with low-fat nutrition support in burn patients. J Parenter Enteral Nutr 19:482–491

Gershwin ME, Borchers AT, Keen CL, Hendler S, Hagie F, Greenwood MR (2010) Public safety and dietary supplementation. Ann N Y Acad Sci 1190:104–117

Ghosh S, Novak EM, Innis SM (2007) Cardiac proinflammatory pathways are altered with different dietary n-6 linoleic to n-3 α-linolenic acid ratios in normal, fat-fed pigs. Am J Physiol Heart Circ Physiol 293:H2919–H2927

Giri RK, Selvaraj SK, Kalra VK (2003) Amyloid peptide-induced cytokine and chemokine expression in THP-1 monocytes is blocked by small inhibitory RNA duplexes for early growth response-1 messenger RNA. J Immunol 170:5281–5294

Gota VS, Maru GB, Soni TG, Gandhi TR, Kochar N, Agarwal MG (2010) Safety and pharmacokinetics of a solid lipid curcumin particle formulation in osteosarcoma patients and healthy volunteers. J Agric Food Chem 58:2095–2099

Haffner SM (2006) The metabolic syndrome: inflammation, diabetes mellitus, and cardiovascular disease. Am J Cardiol 97:3A–11A

Han YS, Zheng WH, Bastianetto S, Chabot JG, Quirion R (2004) Neuroprotective effects of resveratrol against beta-amyloid-induced neurotoxicity in rat hippocampal neurons: involvement of protein kinase C. Br J Pharmacol 141:997–1005

Hashimoto M, Hossain S, Shimada T, Shido O (2006) Docosahexaenoic acid-induced protective effect against impaired learning in amyloid β-infused rats is associated with increased synaptosomal membrane fluidity. Clin Exp Pharmacol Physiol 33:934–939

Hashimoto M, Shahdat HM, Yamashita S, Katakura M, Tanabe Y, Fujiwara H, Gamoh S, Miyazawa T, Arai H, Shimada T, Shido O (2008) Docosahexaenoic acid disrupts in vitro amyloid beta fibrillation and concomitantly inhibits amyloid levels in cerebral cortex of Alzheimer's disease model rats. J Neurochem 107:1634–1646

He K, Rimm EB, Merchant A, Rosner BA, Stampfer MJ, Willett WC, Ascherio A (2002) Fish consumption and risk of stroke in men. JAMA 288:3130–3136

Heine PA, Taylor JA, Iwamoto GA, Lubahn DB, Cooke PS (2000) Increased adipose tissue in male and female estrogen receptor-α knockout mice. Proc Natl Acad Sci USA 97:12729–12734

Henry LA, Witt DM (2002) Resveratrol: phytoestrogen effects on reproductive physiology and behavior in female rats. Horm Behav 41:220–228

Higdon JV, Frei B (2003) Tea catechins and polyphenols: health effects, metabolism, and antioxidant functions. Crit Rev Food Sci Nutr 43:89–143

Hong JT, Ryu SR, Kim HJ, Lee JK, Lee SH, Yun YP, Lee BM, Kim PY (2001) Protective effect of green tea extract on ischemia/reperfusion-induced brain injury in Mongolian gerbils. Brain Res 888:11–18

Hossain MS, Hashimoto M, Masumura S (1998) Influence of docosahexaenoic acid on cerebral lipid peroxide level in aged rats with and without hypercholesterolemia. Neurosci Lett 244:157–160

Hossain MS, Hashimoto M, Gamoh S, Masumura S (1999) Antioxidative effects of docosahexaenoic acid in the cerebrum versus cerebellum and brainstem of aged hypercholesterolemic rats. J Neurochem 72:1133–1138

Huynh NN, Chin-Dusting J (2006) Amino acids, arginase and nitric oxide in vascular health. Clin Exp Pharmacol Physiol 33:1–8

Iso H, Rexrode KM, Stampfer MJ, Manson JE, Colditz GA, Speizer FE, Hennekens CH, Willett WC (2001) Intake of fish and omega-3 fatty acids and risk of stroke in women. JAMA 285:304–312

Jang JH, Surh YJ (2003) Protective effect of resveratrol on beta-amyloid-induced oxidative PC12 cell death. Free Radic Biol Med 34:1100–1110

Joseph JA, Shukitt-Hale B, Willis LM (2009) Grape juice, berries, and walnuts affect brain aging and behavior. J Nutr 139:1813S–1817S

Kaplan S, Bisleri G, Morgan JA, Cheema FH, Oz MC (2005) Resveratrol, a natural red wine polyphenol, reduces ischemia-reperfusion-induced spinal cord injury. Ann Thorac Surg 80:2242–2249

Kasper S, Schubert H (2009) Ginkgo-Spezialextrakt EGb 761® in der Behandlung der Demenz: Evidenz für Wirksamkeit und Verträglichkeit [Ginkgo biloba extract EGb 761® in the treatment of dementia: evidence of efficacy and tolerability]. Fortschr Neurol Psychiatr 77:494–506

Khan A, Safdar M, Ali Khan MM, Khattak KN, Anderson RA (2003) Cinnamon improves glucose and lipids of people with type 2 diabetes. Diabetes Care 26:3215–3218

Khan MM, Ahmad A, Ishrat T, Khan MB, Hoda MN, Khuwaja G, Raza SS, Khan A, Javed H, Vaibhav K, Islam F (2010) Resveratrol attenuates 6-hydroxydopamine-induced oxidative damage and dopamine depletion in rat model of Parkinson's disease. Brain Res 1328:139–151

Kim D, Nguyen MD, Dobbin MM, Fischer A, Sananbenesi F, Rodgers JT, Delalle I, Baur JA, Sui G, Armour SM, Pulgserver P, Sinclair DA, Tsai LH (2007) SIRT1 deacetylase protects against neurodegeneration in models for Alzheimer's disease and amyotrophic lateral sclerosis. EMBO J 26:3169–3179

Kim J, Lee HJ, Lee KW (2010) Naturally occurring phytochemicals for the prevention of Alzheimer's disease. J Neurochem 112:1415–1430

Kimura Y, Kimura H (2004) Hydrogen sulfide protects neurons from oxidative stress. FASEB J 18:1165–1167

King VR, Huang WL, Dyall SC, Curran OE, Priestley JV, Michael-Titus AT (2006) Omega-3 fatty acids improve recovery, whereas omega-6 fatty acids worsen outcome, after spinal cord injury in the adult rat. J Neurosci 26:4672–4680

Knoops KT, de Groot LC, Kromhout D, Perrin AE, Moreiras-Varela O, Menotti A, van Staveren WA (2004) Mediterranean diet, lifestyle factors, and 10-year mortality in elderly European men and women: the HALE project. JAMA 292:1433–1439

Koch HP, Lawson LD (1996) Garlic: the science and therapeutic application of *Allium sativum* L. and related species, 2nd edn. Williams & Wilkins, Baltimore

Koh SH, Kim SH, Kwon H, Kim JG, Kim JH, Yang KH, Kim J, Kim SU, Yu HJ, Do BR, Kim KS, Jung HK (2004) Phosphatidylinositol-3 kinase/Akt and GSK-3 mediated cytoprotective effect of epigallocatechin gallate on oxidative stress-injured neuronal differentiated N18D3 cells. Neurotoxicology 25:793–802

Költringer P, Langsteger W, Eber O (1995) Dose-dependent hemorheological effects and microcirculatory modifications following intravenous administration of Ginkgo biloba special extract EGb 761®. Clin Hemorheol 15:649–656

Kulkarni SK, Bhutani MK, Bishnoi M (2008) Antidepressant activity of curcumin: Involvement of serotonin and dopamine system. Psychopharmacology (Berl) 201:435–442

Kulkarni S, Dhir A, Akula KK (2009) Potentials of curcumin as an antidepressant. ScientificWorld J 9:1233–1241

Kuner P, Schubenel R, Hertel C (1998) Beta-amyloid binds to p57NTR and activates NFkappaB in human neuroblastoma cells. J Neurosci Res 54:798–804

Lagouge M, Argmann C, Gerhart-Hines Z, Meziane H, Lerin C, Daussin F, Messadeq N, Milne J, Lambert P, Elliott P, Geny B, Laakso M, Puigserver P, Auwerx J (2006) Resveratrol improves mitochondrial function and protects against metabolic disease by activating SIRT1 and PGC-1alpha. Cell 127:1109–1122

Lander HM, Tauras JM, Ogiste JS, Hori O, Moss RA, Schmidt AM (1997) Activation of the receptor foe advanced glycation end products triggers a p21(ras)-dependent mitogen-activated protein kinase pathway regulated by oxidant stress. J Biol Chem 272:17810–14

Lang-Lazdunski L, Biondeau N, Jarretou G, Heurteaux C (2003) Linolenic acid prevents neuronal cell death and paraplegia after transient spinal cord ischemia in rats. J Vasc Surg 38:564–575

Lankinen M, Schwab U, Kolehmainen M, Paananen J, Poutanen K, Mykkänen H, Seppänen-Laakso T, Gylling H, Uusitupa M, Orešič M (2011) Whole grain products, fish and bilberries alter glucose and lipid metabolism in a randomized, controlled trial: the Sysdimet study. PLoS One 6:e22646

Lê KA, Tappy L (2006) Metabolic effects of fructose. Curr Opin Clin Nutr Metab Care 9(4):469–475

Lee S, Suh S, Kim S (2000) Protective effects of the green tea polyphenol (−)-epigallocatechin gallate against hippocampal neuronal damage after transient global ischemia in gerbils. Neurosci Lett 287:191–194

Levites Y, Weinreb O, Maor G, Youdim MB, Mandel S (2001) Green tea polyphenol (−)-epigallocatechin-3-gallate prevents N-methyl-4-phenyl-1,2,3,6-tetrahydropyridine-induced dopaminergic neurodegeneration. J Neurochem 78:1073–1082

Levites Y, Amit T, Mandel S, Youdim MB (2003) Neuroprotection and neurorescue against Abeta toxicity and PKC-dependent release of nonamyloidogenic soluble precursor protein by green tea polyphenol (−)-epigallocatechin-3-gallate. FASEB J 17:952–954

Li Z, Kaplan ML, Hatchey DL (2000) Hepatic microsomal and peroxisomal docosahexaenoate biosynthesis during piglet development. Lipids 35:1325–1333

Li L, Braiteh FS, Kurzrock R (2005) Liposome-encapsulated curcumin: *in vitro* and *in vivo* effects on proliferation, apoptosis, signaling, and angiogenesis. Cancer 104:1322–1331

Li C, Allen A, Kwagh J, Doliba NM, Qin W, Najafi H, Collins HW, Matschinsky FM, Stanley CA, Smith TJ (2006) Green tea polyphenols modulate insulin secretion by inhibiting glutamate dehydrogenase. J Biol Chem 281:10214–10221

Liu A, Lou H, Zhao L, Fan P (2006) Validated LC/MS/MS assay for curcumin and tetrahydrocurcumin in rat plasma and application to pharmacokinetic study of phospholipid complex of curcumin. J Pharm Biomed Anal 40:720–727

Liu C, Shi Z, Fan L, Zhang C, Wang K, Wang B (2011) Resveratrol improves neuron protection and functional recovery in rat model of spinal cord injury. Brain Res 1374:100–109

Lukiw WJ, Cui JG, Marcheselli VL, Bodker M, Botkjaer A, Gotlinger K, Serhan CN, Bazan NG (2005) A role for docosahexaenoic acid-derived neuroprotectin D1 in neural cell survival and Alzheimer disease. J Clin Invest 115:2774–2783

Ma QL, Yang F, Rosario ER, Ubeda OJ, Beech W, Gant DJ, Chen PP, Hudspeth B, Chen C, Zhao Y, Vinters HV, Frautschy SA, Cole GM (2009a) Beta-amyloid oligomers induce phosphorylation of tau and inactivation of insulin receptor substrate via c-Jun N-terminal kinase signaling: suppression by omega-3 fatty acids and curcumin. J Neurosci 29:9078–9089

Ma QL, Yang F, Rosario ER, Ubeda QJ, Beech W, Gant DJ, Chen PP, Hudspeth B, Chen C, Zhao Y, Vinters HV, Frautchy SA, Cole GM (2009b) Beta-amyloid oligomers induce phosphorylation of tau and inactivation of insulin receptor substrate via c-Jun N-terminal kinase signaling: suppression by omega-3 fatty acids and curcumin. J Neurosci 29:9078–9089

Mandel SA, Amit T, Kalfon L, Reznichenko L, Youdim MB (2008) Targeting multiple neurodegenerative diseases etiologies with multimodal-acting green tea catechins. J Nutr 138:1578S–1583S

Mattson MP, Cheng A (2006) Neurohormetic phytochemicals: Low-dose toxins that induce adaptive neuronal stress responses. Trends Neurosci 29:632–639

Mattson MP, Son TG, Camandola S (2007) Viewpoint: mechanisms of action and therapeutic potential of neurohormetic phytochemicals. Dose Response 5:174–186

Mattson MP (2008) Hormesis defined. Ageing Res Rev 7:1–7

McEwen BS (2001) Plasticity of the hippocampus: adaptation to chronic stress and allostatic load. Ann NY Acad Sci 933:265–277

Michael-Titus AT (2007) Omega-3 fatty acids and neurological injury. Prost Leukot Essent Fatty Acids 77:295–300

Milne JC, Lambert PD, Schenk S, Carney DP, Smith JJ, Gagne DJ, Jin L, Boss O, Perni RB, Vu CB, Bemis JE, Xie R, Disch JS, Ng PY, Nunes JJ, Lynch AV, Yang H, Galonek H, Israelian K, Choy W, Iffland A, Lavu S, Medvedik O, Sinclair DA, Olefsky JM, Jirousek MR, Elliott PJ, Westphal CH (2007) Small molecule activators of SIRT1 as therapeutics for the treatment of type 2 diabetes. Nature 450:712–716

Minamishima S, Bougaki M, Sips PY, Yu JD, Minamishima YA, Elrod JW, Lefer DJ, Bloch KD, Ichinose F (2009) Hydrogen sulfide improves survival after cardiac arrest and cardiopulmonary resuscitation via a nitric oxide synthase 3-dependent mechanism in mice. Circulation 120:888–896

Mira L, Fernandez MT, Santos M, Rocha R, Florêncio MH, Jennings KR (2002) Interactions of flavonoids with iron and copper ions: a mechanism for their antioxidant activity. Free Radic Res 36:1199–1208

Moon HS, Lee HG, Choi YJ, Kim TG, Cho CS (2007) Proposed mechanisms of (−)-epigallocatechin-3-gallate for anti-obesity. Chem Biol Interact 167:85–89

Munch G, Thome J, Foley P, Schinzel R, Riederer P (2000) Advanced glycation end products in ageing and Alzheimer's disease. Brain Res Rev 23:134–43

Murase T, Misawa K, Haramizu S, Hase T (2009) Catechin-induced activation of the LKB1/AMP-activated protein kinase pathway. Biochem Pharmacol 78:78–84

Nakagawa T, Hu H, Zharikov S, Tuttle KR, Short RA, Glushakova O, Ouyang X, Feig DI, Block ER, Acosta JH, Patel JM, Johnson RJ (2006) A causal role for uric acid in fructose-induced metabolic syndrome. Am J Physiol 2006(290):625–631

Okada M, Amamoto T, Tomonaga M, Kawachi A, Yazawa K, Mine K, Fujiwara M (1996) The chronic administration of docosahexaenoic acid reduces the spatial cognitive deficit following transient forebrain ischemia in rats. Neuroscience 71:17–25

Ono K, Hasegawa K, Naiki H, Hamada M (2004) Curcumin has potent anti-amyloidogenic effects for Alzheimer's beta-amyloid fibrils in vitro. J Neurosci Res 75:742–750

Padiya R, Khatua TN, Bague R, Kuncha M, Banerjee SK (2011) Garlic improves insulin sensitivity and associated metabolic syndromes in fructose fed rats. Nutr Metab (Lond) 8:53

Pan T, Fei J, Zhou X, Jankovic J, Le W (2003) Effects of green tea polyphenols on dopamine uptake and on $MPP^+$-induced dopamine neuron injury. Life Sci 72:1073–1083

Panickar KS, Polansky MM, Graves DJ, Urban JF Jr, Anderson RA (2012) A procyanidin type A trimer from cinnamon extract attenuates glial cell swelling and the reduction in glutamate uptake following ischemia-like injury in vitro. Neuroscience 202:87–98

Papaioannou N, Tooten PC, van Ederen AM, Bohl JR, Rofina J, Tsangaris T, Gruys E (2001) Immunohistochemical investigation of the brain of aged dogs. I. Detection of neurofibrillary tangles and of 4-hydroxynonenal protein, an oxidative damage product, in senile plaques. Amyloid 8:11–21

Park CE, Kim MJ, Lee JH, Min BI, Bae H, Choe W, Kim SS, Ha J (2007) Resveratrol stimulates glucose transport in C2C12 myotubes by activating AMP-activated protein kinase. Exp Mol Med 39:222–229

Patel J, Iyer A, Brown L (2009) Evaluation of the chronic complications of diabetes in a high fructose diet in rats. Ind J Biochem Biophys 46:66–72

Peterson DW, George RC, Scaramozzino F, LaPointe NE, Anderson RA, Graves DJ, Lew J (2009) Cinnamon extract inhibits tau aggregation associated with Alzheimer's disease in vitro. J Alzheimers Dis 17:585–597

Potenza MA, Marasciulo FL, Tarquinio M, Tiravanti E, Colantuono G, Federici A, Kim JA, Quon MJ, Montagnani M (2007) EGCG, a green tea polyphenol, improves endothelial function and insulin sensitivity, reduces blood pressure, and protects against myocardial I/R injury in SHR. Am J Physiol Endocrinol Metab 292:E1378–E1387

Qin B, Nagasaki M, Ren M, Bajotto G, Oshida Y, Sato Y (2004) Cinnamon extract prevents the insulin resistance induced by a high-fructose diet. Horm Metab Res 36:119–125

Qin B, Polansky MM, Anderson RA (2009) Cinnamon extract regulates plasma levels of adipose-derived factors and expression of multiple genes related to carbohydrate metabolism and lipogenesis in adipose tissue of fructose-fed rats. Horm Metab Res 42:187–193

Qin B, Panickar KS, Anderson RA (2010) Cinnamon: potential role in the prevention of insulin resistance, metabolic syndrome, and type 2 diabetes. J Diabetes Sci Technol 4:685–693

Rahman K, Lowe GM (2006) Garlic and cardiovascular disease: a critical review. J Nutr 136:736S–740S

Ramassamy C, Christen Y, Clostre F, Costentin J (1992) The Ginkgo biloba extract, EGb 761®, increases synaptosomal uptake of 5-hydroxytryptamine: in-vitro and ex-vivo studies. J Pharm Pharmacol 44:943–945

Ray B, Chauhan NB, Lahiri DK (2011a) The "aged garlic extract:" (AGE) and one of its active ingredients S-allyl-L-cysteine (SAC) as potential preventive and therapeutic agents for Alzheimer's disease (AD). Curr Med Chem 18:3306–3313

Ray B, Chauhan NB, Lahiri DK (2011b) Oxidative insults to neurons and synapse are prevented by aged garlic extract and S-allyl-L-cysteine treatment in the neuronal culture and APP-Tg mouse model. J Neurochem 117:388–402

Reisin E (2009) The benefit of the Mediterranean-style diet in patients with newly diagnosed diabetes. Curr Hypertens Rep 12:56–8

Ren J, Fan C, Chen N, Huang J, Yang Q (2011) Resveratrol pretreatment attenuates cerebral ischemic injury by upregulating expression of transcription factor Nrf2 and HO-1 in rats. Neurochem Res 36:2352–2362

Rezai-Zadeh K, Arendash GW, Hou H, Fernandez F, Jensen M, Runfeldt M, Shytle RD, Tan J (2008) Green tea epigallocatechin-3-gallate (EGCG) reduces beta-amyloid mediated cognitive impairment and modulates tau pathology in Alzheimer transgenic mice. Brain Res 1214:177–187

Riviere C, Richard T, Quentin L, Krisa S, Merillon JM, Monti JP (2007) Inhibitory activity of stilbenes on Alzheimer's beta-amyloid fibrils in vitro. Bioorg Med Chem 15:1160–1167

Roberts CK, Barnard RJ, Sindhu RK, Jurczak M, Ehdaie A, Vaziri ND (2006) Oxidative stress and dysregulation of NAD(P)H oxidase and antioxidant enzymes in diet-induced metabolic syndrome. Metabolism 55:928–934

Rojas P, Serrano-García N, Medina-Campos ON, Pedraza-Chaverri J, Maldonado PD, Ruiz-Sánchez E (2011) S-Allylcysteine, a garlic compound, protects against oxidative stress in 1-methyl-4-phenylpyridinium-induced parkinsonism in mice. J Nutr Biochem 22:937–944

Roussel AM, Hininger I, Benaraba R, Ziegenfuss TN, Anderson RA (2009) Antioxidant effects of a cinnamon extract in people with impaired fasting glucose that are overweight or obese. J Am Coll Nutr 28:16–21

Ruan H, Yang Y, Zhu X, Wang X, Chen R (2009) Neuroprotective effects of (+/–)-catechin against 1-methyl-4-phenyl-1,2,3,6-tetrahydropyridine (MPTP)-induced dopaminergic neurotoxicity in mice. Neurosci Lett 450:152–157

Saiko P, Szakmary A, Jaeger W, Szekeres T (2008) Resveratrol and its analogs: defense against cancer, coronary disease and neurodegenerative maladies or just a fad? Mutat Res 658:68–94

Saleem S, Ahmad M, Ahmad AS, Yousuf S, Ansari MA, Khan MB, Ishrat T, Islam F (2006) Behavioral and histologic neuroprotection of aqueous garlic extract after reversible focal cerebral ischemia. J Med Food 9:537–544

Sanmukhani J, Anovadiya A, Tripathi CB (2011) Evaluation of antidepressant like activity of curcumin and its combination with fluoxetine and imipramine: an acute and chronic study. Acta Pol Pharm 68:769–775

Sasaki N, Toki S, Chowei H, Saito T, Nakano N, Hayashi Y, Takeuchi M, Makita Z (2001) Immunohistochemical distribution of the receptor for advanced glycation end products in neurons and astrocytes in Alzheimer's disease. Brain Res 888:256–262

Savaskan E, Olivieri G, Meier F, Seifritz E, Wirz-Justice A, Muller-Spahn F (2003) Red wine ingredient resveratrol protects from beta-amyloid neurotoxicity. Gerontology 49:380–383

Scapagnini G, Sonya V, Nader AG, Calogero C, Zella D, Fabio G (2011) Modulation of Nrf2/ARE pathway by food polyphenols: a nutritional neuroprotective strategy for cognitive and neurodegenerative disorders. Mol Neurobiol 44:192–201

Schafer DP, Lehrman EK, Stevens B (2012) The "quad-partite" synapse: Microglia–synapse interactions in the developing and mature CNS. Glia 61:24–36

Scholl J (2012) Traditional dietary recommendations for the prevention of cardiovascular disease: do they meet the needs of our patients? Cholesterol 2012:367898

Scott BL, Bazan NG (1989) Membrane docosahexaenoate is supplied to the developing brain and retina by the liver. Proc Natl Acad Sci USA 86:2903–2907

Sharma S, Zhuang Y, Ying Z, Wu A, Gomez-Pinilla F (2009) Dietary curcumin supplementation counteracts reduction in levels of molecules involved in energy homeostasis after brain trauma. Neuroscience 161:1037–1044

Sheng X, Zhang Y, Gong Z, Huang C, Zang YQ (2008) Improved insulin resistance and lipid metabolism by cinnamon extract through activation of peroxisome proliferator-activated receptors. PPAR Res 2008:581348

Shenouda SM, Vita JA (2007) Effects of flavonoid-containing beverages and EGCG on endothelial function. J Am Coll Nutr 26:366S–372S

Shoba G, Joy D, Joseph T, Majeed M, Rajendran R, Srinivas PS (1998) Influence of piperine on the pharmacokinetics of curcumin in animals and human volunteers. Planta Med 64:353–356

Singh M, Arseneault M, Sanderson T, Murthy V, Ramassamy C (2008) Challenges for research on polyphenols from foods in Alzheimer's disease: bioavailability, metabolism, and cellular and molecular mechanisms. J Agric Food Chem 56:4855–4873

Sloane JA, Hollander W, Moss MB, Rosene DL, Abraham CR (1999) Increased microglial activation and protein nitration in white matter of the aging monkey. Neurobiol Aging 20:395–405

Sohal RS, Agarwal S, Candas M, Forster MJ, Lal H (1994) Effect of age and caloric restriction on DNA oxidative damage in different tissues of C57BL/6 mice. Mech Ageing Dev 76:215–224

Son TG, Camandola S, Mattson MP (2008) Hormetic dietary phytochemicals. Neuromolecular Med 10:236–246

Sönmez Ü, Sönmez A, Erbil G, Tekmen I, Baykara B (2007) Neuroprotective effects of resveratrol against traumatic brain injury in immature rats. Neurosci Lett 420:133–137

Spencer JP (2009) The impact of flavonoids on memory: physiological and molecular considerations. Chem Soc Rev 38:1152–1161

Storz P (2011) Forkhead homeobox type O transcription factors in the responses to oxidative stress. Antioxid Redox Signal 14:593–605

Su HM, Bernardo L, Mirmiran M, Ma XH, Corso TN, Nathanielsz PW, Brenna JT (1999a) Bioequivalence of dietary alpha-linolenic and docosahexaenoic acids as sources of docosahexaenoate accretion in brain and associated organs of neonatal baboons. Pediatr Res 45:87–93

Su HM, Bernardo L, Mirmiran M, Ma XH, Nathanielsz PW, Brenna JT (1999b) Dietary 18:3n-3 and 22:6n-3 as sources of 22:6n-3 accretion in neonatal baboon brain and associated organs. Lipids 34(Suppl):S347–S350

Takeuchi M, Bucala R, Suzuki T, Ohkubo T, Yamazaki M, Koike T, Kameda Y, Makita Z (2000) Neurotoxicity of advanced glycation end-products for cultured cortical neurons. J Neuropathol Exp Neurol 59:1094–1105

Tamura Y, Ogihara T, Uchida T, Ikeda F, Kumashiro N, Nomiyama T, Sato F, Hirose T, Tanaka Y, Mochizuki H, Kawamori R, Watada H (2007) Amelioration of glucose tolerance by hepatic inhibition of nuclear factor kappa B in db/db mice. Diabetologia 50:131–141

Tattelman E (2005) Health effects of garlic. Am Fam Physician 72:103–106

Taubes G (2008) Good calories. Fats, Carbs, and the Controversial Science of Diet and Health, AA Knopf, New York, Bad Calories

Urakawa H, Katsuki A, Sumida Y, Gabazza EC, Murashima S, Morioka K, Maruyama N, Kitagawa N, Tanaka T, Hori Y, Nakatani K, Yano Y, Adachi Y (2003) Oxidative stress is associated with adiposity and insulin resistance in men. J Clin Endocrinol Metab 88:4673–4676

van der Heide LP, Smidt MP (2005) Regulation of FoxO activity by CBP/p300-mediated acetylation. Trends Biochem Sci 30:81–86 van der Horst A, Burgering BM (2007) Stressing the role of FoxO proteins in lifespan and disease. Nat Rev Mol Cell Biol 8:440–450

Velayutham P, Babu A, Liu D (2008) Green tea catechins and cardiovascular health: an update. Curr Med Chem 15:1840–1850

Vicente Miranda H, Outeiro TF (2010) The sour side of neurodegenerative disorders: the effects of protein glycation. J Pathol 221:13–25

Vignes M, Maurice T, Lanté F, Nediar M, Thethi K, Guiramand J, Recasens M (2006) Anxiolytic properties of green tea polyphenol (−)-epigallocatechin gallate (EGCG). Brain Res 1110:102–115

Vlassara H, Fuh H, Donnelly T, Cybulsky M (1995) Advanced glycation endproducts promote adhesion molecule (VCAM-1, ICAM-1) expression and atheroma formation in normal rabbits. Mol Med 1:447–456

Wang Q, Xu J, Rottinghaus GE, Simonyi A, Lubahn D, Sun GY, Sun AY (2002) Resveratrol protects against global cerebral ischemic injury in gerbils. Brain Res 958:439–447

Wang J, Ho L, Zhao Z, Seror I, Humala N, Dickstein DL, Thiyagarajan M, Percival SS, Talcott ST, Pasinetti GM (2006) Moderate consumption of Cabernet Sauvignon attenuates Abeta neuropathology in a mouse model of Alzheimer's disease. FASEB J 20:2313–2320

Wang JG, Anderson RA, Graham GM, Chu MC, Sauer MV, Guarnaccia MM, Lobo RA (2007) The effect of cinnamon extract on insulin resistance parameters in polycystic ovary syndrome: a pilot study. Fertil Steril 88:240–243

Wang R, Li YB, Li YH, Xu Y, Wu HL, Li XJ (2008) Curcumin protects against glutamate excitotoxicity in rat cerebral cortical neurons by increasing brain-derived neurotrophic factor level and activating TrkB. Brain Res 1210:84–91

Ward RE, Huang W, Curran OE, Priestly JV, Michael-Titus AT (2010) Docosahexaenoic acid prevents white matter damage following spinal cord injury. J Neurotrauma 27:1–12

Wautier JL, Zoukourian C, Chappey O, Wautier MP, Guillausseau PJ, Cao R, Hori O, Stern D, Schmidt AM (1996) Receptor-mediated endothelial cell dysfunction in diabetic vasculopathy: soluble receptor for advanced glycation end products blocks hyperpermeability in diabetic rats. J Clin Invest 97:238–243

Whiteman M, Gooding KM, Whatmore JL, Ball CI, Mawson D, Skinner K, Tooke JE, Shore AC (2010) Adiposity is a major determinant of plasma levels of the novel vasodilator hydrogen sulphide. Diabetologia 53:1722–1726

Willett WC (2006) The Mediterranean diet: science and practice. Public Health Nutr 9:105–110

Wolfram S, Raederstorff D, Preller M, Wang Y, Teixeira SR, Riegger C, Weber P (2006) Epigallocatechin gallate supplementation alleviates diabetes in rodents. J Nutr 136:2512–2518

Wu A, Molteni R, Ying Z, Gomez-Pinilla F (2003) A saturated-fat diet aggravates the outcome of traumatic brain injury on hippocampal plasticity and cognitive function by reducing brain-derived neurotrophic factor. Neuroscience 119:365–375

Wu A, Ying Z, Gomez-Pinilla F (2004) Dietary omega-3 fatty acids normalize BDNF levels, reduce oxidative damage, and counteract learning disability after traumatic brain injury in rats. J Neurotrauma 21:1457–1467

Wu A, Ying Z, Gomez-Pinilla F (2005) Omega-3 fatty acids supplementation restores homeostatic mechanisms disrupted by traumatic brain injury. J Neurotrauma 22:1212

Wu A, Ying Z, Gomez-Pinilla F (2006) Dietary curcumin counteracts the outcome of traumatic brain injury on oxidative stress, synaptic plasticity, and cognition. Exp Neurol 197:309–317

Wu Y, Li X, Zhu JX, Xie W, Le W, Fan Z, Jankovic J, Pan T (2011) Resveratrol-Activated AMPK/SIRT1/Autophagy in Cellular Models of Parkinson's Disease. Neurosignals 19:163–174

Xu Y, Ku BS, Yao HY, Lin YH, Ma X, Zhang YH, Li XJ (2005) The effects of curcumin on depressive-like behaviors in mice. Eur J Pharmacol 518:40–46

Yan SD, Stern D, Kane MD, Kuo YM, Lampert HC, Roher AE (1998) RAGE-Aβ interactions in the pathophysiology of Alzheimer's disease. Restor Neurol Neurosci 12:167–173

Yao D, Brownlee M (2010) Hyperglycemia-induced reactive oxygen species increase expression of the receptor for advanced glycation end products (RAGE) and RAGE ligands. Diabetes 59:249–55

Yoshitake T, Yoshitake S, Kehr J (2010) The Ginkgo biloba extract EGb 761® and its main constituent flavonoids and ginkgolides increase extracellular dopamine levels in the rat prefrontal cortex. Br J Pharmacol 159:659–668

Yousuf S, Atif F, Ahmad M, Hoda N, Ishrat T, Khan B, Islam F (2009) Resveratrol exerts its neuroprotective effect by modulating mitochondrial dysfunctions and associated cell death during cerebral ischemia. Brain Res 1250:242–253

Zhao J, Yu S, Zheng W, Feng G, Luo G, Wang L, Zhao Y (2010) Curcumin improves outcomes and attenuates focal cerebral ischemic injury via anti-apoptotic mechanisms in rats. Neurochem Res 35:374–379

Zhong L, Furne JK, Levitt MD (2006) An extract of black, green, and mulberry teas causes malabsorption of carbohydrate but not of triacylglycerol in healthy volunteers. Am J Clin Nutr 84:551–555

Ziegenfuss TN, Hofheins JE, Mendel RW, Landis J, Anderson RA (2006) Effects of a water-soluble cinnamon extract on body composition and features of the metabolic syndrome in pre-diabetic men and women. J Int Soc Sports Nutr 3:45–53

# Chapter 7
# Metabolic Syndrome as a Risk Factor for Stroke

## 7.1 Introduction

The metabolic syndrome (MetS) is a complex condition which is characterized by insulin resistance (a reduced physiological response of peripheral tissues to the action of insulin) and a clustering of other cardiovascular risk factors, such as hypertension, obesity, increased levels of triacylglycerols, the presence of small dense low-density lipoprotein (LDL), a low level of high-density lipoprotein (HDL) cholesterol, and hypercoagulability (Shulman 2000; Borch-Johnsen 2007). MetS patients are essentially at twice the risk of cardiovascular disease compared to those without the syndrome. MetS also elevates the risk of type II diabetes mellitus by about fivefold. In most countries, about 20–30 % of the adult population is predisposed to MetS (Grundy 2008). The MetS is not a discrete entity known to be caused by a single factor. It shows considerable variation in the above mentioned risk factors (hypertension, abdominal obesity, and increased levels of triacylglycerols) (Fig. 7.1). This variation is even greater among different racial and ethnic groups (Grundy et al. 2005). The MetS is not restricted to adults only; the predisposition of MetS however starts much earlier in life especially during adolescence and young age (Smith and Essop 2009; Tailor et al. 2010; Ford et al. 2002). The predisposition of MetS depends on genetic as well as environmental factors that allow its progression. An abundant high-calorie, high-fat diet and sedentary lifestyle are considered central to the development of MetS in individuals at risk. The molecular mechanism associated with the development of MetS is not fully understood. However, excessive food intake along with obesity and induction of oxidative stress and low grade inflammation may be early events in the development and progression of MetS (Roberts and Sindhu 2009). Oxidative stress plays an important role in the etiology of various risk factors of MetS-related manifestations, including atherosclerosis, hypertension, type II diabetes, adiposity, and insulin resistance (Roberts and Sindhu 2009). Thus, oxidative stress, a cytotoxic process that occurs in the cell when antioxidant mechanisms are overwhelmed by reactive oxygen species (ROS), induces alterations in insulin signaling by targeting insulin receptor and insulin receptor

A.A. Farooqui, *Metabolic Syndrome: An Important Risk Factor for Stroke, Alzheimer Disease, and Depression*, DOI 10.1007/978-1-4614-7318-3_7,

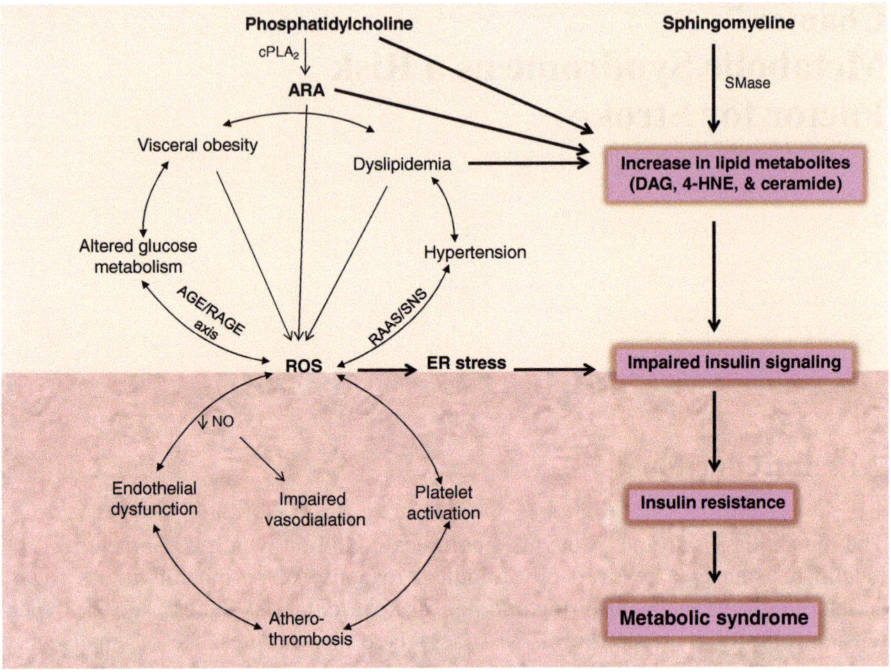


**Fig. 7.1** Oxidative stress connects various components (visceral obesity, dyslipidemia, hypertension) of MetS and development of insulin resistance. Arachidonic acid (ARA); sphingomyelinase (SMase); reactive oxygen species (ROS); nitric oxide (NO); diacylglycerol (DAG); 4-hydroxynonenal (4-HNE); advanced glycation products (AGEs); receptor for advanced glycation end products (RAGE); renin–angiotensin–aldosterone system (RAAS); and sympathetic nervous system (SNS)

substrates (IRSs) or through the activation of ser/thr kinases that regulate insulin signaling (Rains and Jain 2011). In this scenario, the ROS-mediated changes in the insulin signaling pathway not only support, but also advance insulin resistance with subsequent development of diabetes. PtdIns 3-kinase and the MAP kinases (ERK1/2) are major components of insulin-mediated signaling. Insulin mediates its effects through interaction and activation of insulin receptor tyrosine kinase and phosphorylation of the IRS proteins (e.g., IRS-1 and IRS-2) (Taniguchi et al. 2006). As stated above, insulin signaling network involves two major pathways: the phosphatidylinositol 3-kinase (PtdIns 3K) and mitogen-activated protein (MAP) pathways (Gallagher et al. 2008; Chang et al. 2004). PtdIns 3K is the pathway associated with the metabolic effects of insulin, whereas MAP plays a role in growth and mitogenesis (Gallagher et al. 2008; Zick 2001). In the PtdIns 3K pathway, IRS is tyrosine phosphorylated. Phosphorylated IRS interacts with the p85 subunit of PtdIns 3K, resulting in the synthesis of PtdIns(3,4,5)$P_3$ (phosphatidylinositol 3,4,5-trisphosphate) (Gallagher et al. 2008). PtdIns(3,4,5)$P_3$ recruits the serine kinases phosphoinositide-dependent kinase (PDK-1), PKB/Akt, and PKC to the plasma membrane via their pleckstrin homology (PH) domains. The activation of

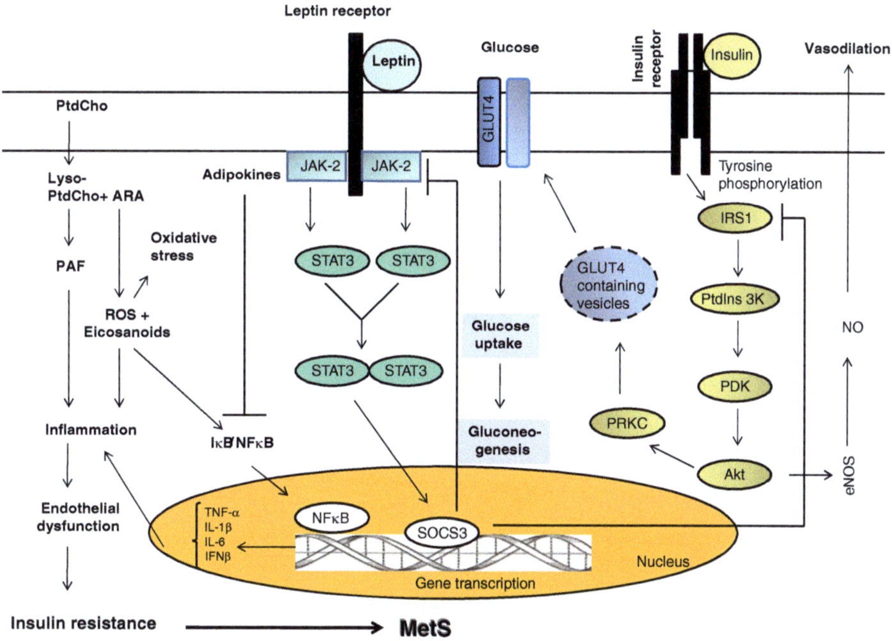


**Fig. 7.2** Interactions between leptin and insulin signaling. Phosphatidylcholine (PtdCho); lyso-phosphatidylcholine (lyso-PtdCho); arachidonic acid (ARA); platelet-activating factor (PAF); reactive oxygen species (ROS); insulin receptor substrate-2 (IRS-2); serine/threonine protein kinase (Atk/PKB); Janus kinase 2 (JAK2); signal transducer and activator of transcription 3 (STAT3); suppressor of cytokine signaling (SOCS); phosphoinositide-dependent kinase (PDK); phosphatidylinositol 3-kinase (PtdIns 3K); tumor necrosis factor-α (TNF-α); interleukin-1β (IL-1β); interleukin-6 (IL-6); nuclear factor-κB (NF-κB); and glucose transporter type 4 (GLUT4)

these kinases leads to several insulin responses, such as GLUT4 translocation to the membrane (Fig. 7.2) (Gallagher et al. 2008; Khan and Pessin 2002; Chang et al. 2004). Akt's role related to insulin includes glucose transport and storage, protein synthesis, and stopping lipid degradation (Gallagher et al. 2008). PtdIns pathway also mediates the regulation of nitric oxide (NO) production from the endothelium and vascular smooth muscle cells (VSMC) (Kim et al. 2006). The bioavailability of NO is a critical factor for maintaining normal vascular functions including vasomotor reactivity, antithrombosis state, barrier function, and nonadhesive state to inflammation cells (Yang and Ming 2006). In the endothelium, endothelial nitric oxide synthase (eNOS) converts L-arginine to L-citrulline and NO. Active NO levels are largely regulated by eNOS gene expression or its activity (Bredt 1999). Many cardiovascular and cerebrovascular risk factors may negatively influence NO levels by different mechanisms. ROS including superoxide anions react with NO and reduce NO bioavailability (Harrison et al. 2003). A decrease in the relative bioavailability of NO not only impairs endothelium-dependent vasorelaxation but also activates other mechanisms that play an important role in atherogenesis (Quyyumi 1998).

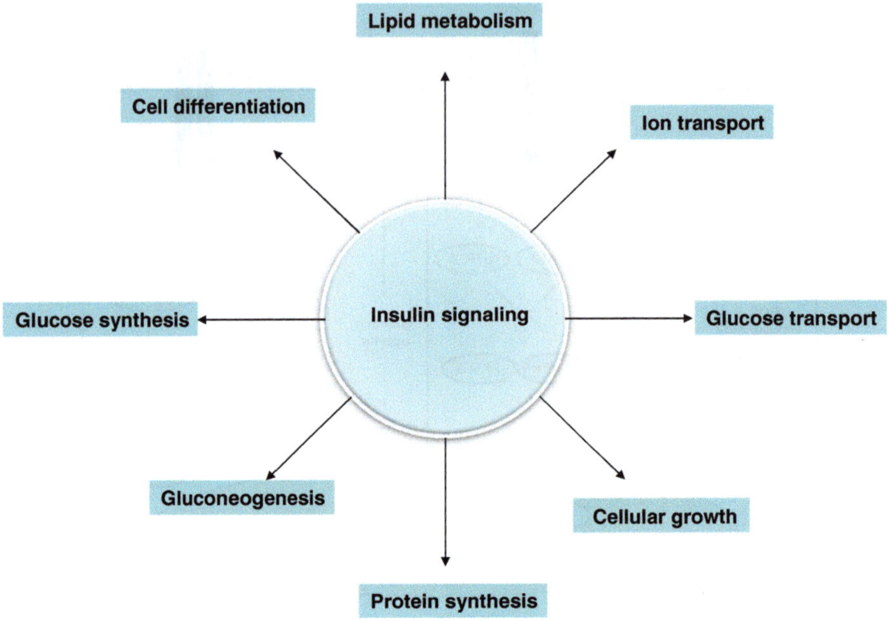


**Fig. 7.3** Role of insulin signaling in the brain metabolism

Insulin resistance is accompanied by a specific impairment in PtdIns 3K-dependent signaling pathways, whereas MAPK-dependent pathway remains unaffected (Mather et al. 2004). However, MARK-dependent pathway leads to the gene expression of various cellular proliferation or differentiation components. After phosphorylation of IRS-1/2, the adaptor proteins GRB-2 and SOS are recruited and work together with a stimulated tyrosine phosphatase, SHP-2, to activate membrane-bound Ras. Activated Ras induces a kinase cascade, allowing extracellular signal-regulated kinase (ERK) to translocate to the nucleus for gene expression (Saltiel and Kahn 2001). Insulin signaling is associated with many biochemical processes, including glucose transport, gluconeogenesis, lipid metabolism, and cell growth (Fig. 7.3).

## 7.2 Adipocytokines and Their Contribution in the Pathogenesis of MetS

Adipose tissue is a metabolically active endocrine organ that secretes adipocytokines, such as adiponectin, resistin, visfatin, retinol-binding protein-4 (RBP-4) and leptin (Table 7.1) (Ahima and Flier 2000; Frühbeck 2004; Graham et al. 2006; Dodson et al. 2011; Ye 2009). "Adipocytokines" (a) control energy balance and appetite and influence insulin sensitivity via endocrine mechanisms and (b)

**Table 7.1** Properties and roles of adipocytokines

| Name | Molecular mass (kDa) | Role | Reference |
|---|---|---|---|
| Adiponectin | 30 | Anti-inflammatory, insulin sensitizer | Bakhai (2008) |
| Resistin | 12.5 | Proinflammatory, induces insulin resistance | Heidemann et al. (2008) |
| Visfatin | 100 | | Körner et al. (2007) |
| RBP-4 | 21 | Carrier protein for vitamin A in plasma | Frey et al. (2008) |
| Leptin | 16 | Regulation of food intake, weight, and energy balance, regulation of immune response | Shpilman et al. (2011) |
| TNF-α | 51 | Proinflammatory, impairs insulin signaling, mediates insulin resistance | Utsumi et al. (1992) |
| IL-6 | 21–28 | Releases CRP from the liver, impairs insulin signaling, produces insulin resistance | Tsunoda et al. (2000) |
| Angiotensinogen | 56.8 | Contributes to hypertension | Tewsbury et al. (1978) |

modulate adipocyte size/number and angiogenesis via paracrine mechanisms, thus exerting a major role in the regulation of fat mass (Waki and Tontonoz 2007; Matsuzawa 2006). Furthermore, they can also exert a role in the control of blood pressure, lipoprotein metabolism, coagulation, immunity, and inflammation (Reaven et al. 2004; Matsuzawa 2006). Under basal conditions, only low levels of adipocytokines are released into the systemic circulation. Targets for adipocytokines include the brain, liver, and skeletal system (Rabe et al. 2008), where they also regulate lipid metabolism, insulin sensitivity, and angiogenesis (Trayhurn and Beattie 2001; Rajala and Scherer 2003).

Under pathological conditions, such as MetS and inflammatory bowel diseases, there occurs an increase in levels of adipocytokines resulting in abnormalities in signal transduction processes (Karmiris et al. 2006; Hallikainen et al. 2007; Norata et al. 2007; Pini et al. 2008). Thus, abnormal production of adipocytokines by expanded visceral fat during obesity contributes to a proinflammatory state and development of MetS and inflammatory bowel diseases (Hu et al. 2004). Many of the peripheral actions of adipocytokines released from adipose tissue are mediated through the activation NF-κB, and its action is further promoted by the effects of other hormones, metabolites, and inflammatory cytokines present in MetS. The activation of NF-κB not only leads to oxidative stress but also provides a link between inflammation and oxidative stress. Both these processes are crucial to the development of MetS. ROS have been implicated in the pathogenesis of MetS, including hypertension, atherosclerosis, diabetes, and even obesity itself (Libby 2003). The inhibition of NF-κB by adiponectin is responsible for attenuation of expression of cell adhesion molecules (Ouedraogo et al. 2007).

In visceral tissues, adipokine signaling inhibits insulin signaling through the involvement of suppressors of cytokine signaling (SOCS) 1 and 3 and nitric oxide (NO). SOCS proteins are induced during inflammation by interleukin (IL)-6 and induce ubiquitinylation and degradation of IRS proteins (see below). NO is an endogenous diffusible signaling molecule that is produced by nitric oxide synthase (NOS), which is inducible by a variety of inflammatory cytokines. In blood vessels NO functions as a vasodilator, while in the brain it acts as a neurotransmitter. However, if generated in excess and in the appropriate redox state, NO can be neurotoxic (Drew and Leeuwenburgh 2002). Thus, NO reacts with superoxide and generates peroxynitrite. In addition, signaling by NO is transduced mainly by targeted modifications of critical cysteine residues in proteins, including S-nitrosylation and S-oxidation, as well as by lipid and tyrosine nitration (Calabrese et al. 2009). It is shown that the PtdIns 3K/Akt signaling pathway associated with the degradation of IRS1 and iNOS activity results in reduced activity of Akt, the main mediator of IRS signaling (Kim et al. 2006).

### *7.2.1 Adiponectin and AMP Kinase in the Metabolic Syndrome*

Adiponectin, a 244 amino acid multimeric protein (30 kDa) solely produced and secreted by the adipose tissue, acts through its seven transmembrane receptors distinct from G protein-coupled receptors (AdipoR1 and AdipoR2), which are expressed throughout the brain (Ahima 2005; Rondinone 2006). It is expressed almost exclusively in mature adipocytes where it is the most abundant adipokine synthesized and released locally and in the blood stream. Adiponectin modulates hypothalamic and brainstem neuronal activity and acts centrally to control peripheral metabolism (Hoyda et al. 2007; Kubota et al. 2007). Adiponectin circulates as trimer, hexamer, and higher-order multimer in serum. Low plasma levels of adiponectin are associated with MetS (Ahima 2005). The deficiency of adiponectin has no apparent metabolic effects in lean mice. However, adiponectin deficiency in mice fed with high-fat diet causes insulin resistance, hyperlipidemia, inflammation, and vascular injury (Ma et al. 2002; Nawrocki et al. 2006). Low levels of adiponectin are found in CSF (Yamauchi et al. 2007; Kusminski et al. 2007). Adiponectin is a reliable biomarker for MetS in a clinical setting. Unlike insulin, glucose, and triacylglycerols, levels of circulating adiponectin do not show diurnal fluctuation (Ahonen et al. 2010; Okamoto 2011; Menzaghi et al. 2006; Sandeep et al. 2007). Levels of adiponectins are decreased in the serum in patients with type II diabetes mellitus (Hotta et al. 2000). Monkeys with reduced plasma levels of adiponectin (before the onset of diabetes) later developed type II diabetes mellitus (Hotta et al. 2001). Adiponectin not only increases insulin sensitivity and ameliorates obesity but also has several anti-inflammatory actions (Fig. 7.4). Studies on the effect of adiponectin have indicated that this adipocytokine not only activates cAMP-dependent protein kinase A but also inhibits endothelial nuclear factor-κβ (NF-κβ) signaling.

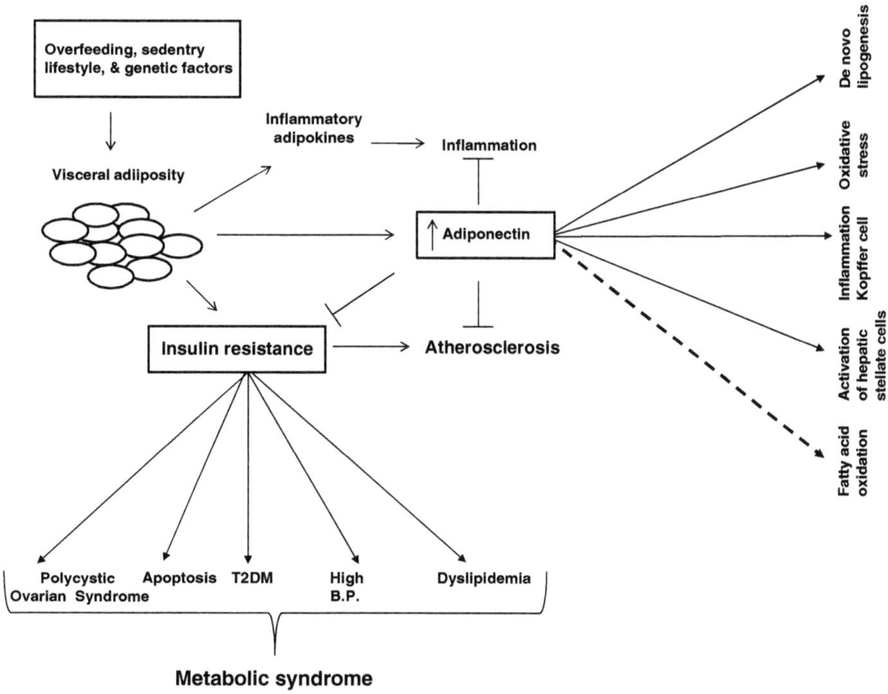


**Fig. 7.4** Link among visceral obesity, insulin resistance, type II diabetes, and atherosclerosis

Adiponectin is a potent anti-inflammatory protein which suppresses TNF-α-induced NF-κB activation and blocks TNF-α release from endothelial cells and macrophages. Thus, activities of adiponectin are inversely proportional to obesity, diabetes, and other insulin-resistant states. Collective evidence suggests that at the molecular level, adiponectin exerts insulin-sensitizing effects by increasing glucose uptake, NO production, and free fatty acid oxidation (Chen et al. 2003; Soodini and Hamdy 2004; Dyck 2009) and shows an anti-inflammatory activity mainly through a cAMP-mediated interference with NF-κB signaling (Ouchi et al. 2000).

Similarly in macrophages, Toll-like receptor-mediated NF-κβ signaling is also inhibited by adiponectin (Yamaguchi et al. 2005). Adiponectin also inhibits vascular smooth muscle migration and proliferation (Okamoto et al. 2002; O'Rourke et al. 2002). It promotes clearance of apoptotic cells by macrophages through a receptor-dependent pathway involving calreticulin. This property of adiponectin is similar to surfactant proteins and C1q, which serve as anti-inflammatory molecules by promoting the clearance of apoptotic cell debris (Vandivier et al. 2002). In addition, adiponectin also forms heterooligomers with additional members of the C1q/TNF-related protein (CTRP) family.

In the liver, adiponectin signaling is mediated primarily through activation of adenosine monophosphate-activated kinase (AMPK) by phosphorylation at $Thr_{172}$. Peroxisome proliferator-activated receptor alpha (PPAR-α) and serine/threonine

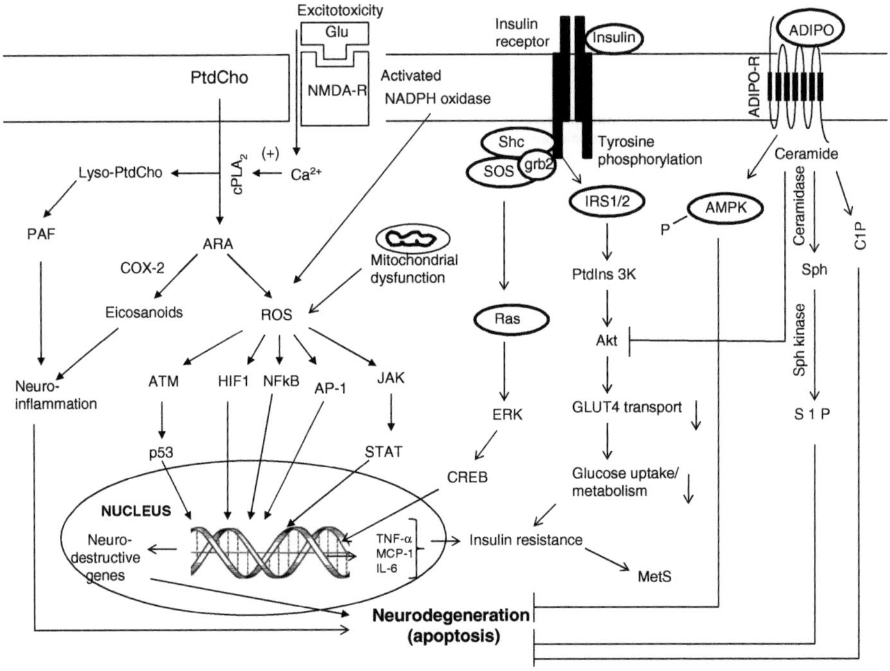


**Fig. 7.5** Interactions among oxidative stress, insulin, and adiponectin signaling. Ataxia telangiectasia mutated (ATM); heat shock transcription factor 1 (HSF1); nuclear factor-κB (NF-κB); Janus protein kinase (JAK); cytosolic phospholipase $A_2$ ($cPLA_2$); cyclooxygenase-2 (COX-2); phosphatidylcholine (PtdCho); lyso-phosphatidylcholine (lyso-PtdCho); arachidonic acid (ARA); platelet-activating factor (PAF); adiponectin (ADIPO); adiponectin receptor (ADPOR); protein kinase A (PKA); glutamate (Glu); *N*-methyl-D-aspartic acid receptor (NMDA-R); insulin receptor substrate-2 (IRS-2); serine/threonine protein kinase (Atk/PKB); phosphatidylinositol 3-kinase (PtdIns 3K); glucose transporter type 4 (GLUT4); sphingosine (Sph); sphingosine kinase (sph kinase), ceramide-1-phosphate (C1P); and sphingosine-1-phosphate (S1P)

kinase 11 (LKB1) also participate in adiponectin signaling (Hardie 2003; Yamaguchi et al. 2007). However, their involvement in the regulation of cellular energy stores and insulin sensitivity through adiponectin signaling has not been completely defined (Hardie 2003). Plasma glucose and FFAs lowering effects of adiponectin may be partly due to adiponectin-mediated AMP-activated protein kinase (AMPK) activation (Fig. 7.5), which in turn stimulates skeletal muscle fatty acid oxidation and glucose uptake. In addition, adiponectin also activates peroxisome proliferator-activated receptor in the liver and skeletal muscle. This results in stimulating the phosphorylation of acetyl-CoA carboxylase and fatty acid oxidation, leading to decrease in tissue triglyceride levels in the muscle and liver. Furthermore, the activation of AMPK by adiponectin suppresses endogenous glucose production, concomitantly inhibiting phosphoenolpyruvate carboxykinase and glucose-6-phosphatase expression. AMPK is an energy-sensing enzyme that is activated by acute increases in the cellular AMP-to-ATP ratio. In the skeletal muscle, AMPK activity is activated by exercise, hypoxia, ischemia, and osmotic stress, all of which reduced cellular energy level (Fujii et al. 2006). When levels of intracellular ATP are decreased, AMPK acts to switch off ATP-consuming pathways, including glycogen, fatty acid, and protein synthesis pathways, and acts to switch on alternative pathways for ATP regeneration (such as glucose transport, glycolysis, and fatty acid oxidation). AMPK may also play a role in enhancing insulin sensitivity and/or responsiveness for glucose transport (Fisher et al. 2002; Jessen et al. 2003; Handy et al. 2010) in the skeletal muscle. The activation of AMPK may produce beneficial effects in subjects with type II diabetes and MetS, such as increased glucose uptake and metabolism by the muscle and other tissues, decreased glucose production by the liver, and decreased synthesis and increased oxidation of fatty acids. Indeed, experiments with animal models of type II diabetes and the MetS have shown that activation of AMPK using 5-aminoimidazole-4-carboxamide 1-β-D-ribofuranoside riboside can reverse many of the metabolic defects of these animals in vivo (Song et al. 2002; Buhl et al. 2002). Because of this reason the AMPK system is a major target of antidiabetic drug metformin and rosiglitazone (Zhou et al. 2001; Fryer et al. 2002). The AMPK is also activated by exercise, which is known to have beneficial effects in the treatment and prevention of type II diabetes. Finally, adiponectin strongly inhibits the expression of adhesion molecules, including intracellular adhesion molecule 1, vascular cellular adhesion molecule 1, and E-selectin.

Mounting evidence suggests that adiponectin enhances AMPK and the PPARα pathway in the liver and skeletal muscle. It increases fatty acid oxidation, which lowers circulating free fatty acids and prevents insulin resistance. Adiponectin not only exerts an antiatherosclerotic effect but inhibits macrophage activation and foam cell accumulation and augments endothelial nitric oxide production and protects the vasculature by reducing platelet aggregation and vasodilation. Adiponectin deficiency has been reported to contribute to coronary heart disease, insulin resistance, nonalcoholic fatty liver disease, and a wide array of cancers (Hardie 2003; Yamauchi et al. 2007).

Sphingolipid metabolism has garnered increasing attention for its roles in insulin resistance. Thus, accumulation of ceramide is crucial for the etiology of obesity-induced insulin resistance (Holland et al. 2007; Holland and Scherer 2009). The systemic administration of recombinant adiponectin to mice fed with a high-fat diet or to leptin-deficient ob/ob mice reduces hepatic ceramide content. By modulating circulating adiponectin levels by genetic approaches, it is shown that elevation or reduction in adiponectin levels decreases or increases hepatic ceramide content, respectively (Holland and Scherer 2009; Holland et al. 2011). Furthermore, in the liver overexpression of adiponectin receptors increases hepatic ceramidase activity and, concomitantly, reduces hepatic ceramide content, supporting the view that there is a strong association among adiponectin levels, hepatic ceramide content, and insulin sensitivity. Adiponectin-mediated AMPK activation can be blocked by a broad-spectrum ceramidase inhibitor implying that sphingosine-1-phosphate (S1P) generation may be required for the activation of AMPK by adiponectin (Fig. 7.5) (Holland et al. 2011; Lancaster and Febbraio 2011). S1P, a potent inducer of proliferation, is known to inhibit apoptosis (Farooqui 2011). Collective evidence suggests that adiponectin exerts multiple effects and plays a key role not only in glucose and lipid metabolism, insulin sensitivity, blood pressure, and angiogenesis but also in decreasing inflammation and promoting cell survival through inhibition of apoptosis.

### 7.2.2 Resistin in Metabolic Syndrome

Resistin is a 12.5 kDa cysteine-rich peptide (108 amino acids), which is predominantly expressed in adipocytes. Resistin regulates glucose homeostasis, adipogenesis, and inflammation (Filková et al. 2009). It induces insulin resistance by regulating adipose tissue deposition through a negative feedback mechanism and exerts proinflammatory effects through activation of the transcription factor NF-κB (Bokarewa et al. 2005). Circulating levels of resistin are increased in obesity, and treatment with recombinant resistin or transgenic overexpression of resistin mediates hepatic insulin resistance in mice (Steppan et al. 2001; Muse et al. 2004; Li et al. 2009). Furthermore, deletion of resistin enhances insulin sensitivity in the livers of mice fed with a high-fat diet and in the muscle and white adipose tissue in leptin-deficient ob/ob mice (Qi et al. 2006). Resistin inhibits adipogenesis, whereas the loss of resistin function increases body weight and fat and enhances insulin sensitivity. Thus, resistin is closely associated with glucose homeostasis. Intracerebroventricular administration of resistin inhibits food intake and induces hepatic insulin resistance (Tovar et al. 2005). In humans resistin is mainly expressed in mononuclear cells, including macrophages (Patel et al. 2003). Human resistin is induced in response to various inflammatory stimuli such as tumor necrosis factor-α (TNF-α) or interleukin-6 (IL-6) and resistin itself (Kaser et al. 2003). It induces proinflammatory activity mediated through the activation of extracellular signal-regulated kinases 1 and 2 (ERK1/2). Administration of resistin improves glucose tolerance and insulin action in mice with normal glucose tolerance. On the other hand, administration of anti-resistin antibody improved blood glucose and insulin action in mice with diet-induced obesity (Steppan et al. 2001). Resistin and adiponectin produce opposing effects on vascular endothelial cells, with resistin enhancing the expression of various adhesion molecules such as ICAM1, while adiponectin downregulates expression of these molecules.

### 7.2.3 Visfatin in Metabolic Syndrome

Visfatin is a novel adipocytokine with insulin mimetic effect (Fukuhara et al. 2005). It is preferentially produced in the visceral adipose tissue with antidiabetic effects similar to insulin. It increases glucose transport and lipogenesis and decreases glucose production by hepatocytes. Visfatin interacts with insulin receptors and improves insulin sensitivity in vivo as evidenced by a reduction of glucose and insulin levels in diabetic mice infused with visfatin. The insulin-sensitizing effect of visfatin is additive to that of insulin (Fukuhara et al. 2005). Upregulation of visfatin mRNA has been observed under inflammatory conditions that occur during development of atherosclerosis and in inflammatory bowel disease. Few studies have been performed on the association of resistin with retinol-binding protein-4 (RBP-4), a newly identified adipokine that was reported to be associated with insulin resistance and MetS (Qi et al. 2007). RBP-4 is involved in the developmental pathology of type II diabetes mellitus and obesity.

### *7.2.4 Leptin in Metabolic Syndrome*

Leptin (16 kDa protein) is produced and secreted by adipocytes and is the product of the *ob* gene (Zhang et al. 1994). It acts as an afferent signal in a negative feedback loop inhibiting the appetite and regulating adiposity (Frühbeck and Gómez-Ambrosi 2003). In addition to modulating satiety, leptin induces many peripheral effects (Anubhuti and Arora 2008). Leptin increases insulin sensitivity in normal and diabetic murine models (Frühbeck and Salvador 2000). This hormone not only increases the basal and insulin-induced glucose uptake in isolated muscles of rodents as well as in cultured cells in a PtdIns 3-kinase (PtdIns 3K)-dependent manner (Yaspelkis et al. 2004) but also activates AMPK and stimulates the gene expression of GLUT4 in the plasma membrane of the skeletal muscle cells by reducing the expression and increasing the phosphorylation of the negative regulators of GLUT4 translocation TBC1D1 (a putative GTPase-activating protein from Rab family) and TBC1D4 (another putative GTPase-activating protein from Rab family) (Holmes et al. 2005; Sáinz et al. 2012). Leptin is highly conserved among species, and its structure resembles that of cytokines, especially interleukin-6 (Madej et al. 1995). Leptin acts as an adiposity signal to the brain and serves as a feedback mechanism to limit weight gain by decreasing orexigenic and increasing anorexigenic peptide expression in the hypothalamus (Ahima 2005). Thus, leptin regulates body weight not only through the suppression of appetite and decreasing food intake but also by stimulating energy expenditure and controlling insulin secretion. Plasma levels of leptin are directly correlated with the amount of body fat, and obese people have high circulating levels of leptin, indicating the occurrence of resistance to its action (Jéquier 2002). Leptin enters the brain through a saturable transport system. Leptin receptors are found both in the hypothalamus and in extrahypothalamic tissues, and their main role is to inform the brain regarding the amount of stored fat, thus primarily regulating food intake and energy expenditure (Campfield et al. 1995); however, in obese humans increased leptin levels are unable to induce weight loss: this phenomenon is attributed to a selective resistance to its metabolic actions (Heymsfield et al. 1999). Interactions of leptin with long leptin receptor (LRb) result in autophosphorylation of JAK2 and activation of STAT3 (Munzberg and Myers 2005) (Fig. 7.2). These events are followed by the migration of STAT3 into the nucleus and transcription regulation of neuropeptides and various leptin target genes (Munzberg and Myers 2005) (Fig. 7.2). Leptin stops its own action through phosphorylation of Tyr985 and induction of suppressor of cytokine signaling 3 (SOCS3) (Fig. 7.2). Protein tyrosine phosphatase 1B, a well-known inhibitor of insulin action, also terminates leptin signaling through inactivation of JAK2 (Zabolotny et al. 2002). Leptin acting through LRb not only regulates insulin receptor substrate-1 and insulin receptor substrate-2 (IRS-1 and IRS-2) and mitogen-activated protein kinase but modulates extracellular-regulated kinase, Akt, and phosphatidylinositol 3-kinase in the hypothalamus, supporting the possibility of cross talk between leptin and insulin (Niswender et al. 2004). Both in vivo and in vitro studies indicate that leptin and insulin signaling pathways are connected at IRS and PtdIns 3K levels (Kim et al. 2000). Leptin mimics the ability of insulin to facilitate glucose uptake and glycogen storage through activation of Janus kinase 2 (JAK2) that in turn phosphorylates IRS2 to activate PtdIns 3K (Kellerer et al. 1997). Moreover, intracerebroventricular administration of PtdIns 3K inhibitors retards the ability of leptin (Niswender et al. 2001) and insulin (Niswender et al. 2003), but not other anorexigenic compounds, to reduce food intake, reinforcing the concept that the PtdIns 3K signaling pathway is closely associated with the central actions of these hormones. Under low energy situation, a fall in leptin level is not only accompanied by the decrease in thermogenesis and increase in fatty acid oxidation, but also in the suppression of hypothalamic pituitary adrenal and gonadal axes activity. Leptin also modulates the activities of the hematopoietic (Bennett et al. 1996) and immune (Matarese et al. 2005) systems, as well as angiogenesis (Bouloumie et al. 1998). Collective evidence suggests that insulin and leptin mediate their signals through the autophosphorylation of tyrosine kinase family of receptors (Hadari et al. 1992; Baumann et al. 1996; Tartaglia 1997; Koch et al. 1991). An important difference in insulin- and leptin-mediated signaling lies in the manner of activation upon ligand binding—the insulin receptor has intrinsic activity in that insulin binding stimulates autophosphorylation along with activation of downstream signaling targets, whereas leptin receptor requires JAK–STAT binding for full activation (Baumann et al. 1996; Tartaglia 1997). Once activated, both insulin and leptin activate IRS and their downstream targets (Tartaglia 1997; Koch et al. 1991).

Insulin, glucocorticoids, and cytokines, e.g., tumor necrosis factor-α and interleukin-6, increase leptin, whereas cold exposure and adrenergic stimulation decrease leptin. Leptin also contributes to inflammatory mechanism and immune responses. Leptin stimulates the production of proinflammatory cytokines from cultured monocytes and increases the production of Th1-type cytokines, downregulating Th2-type cytokines, thus polarizing T cells toward a Th1 phenotype (Montecucco et al. 2006). Thus, leptin is a proinflammatory cytokine linking the immune and inflammatory processes to the neuroendocrine system (Otero et al. 2006). It not only stimulates energy expenditure through activation of the sympathetic nervous system but also promotes lipolysis and apoptosis of adipocytes. Leptin also suppresses the expression and activity of the collagen-degrading matrix metalloproteinase-1 (MMP-1) (Lin et al. 2006; Handy et al. 2010) and promotes expression of tissue inhibitor of metalloproteinase-1 (TIMP-1) (Lin et al. 2006), an important negative regulator of this MMP. Leptin also promotes the maintenance of hepatic stellate cell in the "activated" phenotype by stimulating their proliferation and suppressing apoptosis (Saxena et al. 2004). Leptin also induces monocytes to produce nitric oxide, leukotriene B4 (LTB4), cholesterol acyl-transferases-1 (ACAT-1), and cyclooxygenase-2 (COX-2) (De Rosa et al. 2007). Thus, leptin and adiponectin exert an insulin-sensitizing effect through fatty acid oxidation, and, in addition, adiponectin is associated with antiatherogenic, antidiabetic, and anti-inflammatory properties. In obesity, insulin resistance is linked not only to the leptin resistance and elevated leptin levels but also with low adiponectin levels, which are associated with higher cardiovascular disease risk (Guerre-Millo 2004; Inadera 2008; Bakhai 2008).

Inefficient leptin action is associated with hyperphagia, decrease in fat oxidation, increase in tissue triglyceride levels, insulin resistance, and overweight. Subsequently, plasma leptin levels rise. The majority of obese individuals with high leptin levels show a leptin insensitivity or "resistance," which occurs at the leptin receptor level (Zimmet et al. 1999). The molecular mechanism associated with "leptin resistance" is not fully understood. However, it is proposed that "leptin resistance" may be associated with the impairment in transporter, receptor, post-receptor, and downstream neuronal circuitry functions in animal models of obesity. Leptin is transported across the blood–brain barrier (BBB) by a saturable transporter (Banks et al. 1996), and impairment in transport may be caused by above mentioned parameters (Banks and Farrell 2003). The relation between cerebrospinal fluid and serum levels of leptin in obese humans indicates that defective BBB transport accounts for the overall resistance to leptin than the receptor/post-receptor defects (Banks 2003). Levels of serum triglycerides inhibit the transport of leptin across the BBB, thus attenuating the leptin signal across the BBB and providing a mechanism for peripheral leptin resistance (Banks et al. 2004). Triglycerides are elevated in both starvation and obesity. It is postulated that hypertriglyceridemia evolved as a starvation signal to the brain that acts in part to inhibit the transport of the leptin across the BBB. The hypertriglyceridemia in obesity invokes this aspect of the starvation response, inducing leptin resistance at the BBB. Thus, the BBB plays important roles in leptin resistance, obesity, and starvation (Banks 2008). In animal models of MetS, leptin resistance and leptin deficiency upregulate the hepatic expression of SREBP-1c mRNA. This may stimulate the lipid synthesis via de novo lipogenesis. Thus, "leptin resistance" may lead to obesity and MetS (Unger 2003).

Infusion of leptin for 12 days increases arterial pressure and heart rate, despite a decrease in body weight that is expected to decrease arterial pressure (Shek et al. 1998). Leptin-mediated increase in arterial pressure is probably due to the central action of the leptin because intracerebroventricular (ICV) administration of leptin mimics the effects of systemic administration (Correia et al. 2001). The substantial dose-dependent increase in heart rate and the greater response to air-jet stress observed in leptin-treated rats give support for centrally mediated activation of the sympathetic nervous system (Correia et al. 2001). Moreover, blocking the adrenergic system inhibits the pressor response to leptin (Carlyle et al. 2002).

Leptin not only produces neurotrophic effects by reducing cell death in serum and growth factor-deprived neuronal cultures (Guo et al. 2008) but also plays a role in synaptogenesis, induction of long-term potentiation (LTP) in the hippocampus (Shanley et al. 2001), and enhanced synaptic protein expression (Walker et al. 2007). Leptin acts by modulating NMDA receptor-dependent hippocampal synaptic plasticity as well as hippocampal-dependent learning and memory (Farr et al. 2006; Harvey 2007). Rodents with dysfunctional leptin receptors show impaired hippocampal synaptic plasticity and deficits in spatial memory tasks (Li et al. 2002). Direct application of leptin into the hippocampus also improves performance in specific memory tasks (Farr et al. 2006). Under enhanced excitability conditions, leptin produces a novel form of NMDA receptor-dependent LTD (Durakoglugil et al. 2005). In addition, leptin also promotes rapid changes in neuronal structure including alterations in dendritic morphology and synaptic density, which are likely to contribute to leptin-driven changes in synaptic efficacy (O'Malley et al. 2007). Finally, leptin regulates neurogenesis in the dentate gyrus of adult mice as well as in cultured adult hippocampal progenitor cells (Garza et al. 2008). Chronic administration of leptin in adult mice increases cell proliferation, but has no significant effects on the differentiation and the survival of newly proliferated cells in the dentate gyrus. Analysis of signal transduction pathways indicates that leptin stimulates the phosphorylation of Akt and STAT3, but not ERK1/2. In addition, pretreatment of the cells with specific inhibitors of Akt or STAT3 attenuates leptin-mediated cell proliferation in a dose-dependent manner supporting the view that leptin modulates neurogenesis in adult hippocampus through the involvement of the Akt and STAT3 signaling pathways (Garza et al. 2008).

### 7.2.5 *TNF-α in Metabolic Syndrome*

Chronic inflammation is a key component of MetS. It is characterized by elevated levels of circulating inflammatory cytokines (Dandona et al. 2005). The molecular mechanism involved in elevation of cytokines in MetS is not fully understood. However, it is proposed that overnutrition-mediated stimulation of IKKβ/NF-κB in hypothalamic neurons modulates chronic inflammation in the peripheral tissues (metabolic inflammation), and therefore causes various metabolic defects in those tissues which contribute to type II diabetes, the key component of MetS (Hotamisligil 2006; Zhang et al. 2008). NF-κB resides in the cytoplasm in an inactive form through binding to the inhibitory protein IκB. Activation of IKKβ by phosphorylation at S177 and S181 induces phosphorylation of its substrate IκBα at S32 and S36, ubiquitination, and subsequent proteasomal degradation. The migration of IκBα releases NF-κB to translocate into the nucleus where it mediates the transcription of proinflammatory cytokine (TNF-α, IL-1β, IL-6, etc.) genes. These cytokines contribute to chronic peripheral inflammation. It is also reported that activation of IKKβ/NF-κB can also be mimicked without obesity by an acute but extreme oversupply of various nutrients to the brain (Zhang et al. 2008). Persistent signals to the brain under conditions of overnutrition also stimulate an innate immunity-like response directed by IKKβ/NF-κB in the hypothalamic neurons; this process affects the neuronal

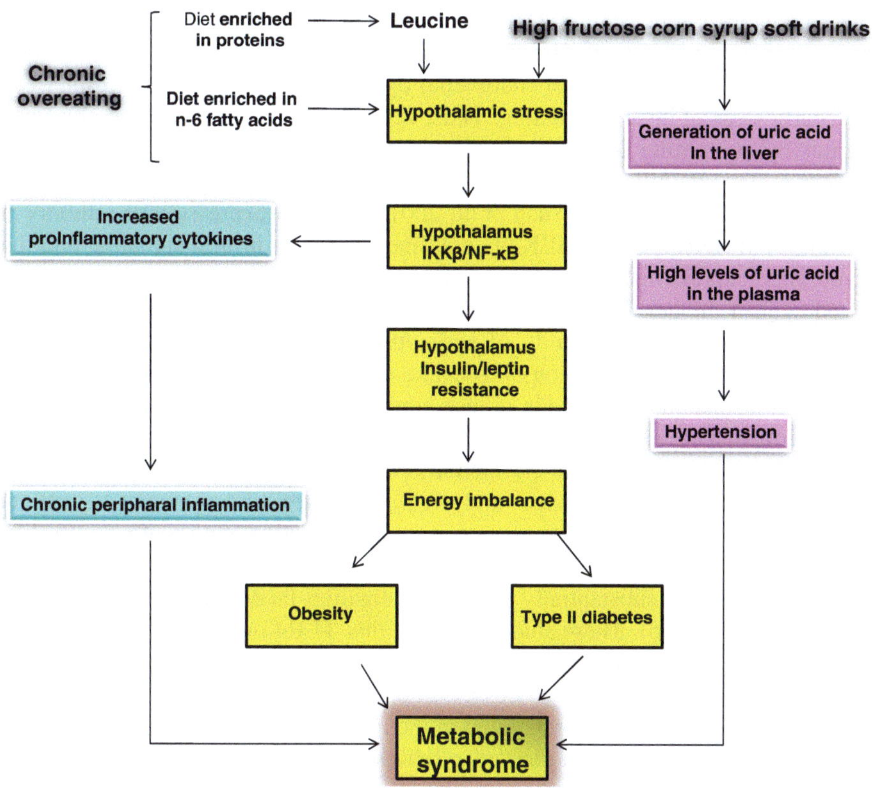


**Fig. 7.6** Effect of chronic nutrition on IKKβ/NF-κB in hypothalamic neurons causing insulin/leptin resistance

regulation of energy balance (Fig. 7.6). Increase in proinflammatory cytokines production causes an imbalance among cytokine levels and contribute to dysfunction of cytokine network. Consequently, MetS is presently viewed not only as a metabolic disorder but also as an inflammatory disease affecting both innate and acquired immune systems (Schmidt and Duncan 2003). Chronic, low-grade inflammation in MetS leads to endothelial dysfunction and the release of inflammatory cytokines (TNF-α, IL-1β, and IL-6), which induce insulin resistance in the adipose tissue and muscle (Hanley et al. 2004; Shulman 2000; Hotamisligil 2006) (Fig. 7.4). TNF-α is a pleiotropic cytokine with a central role in inflammation. TNF-α is produced by adipocytes and fat-infiltrating macrophages and is overexpressed in the plasma and adipose tissue of obese human and animal models. TNF-α promotes insulin resistance by decreasing tyrosine kinase activity of insulin receptor and insulin signaling via MAPK pathways in vitro and in vivo, thereby reducing insulin activity (Hotamisligil et al. 1994). Studies in mice suggest that the attenuation of TNF-α activity improves glucose homeostasis (Uysal et al. 1997. In type II diabetes patients, blocking of TNF-α secretion reduces plasma levels of inflammatory markers and ameliorates insulin resistance (Dominguez et al. 2005), suggesting that TNF-α has an important role in inducing insulin resistance mediated by inflammation. In addition TNF-α also contributes to endothelial dysfunction in MetS. TNF-α interferes with early steps of insulin signaling. Thus, TNF-α inhibits IRS1 tyrosine phosphorylation by promoting its serine phosphorylation. TNF-α induces expression of nuclear factor-κB (NF-κB), a transcription factor closely associated with TNF-α-mediated insulin resistance (Yuan et al. 2001). The overexpression of inhibitor of IκB kinase-β (IKKβ) in the HEK293 cell line attenuates insulin signaling and that ob/ob mice expressing only one copy of the gene encoding IKKβ (*Ikbkb*) are protected against the development of insulin resistance (Yuan et al. 2001). The c-Jun N-terminal kinase (JNK) family of serine/threonine protein kinases—which are activated by TNF-α and ligation of Toll-like receptors (TLRs)—are also important regulators of insulin resistance in mouse models of obesity (Hirosumi et al. 2002). Fatty acids, in particular saturated fatty acids, activate macrophage-like cells via the TLR4. FFA concentrations are elevated in obesity and can directly mediate inflammatory responses in macrophages and even adipocytes. FFAs activate NF-κB and cytokine production in adipocytes and macrophages, and induction of inflammatory signaling is retarded in the absence of functional TLR4. Moreover, mice lacking functional TLR4 are substantially protected from high-fat diet-induced insulin resistance (Shi et al. 2006). These studies suggest that TLR4 on adipocytes and macrophages may be a sensor of elevated FFA concentrations, which initiates inflammatory and thus insulin-desensitizing processes. In both genetic and dietary animal models of obesity, JNK activity is upregulated in the liver, muscle, and adipose tissue, and loss of JNK1 prevents insulin resistance (Hirosumi et al. 2002). NF-κB also modulates the expression of adhesion molecules, macrophage chemoattractive protein-1, and macrophage colony-stimulating factor in the endothelial and VSMC (Lyon et al. 2003) (Fig. 7.2).

### *7.2.6 IL-6 in Metabolic Syndrome*

In the viscera, IL-6 is secreted by a wide variety of cells such as endothelial cells, myocytes, adipocytes, β-pancreatic cells, monocytes, and macrophages. This cytokine is associated with the decrease in inflammatory processes by promoting the synthesis of anti-inflammatory cytokines and by negatively regulating inflammatory targets. Therefore, this protein has been classified as both a pro- and anti-inflammatory; at certain level it acts as a defense mechanism but in chronic inflammation it has rather proinflammatory properties (Steensberg et al. 2003). In humans, higher circulating IL-6 levels are not only associated with obesity and visceral fat deposition (Qi et al. 2007), increased risk of impaired glucose tolerance, and type II diabetes (Qi et al. 2007; Stephens et al. 2007) but also with hypertension (Fernandez-Real et al. 2001). IL-6 is closely associated with the production of CRP (Stephens et al. 2004). IL-6 is preferentially released from visceral fat and upregulates suppressor of cytokine signaling 3 (SOCS3) in the liver that causes hepatic insulin resistance. Visceral adipose tissue secretes about two to three times more IL-6 than subcutaneous tissue secreting also other molecules that stimulate further IL-6 expression. Mounting evidence suggests that variation in IL-6 gene is associated with the modulation of cytokine expression, which subsequently may play an important role in impaired glucose and lipid homeostasis, a process which is closely associated with response to fat intake and pathogenesis of cardiometabolic diseases. It is also reported that TNF-α, IL-1β, and IL-6 promote the attachment and migration of monocytes into the vessel wall and their conversion into macrophages, where they phagocytose oxidized LDL and form lipid-laden foam cells. This is a key early stage in the development of atherosclerosis—as the foam cells accumulate in the vessel wall, they form fatty streaks that ultimately develop into atherosclerotic plaques (Lyon et al. 2003).

### *7.2.7 IL-1β in Metabolic Syndrome*

Interleukin-1β (IL-1β) acts through IL-1β receptor and plays a central role in the regulation of immune responses and induction of inflammatory processes. Animal studies have shown that central administration of IL-1β produces several effects, including activation of the hypothalamic–pituitary–adrenal (HPA) axis (Shintani et al. 1995), inhibition of hippocampal long-term potentiation (Murray and Lynch 1998), downregulation of hippocampal brain-derived neurotrophic factor (Barrientos et al. 2003), and impairment in hippocampal-dependent contextual fear conditioning (Pugh et al. 1999). In contrast, blockade of the receptor by IL-1RI Ra inhibits these stress-like effects (Barrientos et al. 2003; Pugh et al. 1999) and blocks the antiproliferative effects of INF-α in the hippocampus (Kaneko et al. 2006). Elevated levels of IL-1β and interleukin-1 receptor antagonist (IL-1Ra) have been detected in subjects with hypertension, which is known to contribute to the development of insulin resistance and MetS (Jager et al. 2007). The IL-1Ra level has been shown to be the most sensitive marker for cytokine response in the prediabetic state. IL-1β regulates the production of a variety of inflammatory mediators, such as IL-6, intercellular adhesion molecule (ICAM)-1, and E-selectin (Zhang et al. 1995), and it contributes to metabolic disturbances in obese subjects such as dyslipidemia and insulin resistance (Jager et al. 2007).

## 7.3 Stroke and it Neurochemical Consequences in the Brain

Stroke is a highly dynamic multifactorial metabolic insult caused by severe reduction or blockade in cerebral blood flow due to the formation of a clot. This blockade not only decreases oxygen and glucose delivery to the brain tissue but also results in the breakdown of blood–brain barrier (BBB) and buildup of potentially toxic products in the brain (Farooqui 2010). In the Western world, stroke is caused by ischemic insult (formation of clot) where as in Asian countries majorities of strokes are mediated by intracerebral hemorrhagic insult. Approximately 12 % of strokes are hemorrhagic (rupture of a cerebral blood vessel), whereas the remaining 88 % are ischemic and result from occlusion of a cerebral artery (either thrombotic or embolic). Blockage of a cerebral artery results in interruption of the blood flow and supply of nutrients, glucose, and oxygen to the brain.

Stroke is the third leading cause of death and the major cause of long-term disability, striking almost 800,000 people in the USA each year (Lloyd-Jones et al. 2009). On average stroke in the USA strikes once every 40 s and causes death every 4 min, with an estimated 41.6 % death rate in 2007 (Lloyd-Jones et al. 2009). The estimated cost for stroke is 73.7 billion dollars in 2010 (USA) and projected to be 1.52 trillion dollars in 2050 (Lloyd-Jones et al. 2009). There are many modifiable and non-modifiable risk factors associated with the pathogenesis of stroke. The etiology of stroke is complex and multifactorial, with hypertension, diabetes, smoking, atrial fibrillation, obesity, and carotid artery disease identified as well-known potentially modifiable risk factors (Bousser 2012). While these predisposing risk factors are closely involved in the etiology of stroke, several large population-based cohort studies have indicated that age is the single most important risk factor for stroke, as the stroke rate more than doubles for men and women for every 10 years after the age of 55 (Rojas et al. 2007; Roger et al. 2011). Moreover, advanced age is associated with adverse outcomes and delayed recovery after the initial acute cerebrovascular accident (Black-Schaffer and Winston 2004). The onset of MetS increases the chances of stroke in patients of all ages, ranging from neonates to the elderly. The annual incidence of first-time stroke for patients of all ages is 158 per 100,000 (Kleindorfer et al. 2006). Patients younger than 50 years of age are ten times less likely to have a stroke, with an estimated annual incidence of approximately 10 per 100,000 in this age group (Putaala et al. 2009). The risk of stroke is high within 5 years of diagnosis for type 2 diabetes with 21 % mortality. Stress is associated with high blood pressure and atherosclerosis. Both of these are strong risk factors for stroke (Farooqui 2010). Those humans who experience more stress and are less able to cope with it emotionally and physically are more at risk for stroke. Our body responds to stress by expanding blood vessels and increasing blood flow. Repeated or long-lasting blood pressure spikes may cause vascular wall damage and formation of clot leading to blockade of blood vessel ultimately resulting in ischemic stroke (Farooqui 2010). Stroke-induced injury affects multiple different neuronal population and phenotypes. For example, an infarct might involve the thalamus, hippocampus, and striate visual cortex, affecting three or more very different neuronal populations including neurons, oligodendrocytes, astrocytes, and endothelial cells (Savitz et al. 2003, 2004). The size and position of the affected region depends upon, which vessel is occluded. This is in contrast to neurodegenerative diseases, where homogeneous population of neurons dies in a specific brain region (Farooqui 2010). The brain is particularly vulnerable to oxidative damage not only due to high oxygen consumption and the presence of polyunsaturated fatty acids in neural membrane phospholipids but also because of high amounts of redox-active transition metals and relatively low activities of antioxidant enzymes (Farooqui 2010). Interruption in glucose and oxygen supply to the brain may result

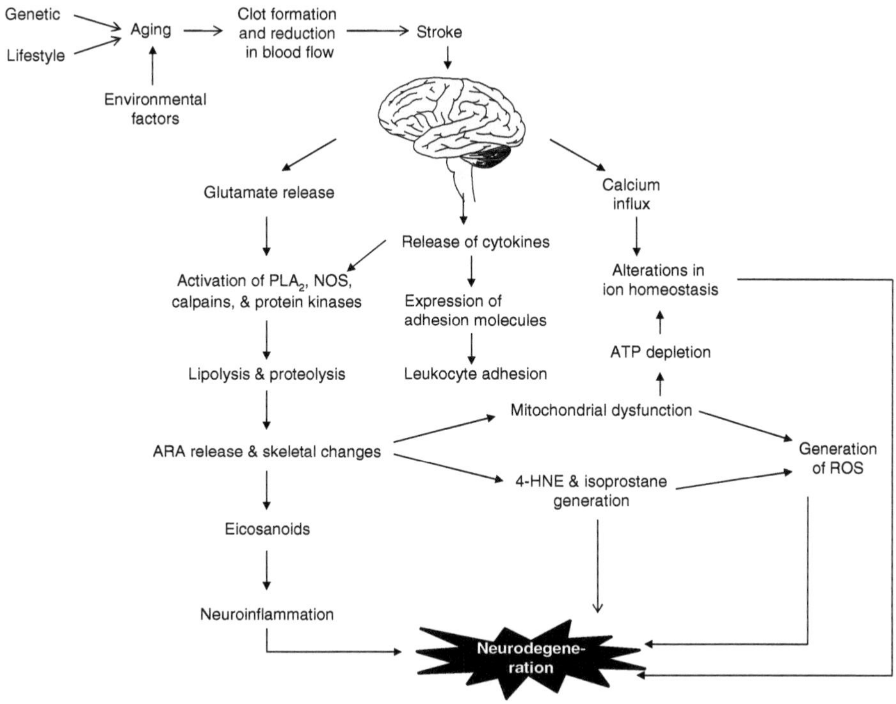


**Fig. 7.7** Neurochemical changes associated with ischemic injury

in compromised ATP generation; loss of ion homeostasis; mitochondrial dysfunction; production of ROS, such as superoxide and hydroxyl anion, and reactive nitrogen species (RNS), such as NO and $ONOO^-$; and changes in redox status of neural cells. These processes also contribute to cerebral edema, which is the primary cause of patient mortality after stroke. The initial response to a transient insufficiency of energy is depolarization, which causes $Na^+$ influx into axons. Prolonged depletion of ATP produces a massive influx of $Ca^{2+}$ that facilitates neural cell death (Fig. 7.7) (Farooqui 2010).

Stroke-induced injury is mediated by excitotoxic insult, which involves not only the release of excess glutamate in the extracellular space and overactivation of glutamate receptors but also the dramatic increase of intracellular calcium ions, which in turn activates multiple intracellular death pathways as well as expression and release of cytokines from activated microglial cells and astrocytes (Farooqui and Horrocks 1994; Farooqui 2010). Thus, stroke triggers a complex series of biochemical and molecular mechanisms that impair the neurological functions through the breakdown of cellular and subcellular integrity, alterations in ionic balance and redox status, generation of free radicals, induction of proinflammatory cytokine, and development of neuroinflammation. These processes are mediated by the activation of phospholipases $A_2$, C, and D ($PLA_2$, PLC, and PLD), cyclooxygenases,

**Table 7.2** Status of excitotoxicity, ARA-derived enzymic and nonenzymic lipid mediators, oxidative stress, and neuroinflammation following ischemic injury

| Neurochemical parameter | Ischemic injury | Reference |
|---|---|---|
| Glycerophospholipid metabolism | Altered | Farooqui (2010, 2011) |
| Excitotoxicity | Increased | Farooqui (2010, 2011) |
| Arachidonic acid | Increased | Farooqui (2010, 2011) |
| Prostaglandins | Increased | Farooqui (2010, 2011) |
| Leukotrienes | Increased | Farooqui (2010, 2011) |
| Thromboxanes | Increased | Farooqui (2010, 2011) |
| 4-Hydroxynonenal | Increased | Farooqui (2010, 2011) |
| Isoprostane | Increased | Farooqui (2010, 2011) |
| Oxidative stress | Increased | Farooqui (2010, 2011) |
| Inflammation | Increased | Farooqui (2010, 2011) |
| Neurodegeneration | Increased | Farooqui (2010, 2011) |

lipoxygenases, epoxygenases, calcium/calmodulin-dependent kinases (CaMKs), mitogen-activated protein kinases (MAPKs)—such as extracellular signal-regulated kinase (ERK), p38, and c-Jun N-terminal kinase (JNK)—nitric oxide synthases (NOSs), calpains, calcinurin, and endonucleases. Activation of these enzymes generates high levels of a variety of arachidonic acid (ARA)-derived enzymic lipid mediators [prostaglandins (PGs), leukotrienes (LTs), thromboxanes (TXs)] as well as nonenzymic lipid mediators [4-hydroxynonenal (4-HNE), isoprostanes (IsoPs), acrolein, and malonaldehyde], which contribute to neurodegeneration in stroke-mediated neuronal injury (Table 7.2) (Farooqui 2011).

Stroke-mediated neuronal injury is accompanied by the induction of neuroinflammation, which is accompanied by the activation of microglia and astrocytes, which migrate rapidly to the site where neurodegenerative process is taking place. Microglia, which are resident immune cells of the central nervous system (CNS), undergo significant functional changes with aging ranging from altered iron storage to cytokine production and accumulation of lipofuscin (Brown 2009). Microglia not only secrete and release detrimental proinflammatory cytokines (TNF-α, IL-1β, IL-6), chemokines, superoxide, NO, and proteases but also engulf dead cells and clear cellular debris. Microglial cells also promote the death of neurons that are engaged in synaptogenesis. Thus, activated microglial cells act as immunocompetent macrophage-like cell in the injured brain. In the healthy brain, microglial cells are in intimate contact with neurons. They provide developmental support and promote the clearance of aberrant proteins (Napoli and Neumann 2009). Healthy neurons maintain microglia in an inactive state via secreted and membrane-bound signals, including CD200, CX3CL1 (fractalkine), neurotransmitters and neurotrophins (Biber et al. 2007). Signaling through purinergic and Toll-like receptor systems plays an important role in chronic microglial activation with significant functional consequences. Like microglial cells, astrocytes also release proinflammatory cytokines, chemokines, ROS, PGs, LTs, TXs, and PAF, which promote and support acute neuroinflammation (Fig. 7.8).

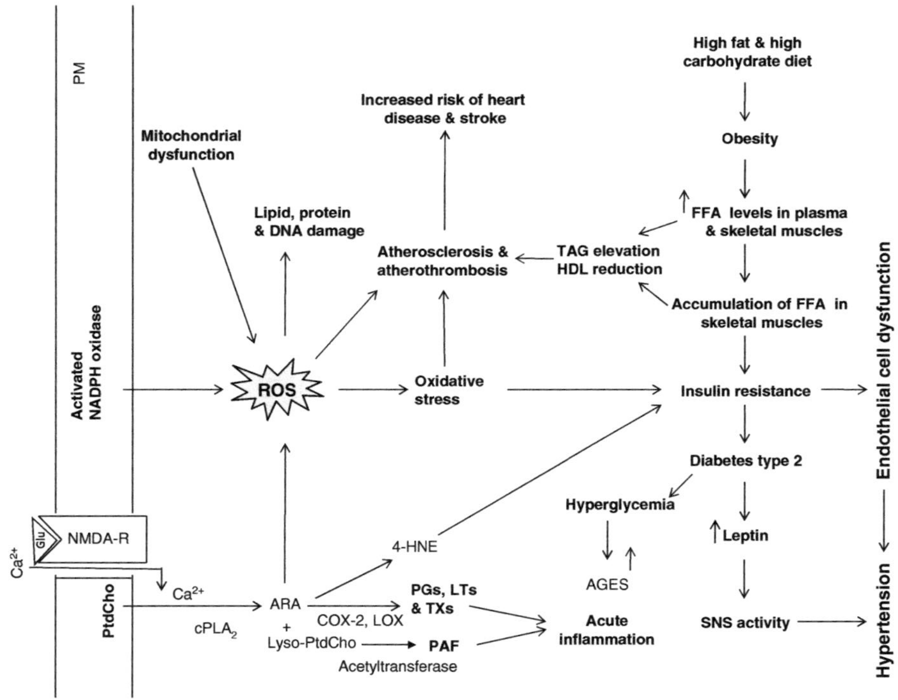


**Fig. 7.8** Hypothetical diagram showing commonality between biochemical reactions of stroke and MetS. Cytosolic phospholipase $A_2$ ($cPLA_2$); cyclooxygenase-2 (COX-2); phosphatidylcholine (PtdCho); lyso-phosphatidylcholine (lyso-PtdCho); arachidonic acid (ARA); platelet-activating factor (PAF); glutamate (Glu); *N*-methyl-D-aspartic acid receptor (NMDA-R); triacylglycerol (TAG); free fatty acid (FFA); sympathetic nervous system activity (SNS activity); reactive oxygen species (ROS); and advanced glycation end products (AGES)

Astrocytes have been implicated in many functions associated with injury pathophysiology such as buffering of potassium ions and maintaining the integrity of the BBB (Bitto et al. 2010). Despite increased understanding of astrocyte functions pre- and post-injury, the effect of age on astrocyte function is not entirely clear. However, inflammatory cytokines alter astrocytic ability to respond to injury. Astrocytes have also been shown to express markers associated with aging in both aged humans and rodents. These markers include increases not only in glial fibrillary acidic protein (GFAP) and cytokine release but also in aggregation of proteins (Aβ, α-synuclein, and huntingtin) associated with cellular death (Salminen et al. 2011). These studies support the view that roles of astrocyte in the brain change with aging and may influence both normal homeostatic mechanisms as well as the response to neural cell injury. It is also shown that astrocytes play essential roles in orchestrating the injury response as well as in regulating neuroinflammation and repair in a manner that markedly impacts functional and clinical outcomes of stroke-mediated neural injury. Emerging evidence suggests that activated microglia and astrocytes take part in neuroinflammation, which is different from nonneural (peripheral) inflammation due to the involvement of a complex network of neural cells, signaling molecules, and lipid mediators that occur within the brain (Farooqui 2010).

Age not only is a risk factor for stroke but also has a profound impact on recovery (Buga et al. 2011). Following injury, aged animals initiate the development of glial scar and increase in expression and release of injury associated signaling molecules, such as proinflammatory cytokines and adhesion molecules leading to neuroinflammation and further increase in oxidative stress. Synergistic effect of neuroinflammation and oxidative stress results in a more deleterious response after injury (Gemma et al. 2010).

## 7.4 Stroke and Induction of Cytokines and Chemokines

The cytokines and chemokines are important inflammatory mediators, which are upregulated within the cerebral tissue during the acute phase of stroke. Cytokines and chemokines are not only expressed by cells of the immune system, but cytokines are also produced endogenously by the resident brain cells (microglia and neurons). Cytokines possess both pro- and anti-inflammatory properties, which play an important role in the progression of the cerebral infarct (Farooqui 2010). Ischemic/perfusion injury induces the expression of four major cytokines, namely, tumor necrosis factor-α (TNF-α), interleukin (IL)-1, IL-8, and IL-6 in different regions of the rat brain as well as in cell culture experiments (Table 7.3) (Al-Bahrani et al. 2007; Tuttolomondo et al. 2008; Brea et al. 2009). Many of the abovementioned cytokines, such as TNF-α, IL-1β, IL-6, and IL-18, are synthesized as inactive precursor proteins that are processed by enzymic cleavage into the final mature and biologically active form. All neural cells (neurons, astrocytes, microglia, and oligodendrocytes) synthesize and secrete inflammatory cytokines, which contribute to cellular intercommunication through autocrine, paracrine, or endocrine

**Table 7.3** Effect of ischemic injury on cytokines, chemokines, and adhesion molecules

| Target | Effect | Reference |
|---|---|---|
| TNF-1α | Upregulation | Al-Bahrani et al. (2007) |
| IL-1β, IL-8, IL-6 | Upregulation | Al-Bahrani et al. (2007); Tuttolomondo et al. (2008) |
| MCP-1 | Upregulation | Terao et al. (2009) |
| MIP-1α (protein-3α) | Upregulation | Terao et al. (2009) |
| Adhesion molecules | Upregulation | Wen et al. (2006) |

mechanisms. Their actions involve a complex network linked to feedback loops and cascades. Cytokines mediate their effects by interacting with specific membrane receptors that are composed of an extracellular ligand-binding region, a membrane-spanning region, and an intracellular region that is activated by binding of cytokines and hence delivering a signal to the nucleus (Rothwell and Relton 1993). Cytokine receptors are expressed constitutionally throughout the brain tissue at low levels. Cytokines play an important role not only in neuronal development, maturation, survival, and regeneration but also in recovery process following neural insult (Rothwell and Relton 1993). In addition, cytokines contribute to interconnection between the brain and the immune system through hormonal cascades and cell-to-cell interactions. Indeed, the balance between pro- and anti-inflammatory cytokines not only determines the prowess of the immunological response but also influences the fate of the injured neurons following ischemic insult (Farooqui et al. 2007). In addition to direct signaling for a proinflammatory event, cytokines also modulate numerous downstream regulatory processes, such as induction of proteins that inhibit signal transduction pathways (e.g., SOCS proteins), induction of transcriptional repressors and transrepressors (e.g., ATF3 and Nurr1), and the production of soluble or cell surface mediators with anti-inflammatory activities (e.g., IL-10, transforming growth factor (TGF)-β, resolvins, and ligands for TAM receptors) (Kraft and Harry 2011). The diverse and multifunctional capacity of cytokine network present many challenges in understanding neural cell dynamics and impact on the elevated expression of other cytokines. For example, IL-1α, IL-1β, and the IL-1R antagonist (IL-1RA) all work via activation (or antagonism) of IL-1 receptor 1 (IL-1R1), yet IL-1α and IL-1β can also elicit disparate IL-1R1-independent signaling events (Kraft and Harry 2011). IL-1β is a major regulator of the expression of several MMPs. This cytokine not only induces the production of NO but also blocks glutamate uptake. In addition, it also facilitates the production of growth factors and increases the deposition of extracellular matrix molecules, laminins, and chondroitin sulfate proteoglycans (Liberto et al. 2004; Herx et al. 2000).

In the brain, the major contributors of neuroinflammation are eicosanoids and platelet-activating factor, which are generated through $PLA_2$/cyclooxygenase cascade and lysophospholipid acetyltransferase reactions, respectively (Farooqui and Horrocks 2007). Cytokines act by stimulating $cPLA_2$/cyclooxygenase pathway leading to the generation of prostaglandins, leukotrienes, and thromboxanes. The stimulation of $cPLA_2$ by TNF-α and IL-1β involves the phosphorylation of $cPLA_2$

by mitogen-activated protein kinase in the presence of agents that mobilize intracellular $Ca^{2+}$ (Clark et al. 1995). In addition, it is well known that TNF-α stimulates caspase-3, which through its proteolytic action cleaves $cPLA_2$ leading to its activation (Wissing et al. 1997). Acetyl-Asp-Glu-Val-Asp-aldehyde, a tetrapeptide inhibitor of caspase-3, prevents the proteolytic cleavage and activation of $cPLA_2$ indicating that caspase-3-mediated $cPLA_2$ proteolysis results in the activation of this enzyme (Wissing et al. 1997). Thus, the stimulation of $cPLA_2$ and caspase-3 along with induction of cyclooxygenase results in oxidative stress, mitochondrial dysfunction, and calcium ion overload along with the release of cytochrome c and the activation of downstream caspase-9 and caspase-3 resulting in cell death (Farooqui 2009). In addition, cytokines also facilitate the expression of other enzymes, such as cyclooxygenase-2, inducible nitric oxide synthase, and myeloperoxidase. These enzymes promote neuroinflammation following ischemic injury (Phillis et al. 2006). In the brain, IL-6 plays an important role in the development and differentiation of neurons in both peripheral and CNS (Gadient and Otten 1997). It not only promotes the activation of microglia (Heyser et al. 1997) but also induces the synthesis of acute-phase proteins (Castell et al. 1989) and phosphorylation of tau protein in neurons (Quintanilla et al. 2004). In AD brain, microglia and astrocytes are stimulated by IL-6 and are recruited to release other proinflammatory cytokines and acute-phase proteins, such as C-reactive protein (CRP) (Querfurth and LaFerla 2010). Therefore, IL-6 plays a pivotal role in brain inflammation that may contribute to the pathogenesis of AD. It should be noted that increasing serum levels of both IL-6 and CRP, but not TNF-α, are negatively associated with performance-based mobility function, such as longer time to complete a 4-mile walk and lower grip strength (Brinkley et al. 2009). It is also proposed that elevated inflammatory cytokine concentrations are markers of increased disease severity or other underlying factors that may be involved in association between inflammation and physical function in patients with chronic diseases, such as cardiovascular and cerebrovascular diseases and neurodegenerative diseases.

Chemokines are a large family of structurally related small cytokines (8–10 kDa) originally identified as factors regulating the migration of leukocytes in inflammatory and immune responses (Minami and Satoh 2000, 2003). Examples of chemokines are monocyte chemoattractant protein-1 (MCP-1), macrophage inflammatory protein-1α (MIP-1α), and cytokine-induced neutrophil chemoattractants (CINC). Some chemokines such as SDF-1 and fractalkine are constitutively produced in the brain, and they are associated with maintenance of brain homeostasis or determination of the patterning of neurons and/or glial cells in the developing brain and normal adult brain (Minami and Satoh 2003). Ischemic injury not only stimulates the generation and release of chemokines but also increases the number of chemokine receptors in the brain. It is shown that mRNA expression for MCP-1 and MIP-1α is induced in the rat brain after focal cerebral ischemia. Chemokine signaling is associated with the postischemic inflammatory response. Overlapping pathways involving ROS, Toll-like receptor (TLR) activation, and the nuclear factor-κB system mediate both CXC and CC chemokines in ischemic tissues. Reperfusion accentuates chemokine expression promoting an intense inflammatory reaction (Frangogiannis 2007). ELR-containing CXC chemokines regulate neutrophil infiltration in the ischemic area, whereas CXCR3 ligands may mediate recruitment of Th1 cells. CC chemokines, on the other hand, mediate mononuclear cell infiltration and macrophage activation. Accumulating evidence demonstrates that chemokine signaling mediates actions beyond leukocyte chemotaxis and activation, regulating angiogenesis and fibrous tissue deposition (Frangogiannis 2007).

## 7.5 Neuroprotective Role of Leptin in the Ischemic Injury

As stated above, leptin is a 16 kDa non-glycosylated protein that is primarily produced and secreted by white adipocytes. It acts through binding with leptin receptor and is involved in regulation of food intake and energy balance (Friedman and Halaas 1998; Ahima 2005).

In the brain, leptin receptors are abundantly expressed in the hypothalamus and less abundantly expressed in the cerebral cortex, hippocampus, and cerebellum (Mercer et al. 1996; Guo et al. 2008). The interactions between leptin and leptin receptors activate Janus tyrosine kinase 2, leading to tyrosine residue phosphorylation of the intracellular domains of the receptor. The receptors then serve as docking sites for subsequent signaling events, including the phosphorylation of signal transducer and activator of transcription 3 (STAT3), extracellular signal-related kinase 1/2 (ERK1/2), and Akt (Banks et al. 2000; Kloek et al. 2002). Leptin functions via both central and peripheral mechanisms. In the central mechanism, the regulation of body weight and energy balance by leptin is mediated by neurons (Friedman and Halaas 1998), whereas in the peripheral mechanism, leptin acts by regulating blood pressure via endothelial cells in the cardiovascular and cerebrovascular systems (Lembo et al. 2000; Fortuno et al. 2002). Studies on the systemic administration of leptin indicate that it reduces infarct volume produced by focal cerebral ischemic injury in mice and this effect is abolished by the administration of PD98059, a mitogen-activated protein kinase 1 (MEK1) inhibitor (Zhang et al. 2007). Similarly, intracerebral ventricle infusion of leptin protects neurons and markedly increases the numbers of surviving neurons in the CA1 area of the hippocampus in a dose-dependent manner. Infusion of a specific leptin antagonist 10 min prior to transient global ischemia blocks the pro-survival effects of leptin, suggesting that leptin receptors play a very important role in mediating this neuroprotection. The molecular mechanism of leptin-mediated neuroprotection is not fully understood. However, based on detailed investigations, it is proposed that phosphorylation of Akt and extracellular signal-related kinase 1/2 (ERK1/2) signaling pathways may be closely associated with neuroprotective process in the CA1 area of hippocampus (Zhang and Chen 2008). Similarly, in permanent middle cerebral artery occlusion model of ischemia, the neuroprotective effect of leptin is mediated through the modulation of STAT3 phosphorylation in the ischemic cortex (Amantea et al. 2011). In this mechanism, the binding of leptin and its receptor activates the Janus tyrosine kinase (JAK)-2, which in turn recruits signal transducer and activator of transcription (STAT)-3 proteins to the receptor. Upon phosphorylation, STAT3 dimerizes and translocates to the nucleus where it regulates the expression of neuroprotective genes, such as *bcl-2*, *bcl-xL*, and *TIMP-1* (Vaisse et al. 1996). Thus, systemic acute administration of leptin provides significant reduction of the brain infarct volume and neurological deficit up to 7 days after the induction of ischemia. Detailed investigations indicate that early enhancement of nuclear phospho-STAT3 mediated by leptin in the astrocytes of the ischemic penumbra may contribute to a beneficial effect on the evolution of tissue damage. In addition, the elevation of phospho-STAT3 induced by leptin in the neurons after 24 h permanent middle cerebral artery occlusion is associated with an increased expression of tissue inhibitor of matrix metalloproteinases-1 in the cortex, supporting the view that leptin-mediated neuroprotection involves the phosphorylation of STAT3 (Amantea et al. 2011).

## 7.6 Metabolic Link Between MetS and Stroke

The prevalence of MetS in patients with a history of stroke is 44 % compared to 23 % without stroke (Boden-Albala et al. 2008; Arenillas et al. 2007). MetS increases the risk of vascular disease and stroke through the involvement of hypertension, insulin insensitivity, and dyslipidemia (Li et al. 2010; Abarquez 2003). These processes lead to atherogenesis and the prothrombotic state, which increases the risk of stroke (Uchino et al. 2010). Each component of the MetS may increase the risk of vessel blockage leading to stroke. Involvement of more components of MetS results in the higher risk of developing stroke (Koren-Morag et al. 2005; Dekker et al. 2005). The association between MetS and stroke is not attenuated after adjustment for the presence of established cardiovascular and cerebrovascular risk factors. The prevalence of the MetS as well as stroke increases with the age (Ninomiya et al. 2004). Like stroke patients, MetS patients show the inflammatory activity, increase in the generation of free radicals, changes of neurotrophic factors, and reduction of insulin transport into the brain (Lusis et al. 2008). Another commonality between the MetS and stroke is the hyperglycemia which is detrimental to cognition and other brain functions. MetS in stroke patients involves several metabolic disturbances, such as high levels of low-density lipoproteins, fibrinogen, insulin resistance, and uric acid. These risk factors are closely associated with the pathogenesis of ischemic stroke (Milionis et al. 2000; Maresca et al. 1999; Lehto et al. 1998). Increase in serum uric acid levels (hyperuricemia), an important risk factor of stroke, is linked to obesity (Lee et al. 1995), hypertension (Selby et al. 1990), reduced HDL cholesterol (Wilson et al. 1983), hypertriglyceridemia (Zavaroni et al. 1993), hyperinsulinemia, and reduced insulin sensitivity (Facchini et al. 1991). Similarly, dyslipidemia, a hallmark of the MetS, which is characterized by elevated triglyceride levels and low HDL-C levels, is strongly correlated with ischemic stroke among all the MetSyn components. Although there is a controversy regarding the association between serum triglyceride levels in stroke (Rizos and Mikhailidis 2001), the Copenhagen City Heart Study has indicated that there is a log-linear association between serum triglyceride levels and nonhemorrhagic stroke in both sexes (Lindenstrom et al. 1994). These parameters are components of the MetS in humans.

Another risk factor for stroke and MetS is hyperhomocysteinemia, which induces endothelial damage, mitochondrial disintegration, swelling of pericytes, basement membrane thickening, and perivascular detachment (Troen 2005). The intracellular effects of homocysteine are very divergent. Hyperhomocysteinemia not only induces the activation of caspase-8 and subsequent apoptosis and stimulates monocyte chemoattractant protein-1/interleukin-8 and subsequent neuroinflammation but also increases oxidative stress, retards endothelial nitric oxide synthase activity, and produces peroxynitrite resulting into cell death (Skurk and Walsh 2004). In addition, homocysteine also inhibits capillary endothelial nitric oxide synthase (Faraci 2003) and glucose transporter and transiently alters different cell adhesion molecules (Lee et al. 2004). Homocysteine has also been reported to directly induce cell death of cerebrocortical neurons through the activation of NMDA (Lipton et al. 1997). Chronic hyperhomocysteinemia also enhances lipid peroxidation and decreases glutathione, suggesting the involvement of oxidative stress leading to cognitive impairment (Baydas et al. 2005).

Although MetS is traditionally regarded as a peripheral metabolic disease, recent noninvasive brain imaging studies indicate that diabetes (a major component of MetS) is associated with functional and structural abnormalities in the brain. Thus, computed tomography (CT) and magnetic resonance imaging (MRI) studies on diabetic subjects indicate a relationship among diabetes, cerebral atrophy, and lacunar infarcts (van Harten et al. 2006). Similarly, positron emission tomography (PET) and single-photon emission CT (SPECT) studies also show regional alterations in cerebral blood flow (van Harten et al. 2006). Cerebral edema, cerebral hemorrhage, or intracranial thrombosis is observed in children with type I diabetes and diabetic ketoacidosis (Wootton-Gorges and Glaser 2007). Patients with type II diabetes often show white matter hyperintensities which are detectable with brain imaging techniques (Jongen and Biessels 2008). Studies on chronic cerebrovascular disease patients indicate that all the patients have an inadequate endothelial reaction—the antiaggregatory, fibrinolytic, and anticoagulant activities of the endothelium are lowered. Furthermore, there is endothelium-dependent hyperproduction of intercellular adhesion molecules supporting the view that MetS has considerable impact on the formation of a significant procoagulant state of the hemorrheological and hemostatic systems in patients with chronic cerebrovascular diseases (Tanashian et al. 2010). In addition, impaired cerebral vasomotor reactivity may be another mediator of stroke in patients with MetS. The abovementioned studies suggest that diabetes influences brain function. Thus, the simultaneous presence of multiple risk factors (insulin resistance, hypertension, obesity, and increased levels of triacylglycerols) in MetS supports the view that MetS substantially increases the risk of stroke even after extensive adjustment for various cardiovascular risk factors. Serum uric acid remains an independent risk factor for stroke. Studies on the effect of ischemic injury in adult diabetic mice (db/db) indicate that db/db mice have four times higher blood sugar than its normoglycemic genetic control (db/+ mice) (Tureyen et al. 2011). Following transient middle cerebral artery occlusion (MCAO), the db/db mice not only show significantly higher mortality and bigger infarcts but also increase in cerebral edema and worsening neurological outcome compared to db/+ mice. The db/db mice also show significantly higher postischemic inflammatory markers ($ICAM1^+$ capillaries, extravasated macrophages/neutrophils, and exacerbated proinflammatory gene expression) compared to db/+ mice. In addition, postischemic induction of HSP70, HSP27, and HSP32/HO1 is significantly curtailed in the brains of db/db mice indicating a disability to prevent cell death and failure to induce plasticity in the diabetic brain (Tureyen et al. 2011). Similarly, in the db/db mice cerebral inflammatory response initiating recovery is also delayed and diminished in the diabetic mice (db/db) compared with the nondiabetic db/+ mice (Kumari et al. 2007).

It is obvious that the brain is involved in regulating glucose homeostasis through the modulation of hepatic gluconeogenesis and glycogenolysis and pancreatic function. These activities are largely mediated by central regulation of the autonomic nervous system, which acts in concert with the HPA axis to regulate metabolic responses to changes in energy requirements and plasma glucose concentration (Eikelis and Esler 2005). Disruption of blood flow and alterations in BBB are closely associated with cellular damage in acute and chronic cerebral ischemic injury (Farooqui 2010). During ischemic injury, neurovascular unit (NVU), which forms the interface between the blood and brain tissue, is attacked by free radical (ROS and RNS) and proteases leading to disruption of BBB and degradation of the tight junction proteins in endothelial cells. Thus, disruption of BBB allows the migration of neutrophils from the blood into the brain parenchyma within hours after reperfusion, followed by the entry of macrophages and monocytes (Stanimirovic and Friedman 2012). Macrophages and monocytes, brain endothelial cells, astrocytes, pericytes, microglia, and neurons interact with acellular (basal lamina) components of the NVU leading in increased BBB permeability, brain edema, neurovascular uncoupling, and neuronal dysfunction and damage (Stanimirovic and Friedman 2012). It should be noted that the basic principles of stroke prevention are the same in patients with diabetes and/or the MetS as in those without (Rothwell 2005). Lowering of hypertension is highly effective in both primary and secondary prevention of stroke in diabetics, and there is no evidence to suggest that the benefits of lipid-lowering therapy are any less. Antiplatelet agents produce beneficial effects in asymptomatic diabetic patients, who have a substantially increased risk of acute vascular events. These studies support the view that MetS is a major risk factor for stroke.

## 7.7 Neuroprotective Role of Adiponectin in Stroke

Adiponectin has been shown to produce beneficial effects in cardio- and cerebrovascular disorders (Qi et al. 2004; Chen et al. 2009; Lee et al. 2011). Although hypoadiponectinemia is associated with ischemic cerebrovascular disease, very little is known about the molecular mechanism of cerebroprotective action of adiponectin. Following ischemia/reperfusion injury, administration of adiponectin through the jugular vein not only decreases cerebral infarct size, reduces neurological deficits, and decreases the expression of endogenous immunoglobulin G and matrix metalloproteinase-9 but also significantly inhibits the expression of myeloperoxidase, a distinct indicator of inflammatory cell infiltration, and inflammatory cytokines, interleukin-1β (IL-1β), tumor necrosis factor-α, and IL-8 in response to ischemia/reperfusion injury, supporting the view that adiponectin exerts potent anti-inflammatory actions (Chen et al. 2009; Lee et al. 2011). In addition, ischemia/reperfusion injury facilitates migration of NF-κB (p65 subunit) to the nucleus. Administration of adiponectin prevents the translocation of NF-κB from the cytoplasm into the nucleus (Chen et al. 2009). These results are supported by the results on ischemic injury in adiponectin-deficient mice (Nishimura et al. 2008). In these mice, it is shown that beneficial effects of adiponectin on ischemic injury are mediated through the involvement of eNOS-dependent mechanisms in the vasculature. It is also reported that eNOS contributes to vascular protection and plays an important protective role in the regulation of brain damage after ischemia (Huang et al. 1996; Jeon et al. 2009). In addition, intracerebroventricularly injected adiponectin promotes phosphorylation of AMPK in the rat hypothalamus suggesting a role of adiponectin in modulating energy homeostasis (Guillod-Maximin et al. 2009). Based on these results, it is suggested that the adiponectin–eNOS regulatory signaling axis functions to modulate vascular function under ischemic conditions and it is closely involved in neuroprotection against cerebral ischemic injury.

## 7.8 Molecular Mechanism Associated with Metabolic Link Between MetS and Stroke

The molecular mechanism associated with increased risk of stroke in MetS is not fully understood. However, it is becoming increasingly evident that leptin not only stimulates brain metabolic rate (Morrison 2009; Villanueva and Myers 2008) but also mediates the activation of the pituitary adrenal sympathetic system and the renin–angiotensin system (Shek et al. 1998; Esler et al. 2001). Increased plasma norepinephrine (noradrenaline) and angiotensin II produces vasoconstriction through the involvement of NADPH oxidase (Kazama et al. 2003) which also increases oxygen demand through activation of uncoupling of mitochondrial oxidative phosphorylation (Argiles et al. 2002). This sequence of events may ultimately lead to oxidative stress and stroke (Uchino et al. 2010).

Patients with MetS have higher risk of stroke and increased risk of morbidity and mortality after another stroke (Kernan et al. 2003; Ninomiya et al. 2004). It is reported that cortical neurons develop insulin resistance, which decreases neuroprotection via circulating insulin and insulin-like growth factor-I (IGF-I) (Kim et al. 2011). Acute insulin treatment of primary embryonic cortical neurons results in activation of insulin signaling including downstream signaling that involves activation of extracellular signal-regulated kinase (ERK), Akt, p70S6K, and glycogen synthase kinase 3β (GSK3β) (Fig. 7.9). Precise regulation of Akt phosphorylation is critical for neuronal survival during brain ischemia (Fukunaga and Kawano 2003; Ouyang et al. 1999; Zhao et al. 2006). Decrease in responsiveness of Akt phosphorylation due to hyperinsulinemia explains the increase in neuronal damage reported in both experimental models of diabetes, diabetic patients, and patients with MetS (Allen and Bayraktutan 2008; Meeuwisse-Pasterkamp et al. 2008; Quinn and Lees 2009). Collective evidence suggests that prolonged hyperinsulinemia leads to insulin resistance in cortical neurons. Decreased sensitivity to neuroprotective ligands may explain the increased neuronal damage observed in both experimental models of diabetes and diabetic patients after ischemia/reperfusion injury.

Another mechanism that links MetS with stroke involves c-Jun N-terminal kinase (JNK), a family of serine/threonine protein kinases, which not only is activated by oxidative stress and ischemia but also interferes with insulin action in cultured cells and is activated by inflammatory cytokines and free fatty acid molecules. All these mediators have been implicated in the pathogenesis of MetS and stroke (Hirosumi et al. 2002). Feeding mice with a high-fat diet results in activation of JNK. Moreover, JNK knockout mice are protected against the effects of high-fat diet-induced insulin resistance (Hirosumi et al. 2002). These observations support the view that JNK plays a critical role in the metabolic stress response of obesity as well as hypoxic/ischemic injury. It is also known that JNK is activated by tumor necrosis factor-α, free fatty acids, and ROS (Hirosumi et al. 2002). It is hypothesized that being overweight aggravates apoptosis, microglia activation, and BBB leakage after hypoxic injury and thereby worsens brain damage through JNK hyperactivation in neonatal brains (Tu et al. 2011) (Fig. 7.10). Thus, overweight pups have significantly higher levels of p-JNK levels before and after hypoxic injury compared to the control pups suggesting that an excess volume of fat in the overweight pups may contribute to JNK hyperactivation (Tu et al. 2011). It is interesting to note that JNK is hyperactivated in the neurons, microglia, and vascular endothelial cells post-hypoxic injury in the overweight pups, and inhibition of JNK activation not only reduces hypoxic injury-mediated neuronal apoptosis, and reduces microglia activation, but also attenuates BBB damage in the overweight pups. These findings suggest that overweight pups may induce a programming effect on the neurons, microglia, and vascular endothelial cells of the neonatal brain through JNK hyperactivation after hypoxic injury (Tu et al. 2011). Collective evidence suggests that overweight neonates and rat pups show the increased risk of hypoxia-induced neuronal apoptosis, microglial activation, and BBB damage through JNK hyperactivation.

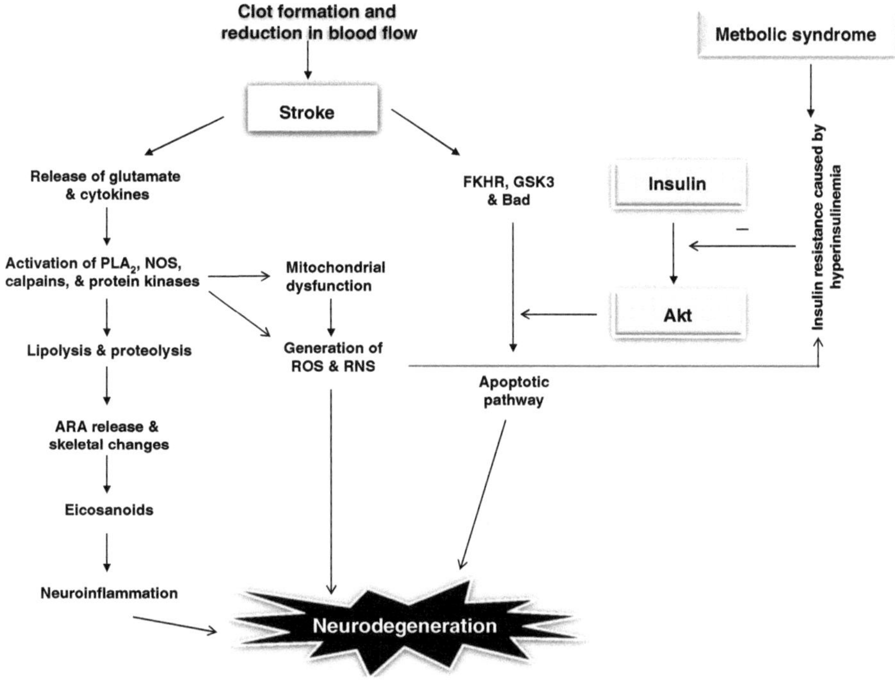


**Fig. 7.9** Hypothetical model showing contribution of metabolic syndrome in stroke-induced neurodegeneration. Cytosolic phospholipase $A_2$ ($cPLA_2$); phosphatidylcholine (PtdCho); nitric oxide synthase (NOS); serine/threonine kinase (Akt,); glycogen synthase kinase 3**β** (GSK3**β**); reactive oxygen species (ROS); reactive nitrogen species (RNS); forkhead transcription factor (FKHR); and Bcl-2-associated death promoter (BAD) protein is a proapoptotic member of the Bcl-2 gene family

## 7.9 Conclusion

MetS is characterized by the clustering of a number of metabolic abnormalities, such as hypertension, dyslipidemia, hyperglycemia, insulin resistance, low HDL cholesterol, higher low-density lipoprotein cholesterol, elevated triglycerides, and increased waist circumference. MetS is associated with a higher risk of cardiovascular (heart disease) and cerebrovascular (stroke) diseases. The dominant underlying risk factors for MetS are abdominal obesity and insulin resistance. Insulin mediates its effects through a signaling network composed of a number of molecules, including the activation of insulin receptor tyrosine kinase and phosphorylation of IRS-1 and IRS-2. Insulin signaling network also requires PtdIns 3K, which activates downstream serine/threonine kinases including Akt, regulating most of the metabolic actions of insulin such as suppression of hepatic glucose production and activation of glucose transport in the muscle and adipocytes. Impairment in insulin signaling network at the enzyme and transcription factor levels may contribute to obesity and insulin resistance. Additional risk factors for MetS include physical

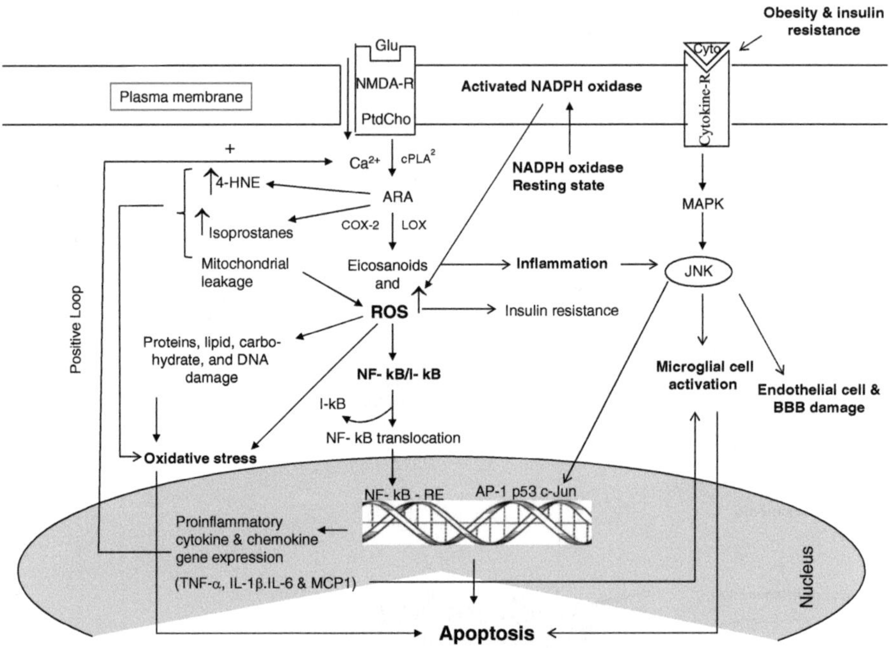


**Fig. 7.10** Contribution of oxidative stress and cytokine signaling in apoptotic cell death. Cytosolic phospholipase $A_2$ ($cPLA_2$); phosphatidylcholine (PtdCho); cyclooxygenase-2 (COX-2); lipoxygenase (LOX); cytokines (Cyto); cytokine receptor (cytokine-R); tumor necrosis factor-α (TNF-α); interleukin-1β (IL-1β); *N*-methyl-D-aspartic acid receptor (NMDA-R); c-Jun $NH_2$-terminal kinase (JNK); and activator protein-1 (AP-1). The expression of cytokines upregulates activities of $cPLA_2$ through a positive loop type of mechanism in the cytoplasm and neural membranes

inactivity, aging, and polycystic ovarian disease. Acquired risk factors for MetS include central adiposity and physical inactivity, both of which can elicit insulin resistance and the MetS. At the molecular level, it is shown that increase in the release of free fatty acids from adipocytes blocks the insulin signaling pathway and induces endothelial dysfunction due to increase in the production of ROS. High levels of triglycerides and low concentrations of HDLs contribute to dyslipidemia and proinflammatory state. Inflammation, the key pathogenic component of atherosclerosis, facilitates thrombosis, a process that is closely associated with stroke, an ischemia-mediated metabolic injury, which is caused by an interruption of cerebral blood flow. It can lead to vascular leakage, inflammation, tissue injury, and necrosis. Changes associated with ischemia include impairment of metabolism, energy failure, ROS production, excitotoxicity, alterations in calcium homeostasis, and activation of proteases, phospholipases, nitric oxide synthases, and protein kinases.

An increase in adipose tissue mass in MetS is accompanied by increased secretion of several adipocytokines such as TNF-α, IL-1β, IL-6, and resistin. These adipocytokines not only initiate and maintain inflammation but also cause endothelial cell dysfunction. Similarly, stroke is accompanied by increase in free fatty acid release; elevation in ROS, diacylglycerol, and ceramide; increase in expression of proinflammatory cytokines; and activation of microglial cells. Interactions among adipokines, insulin signaling, endothelial cell dysfunction, and platelet aggregation in MetS and stroke along with alterations in the expression of transcription factors may play a crucial role in the vascular damage closely associated with the pathogenesis of MetS and stroke.

## References

Abarquez RF Jr (2003) Microvascular disease relevance in the hypertension syndrome. Clin Hemorheol Microcirc 29:295–300
Ahima RS (2005) Central actions of adipocyte hormones. Trends Endocrinol Metab 16:307–313
Ahima R, Flier JS (2000) Adipose tissue as an endocrine organ. Trends Endocrinol Metab 11:327–332
Ahonen TM, Saltevo JT, Kautiainen HJ, Kumpusalo EA, Vanhala MJ (2010) The association of adiponectin and low-grade inflammation with the course of metabolic syndrome. Nutr Metab Cardiovasc Dis 22:285–291
Al-Bahrani A, Taha S, Shaath H, Bakhiet M (2007) TNF-alpha and IL-8 in acute stroke and the modulation of these cytokines by antiplatelet agents. Curr Neurovasc Res 4:31–37
Allen CL, Bayraktutan U (2008) Risk factors for ischaemic stroke. Int J Stroke 3:105–116
Amantea D, Tassorelli C, Russo R, Petrelli F, Morrone LA, Bagetta G, Corasaniti MT (2011) Neuroprotection by leptin in a rat model of permanent cerebral ischemia: effects on STAT3 phosphorylation in discrete cells of the brain. Cell Death Dis 2:e238
Anubhuti V, Arora S (2008) Leptin and its metabolic interactions—an update. Diab Obes Metab 10:973–993
Arenillas JF, Moro MA, Davalos A (2007) The metabolic syndrome and stroke. Stroke 38:2196–2203
Argiles JM, Busquets S, Lopez-Soriano F (2002) The role of uncoupling proteins in pathophysiological states. Biochem Biophys Res Commun 293:1145–1152
Bakhai A (2008) Adipokines-targeting a root cause of cardiometabolic risk. Q J Med 101:767–776
Banks WA (2003) Is obesity a disease of the blood–brain barrier? Physiological, pathological, and evolutionary considerations. Curr Pharm Des 9:801–809
Banks WA (2008) The blood–brain barrier as a cause of obesity. Curr Pharm Des 14:1606–1614
Banks WA, Farrell CL (2003) Impaired transport of leptin across the blood–brain barrier in obesity is acquired and reversible. Am J Physiol 285:E10–E15
Banks WA, Kastin AJ, Huang W, Jaspan JB, Maness LM (1996) Leptin enters the brain by a saturable system independent of insulin. Peptides 17:305–311
Banks AS, Davis SM, Bates SH, Myers MG Jr (2000) Activation of downstream signals by the long form of the leptin receptor. J Biol Chem 275:14563–14572
Banks WA, Coon AB, Robinson SM, Moinuddin A, Shultz JM, Nakaoke R, Morley JE (2004) Triglycerides induce leptin resistance at the blood–brain barrier. Diabetes 53:1253–1260
Barrientos RM, Sprunger DB, Campeau S, Higgins EA, Watkins LR, Rudy JW, Maier SF (2003) Brain-derived neurotrophic factor mRNA downregulation produced by social isolation is blocked by intrahippocampal interleukin-1 receptor antagonist. Neuroscience 121:847–853
Baumann H, Morella KK, White DW, Dembski M, Bailon PS, Kim H, Lai CF, Tartaglia LA (1996) The full-length leptin receptor has signaling capabilities of interleukin 6-type cytokine receptors. Proc Natl Acad Sci USA 93:8374–8378

Baydas G, Ozer M, Yasar A, Tuzcu M, Koz ST (2005) Melatonin improves learning and memory performances impaired by hyperhomocysteinemia in rats. Brain Res 1046:187–194

Bennett BD, Solar GP, Yuan JQ, Mathias J, Thomas GR, Matthews W (1996) A role for leptin and its cognate receptor in hematopoiesis. Curr Biol 6:1170–1180

Biber K, Neumann H, Inoue K, Boddeke HW (2007) Neuronal "On" and "Off" signals control microglia. Trends Neurosci 30:596–602

Bitto A, Sell C, Crowe E, Lorenzini A, Malaguti M, Hrelia S, Torres C (2010) Stress-induced senescence in human and rodent astrocytes. Exp Cell Res 316:2961–2968

Black-Schaffer RM, Winston C (2004) Age and functional outcome after stroke. Top Stroke Rehabil 11:23–32

Boden-Albala B, Sacco RL, Lee HS, Grahame-Clarke C, Rundek T, Elkind MV, Wright C, Giardina EG, DiTullio MR, Homma S, Paik MC (2008) Metabolic syndrome and ischemic stroke risk: Northern Manhattan Study. Stroke 39:30–35

Bokarewa M, Nagaev I, Dahlberg L, Smith U, Tarkowski A (2005) Resistin, an adipokine with potent proinflammatory properties. J Immunol 174:5789–5795

Borch-Johnsen K (2007) The metabolic syndrome in a global perspective. The public health impact-secondary publication. Dan Med Bull 54:157–159

Bouloumie A, Drexler HC, Lafontan M, Busse R (1998) Leptin, the product of Ob gene, promotes angiogenesis. Circ Res 83:1059–1066

Bousser MG (2012) Stroke prevention: an update. Front Med 6:22–34

Brea D, Sobrino T, Ramos-Cabrer P, Castillo J (2009) Inflammatory and neuroimmunomodulatory changes in acute cerebral ischemia. Cerebrovasc Dis 27(Suppl 1):48–64

Bredt DS (1999) Endogenous nitric oxide synthesis: biological functions and pathophysiology. Free Radic Res 31:577–596

Brinkley TE, Leng X, Miller ME, Kitzman DW, Pahor M, Berry MJ, Marsh AP, Kritchevsky SB, Nicklas BJ (2009) Chronic inflammation is associated with low physical function in older adults across multiple comorbidities. J Gerontol A Biol Sci Med Sci 64:455–461

Brown DR (2009) Role of microglia in age-related changes to the nervous system. ScientificWorldJournal 9:1061–1071

Buga AM, Vintilescu R, Pop OT, Popa-Wagner A (2011) Brain aging and regeneration after injuries: an organismal approach. Aging Dis 2:64–79

Buhl ES, Jessen N, Pold R, Ledet T, Flyvbjerg A, Pedersen SB, Pedersen O, Schmitz O, Lund S (2002) Long-term AICAR administration reduces metabolic disturbances and lowers blood pressure in rats displaying features of the insulin resistance syndrome. Diabetes 51:2199–2206

Calabrese V, Cornelius C, Rizzarelli E, Owen JB, Dinkova-Kostova AT, Butterfield DA (2009) Nitric oxide in cell survival: a janus molecule. Antioxid Redox Signal 11:2717–2739

Campfield LA, Smith FJ, Guisez Y, Devos R, Burn P (1995) Recombinant mouse OB protein: evidence for a peripheral signal linking adiposity and central neural networks. Science 269:546–549

Carlyle M, Jones OB, Kuo JJ, Hall JE (2002) Chronic cardiovascular and renal actions of leptin: role of adrenergic activity. Hypertension 39:496–501

Castell JV, Andus T, Kunz D, Heinrich PC (1989) Interleukin-6: the major regulator of acute-phase protein synthesis in man and rat. Ann NY Acad Sci 557:87–99

Chang L, Chian S, Saltiel AR (2004) Insulin signaling and regulation of glucose transport. Mol Med 10:65–71

Chen H, Montagnani M, Funahashi T, Shimomura I, Quon MJ (2003) Adiponectin stimulates production of nitric oxide in vascular endothelial cells. J Biol Chem 278:45021–45026

Chen B, Liao WQ, Xu N, Xu H, Wen JY, Yu CA, Liu XY, Li CL, Zhao SM, Campbell W (2009) Adiponectin protects against cerebral ischemia-reperfusion injury through anti-inflammatory action. Brain Res 1273:129–137

Clark JD, Schievella AR, Nalefski EA, Lin L-L (1995) Cytosolic phospholipase $A_2$. J Lipid Mediat Cell Signal 12:83–117

Correia ML, Morgan DA, Sivitz WI, Mark AL, Haynes WG (2001) Leptin acts in the central nervous system to produce dose-dependent changes in arterial pressure. Hypertension 37:936–942
Dandona P, Aljada A, Chaudhuri A, Mohanty P, Garg R (2005) Metabolic syndrome: a comprehensive perspective based on interactions between obesity, diabetes, and inflammation. Circulation 111:1448–1454
De Rosa V, Procaccini C, Calì G, Pirozzi G, Fontana S, Zappacosta S, La Cava A, Matarese G (2007) A key role of leptin in the control of regulatory T cell proliferation. Immunity 26:241–255
Dekker JM, Girman C, Rhodes T, Nijpels G, Stehouwer CD, Bouter LM, Heine RJ (2005) Metabolic syndrome and 10-year cardiovascular disease risk in the Hoorn study. Circulation 112:666–673
Dodson MV, Mir PS, Hausman GJ et al (2011) Obesity, metabolic syndrome, and adipocytes. J Lipids 2011:721686
Dominguez H, Storgaard H, Rask-Madsen C, Steffen Hermann T, Ihlemann N, Baunbjerg Nielsen D, Spohr C, Kober L, Vaag A, Torp-Pedersen C (2005) Metabolic and vascular effects of tumor necrosis factor-α blockade with etanercept in obese patients with type 2 diabetes. J Vasc Res 42:517–525
Drew B, Leeuwenburgh C (2002) Aging and the role of reactive nitrogen species. Ann N Y Acad Sci 959:66–81
Durakoglugil M, Irving AJ, Harvey J (2005) Leptin induces a novel form of NMDA receptor-dependent long-term depression. J Neurochem 95:396–405
Dyck DJ (2009) Adipokines as regulators of muscle metabolism and insulin sensitivity. Appl Physiol Nutr Metab 34:396–402
Eikelis N, Esler M (2005) The neurobiology of human obesity. Exp Physiol 90:673–682
Esler M, Rumantir M, Wiesner G, Kaye D, Hastings J, Lambert G (2001) Sympathetic nervous system and insulin resistance: from obesity to diabetes. Am J Hypertens 14:304S–309S
Facchini F, Chen YDI, Hollenbeck CB, Reaven GM (1991) Relationship between resistance to insulin-mediated glucose uptake, urinary uric acid clearance, and plasma uric acid concentration. JAMA 266:3008–3011
Faraci FM (2003) Hyperhomocysteinemia, a million ways to lose control. Arterioscler Thromb Vasc Biol 23:371–373
Farooqui AA (2009) Hot topics in neural membrane lipidology. Springer, New York
Farooqui AA (2010) Neurochemical aspects of neurotraumatic and neurodegenerative diseases. Springer, New York
Farooqui AA (2011) Lipid mediators and their metabolism in the brain. Springer, New York
Farooqui AA, Horrocks LA (1994) Excitotoxicity and neurological disorders: involvement of membrane phospholipids. Int Rev Neurobiol 36:267–323
Farooqui AA, Horrocks LA (2007) Glycerophospholipids in brain. Springer, New York
Farooqui AA, Horrocks LA, Farooqui T (2007) Modulation of inflammation in brain: a matter of fat. J Neurochem 101:577–599
Farr SA, Banks WA, Morley JE (2006) Effects of leptin on memory processing. Peptides 27:1420–1425
Fernandez-Real JM, Vayreda M, Richart C, Gutierrez C, Broch M, Vendrell J, Ricart W (2001) Circulating interleukin 6 levels, blood pressure, and insulin sensitivity in apparently healthy men and women. J Clin Endocrinol Metab 86:1154–1159
Filková M, Haluzík M, Gay S, Šenolt L (2009) The role of resistin as a regulator of inflammation: implications for various human pathologies. Clin Immunol 133:157–170
Fisher JS, Gao J, Han DH, Holloszy JO, Nolte LA (2002) Activation of AMP kinase enhances sensitivity of muscle glucose transport to insulin. Am J Physiol Endocrinol Metab 282: E18–E23
Ford ES, Giles WH, Dietz WH (2002) Prevalence of the metabolic syndrome among US adults: findings from the Third National Health and Nutrition Examination Survey. JAMA 287:356–359

Fortuno A, Rodriguez A, Gomez-Ambrosi J, Muniz P, Salvador J, Diez J, Fruhbeck G (2002) Leptin inhibits angiotensin II-induced intracellular calcium increase and vasoconstriction in the rat aorta. Endocrinology 143:3555–3560

Frangogiannis NG (2007) Chemokines in ischemia and reperfusion. Thromb Haemost 97:738–747

Frey SK, Nagl B, Henze A, Raila J, Schlosser B, Berg T, Tepel M, Zidek W, Weickert MO, Pfeiffer AF, Schweigert FJ (2008) Isoforms of retinol binding protein 4 (RBP4) are increased in chronic diseases of the kidney but not of the liver. Lipids Health Dis 7:29

Friedman JM, Halaas JL (1998) Leptin and the regulation of body weight in mammals. Nature 395:763–770

Frühbeck G (2004) The adipose tissue as a source of vasoactive factors. Curr Med Chem Cardiovasc Hematol Agents 2:197–208

Frühbeck G, Gómez-Ambrosi J (2003) Control of body weight: a physiologic and transgenic perspective. Diabetologia 46:143–172

Frühbeck G, Salvador J (2000) Relation between leptin and the regulation of glucose metabolism. Diabetologia 43:3–12

Fryer LG, Parbu-Patel A, Carling D (2002) The anti-diabetic drugs rosiglitazone and metformin stimulate AMP-activated protein kinase through distinct pathways. J Biol Chem 277: 25226–25232

Fujii N, Jessen N, Goodyear LJ (2006) AMP-activated protein kinase and the regulation of glucose transport. Am J Physiol Endocrinol Metab 291:E867–E877

Fukuhara A, Matsuda M, Nishizawa M, Segawa K, Tanaka M, Kishimoto K, Matsuki Y, Murakami M, Ichisaka T, Murakami H, Watanabe E, Takagi T, Akiyoshi M, Ohtsubo T, Kihara S, Yamashita S, Makishima M, Funahashi T, Yamanaka S, Hiramatsu R, Matsuzawa Y, Shimomura I (2005) Visfatin: a protein secreted by visceral fat that mimics the effects of insulin. Science 307:426–430

Fukunaga K, Kawano T (2003) Akt is a molecular target for signal transduction therapy in brain ischemic insult. J Pharmacol Sci 92:317–327

Gadient RA, Otten UH (1997) Interleukin-6 (IL-6)—a molecule with both beneficial and destructive potentials. Prog Neurobiol 52:379–390

Gallagher EJ, LeRoith D, Karnieli E (2008) The metabolic syndrome—from insulin resistance to obesity and diabetes. Endocrinol Metab Clin N Am 37:559–579

Garza JC, Guo M, Zhang W, Lu XY (2008) Leptin increases adult hippocampal neurogenesis in vivo and in vitro. J Biol Chem 283:18238–18247

Gemma C, Bachstetter AD, Bickford PC (2010) Neuron-microglia dialogue and hippocampal neurogenesis in the aged brain. Aging Dis 1:232–244

Graham TE, Yang Q, Bluher M, Hammarstedt A, Ciaraldi TP, Henry RR, Wason CJ, Oberbach A, Jansson PA, Smith U, Kahn B (2006) Retinol-binding protein 4 and insulin resistance in lean, obese, and diabetic subjects. New Engl Med 354:2552–2563

Grundy S (2008) Metabolic syndrome pandemic. Arterioscler Thromb Vasc Biol 28:629–636

Grundy SM, Cleeman JI, Daniels SR, Donato KA, Eckel RH, Franklin BA, Gordon DJ, Krauss RM, Savage PJ, Smith SC Jr, Spertus JA, Costa F (2005) Diagnosis and management of the metabolic syndrome: an American Heart Association/National Heart, Lung, and Blood Institute Scientific Statement (Executive Summary). Circulation 112:e285–e290

Guerre-Millo M (2004) Adipose tissue and adipokines: for better or worse. Diabetes Metab 30:13–19

Guillod-Maximin E, Roy AF, Vacher CM, Aubourg A, Bailleux V, Lorsignol A, Pénicaud L, Parquet M, Taouis M (2009) Adiponectin receptors are expressed in hypothalamus and colocalized with proopiomelanocortin and neuropeptide Y in rodent arcuate neurons. J Endocrinol 200:93–105

Guo Z, Jiang H, Xu X, Duan W, Mattson MP (2008) Leptin-mediated cell survival signaling in hippocampal neurons mediated by JAK STAT3 and mitochondrial stabilization. J Biol Chem 283:1754–1763

Hadari YR, Tzahar E, Nadiv O, Rothenberg P, Roberts CT Jr, LeRoith D, Yarden Y, Zick Y (1992) Insulin and insulinomimetic agents induce activation of phosphatidylinositol 3′-kinase upon its association with pp 185 (IRS-1) in intact rat livers. J Biol Chem 267:17483–17486

Hallikainen M, Kolehmainen M, Schwab U, Laaksonen DE, Niskanen L, Rauramaa R et al (2007) Serum adipokines are associated with cholesterol metabolism in the metabolic syndrome. Clin Chim Acta 383:126–132

Handy JA, Saxena NK, Fu P, Lin S, Mells JE, Gupta NA, Anania FA (2010) Adiponectin activation of AMPK disrupts leptin-mediated hepatic fibrosis via suppressors of cytokine signaling (SOCS-3). J Cell Biochem 110:1195–1207

Hanley AJ, Festa A, D'Agostino RB Jr, Kempf J, Zinman B, Haffner SM, Insulin resistance atherosclerosis study (2004) Metabolic and inflammation variable clusters and prediction of type 2 diabetes: factor analysis using directly measured insulin sensitivity. Diabetes 53:1773–1781

Hardie DG (2003) Minireview: the AMP-activated protein kinase cascade: the key sensor of cellular energy status. Endocrinology 144:5179–5183

Harrison D, Griendling KK, Landmesser U, Hornig B, Drexler H (2003) Role of oxidative stress in atherosclerosis. Am J Cardiol 91:7A–11A

Harvey J (2007) Leptin regulation of neuronal excitability and cognitive function. Curr Opin Pharmacol 7:643–647

Heidemann C, Sun Q, van Dam RM, Meigs JB, Zhang C, Tworoger SS, Mantzoros CS, Hu FB (2008) Molecular characteristics of serum visfatin and differential detection by immunoassays. Ann Intern Med 149:307–316

Herx LM, Rivest S, Yong VW (2000) Central nervous system-initiated inflammation and neurotrophism in trauma: IL-1 beta is required for the production of ciliary neurotrophic factor. J Immunol 165:2232–2239

Heymsfield SB, Greenberg AS, Fujioka K et al (1999) Recombinant leptin for weight loss in obese and lean adults: a randomized, controlled, dose-escalation trial. J Am Med Assoc 282:1568–1575

Heyser CJ, Masliah E, Samimi A, Campbell IL, Gold LH (1997) Progressive decline in avoidance learning paralleled by inflammatory neurodegeneration in transgenic mice expressing interleukin 6 in the brain. Proc Natl Acad Sci USA 94:1500–1505

Hirosumi J, Tuncman G, Chang L, Görgün CZ, Uysal KT, Maeda K, Karin M, Hotamisligil GS (2002) A central role for JNK in obesity and insulin resistance. Nature 420:333–336

Holland WL, Scherer PE (2009) PAQRs: a counteracting force to ceramides? Mol Pharmacol 75:740–743

Holland WL, Brozinick JT, Wang LP, Hawkins ED, Sargent KM, Liu Y, Narra K, Hoehn KL, Knotts TA, Siesky A, Nelson DH, Karathanasis SK, Fontenot GK, Birnbaum MJ, Summers SA (2007) Inhibition of ceramide synthesis ameliorates glucocorticoid-, saturated-fat-, and obesity-induced insulin resistance. Cell Metab 5:167–179

Holland WL, Miller RA, Wang ZV, Sun K, Barth BM, Bui HH, Davis KE, Bikman BT, Halberg N, Rutkowski JM, Wade MR, Tenorio VM, Kuo MS, Brozinick JT, Zhang BB, Birnbaum MJ, Summers SA, Scherer PE (2011) Receptor-mediated activation of ceramidase activity initiates the pleiotropic actions of adiponectin. Nat Med 17:55–63

Holmes BF, Sparling DP, Olson AL, Winder WW, Dohm GL (2005) Regulation of muscle GLUT4 enhancer factor and myocyte enhancer factor 2 by AMP-activated protein kinase. Am J Physiol Endocrinol Metab 289:E1071–E1076

Hotamisligil GS (2006) Inflammation and metabolic disorders. Nature 444:860–867

Hotamisligil GS, Budavari A, Murray D, Spiegelman BM (1994) Reduced tyrosine kinase activity of the insulin receptor in obesity-diabetes. Central role of tumor necrosis factor-alpha. J Clin Invest 94:1543–1549

Hotta K, Funahashi T, Arita Y, Takahashi M, Matsuda M, Okamoto Y, Iwahashi H, Kuriyama H, Ouchi N, Maeda K, Nishida M, Kihara S, Sakai N, Nakajima T, Hasegawa K, Muraguchi M, Ohmoto Y, Nakamura T, Yamashita S, Hanafusa T, Matsuzawa Y (2000) Plasma concentrations of a novel, adipose-specific protein, adiponectin, in type 2 diabetic patients. Arterioscler Thromb Vasc Biol 20:1595–1599

Hotta K, Funahashi T, Bodkin NL, Ortmeyer HK, Arita Y, Hansen BC, Matsuzawa Y (2001) Circulating concentrations of the adipocyte protein adiponectin are decreased in parallel with reduced insulin sensitivity during the progression to type 2 diabetes in rhesus monkeys. Diabetes 50:1126–1133

Hoyda TD, Fry M, Ahima RS, Ferguson AV (2007) Adiponectin selectively inhibits oxytocin neurons of the paraventricular nucleus of the hypothalamus. J Physiol 585:805–816

Hu FB, Meigs JB, Li TY, Rifai N, Manson JE (2004) Inflammatory markers and risk of developing type 2 diabetes in women. Diabetes 53:693–700

Huang Z, Huang PL, Ma J, Meng W, Ayata C, Fishman MC, Moskowitz MA (1996) Enlarged infarcts in endothelial nitric oxide synthase knockout mice are attenuated by nitro-L-arginine. J Cereb Blood Flow Metab 16:981–987

Inadera H (2008) The usefulness of circulating adipokine levels for the assessment of obesity-related health problems. Int J Med Sci 5:248–262

Jager J, Grémeaux T, Cormont M, Le Marchand-Brustel Y, Tanti J-F (2007) Interleukin-1β-induced insulin resistance in adipocytes through down-regulation of insulin receptor substrate-1 expression. Endocrinology 148:241–251

Jeon BT, Shin HJ, Kim JB, Kim YK, Lee DH, Kim KH, Kim HJ, Kang SS, Cho GJ, Choi WS, Roh GJ (2009) Adiponectin protects hippocampal neurons against kainic acid-induced excitotoxicity. Brain Res Rev 61:81–88

Jéquier E (2002) Leptin signaling, adiposity, and energy balance. Ann N Y Acad Sci 967:379–388

Jessen N, Pold R, Buhl ES, Jensen LS, Schmitz O, Lund S (2003) Effects of AICAR and exercise on insulin-stimulated glucose uptake, signaling, and GLUT-4 content in rat muscles. J Appl Physiol 94:1373–1379

Jongen C, Biessels GJ (2008) Structural brain imaging in diabetes: a methodological perspective. Eur J Pharmacol 585:208–218

Kaneko N, Kudo K, Mabuchi T, Takemoto K, Fujimaki K, Wati H, Iguchi H, Tezuka H, Kanba S (2006) Suppression of cell proliferation by interferon-alpha through interleukin-1 production in adult rat dentate gyrus. Neuropsychopharmacology 31:2619–2626

Karmiris K, Koutroubakis IE, Xidakis C, Polychronaki M, Voudouri T, Kouroumalis EA (2006) Circulating levels of leptin, adiponectin, resistin, and ghrelin in inflammatory bowel disease. Inflamm Bowel Dis 12:100–105

Kaser S, Kaser A, Sandhofer A, Ebenbichler CF, Tilg H, Patsch JR (2003) Resistin messenger-RNA expression is increased by proinflammatory cytokines in vitro. Biochem Biophys Res Commun 309:286–290

Kazama K, Wang G, Frys K, Anrather J, Iadecola C (2003) Angiotensin II attenuates functional hyperemia in the mouse somatosensory cortex. Am J Physiol Heart Circ Physiol 285:H1890–H1899

Kellerer M, Koch M, Metzinger E, Mushack J, Capp E, Häring HU (1997) Leptin activates PI-3 kinase in C2C12 myotubes via janus kinase-2 (JAK-2) and insulin receptor substrate-2 (IRS-2) dependent pathways. Diabetologia 40:1358–1362

Kernan WN, Inzucchi SE, Viscoli CM, Brass LM, Bravata DM, Shulman GI, McVeety JC, Horwitz RI (2003) Impaired insulin sensitivity among nondiabetic patients with a recent TIA or ischemic stroke. Neurology 60:1447–1451

Khan AH, Pessin JE (2002) Insulin regulation of glucose uptake: a complex interplay of intracellular signalling pathways. Diabetologia 45:1475–1483

Kim YB, Uotani S, Pierroz DD, Flier JS, Kahn BB (2000) In vivo administration of leptin activates signal transduction directly in insulin-sensitive tissues: overlapping but distinct pathways from insulin. Endocrinology 141:2328–2339

Kim JA, Montagnani M, Koh KK, Quon MJ (2006) Reciprocal relationships between insulin resistance and endothelial dysfunction: molecular and pathophysiological mechanisms. Circulation 113:1888–1904

Kim B, Sullivan KA, Backus C, Feldman EL (2011) Cortical neurons develop insulin resistance and blunted Akt signaling: a potential mechanism contributing to enhanced ischemic injury in diabetes. Antioxid Redox Signal 14:1829–1839

Kleindorfer D, Broderick J, Khoury J, Flaherty M, Woo D, Alwell K, Moomaw CJ, Schneider A, Miller R, Shukla R, Kissela B (2006) The unchanging incidence and case-fatality of stroke in the 1990s: a population-based study. Stroke 37:2473–2478

Kloek C, Haq AK, Dunn SL, Lavery HJ, Banks AS, Myers MG Jr (2002) Regulation of Jak kinases by intracellular leptin receptor sequences. J Biol Chem 277:41547–41555

Koch CA, Anderson D, Moran MF, Ellis C, Pawson T (1991) SH2 and SH3 domains: elements that control interactions of cytoplasmic signaling proteins. Science 252:668–674

Koren-Morag N, Goldbourt U, Tanne D (2005) Relation between the metabolic syndrome and ischemic stroke or transient ischemic attack: a prospective cohort study in patients with atherosclerotic cardiovascular disease. Stroke 36:1366–1371

Körner A, Garten A, Blüher M, Tauscher R, Kratzsch J, Kiess W (2007) Molecular characteristics of serum visfatin and differential detection by immunoassays. J Clin Endocrinol Metab 92:4783–4791

Kraft AD, Harry GJ (2011) Features of microglia and neuroinflammation relevant to environmental exposure and neurotoxicity. Int J Environ Res Public Health 8:2980–3018

Kubota N, Yano W, Kubota T, Yamauchi T, Itoh S, Kumagai H, Kozono H, Takamoto I, Okamoto S, Shiuchi T, Suzuki R, Satoh H, Tsuchida A, Moroi M, Sugi K, Noda T, Ebinuma H, Ueta Y, Kondo T, Araki E (2007) Adiponectin stimulates AMP-activated protein kinase in the hypothalamus and increases food intake. Cell Metab 6:55–68

Kumari R, Willing LB, Krady JK, Vannucci SJ, Simpson IA (2007) Impaired wound healing after cerebral hypoxia-ischemia in the diabetic mouse. J Cereb Blood Flow Metab 27:710–718

Kusminski CM, McTernan PG, Schraw T, Kos K, O'Hare JP, Ahima R, Kumar S, Scherer PE (2007) Adiponectin complexes in human cerebrospinal fluid: distinct complex distribution from serum. Diabetologia 50:634–642

Lancaster GI, Febbraio MA (2011) Adiponectin sphings into action. Nat Med 17:37–38

Lee J, Sparrow D, Vokonas PS, Landsberg L, Weiss ST (1995) Uric acid and coronary heart disease risk: evidence for a role of uric acid in the obesity-insulin resistance syndrome: the Normative Aging Study. Am J Epidemiol 142:288–294

Lee H, Kim H-J, Kim J-M, Chang N (2004) Effects of dietary folic acid supplementation on cerebrovascular endothelial dysfunction in rats with induced hyperhomocysteinemia. Brain Res 996:139–147

Lee EB, Warmann G, Dhir R, Ahima RS (2011) Metabolic dysfunction associated with adiponectin deficiency enhances kainic acid-induced seizure severity. J Neurosci 31:14361–14366

Lehto S, Niscanen L, Ronnemaa T, Laasko M (1998) Serum uric acid is a strong predictor of stroke in patients with non-insulin-dependent diabetes mellitus. Stroke 29:635–639

Lembo G, Vecchione C, Fratta L, Marino G, Trimarco V, d'Amati G, Trimarco B (2000) Leptin induces direct vasodilation through distinct endothelial mechanisms. Diabetes 49:293–297

Li B, Chen N, Luo T, Otsu Y, Murphy TH, Raymond LA (2002) Differential regulation of synaptic and extra-synaptic NMDA receptors. Nat Neurosci 5:833–834

Li FP, He J, Li ZZ, Luo ZF, Yan L, Li Y (2009) Effects of resistin expression on glucose metabolism and hepatic insulin resistance. Endocrine 35:243–251

Li R, Zhang H, Wang W, Wang X, Huang Y, Huang C (2010) Vascular insulin resistance in prehypertensive rats: role of PI3-kinase/Akt/eNOS signaling. Eur J Pharmacol 628:140–147

Libby P (2003) Inflammation in atherogenesis. Nature 420:868–874

Liberto CM, Albrecht PJ, Herx LM, Yong VW, Levison SW (2004) Pro-regenerative properties of cytokine-activated astrocytes. J Neurochem 89:1092–1100

Lin S, Saxena NK, Ding X, Stein LL, Anania FA (2006) Leptin increases tissue inhibitor of metalloproteinase I (TIMP-1) gene expression by a specificity protein 1/signal transducer and activator of transcription 3 mechanism. Mol Endo (Baltimore, MD) 20:3376–3388

Lindenstrom E, Boysen G, Nyboe J (1994) Influence of total cholesterol, high-density lipoprotein cholesterol, and triglycerides on risk of cerebrovascular disease: the Copenhagen City Heart Study. BMJ 309:11–15

Lipton SA, Kim W-K, Choi Y-B, Kumar S, D'Emilia DM, Rayudu PV, Arnelle DR, Stamler JS (1997) Neurotoxicity associated with dual actions of homocysteine at the *N*-methyl-D-aspartate receptor. Proc Natl Acad Sci USA 94:5923–5928

Lloyd-Jones D, Adams R, Carnethon M, De Simone G, Ferguson TB, Flegal K, Ford E, Furie K, Go A, Greenlund K, Haase N, Hailpern S, Ho M, Howard V, Kissela B, Kittner S, Lackland D, Lisabeth L, Marelli A, McDermott M, Meigs J, Mozaffarian D, Nichol G, O'Donnell C, Roger V, Rosamond W, Sacco R, Sorlie P, Stafford R, Steinberger J, Thom T, Wasserthiel-Smoller S, Wong N, Wylie-Rosett J, Hong Y, American Heart Association Statistics Committee and Stroke Statistics Subcommittee (2009) Heart disease and stroke statistics—2009 update: a report from the American Heart Association statistics committee and stroke statistics subcommittee. Circulation 119:480–486

Lusis AJ, Attie AD, Reue K (2008) Metabolic syndrome: from epidemiology to systems biology. Nat Rev Genet 9:819–830

Lyon CJ, Law RE, Hsueh WA (2003) Minireview: adiposity, inflammation, and atherogenesis. Endocrinology 144:2195–2200

Ma K, Cabrero A, Saha PK, Kojima H, Li L, Chang BH, Paul A, Chan L (2002) Increased beta-oxidation but no insulin resistance or glucose intolerance in mice lacking adiponectin. J Biol Chem 277:34658–34661

Madej T, Boguski MS, Bryant SH (1995) Threading analysis suggests that the obese gene product may be a helical cytokine. FEBS Lett 373:13–18

Maresca G, Di Blasio A, Marchioli R, Di Minno G (1999) Measuring plasma fibrinogen to predict stroke and myocardial infarction. Arterioscler Thromb Vasc Biol 19:1368–1377

Matarese G, Moschos S, Mantzoros CS (2005) Leptin in immunology. J Immunol 174:3137–3142

Mather KJ, Lteif A, Steinberg HO, Baron AD (2004) Interactions between endothelin and nitric oxide in the regulation of vascular tone in obesity and diabetes. Diabetes 53:2060–2066

Matsuzawa Y (2006) Therapy insight: adipocytokines in metabolic syndrome and related cardiovascular disease. Nat Clin Pract Cardiovas Med 3:35–42

Meeuwisse-Pasterkamp SH, van der Klauw MM, Wolffenbuttel BH (2008) Type 2 diabetes mellitus: prevention of macrovascular complications. Expert Rev Cardiovasc Ther 6:323–341

Menzaghi C, Coco A, Salvemini L, Thompson R, De Cosmo S, Doria A, Trischitta V (2006) Heritability of serum resistin and its genetic correlation with insulin resistance-related features in nondiabetic Caucasians. J Clin Endocrinol Metab 91:2792–2795

Mercer JG, Hoggard N, Williams LM, Lawrence CB, Hannah LT, Trayhurn P (1996) Localization of leptin receptor mRNA and the long form splice variant (Ob-Rb) in mouse hypothalamus and adjacent brain regions by in situ hybridization. FEBS Lett 387:113–116

Milionis HJ, Winder AF, Mikhailidis DP (2000) Lipoprotein (a) and stroke. J Clin Pathol 53:487–496

Minami M, Satoh M (2000) Chemokines as mediators for intercellular communication in the brain. Nippon Yakurigaku Zasshi 115:193–200

Minami M, Satoh M (2003) Chemokines and their receptors in the brain: pathophysiological roles in ischemic brain injury. Life Sci 74:321–327

Montecucco F, Bianchi G, Gnerre P, Bertolotto M, Dallegri F, Ottonello L (2006) Induction of neutrophil chemotaxis by leptin: crucial role for p38 and Src kinases. Ann N Y Acad Sci 1069:463–471

Morrison CD (2009) Leptin signaling in brain: a link between nutrition and cognition? Biochim Biophys Acta 1792:401–408

Munzberg H, Myers MG Jr (2005) Molecular and anatomical determinants of central leptin resistance. Nat Neurosci 8:566–570

Murray CA, Lynch MA (1998) Evidence that increased hippocampal expression of the cytokine interleukin-1 beta is a common trigger for age- and stress-induced impairments in long-term potentiation. J Neurosci 18:2974–2981

Muse ED, Obici S, Bhanot S, Monia BP, McKay RA, Rajala MW, Scherer PE, Rossetti L (2004) Role of resistin in diet-induced hepatic insulin resistance. J Clin Invest 114:232–239

Napoli I, Neumann H (2009) Microglial clearance function in health and disease. Neuroscience 158:1030–1038

Nawrocki AR, Rajala MW, Tomas E, Pajvani UB, Saha AK, Trumbauer ME, Pang Z, Chen AS, Ruderman NB, Chen H, Rossetti L, Scherer PE (2006) Mice lacking adiponectin show decreased hepatic insulin sensitivity and reduced responsiveness to peroxisome proliferator-activated receptor gamma agonists. J Biol Chem 281:2654–2660

Ninomiya JK, L'Italien G, Criqui MH, Whyte JL, Gamst A, Chen RS (2004) Association of the metabolic syndrome with history of myocardial infarction and stroke in the third National Health and Nutrition Examination Survey. Circulation 109:42–46

Nishimura M, Izumiya Y, Higuchi A, Shibata R, Qiu J, Kudo C, Shin HK, Moskowitz MA, Ouchi N (2008) Adiponectin prevents cerebral ischemic injury through endothelial nitric oxide synthase dependent mechanisms. Circulation 117:216–223

Niswender KD, Morton GJ, Stearns WH, Rhodes CJ, Myers MG Jr, Schwartz MW (2001) Intracellular signalling. Key enzyme in leptin-induced anorexia. Nature 413:794–795

Niswender KD, Morrison CD, Clegg DJ, Olson R, Baskin DG, Myers MG Jr, Seeley RJ, Schwartz MW (2003) Insulin activation of phosphatidylinositol 3-kinase in the hypothalamic arcuate nucleus: a key mediator of insulin-induced anorexia. Diabetes 52:227–231

Niswender KD, Baskin DG, Schwartz MW (2004) Insulin and its evolving partnership with leptin in the hypothalamic control of energy homeostasis. Trends Endocrinol Metab 15:362–369

Norata GD, Ongari M, Garlaschelli K, Raselli S, Grigore L, Catapano AL (2007) Plasma resistin levels correlate with the determinants of metabolic syndrome. Eur J Endocrinol 156:279–284

O'Malley D, MacDonald N, Mizielinska S, Connolly CN, Irving AJ, Harvey J (2007) Leptin promotes rapid dynamic changes in hippocampal dendritic morphology. Mol Cell Neurosci 35:559–572

O'Rourke L, Gronning LM, Yeaman SJ, Shepherd PR (2002) Glucose-dependent regulation of cholesterol ester metabolism in macrophages by insulin and leptin. J Biol Chem 277:42557–42562

Okamoto Y (2011) Adiponectin provides cardiovascular protection in metabolic syndrome. Cardiol Res Pract 2011:313179

Okamoto Y, Kihara S, Ouchi N, Nishida M, Arita Y, Kumada M, Ohashi K, Sakai N, Shimomura I, Kobayashi H, Terasaka N, Inaba T, Funahashi T, Matsuzawa Y (2002) Adiponectin reduces atherosclerosis in apolipoprotein E-deficient mice. Circulation 106:2767–2770

Otero M, Lago R, Gomez R, Lago F, Gomez-Reino JJ, Gualillo O (2006) Leptin: a metabolic hormone that functions like a proinflammatory adipokine. Drug News Perspect 19:21–26

Ouchi N, Kihara S, Arita Y, Okamoto Y, Maeda K, Kuriyama H, Hotta K, Nishida M, Takahashi M, Muraguchi M, Ohmoto Y, Nakamura T, Yamashita S, Funahashi T, Matsuzawa Y (2000) Adiponectin, an adipocyte-derived plasma protein, inhibits endothelial NF-κB signaling through a cAMP-dependent pathway. Circulation 102:1296–1301

Ouedraogo R, Gong Y, Berzins B, Wu X, Mahadev K, Hough K, Chan L, Goldstein BJ, Scalia R (2007) Adiponectin deficiency increases leukocyte-endothelium interactions via upregulation of endothelial cell adhesion molecules in vivo. J Clin Invest 117:1718–1726

Ouyang YB, Tan Y, Comb M, Liu CL, Martone ME, Siesjo BK, Hu BR (1999) Survival- and death-promoting events after transient cerebral ischemia: phosphorylation of Akt, release of cytochrome C and Activation of caspase-like proteases. J Cereb Blood Flow Metab 19:1126–1135

Patel L, Buckels AC, Kinghorn IJ et al (2003) Resistin is expressed in human macrophages and directly regulated by PPAR gamma activators. Biochem Biophys Res Commun 300:472–476

Phillis JW, Horrocks LA, Farooqui AA (2006) Cyclooxygenases, lipoxygenases, and epoxygenases in CNS: their role and involvement in neurological disorders. Brain Res Rev 52:201–243
Pini M, Gove ME, Senello JA, van Baal JW, Chan L, Fantuzzi G (2008) Role and regulation of adipokines during zymosan-induced peritoneal inflammation in mice. Endocrinology 149:4080–4085
Pugh CR, Nguyen KT, Gonyea JL, Fleshner M, Wakins LR, Maier SF, Rudy JW (1999) Role of interleukin-1 beta in impairment of contextual fear conditioning caused by social isolation. Behav Brain Res 106:109–118
Putaala J, Metso AJ, Metso TM, Konkola N, Kraemer Y, Haapaniemi E, Kaste M, Tatlisumak T (2009) Analysis of 1008 consecutive patients aged 15 to 49 with first-ever ischemic stroke: the Helsinki young stroke registry. Stroke 40:1195–1203
Qi Y, Takahashi N, Hileman SM, Patel HR, Berg AH, Pajvani UB, Scherer PE, Ahima RS (2004) Adiponectin acts in the brain to decrease body weight. Nat Med 10:524–529
Qi Y, Nie Z, Lee YS, Singhal NS, Scherer PE, Lazar MA, Ahima RS (2006) Loss of resistin improves glucose homeostasis in leptin deficiency. Diabetes 55:3083–3090
Qi L, Zhang C, van Dam RM, Hu FB (2007) Interleukin-6 genetic variability and adiposity: associations in two prospective cohorts and systematic review in 26,944 individuals. J Clin Endocrinol Metab 92:3618–3625
Querfurth HW, LaFerla FM (2010) Alzheimer's disease. N Engl J Med 362:329–344
Quinn TJ, Lees KR (2009) Hyperglycaemia in acute stroke—to treat or not to treat. Cerebrovasc Dis 27(Suppl 1):148–155
Quintanilla RA, Orellana DI, Gonzalez-Billáult C, Maccioni RB (2004) Interleukin-6 induces Alzheimer-type phosphorylation of tau protein by deregulating the cdk5/p35 pathway. Exp Cell Res 295:245–257
Quyyumi AA (1998) Endothelial function in health and disease: new insights into the genesis of cardiovascular disease. Am J Med 105:32S–39S
Rabe K, Lehrke M, Parhofer KG, Broedl UC (2008) Adipokines and insulin resistance. Mol Med 14:741–751
Rains JL, Jain SK (2011) Oxidative stress, insulin signaling, and diabetes. Free Rad Biol Med 50:567–575
Rajala MW, Scherer PE (2003) Minireview: the adipocyte–at the crossroads of energy homeostasis, inflammation, and atherosclerosis. Endocrinology 144:3765–3773
Reaven G, Abbasi F, McLaughlin T (2004) Obesity, insulin resistance, and cardiovascular disease. Recent Progress Hormone Res 59:207–223
Rizos E, Mikhailidis DP (2001) Are high density lipoprotein (HDL) and triglyceride levels relevant in stroke prevention? Cardiovasc Res 52:199–207
Roberts CK, Sindhu KK (2009) Oxidative stress and metabolic syndrome. Life Sci 84:705–712
Roger VL, Go AS, Lloyd-Jones DM, Adams RJ, Berry JD, Brown TM, Carnethon MR, Dai S, de Simone G, Ford ES, Fox CS, Fullerton HJ, Gillespie C, Greenlund KJ, Hailpern SM, Heit JA, Ho PM, Howard VJ, Kissela BM, Kittner SJ, Lackland DT, Lichtman JH, Lisabeth LD, Makuc DM, Marcus GM, Marelli A, Matchar DB, McDermott MM, Meigs JB, Moy CS, Mozaffarian D, Mussolino ME, Nichol G, Paynter NP, Rosamond WD, Sorlie PD, Stafford RS, Turan TN, Turner MB, Wong ND, Wylie-Rosett J, American Heart Association Statistics Committee and Stroke Statistics Subcommittee (2011) Heart disease and stroke statistics–2011 update: a report from the American Heart Association. Circulation 123:e18–e209
Rojas JI, Zurru MC, Romano M, Patrucco L, Cristiano E (2007) Acute ischemic stroke and transient ischemic attack in the very old—risk factor profile and stroke subtype between patients older than 80 years and patients aged less than 80 years. Eur J Neurol 14(8):895–899
Rondinone CM (2006) Adipocyte-derived hormones, cytokines, and mediators. Endocrine 29:81–90
Rothwell PM (2005) Prevention of stroke in patients with diabetes mellitus and the metabolic syndrome. Cerebrovasc Dis 70(Suppl 1):24–34
Rothwell NJ, Relton JK (1993) Involvement of cytokines in acute neurodegeneration in the CNS. Neurosci Biobehav Rev 17:217–227

Sáinz N, Rodríguez A, Catalán V, Becerril S, Ramírez B, Lancha A, Burgos-Ramos E, Gómez-Ambrosi J, Frühbeck G (2012) Leptin reduces the expression and increases the phosphorylation of the negative regulators of GLUT4 traffic TBC1D1 and TBC1D4 in muscle of ob/ob mice. PLoS One 7:e29389

Salminen A, Ojala J, Kaarniranta K, Haapasalo A, Hiltunen M, Soininen H (2011) Astrocytes in the aging brain express characteristics of senescence-associated secretory phenotype. Eur J Neurosci 34:3–11

Saltiel AR, Kahn CR (2001) Insulin signalling and the regulation of glucose and lipid metabolism. Nature 414:799–806

Sandeep S, Velmurugan K, Deepa R, Mohan V (2007) Serum visfatin in relation to visceral fat, obesity, and type 2 diabetes mellitus in Asian Indians. Metabolism 56:565–570

Savitz SL, Malhotra S, Gupta G, Rosenbaum DM (2003) Cell transplants offer promise for stroke recovery. J Cardiovasc Nurs 18:57–61

Savitz SI, Dinsmore JH, Wechsler LR, Rosenbaum DM, Caplan LR (2004) Cell therapy for stroke. NeuroRx 1:406–414

Saxena NK, Titus MA, Ding X, Floyd J, Srinivasan S, Sitaraman SV, Anania FA (2004) Leptin as a novel profibrogenic cytokine in hepatic stellate cells: mitogenesis and inhibition of apoptosis mediated by extracellular regulated kinase (Erk) and Akt phosphorylation. FASEB J 18:1612–1614

Schmidt MI, Duncan BB (2003) Diabesity: an inflammatory metabolic condition. Clin Chem Lab Med 41:1120–1130

Selby JV, Friedman GD, Quesenberry CPJ (1990) Precursors of essential hypertension: pulmonary function, heart rate, uric acid, serum cholesterol, and other serum chemistries. Am J Epidemiol 131:1017–1027

Shanley LJ, Irving AJ, Harvey J (2001) Leptin enhances NMDA receptor function and modulates hippocampal synaptic plasticity. J Neurosci 21:RC186

Shek EW, Brands MW, Hall JE (1998) Chronic leptin infusion increases arterial pressure. Hypertension 31:409–414

Shi H, Kokoeva MV, Inouye K, Tzameli I, Yin H, Flier JS (2006) TLR4 links innate immunity and fatty acid-induced insulin resistance. J Clin Invest 116:3015–3025

Shintani F, Nakaki T, Kanba S, Sato K, Yagi G, Shiozawa M, Aiso S, Kato R, Asai M (1995) Involvement of interleukin-1 in immobilization stress-induced increase in plasma adrenocorticotropic hormone and in release of hypothalamic monoamines in the rat. J Neurosci 15:1961–1970

Shpilman M, Niv-Spector L, Katz M, Varol C, Solomon G, Ayalon-Soffer M, Boder E, Halpern Z, Elinav E, Gertler A (2011) Development and characterization of high affinity leptins and leptin antagonists. J Biol Chem 286:4429–4442

Shulman GI (2000) Cellular mechanisms of insulin resistance. J Clin Invest 106:171–176

Skurk C, Walsh K (2004) Death receptor induced apoptosis, a new mechanism of homocysteine-mediated endothelial cell cytotoxicity. Hypertension 43:1168–1170

Smith C, Essop MF (2009) Gender differences in metabolic risk factor prevalence in a South African student population. Cardiovasc J Afr 20:178–182

Song XM, Fiedler M, Galuska D, Ryder JW, Fernstrom M, Chibalin AV, Wallberg-Henriksson H, Zierath JR (2002) 5-Aminoimidazole-4-carboxamide ribonucleoside treatment improves glucose homeostasis in insulin-resistant diabetic (ob/ob) mice. Diabetologia 45:56–65

Soodini GR, Hamdy O (2004) Adiponectin and leptin in relation to insulin sensitivity. Metab Syndrome Related Disord 2:114–123

Stanimirovic DB, Friedman A (2012) Pathophysiology of the neurovascular unit: disease cause or consequence? J Cereb Blood Flow Metab. doi: 10.1038/jcbfm.2012.25. [Epub ahead of print]

Steensberg A, Fischer CP, Keller C, Møller K, Pedersen BK (2003) IL-6 enhances plasma IL-1ra, IL-10, and cortisol in humans. Am J Physiol 285(2):E433–E437

Stephens JW, Hurel SJ, Cooper JA, Acharya J, Miller GJ, Humphries SE (2004) A common functional variant in the interleukin-6 gene is associated with increased body mass index in subjects with type 2 diabetes mellitus. Mol Genet Metab 82:180–186

Stephens JW, Hurel SJ, Lowe GD, Rumley A, Humphries SE (2007) Association between plasma IL-6, the IL6 -174G>C gene variant and the metabolic syndrome in type 2 diabetes mellitus. Mol Genet Metab 90:422–428
Steppan CM, Bailey ST, Bhat S, Brown EJ, Banerjee RR, Wright CM, Patel HR, Ahima RS, Lazar MA (2001) The hormone resistin links obesity to diabetes. Nature 409:307–312
Tailor AM, Peeters PH, Norat T, Vineis P, Bomaguera D (2010) An update on the prevalence of the metabolic syndrome in children and adolescents. Int J Pediatr Obes 5:202–213
Tanashian MM, Ionova VG, Orlov SV, Omel'chenko NG, Shabalina AA, Kostyreva MV (2010) Chronic cerebrovascular diseases, metabolic syndrome and the status of hemorrheological and hemostatic systems. Ter Arkh 82:19–24
Taniguchi CM, Emanuelli B, Kahn CR (2006) Critical nodes in signaling pathways: insights into insulin action. Nat Rev Mol Cell Biol 7:85–96
Tartaglia LA (1997) The leptin receptor. J Biol Chem 272:6093–6096
Terao Y, Ohta H, Oda A, Nakagaito Y, Kiyota Y, Shintani Y (2009) Macrophage inflammatory protein-3alpha plays a key role in the inflammatory cascade in rat focal cerebral ischemia. Neurosci Res 64:75–82
Tewsbury DA, Frome WL, Dumas ML (1978) Characterization of human angiotensinogen. J Biol Chem 253:3817–3820
Tovar S, Nogueiras R, Tung LY, Castaneda TR, Vazquez MJ, Morris A, Williams LM, Dickson SL, Dieguez C (2005) Central administration of resistin promotes short-term satiety in rats. Eur J Endocrinol 153:R1–R5
Trayhurn P, Beattie JH (2001) Physiological role of adipose tissue: white adipose tissue as an endocrine and secretory organ. Proc Nutr Soc 60:329–339
Troen AM (2005) The central nervous system in animal models of hyperhomocysteinemia. Prog Neuro-Psychopharmacol Biol Psychiatry 29:1140–1151
Tsunoda S, Kamada H, Yamamoto Y, Ishikawa T, Matsui J, Koizumi K, Kaneda Y, Tsutsumi Y, Ohsugi Y, Hirano T, Mayumi T (2000) Molecular design of polyvinylpyrrolidone-conjugated interleukin-6 for enhancement of in vivo thrombopoietic activity in mice. J Control Release 68:335–341
Tu YF, Tsai YS, Wang LW, Wu HC, Huang CC, Ho CJ (2011) Overweight worsens apoptosis, neuroinflammation and blood–brain barrier damage after hypoxic ischemia in neonatal brain through JNK hyperactivation. J Neuroinflammation 8:40
Tureyen K, Bowen K, Liang J, Dempsey RJ, Vemuganti R (2011) Exacerbated brain damage, edema and inflammation in type-2 diabetic mice subjected to focal ischemia. J Neurochem 116:499–507
Tuttolomondo A, Di Raimondo D, di Sciacca R, Pinto A, Licata G (2008) Inflammatory cytokines in acute ischemic stroke. Curr Pharm Des 14:3574–3589
Uchino K, Lin R, Zaidi SF, Kuwabara H, Sashin D, Bircher N, Chang YF, Hammer MD, Reddy V, Jovin TG, Vora N, Jumaa M, Massaro L, Billigen J, Boada F, Yonas H, Nemoto EM (2010) Increased cerebral oxygen metabolism and ischemic stress in subjects with metabolic syndrome-associated risk factors: preliminary observations. Transl Stroke Res 1:178–183
Unger RH (2003) Lipid overload and overflow: metabolic trauma and the metabolic syndrome. Trends Endocrinol Metab 14:398–403
Utsumi T, Hung MC, Klostergaard J (1992) The role of amino functions in recombinant human tumor necrosis factor in expression of biological activity. Mol Immunol 29:77–81
Uysal KT, Wiesbrock SM, Marino MW, Hotamisligil GS (1997) Protection from obesity-induced insulin resistance in mice lacking TNF-α function. Nature 389:610–614
Vaisse C, Halaas JL, Horvath CM, Darnell JE Jr, Stoffel M, Friedman JM (1996) Leptin activation of Stat3 in the hypothalamus of wild-type and ob/ob mice but not db/db mice. Nat Genet 14:95–97
van Harten B, de Leeuw FE, Weinstein HC, Scheltens P, Biessels GJ (2006) Brain imaging in patients with diabetes: a systematic review. Diabetes Care 29:2539–2548

Vandivier RW, Ogden CA, Fadok VA, Hoffmann PR, Brown KK, Botto M, Walport MJ, Fisher JH, Henson PM, Greene KE (2002) Role of surfactant proteins A, D and C1q in the clearance of apoptotic cells in vivo and in vitro: calreticulin and CD91 as a common collectin receptor complex. J Immunol 169:147–152

Villanueva EC, Myers MG Jr (2008) Leptin receptor signaling and the regulation of mammalian physiology. Int J Obes 32:S8–S10

Waki H, Tontonoz P (2007) Endocrine functions of adipose tissue. Ann Rev Path 2:31–56

Walker CD, Long H, Williams S, Richard D (2007) Long-lasting effects of elevated neonatal leptin on rat hippocampal function, synaptic proteins and NMDA receptor subunits. J Neurosci Res 85:816–828

Wen YD, Zhang HL, Oin ZH (2006) Inflammatory mechanism in ischemic neuronal injury. Neurosci Bull 22:171–182

Wilson PWF, Garrison RJ, Abbott RD, Castelli WB (1983) Factors associated with lipoprotein cholesterol levels: the Framingham Study. Arteriosclerosis 3:273–281

Wissing D, Mouritzen H, Egeblad M, Poirier GG, Jäättelä M (1997) Involvement of caspase-dependent activation of cytosolic phospholipase $A_2$ in tumor necrosis factor-induced apoptosis. Proc Natl Acad Sci USA 94:5073–5077

Wootton-Gorges SL, Glaser NS (2007) Imaging of the brain in children with type I diabetes mellitus. Pediatr Radiol 37:863–869

Yamaguchi N, Arueta JG, Mashuiro Y, Kagishita M, Nonaka K, Saito T, Hanazawa S, Yamashita Y (2005) Adiponectin inhibits Toll-like receptor family-induced signaling. FEBS Lett 579:6821–6826

Yamauchi T, Nio Y, Maki T, Kobayashi M, Takazawa T, Iwabu M, Okada-Iwabu M, Kawamoto S, Kubota N, Kubota T, Ito Y, Kamon J, Tsuchida A, Kumagai K, Kozono H, Hada Y, Ogata H, Tokuyama K, Tsunoda M, Ide T, Murakami K, Awazawa M, Takamoto I, Froguel P, Hara K, Tobe K, Nagai R, Ueki K, Kadowaki T (2007) Targeted disruption of AdipoR1 and AdipoR2 causes abrogation of adiponectin binding and metabolic actions. Nat Med 13:332–339

Yang Z, Ming XF (2006) Recent advances in understanding endothelial dysfunction in atherosclerosis. Clin Med Res 4:53–65

Yaspelkis BB, Singh MK, Krisan AD, Collins DE, Kwong CC, Bernard JR, Crain AM (2004) Chronic leptin treatment enhances insulin-stimulated glucose disposal in skeletal muscle of high-fat fed rodents. Life Sci 74:1801–1816

Ye J (2009) Emerging role of adipose tissue hypoxia in obesity and insulin resistance. Int J Obes (Lond) 33:54–66

Yuan M, Konstantopoulos N, Lee J, Hansen L, Li ZW, Karin M, Shoelson SE (2001) Reversal of obesity- and diet-induced insulin resistance with salicylates or targeted disruption of IKKβ. Science 293:1673–1677

Zabolotny JM, Bence-Hanulec KK, Stricker-Krongrad A, Haj F, Wang Y, Minokoshi Y, Kim YB, Elmquist JK, Tartaglia LA, Kahn BB, Neel BG (2002) PTP1B regulates leptin signal transduction in vivo. Dev Cell 2:489–495

Zavaroni I, Mazza S, Fantuzzi M, Dall'Aglio E, Bonora E, Delsignore R, Passeri M, Reaven GM (1993) Changes in insulin and lipid metabolism in males with asymptomatic hyperuricemia. Int J Med 234:25–30

Zhang F, Chen J (2008) Leptin protects hippocampal CA1 neurons against ischemic injury. J Neurochem 107:578–587

Zhang Y, Proenca R, Maffei M, Barone M, Leopold L, Friedman JM (1994) Positional cloning of the mouse obese gene and its human homologue. Nature 372:425–432

Zhang D, Jiang SL, Rzewnicki D, Samols D, Kushner I (1995) The effect of interleukin-1 on C-reactive protein expression in Hep3B cells is exerted at the transcriptional level. Biochem J 310:143–148

Zhang F, Wang S, Signore AP, Chen J (2007) Neuroprotective effects of leptin against ischemic injury induced by oxygen-glucose deprivation and transient cerebral ischemia. Stroke 38:2329–2336

Zhang X, Zhang G, Zhang H, Karin M, Bai H, Cai D (2008) Hypothalamic IKKβ/NF-κB and ER stress link overnutrition to energy imbalance and obesity. Cell 135:61–73

Zhao H, Sapolsky RM, Steinberg GK (2006) Phosphoinositide-3-kinase/akt survival signal pathways are implicated in neuronal survival after stroke. Mol Neurobiol 34:249–270

Zhou G, Myers R, Li Y, Chen Y, Shen X, Fenyk-Melody J, Wu M, Ventre J, Doebber T, Fujii N, Musi N, Hirshman MF, Goodyear LJ, Moller DE (2001) Role of AMP-activated protein kinase in mechanism of metformin action. J Clin Invest 108:1167–1174

Zick Y (2001) Insulin resistance: a phosphorylation-based uncoupling of insulin signaling. Trends Cell Biol 11:437–441

Zimmet P, Boyko EJ, Collier GR, de Courten M (1999) Etiology of the metabolic syndrome: potential role of insulin resistance, leptin resistance, and other players. Ann N Y Acad Sci 892:25–44

# Chapter 8
# Metabolic Syndrome as a Risk Factor for Alzheimer Disease

## 8.1 Introduction

Alzheimer disease (AD) is a highly dynamic age-related neurodegenerative disease, which affects approximately 30 million people worldwide with numbers expected to increase substantially over the next several decades. It is estimated that the worldwide costs for AD and dementia care are $315.4 billion annually (USA; Alzheimer's Association). According to Alzheimer's Association, 5.4 million Americans of all ages suffer from AD in 2011. This figure includes 5.2 million people aged 65 and older and 200,000 individuals under age 65 who have younger-onset AD (Hebert et al. 2003; Alzheimer's Association 2010). Nearly half of people age 85 and older suffer from AD. According to Alzheimer's Association, every 68 s, someone in America develops AD. AD is the sixth leading cause of death in the USA (Alzheimer's Association 2012). In 2010, the annual US health-care cost for the treatment of AD reached about US $144 billion (Alzheimer's Association 2010). Similarly, total health care in 2010, including long-term care payments for AD and other dementia care, was expected to reach US $172 billion (Alzheimer's Association 2010). It is predicted that by 2050, the number of people with AD and other forms of dementia in the USA will be tripled. This increase in AD can be largely attributed to increasing longevity, consumption of processed food, physical inactivity, and the aging of the "baby boomer generation." Given this increase in the population and cost of treatment, it is important to set achievable goals in the near future. Sporadic AD, which has no clear pattern of genetic transmission, accounts of over 90 % of the cases, whereas familial (heritable) forms of AD account for 5–10 % of all cases.

Neuropathologically, AD is characterized by neurodegeneration in the nucleus basalis, hippocampus, and cerebral cortex along the combined presence of two lesions in the brain—extracellular amyloid-β (Aβ) plaques (senile plaques, SP) and intracellular neurofibrillary tangles (NFTs). The extracellular deposits contain aggregated Aβ peptides (Masters et al. 1985), while intraneuronal NFTs are aggregates of hyperphosphorylated forms of the neurofilament-associated protein tau (Querfurth and LaFerla 2010). Clinically, AD is characterized by a gradual decline

A.A. Farooqui, *Metabolic Syndrome: An Important Risk Factor for Stroke, Alzheimer Disease, and Depression*, DOI 10.1007/978-1-4614-7318-3_8,

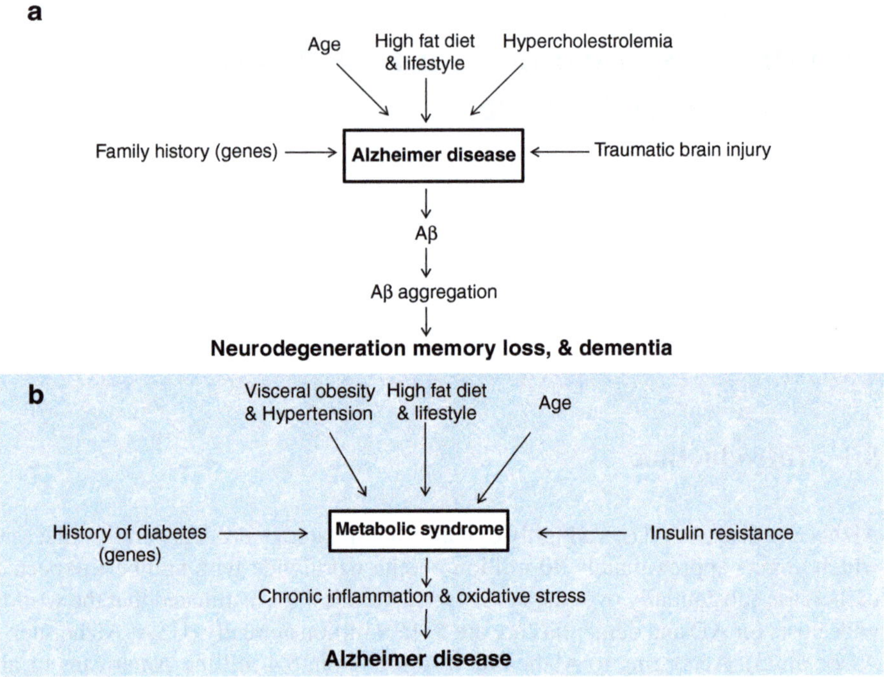


**Fig. 8.1** Risk factors for Alzheimer disease (**a**) and metabolic syndrome (**b**)

in cognition and changes in behavior and personality including difficulty in reasoning, disorientation, and language problems. Neurochemically, AD is characterized by neuroinflammation, oxidative stress, along with enrichment in metabolism of neural membrane phospholipids, sphingolipids, and cholesterol due to the activation of phospholipases $A_2$ ($PLA_2$), sphingomyelinases (SMase), and cholesterol hydroxylases (CYP46). The activation of these enzymes not only increases levels of phospholipid-, sphingolipid, and cholesterol-derived lipid mediators, induces mitochondrial dysfunction, and activates caspases, but also stimulates protein phosphorylation and loss of synapses along with progressive impairment of memory and severe dementia (Barnham et al. 2004; Farooqui 2010; Farooqui et al. 2010). It is reported that humans carry neurodegeneration-linked mutations and develop neurodegenerative diseases during their fifth decade, whereas sporadic neurodegenerative maladies appear during the seventh decade or later (Amaducci and Tesco 1994). Thus, advancing age is the most prevalent risk factor for sporadic AD and the accumulation of Aβ in the brain (Price et al. 2009) possibly because of the decline of cellular protein quality control processes (Powers et al. 2009).

Other risk factors for AD include environmental factors, incidence of head injury, alcohol consumption, smoking, sedentary lifestyle, elevated plasma levels of homocysteine, overeating, obesity, severe adverse stress, and hypercholesterolemia (Fig. 8.1) (Mejia et al. 2003; Farooqui 2010, 2012). Toxic effects of elevated homocysteine and its spontaneous oxidation product, homocysteic acid, are based on their

**Table 8.1** Effect of homocysteine on the brain

| Levels of homocysteine | Effect | Reference |
|---|---|---|
| Elevation | Stimulation of NMDA receptor | Lipton et al. (1997) |
| Elevation | Apoptosis | Kruman et al. (2000) |
| Elevation | Behavioral deficit | Troen and Rosenberg (2005) |
| Elevation | Synaptic plasticity | Garcia and Zanibbi (2004) |

ability to activate NMDA receptors not only through glutamate binding site (as an agonist) but also through partial interactions with glycine-binding site (as an antagonist). This leads to increase in intracellular levels of intracellular calcium and reactive oxygen species (Boldyrev and Johnson 2007). Homocysteine also induces long-lasting behavioral deficits and causes changes in synaptic plasticity (Table 8.1). In addition, vascular effects of homocysteine include increase in proliferation of smooth muscle cells and increase in platelet aggregation (Garcia and Zanibbi 2004). Elevated levels of homocysteine induce atherogenesis by directly increasing formation of reactive oxygen species and by promoting oxidation of low-density lipoprotein. In contrast, hypercholesterolemia causes stimulation of amyloidogenic pathway for Aβ production (Grimm et al. 2007). It should be noted that elevated levels of homocysteine are usually considered as indirect indicator of Vitamin B deficiency (Folstein et al. 2007). Hyperhomocysteinemia has also been associated with depression (Bottiglieri et al. 2000). In spite of above results, more studies are warranted to explore whether there is a causal relationship among hyperhomocysteinemia, AD, and late-life depression. Positive lifestyle factors such as an active and social life, healthy diet, and physical and mental exercise decrease the risk of developing AD and the progression of the symptoms and neuropathological hallmarks after disease onset (Pardon and Rattray 2008). It is proposed that positive lifestyle factors mediate their effects by producing changes in neuroendocrine function and through the induction of mild stress response leading to beneficial effects of positive lifestyle factors on brain function and non-pathological aging.

The pathogenesis of AD is tightly linked to Aβ aggregation and deposition, hyperphosphorylation of tau protein, oxidative stress, accumulation of iron, and loss of synapses, but it remains unclear as to how these factors result in neurodegeneration in AD (Fig. 8.2). It is hypothesized that abnormal amyloid precursor protein (APP) processing, generation, and aggregation of β-amyloid (Aβ) along with mitochondrial dysfunction, accumulation of iron, and alterations in phospholipid-, sphingolipid-, and cholesterol-derived lipid mediators may facilitate reduction in energy production, loss of synapse, alterations in cognitive function, and neurodegeneration in AD (Masliah et al. 2001; Farooqui 2009, 2011; Farooqui et al. 2010; Grünblatt et al. 2011). Thus, levels of phospholipid-derived lipid mediators, such as (eicosanoids, 4-hydroxynonenal (4-HNE), and isoprostanes (IsoPs)), sphingolipid-derived lipid mediator (ceramide), and cholesterol-derived lipid mediator (hydroxycholesterols) are noticeably increased in AD (Fig. 8.3) (Farooqui 2012). This increase in ceramide is accompanied by elevation in acid sphingomyelinase (ASMase), acid ceramidase, and $PLA_2$ expression (Han et al. 2002; Cutler et al. 2004; He et al. 2010; Farooqui 2010).

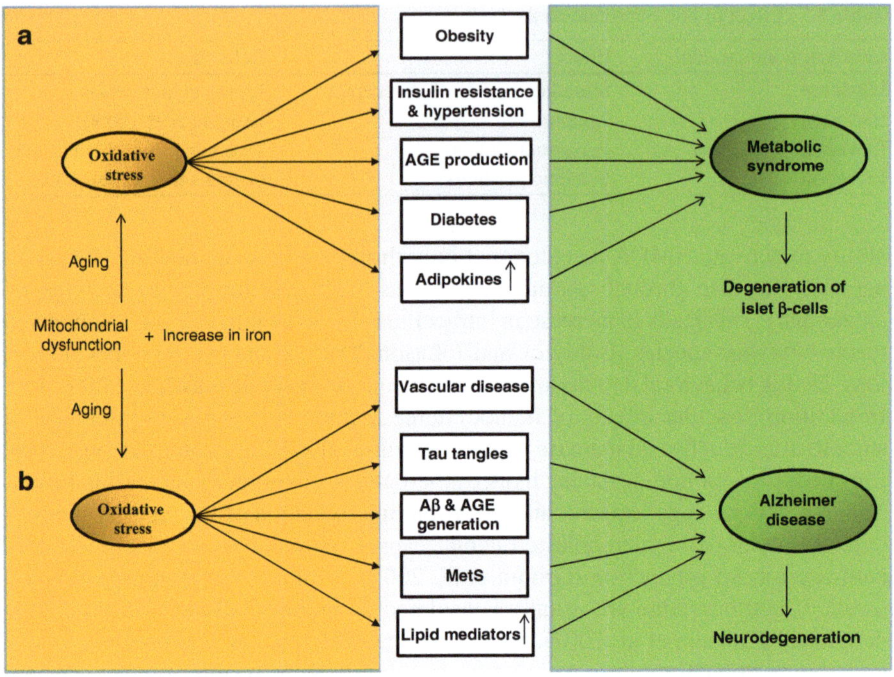


**Fig. 8.2** Effect of oxidative stress on metabolic syndrome (**a**) and Alzheimer disease (**b**)

Importantly, significant correlations are observed between the brain $PLA_2$ and SMase activity and the levels of Aβ peptide and hyperphosphorylated τ protein in AD patients (Kanfer et al. 1998; Farooqui et al. 2003, 2006; Farooqui 2012). Treatment of neural cell cultures with Aβ oligomers not only activates SMase and $PLA_2$ activity, but also increases ceramide, 4-HNE, and IsoP levels and induces apoptotic cell death (He et al. 2010; Farooqui 2012).

Vascular risk factors also contribute to the pathogenesis of AD (Zlokovic 2005; Humpel and Marksteiner 2005). Thus, chronic endothelial dysfunction plays a pivotal role in the pathogenesis of AD. Endothelial cell dysfunction not only contributes to the breakdown of the blood–brain barrier (BBB) and impairs cerebrovascular blood flow but may also promote cerebral amyloid angiopathy, atherosclerosis, or other important secondary-related inflammatory processes (Bell and Zlokovic 2009). In fact, mild chronic cerebrovascular hypoperfusion and hypometabolism caused by above mentioned risk factors may produce decrease in cerebral blood flow, which may lower metabolic rates of glucose, and oxygen utilization. This may be one of the very early events in AD (Iadecola 2004). The reason for brain hypometabolism is unclear but may include defects in glucose transport at the blood–brain barrier, glycolysis, and/or mitochondrial function. Increase in obesity and blood pressure in the middle age may also increase the risk of developing AD by

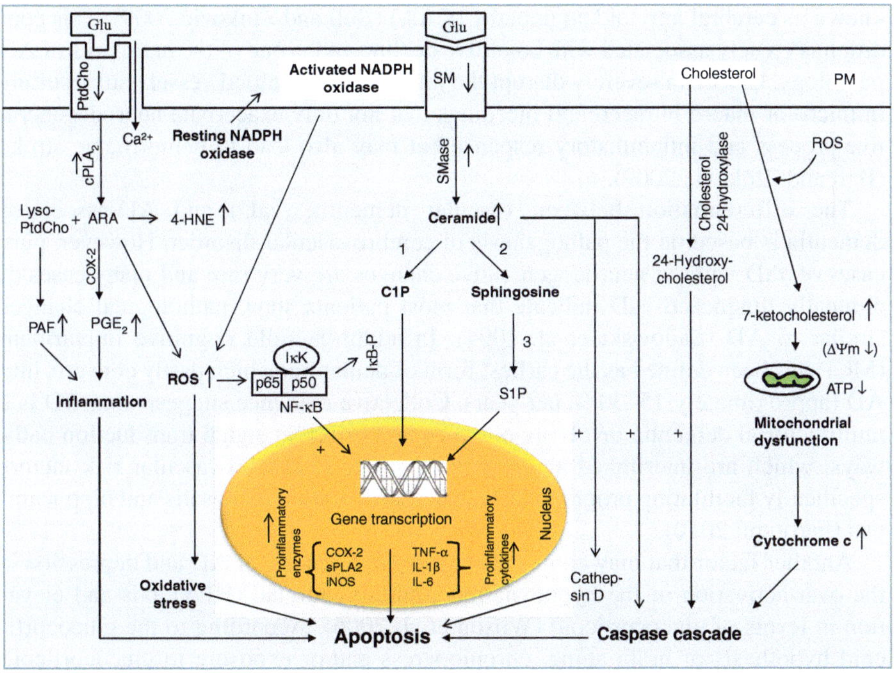


**Fig. 8.3** Hypothetical diagram showing interactions among phospholipid-, sphingolipid-, and cholesterol-derived lipid mediators and contribution of oxidative stress in proinflammatory gene expression. Phosphatidylcholine (PtdCho), sphingomyelin (SM), cytosolic phospholipase $A_2$ ($cPLA_2$), sphingomyelinase (SMase), arachidonic acid (ARA), reactive oxygen species (ROS), cyclooxygenase-2 (COX-2), reactive oxygen species (ROS), ceramide-1-kinase (1), ceramidase (2), sphingosine-1-kinase (3), ceramide-1-phosphate (C1P), sphingosine-1-phosphate (S1P), and plasma membrane (PM). Proinflammatory genes include TNF-α, IL-1β, nitric oxide synthase, cyclooxygenase-2 (COX-2), and secretory phospholipase $A_2$ ($sPLA_2$). *Positive sign* (+) indicates stimulation, *upward arrow* indicates increase, and *downward arrow* indicates decrease in levels of precursors and lipid mediators

decreasing the vascular integrity of the BBB, resulting in protein extravasation into brain tissue (Xu et al. 2011; Kalaria 2010). In turn, protein extravasation may not only cause cellular damage, reduction in synaptic function, and apoptosis but also induce increase in Aβ accumulation, resulting in cognitive impairment and dementia (a progressive decline in the ability to remember, learn, understand, and communicate), which can be distinguished from vascular dementia (vaD) (Deane et al. 2004). As stated above, the decrease in cerebral blood flow negatively affects the synthesis of proteins required for learning and memory and eventually leading to neuritic injury and neuronal death. Impairment in clearance of Aβ from the brain by the cells of the neurovascular unit may lead to its accumulation on blood vessels and in brain parenchyma. The accumulation of Aβ on the cerebral blood vessels is known as cerebral amyloid angiopathy (CAA) (Bell and Zlokovic 2009). This condition is closely associated with cognitive decline and is one of the hallmarks of AD pathology. CAA can severely disrupt the integrity of the blood vessel wall resulting in micro or macro intracerebral bleedings that not only exacerbate neurodegenerative process and inflammatory response but may also lead to hemorrhagic stroke (Bell and Zlokovic 2009).

The differentiation between vascular dementia (vaD) and AD-associated dementia is based on the pathogenesis of cerebrovascular disorder. However, pure cases of vaD without neurodegenerative changes are very rare and many cases of clinically diagnosed vaD indicate that most patients show pathological changes similar to AD (Sadowski et al. 2004). In addition, mild cognitive impairment (MCI) has been defined as the earliest form of dementia, which partly converts into AD (approximately 15–30 % per year). Collective evidence suggests that AD is a multifactorial dementia involving complex and different signal transduction pathways, which are interlinked along with exposure to cerebrovascular risk factors specifically facilitating progressive neuronal loss in nucleus basalis and hippocampus (Farooqui 2010).

Another factor that may contribute to the pathogenesis of AD and depression is the over-activation of the hypothalamic–pituitary–adrenal (HPA) axis and elevation in levels of glucocorticoid (Wilson et al. 2006). According to the glucocorticoid hypothesis of brain aging, chronic stress and/or exposure to glucocorticoid may promote hippocampal aging and even the pathogenesis of AD and depression (Landfield et al. 2007). A basal dysregulation of the HPA axis may mediate abnormal cognitive and pathological responses to adverse environmental stress in AD and depression patients. Even in neurologically normal subjects, excessive adverse stress may contribute to cognitive impairment (Wolf 2009), an effect that is mimicked in experimental rodent models of stress which demonstrate impairments in spatial memory, contextual memory, and object recognition in response to psychosocial or environmental stress (Li et al. 2008). Although it is well known that chronic stress exacerbates Aβ accumulation and memory deficits in AD transgenic (Tg) mice (Green et al. 2006; Lee et al. 2009; Huang et al. 2011), the involvement of a glucocorticoid receptor-dependent pathway remains controversial. Recent studies have also indicated that corticotropin-releasing factor (CRF) promotes the increase and deposition of Aβ in AD (Kang et al. 2007). CRF acts through G protein-coupled receptor type 1 ($CRF_1$) with widespread brain expression to orchestrate the stress response (Bale and Vale 2004). CRF overproduction has also been implicated in depression, anxiety, and posttraumatic stress disorder (Pervanidou and Chrousos 2010; Stengel and Taché 2010). Collective evidence suggests that stress differentially influences the pathogenesis of AD and CRF not only plays an important role in Aβ aggregation and tau pathology, but increased CRF in stress-related psychiatric disorders may predispose animals to AD-like neuropathology. Understanding this emerging link between CRF, AD-related pathology, and stress is of crucial importance in future treatments of not only AD but also neuropsychiatric disorders.

## 8.2 APP Processing, Endoplasmic Reticulum Stress in Alzheimer Disease

Genetic, biochemical, and behavioral studies suggest that APP is a transmembrane glycoprotein with a large extracellular domain and a short cytoplasmic tail. The extracellular domain is composed of two subdomains termed E1 and E2 that are interconnected by a highly acidic domain. APP is closely associated with the pathogenesis of AD. APP is ubiquitously expressed in brain (neurons and glial cells) and peripheral tissues (muscles, epithelial tissue, and circulating cells, such as platelets). APP occurs in three isoforms (APP695, APP751, and APP770), generated from differential splicing of exons 7 and 8 (Sandbrink et al. 1996). Exon 7 is homologous to protease inhibitors of the Kunitz type (KPI domain), while exon 8 is related to the MRC OX-2 antigen in thymocytes (Kitaguchi et al. 1988; Sandbrink et al. 1996). The *APP* gene has been mapped to chromosome 21. APP695 predominantly occurs in neurons (Henriques et al. 2007; Gralle and Ferreira 2007). For the sake of simplicity, I will use APP throughout this monograph. Although physiological roles of APP in normal brain are still unclear, studies on APP transgenic and APP knockout (KO) mouse models indicate that APP is associated with modulation of neurite outgrowth, synaptic activity, regulation of metal homeostasis, synaptic transmission, and synaptic adhesion at the neuromuscular junction (Hoe et al. 2009; Priller et al. 2006; Bellingham et al. 2004; Duce et al. 2010; Yang et al. 2007; Wang et al. 2009).

Two major pathways are involved in APP processing. The amyloidogenic pathway involves the action of β- and γ-secretases to generate Aβ1–40/Aβ1–42 peptide, and the nonamyloidogenic involves the action of α- and γ-secretases. This pathway cleaves APP within the Aβ sequence and releases a soluble secretory amyloid precursor protein (sAPPα) (Fig. 8.4) (Clippingdale et al. 2001; Fahrenholz 2007; Postina 2008; Thinakaran and Koo 2008). Although the monomeric forms of Aβ1–40/Aβ1–42 are not neurotoxic, small Aβ1–40/Aβ1–42 oligomers (soluble) and fibrillar forms of Aβ1–40/Aβ1–42 (insoluble forms) are toxic and are closely associated with neuronal damage in AD (Lambert et al. 1998). In the amyloidogenic pathway, the first cleavage of APP is carried out by β-secretase at 99 amino acids from C-terminus producing a soluble ectodomain fragment (sAPPβ) released into the extracellular space and a CTF of 99 amino acids (C99) retained in the membrane. The C99 fragment begins at residue 1 of the Aβ region. The following cleavage of C99 by γ-secretase leads to the release of the Aβ40/Aβ42 peptide. Recent studies have indicated that extracellular Aβ undergoes phosphorylation by a cell surface-localized or secreted form of protein kinase A (Kumar and Walter 2011). The phosphorylation of serine residue 8 promotes aggregation by stabilization of β-sheet conformation of Aβ and increased formation of oligomeric Aβ aggregates that represent nuclei for fibrillization. Phosphorylated Aβ can be detected not only in the brains of transgenic mice but also in human AD brains. It is suggested that phosphorylation-mediated aggregation of Aβ may be relevant in the pathogenesis of late-onset AD (Kumar and Walter 2011). Aβ1–40/Aβ1–42 oligomer (also known as ADDLs) correlates most strongly with the severity of dementia in AD patients

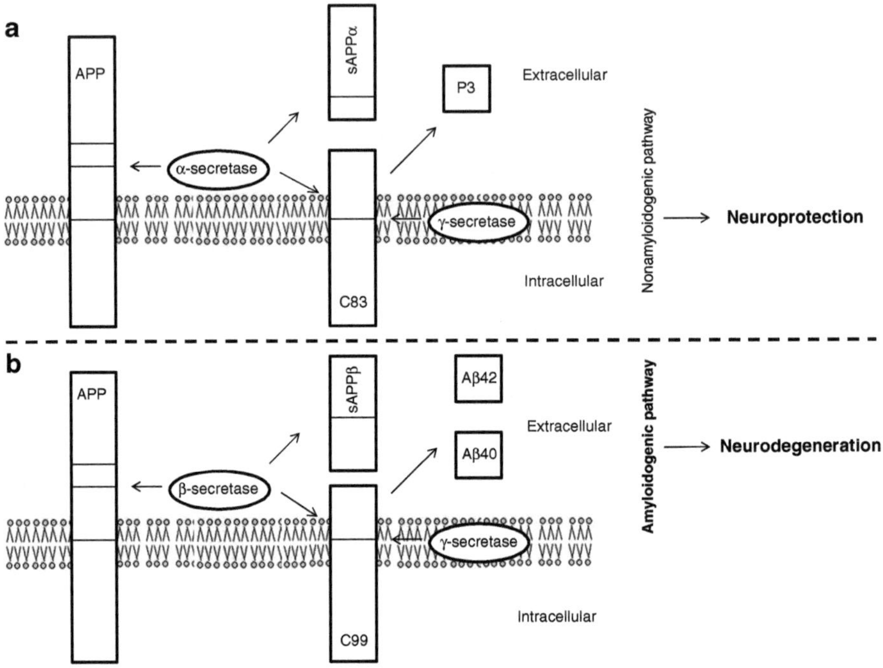


**Fig. 8.4** Nonamyloidogenic (**a**) and amyloidogenic pathways (**b**) for amyloid precursor protein (APP). In nonamyloidogenic pathway, action of α-secretase (ADAM10) leads to the generation of membrane-bound C83 fragment and soluble (sAPPα), which induces neuroprotection. In contrast, cleavage of APP by β-secretase (BACE-1) and γ-secretase (presenilin complex) produces Aβ40 and Aβ42, which are closely associated with the pathogenesis of AD

(McLean et al. 1999). Unlike the insoluble fibrils, ADDLs are diffusible molecules that interact with synapses in hippocampal and entorhinal cortical areas of the brain. ADDLs either create cytotoxic pores or bind to particular membrane targets as highly specific ligands promoting the pathogenesis of AD (Lacor et al. 2004). These interactions of ADDLs with synapses inhibit long-term potentiation (LTP) (Walsh et al. 2002; Selkoe 2008). Interactions between synapses and ADDLs produce AD-like pathology not only through neuronal tau hyperphosphorylation (De Felice et al. 2008) and induction of oxidative stress (De Felice et al. 2007), but also through the synapse deterioration and loss (Roselli et al. 2005). ADDLs also produce down-regulation of plasma membrane insulin receptors (IRs), via a mechanism sensitive to calcium calmodulin-dependent kinase II (CaMKII) and casein kinase II (CK2) inhibition (De Felice et al. 2009). Most significantly, this loss of surface IRs, and ADDL-induced oxidative stress and synaptic spine deterioration, can be completely blocked by insulin. Collective evidence suggests that synapse vulnerability to ADDLs can be mitigated by insulin, supporting the view that diabetes medications may have significant potential to slow or retard AD pathogenesis.

In contrast, APP hydrolysis by α-secretase (ADAM 10) results in production of soluble N-terminal APPα and a 10 kDa C-terminal fragment (α-CTF) that can be further processed by γ-secretase to generate Aβ17–40/Aβ17–42 fragments. The CTF fragment is subsequently cleaved by the γ-secretase with the production of a short fragment called p3. The soluble N-terminal APPα has neuroprotective properties. Sirtuin1 (SIRT1), NAD-dependent deacetylase directly activates transcription of α-secretase, which initiates the activation of notch pathway by cleaving the membrane-bound notch receptor, thus liberating an intracellular domain that activates nuclear genes for neurogenesis (Costa et al. 2005; Hartmann et al. 2001). Studies in neuroblastoma cells show that the suppression of Aβ production by SIRT1 is blocked by normalizing α-secretase levels with specific shRNAs. Based on these observations, it is suggested that the activation of α-secretase by SIRT1 is mechanistically linked to Aβ suppression (Donmez et al. 2010). The molecular mechanism associated with SIRT1-mediated activation of α-secretase is not fully understood. However, it is proposed that deacetylation of the retinoic acid receptor (RARβ) is closely associated with activation of α-secretase (Donmez et al. 2010). Levels of α-secretase (ADAM-10) and levels of APPsα are significantly decreased in the cerebral spinal fluid (CSF) of AD patients compared to controls (Fahrenholz 2007; Postina 2008; O'Brien and Wong 2011).

Compromising α-site cleavage and overwhelming β-site cleavage may result in overproduction of Aβ. Aβ is degraded by either neprilysin or by insulin-degrading enzyme (IDE), which accounts for the homeostasis of Aβ in the brain. The deletion of β-secretase (a novel aspartyl protease) and γ-secretase (a complex of proteins consisting of presenilin1 (PS1) or presenilin2 (PS2), nicastrin (NCT), Aph1, and Pen2) in APP transgenic mice abolishes neuronal production of Aβ and deposition of amyloid plaques (Luo et al. 2001). These observations support the view that changes in activities of β- and γ-secretases are the primary driver for the neurodegeneration and cognitive dysfunction in sporadic AD (Fig. 8.4). Indeed, increase in β-secretase and γ-secretase protein levels and activity has been observed in brains from AD patients (Li et al. 2004). Neuronal β-secretase and γ-secretase levels and activity increase not only with age but also following pathological events such as ischemia and traumatic brain injury along with elevation Aβ levels and an increase in risk of AD (Cole and Vassar 2007). Interestingly, activities of α-, β-, and γ-secretases and APP processing are modulated by lipids. Thus, cholesterol and GM1 have been reported to increase the generation of Aβ (Zha et al. 2004; Wolozin 2004), whereas DHA and sphingomyelin decrease amyloidogenic processing of APP (Grimm et al. 2005, 2011a, b). At low plasma membrane cholesterol concentrations, APP is processed through α-secretase-initiated processing, but as concentration of cholesterol rises in the membrane, APP is processed by β- and γ-secretases resulting in Aβ production (Vassar 2004; Cui et al. 2011). These observations support the view that Aβ metabolism and amyloidogenic pathway are linked with cholesterol homeostasis (Grimm et al. 2007). Collective evidence suggests that cytotoxic effects of Aβ are strongly associated with its ability to interact with membranes of neurons, astrocytes, microglial, and cerebral endothelial cells. Aβ peptides have been reported to directly bind with membrane molecules and alter biophysical membrane properties of membranes (Friedman et al. 2009; Buchsteiner et al. 2010; Eckert et al. 2010). In addition, Aβ also indirectly modulates membrane properties by interacting with membrane receptors and triggering downstream signaling pathways. Moreover, there is evidence that Aβ1–42 oligomers can accelerate the amyloidogenic processing of APP by changing membrane physical properties and interacting with lipid raft-related ganglioside GM-1 (Peters et al. 2009). It is also reported that membrane properties and composition and the presence of metal ions, in turn, play critical roles in Aβ cytotoxicity associated with its conformational changes, aggregation into oligomers and fibrils, and ability to interact with membrane molecules (Lau et al. 2007).

Corticotropin-releasing hormone (CRH), which modulates the stimulation of excitatory neurotransmission, synaptic plasticity, cognition, and other behavioral states, not only induces neuroprotection but also regulates APP processing and tau phosphorylation, via the CRH1 receptor (Bayatti and Behl 2005). These observations support the view that the significant reduction in CRH signaling (Behan et al. 1997) may play a significant role in progression of AD. An imbalance between production and clearance of Aβ aggregative-prone peptides triggers the formation of senile plaques (Querfurth and LaFerla 2010). Even though senile plaques are the most evident AD hallmark, recent reports highlight that Aβ oligomers, because of their potent synaptotoxicity, play a crucial role in AD onset and development (Chiti and Dobson 2006; Walsh and Selkoe 2007). Studies on the effect of *trans* fatty acids indicate that *trans* fatty acids increase amyloidogenic and decrease nonamyloidogenic processing of APP, leading to increase in production of Aβ peptides (Grimm et al. 2011a). Moreover, *trans* fatty acids also promote the oligomerization and aggregation of Aβ peptide initiating the formation of senile plaques. In contrast, docosahexaenoic acid (DHA), an omega-3 fatty acid, reduces amyloidogenic processing by inhibiting β- and γ-secretase activity. However, it has no effect on the expression and protein levels of BACE1 and presenilin1 (Grimm et al. 2011b). In addition, DHA increases protein stability of α-secretase resulting in increase in nonamyloidogenic processing of APP. Collective evidence suggests that DHA directs amyloidogenic processing of APP towards nonamyloidogenic processing by effectively reducing Aβ release (Grimm et al. 2011b). In contrast, gangliosides are among the most potent stimulators of Aβ production. They are enriched in amyloid plaques and they rapidly interact with Aβ (Grimm et al. 2012) indicating that gangliosides are closely associated with the pathology of AD. Alterations in concentration and composition of gangliosides have been reported to occur in the brains of AD patients and in transgenic mouse models of AD (Molander-Melin et al. 2005; Barrier et al. 2007). Ganglioside clusters in neuronal membranes facilitate the formation of amyloid fibrils (Yamamoto et al. 2008). Furthermore, GM1 not only drastically increases Aβ production (Zha et al. 2004) but also binds to Aβ and acts as a seed for Aβ aggregation in amyloid plaques (Wakabayashi et al. 2005; Okada et al. 2007). The underlying mechanisms of above mentioned processes are not fully understood. However, it is recently reported that gangliosides regulate APP processing by modulating the GD3 synthase activity.

It is well known that endoplasmic reticulum (ER) contributes to protein synthesis, calcium signaling, lipid biosynthesis, and protein folding and maturation.

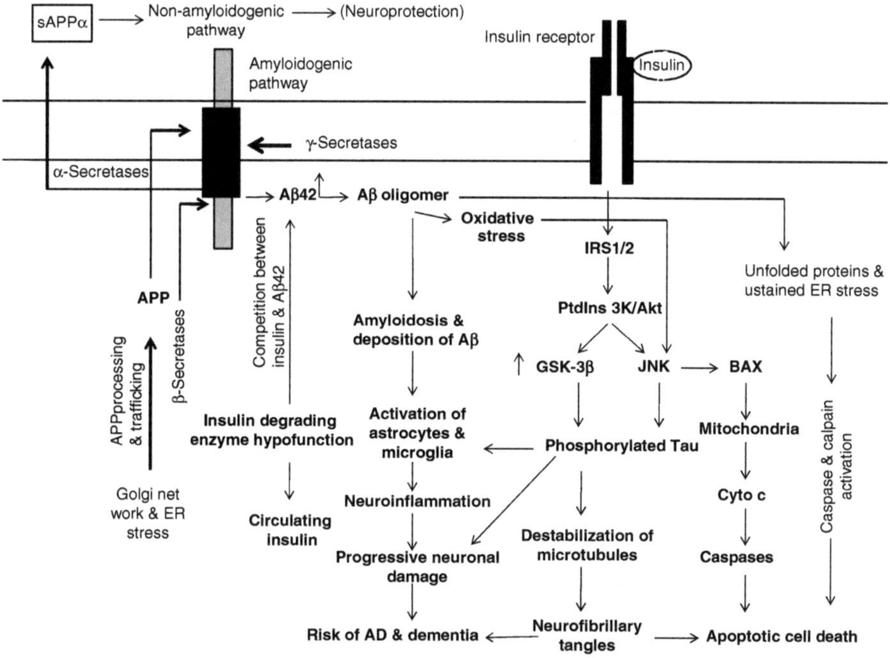


**Fig. 8.5** Altered insulin receptor signaling and increased risk of developing Alzheimer disease. Insulin receptor substrate (IRS), phosphatidylinositol 3-kinase (PtdIns 3K), glycogen synthase kinase 3 (GSK3), amyloid precursor protein (APP), β-amyloid (Aβ), serine/threonine protein kinase (Akt), c-Jun $NH_2$-terminal kinase (JNK), endoplasmic reticulum (ER), cytochrome c (Cyto c), and proapoptotic Bcl-2 family (BAX), and *upward arrow* indicates increase and *downward arrow* indicates decrease

All these processes are driven by glucose metabolism. Perturbations that induce changes in ER homeostasis disrupt protein folding and lead to the accumulation of unfolded proteins and protein aggregates that can be injurious to the cell. ER stress can be provoked by a variety of physiological conditions, such as perturbations in calcium homeostasis, glucose/energy deprivation, redox changes, ischemia, and mutations that impair client protein folding (Fig. 8.5). Mutations in presenilin inhibit activation of endoplasmic transmembrane kinase (IRE1), endoplasmic transmembrane protein (ATF6), and PKR-like ER kinase (PERK), all of which act as signal transducers of ER stress in the ER membrane (Hosoi and Ozawa 2012), thereby making neurons vulnerable to ER stress (Katayama et al. 1999, 2001). Thus, ER stress may also facilitate the onset of AD through the phosphorylation of translation initiation factor eIF2α and elevation in β-secretase activity leading to increase in Aβ production (O'Connor et al. 2008). Collectively, these studies suggest that abnormal APP processing along with increase in ER stress plays a key role in AD pathogenesis as a cause or consequence (Fig. 8.5) (Hoozemans et al. 2009). Statins, the well-known inhibitors of cholesterol synthesis, not only modulate γ-secretase and increase the catabolism of APP but also reduce the risk of AD

(Buxbaum et al. 2002). The proteolytic activity of γ-secretase is stimulated by cerebrosides, anionic glycerophospholipids, and cholesterol, suggesting the involvement of lipid rafts in the modulation of secretases (Kalvodova et al. 2005). As mentioned above, levels of prostaglandins and 4-HNE are also increased in AD (Farooqui 2012) (Fig. 8.3). Detailed investigations using proteomics indicate that 4-HNE attaches itself to many enzymes (ATP synthase, α-enolase, aconitase, aldolase, glutamine synthetase, Mn-superoxide dismutase) and proteins (peroxiredoxin 6, dihydropyrimidinase-related protein-2 and α-tubulin) (Perluigi et al. 2009). These enzymes and proteins are associated with glucose metabolism, maintenance of glutamate levels, antioxidant defense systems, axonal growth, and maintenance of cytoskeleton (Butterfield et al. 2010). 4-HNE also binds to histones and this binding alters the ability of histone to bind DNA. It is proposed that alterations in DNA–histone interactions may contribute to the vulnerability of neurons in AD brain (Drake et al. 2004).

## 8.3 High-Fat Diet and Neurochemical Alterations in AD and Metabolic Syndrome

Lifestyle (high-fat diet and lack of exercise) plays a critical role in the pathogenesis of AD and MetS through the induction of oxidative stress and chronic inflammation in the brain as well as visceral organs (Kalmijn 2000; Grant et al. 2002). Studies on the effect of high-fat diet in the brain and visceral organs indicate that high-fat diet significantly increases ROS generation as well as expression of gp91phox, p22phox, p47phox, and p67phox NADPH oxidase subunits in cerebral cortex (Zhang et al. 2005). In addition, levels of $PGE_2$ and COX-2 expression are also increased in the brain from high-fat-fed animals, supporting the view that excessive intake of high-fat food induces oxidative stress and inflammation along with many signs of MetS. ROS interferes with NF-κB, a transcription factor responsible for regulating the expression of genes involved in cell survival, cell adhesion, inflammation, differentiation, and growth (Fig. 8.3) (O'Neill and Kaltschmidt 1997). It should be noted that the consumption of high-fat diet is a central risk factor for MetS linked to obesity. Adverse effects of high-fat diet on metabolic homeostasis are linked to adipose tissue physiology and are highly influenced by gender (Macotela et al. 2009). The imbalance between caloric intake and energy expenditure may cause hyperplasia and hypertrophy of adipocytes depending not only on the type but also on the amount of dietary fat (Funaki 2009). Many factors, such as peroxisome proliferator-activated receptors gamma (PPARγ) and adipocytokines, modulate the adipose tissue activity (Jeong and Yoon 2011). Dietary fatty acids and their derivatives, which are ligands for PPARγ, may trigger physiological responses such as adipogenesis and adipokine secretion. Thus, it is likely that induction of MetS by high-fat diet may require PPARγ to induce different levels of systemic homeostatic remodeling.

Detailed investigations have revealed that consumption of high-protein and high-fat diet (American diet) upregulates three major pathways important for mammalian target of rapamycin complex1 (mTORC1) activation: (a) increased supply of glucose and fat, (b) stimulation of food-induced insulin/IGF signaling, and (c) abundance of leucine which is supplied by meat and dairy proteins. High levels of glucose and fat increase cellular energy (ATP) levels and thus suppress AMP kinase (AMPK) activity resulting in mTORC1 activation, which represents an essential intracellular target for the actions of hormones and nutrients on food intake and body weight regulation. By being at the crossroads of a nutrient–hormonal signaling network, hypothalamic mTORC1 controls important functions in peripheral organs, such as muscle oxidative metabolism, white adipose tissue differentiation, and β-cell-dependent insulin secretion. Notably, dysregulation of the mTORC1 pathway by high-protein and high-fat diet has been implicated in the development of obesity and obesity-related conditions, such as type II diabetes (Melnik 2012).

In a transgenic mouse model of AD, high-protein and high-fat diet not only produces ER stress, peripheral insulin resistance, reduction in brain insulin, and alterations in insulin receptor signaling, but induces AD-like pathological changes (Ho et al. 2004), suggesting that insulin resistance is closely associated with obesity, type II diabetes, and AD. Although the mechanism involved in ER stress-mediated insulin resistance is not fully understood, activation of stress-activated MAP kinase c-Jun N-terminal kinase (JNK) may be closely associated with the signaling pathways linking ER stress and insulin resistance. In addition, involvement of the redox-sensitive serine kinase JNK, which phosphorylates IRS-1 on serine$^{307}$, has also been proposed (Lim et al. 2009; Ye et al. 2010). In addition, consumption of high fat diet significantly increases the phosphorylation of IκB resulting in the release of NF-κB in the cytoplasm and its translocation to the nucleus, where NF-κB promotes the expression of proinflammatory cytokines (Zhang et al. 2005). This is tempting to suggest that like AD, intracellular oxidative stress in visceral organs may contribute to the pathogenesis of MetS and related diseases.

The induction of oxidative stress in AD and MetS is also supported by recent finding on the accumulation of iron in the same brain regions, which are characterized by Aβ deposition, such as hippocampus, parietal cortex, and motor cortex (Connor et al. 1992; Grünblatt et al. 2011). In addition, iron accumulation is also seen in neurons with neurofibrillary tangles (NFT) (Smith et al. 2010). The binding of ferric iron to the tau protein precedes the aggregation of hyperphosphorylated tau and the subsequent formation of NFT (Yamamoto et al. 2002). Aβ interacts with transition metal ions through three histidine residues located in the N-terminal domain (Nakamura et al. 2007). Similarly, iron plays a pathogenic role in type II diabetes and its complications, such as microangiopathy and atherosclerosis (Swaminathan et al. 2007; Rajpathak et al. 2009). In addition to the induction of oxidative stress, iron may also impede insulin extraction in the liver, impair pancreatic insulin secretion, and interfere with insulin action and glucose uptake in adipocytes.

## 8.4 Insulin Receptor and Insulin Signaling in Alzheimer Disease and Metabolic Syndrome

Glucose homeostasis is maintained by the rapid action of insulin to stimulate glucose uptake and metabolism in peripheral tissues. Skeletal muscle is the primary site of glucose disposal in the insulin-stimulated state (DeFronzo 1997). Insulin resistance in skeletal muscle is a major pathogenic cause of type I and type II diabetes (Petersen and Shulman 2002). The ability of insulin to increase glucose transport in skeletal muscle is mediated through the translocation of glucose transporter 4 (GLUT4) from the intracellular vesicles to the plasma membrane (Kahn 1996). In the muscle of type II diabetic subjects, the expression of the GLUT4 gene is normal, and impaired glucose uptake by insulin action is related to alteration in trafficking or impaired function of GLUT4 (Dohm et al. 1991). Because glucose transport in response to other stimuli that use different signaling pathways is normal in the muscle of type II diabetic subjects (Kahn 1996), the resistance to insulin stimulation may be due to impaired insulin signal transduction (Kim et al. 1999).

Insulin is an anabolic hormone (51 amino acid peptide with mol mass of 5,800 Da), which is produced and secreted by β cells of the islets of Langerhans located in the pancreas, and its serum concentration increases in a direct proportion to the glucose concentration. Insulin plays a crucial role in the metabolism and regulation of carbohydrate and fat in the body by promoting the uptake of glucose and its storage as glycogen in the liver, muscles, and fat cells (Duckworth and Kitabchi 1981). Insulin exerts its effects via the insulin receptor (IR), a transmembrane receptor tyrosine kinase with two α-subunits and two β-subunits. The α-subunits (120–135 kDa) contain the insulin-binding sites. The β-subunits (95 kDa) form transmembrane and intracellular parts of the receptor (White 2003). The intracellular part of the β-subunits contains ATP-binding motifs, autophosphorylation sites, and tyrosine-specific protein kinase activity, which facilitate rapid autophosphorylation upon ligand binding. Insulin is closely related to another polypeptide, insulin-like growth factor I (IGF-1) (Matheny et al. 2010). IGF-1 not only regulates growth, particularly during development, but also exerts anabolic effects on mature organs and tissues. IGF-1 is comprised of 70 amino acids (7,649 Da) in a single chain that contains three intramolecular disulfide bridge (Matheny et al. 2010). IGF-1 is abundantly produced in the liver, and its supply and actions are regulated by interactions with IGF-binding proteins (IGFBPs) (Kuemmerle 2012). Alternative splicing of the IR gene on chromosome 19 leads to two isoforms: isoform A and B. Both isoforms bind to insulin with similar affinity (McClain 1991). Isoform A is expressed in hematopoietic cells, fetal tissue, and the adult nervous system (Moller et al. 1989) and has higher affinity to IGF-1 and IGF-2 compared to isoform B (Yamaguchi et al. 1991; Frasca et al. 1999). Isoform B is found in differentiated tissue of the liver, adipose tissue, and muscle (Moller et al. 1991).

Among visceral tissues, the most important sites for the action of insulin are the liver, adipose tissues, and muscles. The peripheral insulin resistance syndrome occurs when visceral tissues become less responsive to the effects of insulin, not only modulating its actions on cellular glucose uptake but also regulating levels of glucose in the blood. It is typically accompanied by compensatory hyperinsulinemia in the periphery, which has independent toxic deleterious effects upon cells of the liver, adipose tissues, and muscles. In the brain, insulin and IGF signaling regulates a broad array of neuronal and glial activities, including growth, survival, metabolism, gene expression, protein synthesis, cytoskeletal assembly, synapse formation, neurotransmitter function, and plasticity (de la Monte and Wands 2005; D'Ercole and Ye 2008). In addition, insulin and IGF pathways have critical roles in maintaining cognitive function. In the brain, insulin also inhibits the firing of neurons in the hippocampus and hypothalamus, blocks the reuptake of norepinephrine, modulates catecholamine turnover in the hypothalamus, and regulates norepinephrine and dopamine transporter mRNA levels in neurons (Craft 2007; McEwen et al. 2002). At the molecular level, the binding of insulin to IR results in autophosphorylation of tyrosine kinase (Bevan 2001; White 2003). Binding of insulin leads to a rapid autophosphorylation on several tyrosine residues, which provide docking sites for adaptor proteins, such as the insulin receptor substrate proteins (IRS-1, IRS-2, IRS-3, IRS-4, Gab1, and Shc). Following autophosphorylation, at least two signaling pathways are known to occur. In the first pathways, Akt together with protein kinase C promotes translocation of the GLUT4 to the plasma membrane enabling the uptake of glucose in the bloodstream. In the second pathway, phosphorylation initiates the recruitment and activation of Ras/mitogen-activated protein kinase (MAPK). The phosphoinositide 3-kinase/Akt (PtdIns 3K/Akt) signaling pathway is central to proper insulin signaling (Vander Haar et al. 2007). PtdIns 3K acts on the membrane PtdIns 4,5-$P_2$ (phosphatidylinositol 4,5-bisphosphate) turning it into PtdIns $P_3$ (phosphatidylinositol 3,4,5-trisphosphate). PtdIns $P_3$ recruits PKB (protein kinase B, also called Akt), which targets GSK3 (glycogen synthase kinase 3). Phosphorylation of GSK3 by PKB causes its inactivation, reducing the phosphorylation of glycogen synthase (GS). This decreased phosphorylation leads to a more active GS metabolite, which increases the conversion of glucose-6-phosphate to glycogen. Insulin/IR signaling is modulated by cholesterol and glycosphingolipids (GM1 ganglioside) (Simons and Toomre 2000; Hannun and Obeid 2008; Yamamoto et al. 2012). In addition, phosphatase and tensin homologue deleted on chromosome 10 (PTEN), a lipid phosphatase, serves as an important negative modulator for the insulin signaling pathway by hydrolyzing phosphatidylinositol 3,4,5-trisphosphate to phosphatidylinositol 4,5-bisphosphate, antagonizing the PtdIns 3K pathway (Vander Haar et al. 2007). Thus, the physiological regulation of insulin action is controlled by the balance between phosphorylation and dephosphorylation. The degradation of insulin facilitates the cellular response to the hormone by decreasing insulin availability (Duckworth et al. 1998). Liver and kidney are the two primary sites for insulin degradation. Most insulin is removed in the liver during the first transit, while most of the systematically circulating insulin is degraded in the kidney (Castillo et al. 1994). The degradation of insulin normally involves endocytosis of the insulin receptor complex followed by the action of IDE (Duckworth et al. 1998).

Perturbations in insulin and insulin-like growth factor I (IGF-1) signaling and insulin resistance are closely associated with pathogenesis of diabetes (an important component of MetS) and AD (Erol 2008; Talbot et al. 2012). It is known for some time that type I and type II diabetes increase the risk for cognitive dysfunction (Ott et al. 1999; Arvanitakis et al. 2004). Although mechanisms by which insulin resistance mediates alterations in cognitive function are not fully understood, involvement of alterations in neurotransmission and memory formation has been hypothesized. It is well known that impairment in central cholinergic activity may contribute to the pathogenesis of AD (Wenk 2006). Similarly, rats with streptozotocin-induced diabetes also show a decrease in the production and release of acetylcholine compared with control rats (Welsh and Wecker 1991). Administration of glucose not only reverses blocked cholinergic activity induced by scopolamine in mice model (Stone et al. 1988a, b) but also induces changes in behavior by enhancing cholinergic activity (Craft et al. 1996), supporting the view that molecular mechanisms associated with cognitive dysfunction in diabetes and AD may be closely related. In both pathological conditions, insulin resistance may not only involve similar inflammatory mediators and vascular parameters (C-reactive protein, α-1-antichymotrypsin, IL-6, TNF-α, intercellular adhesion molecule 1, etc.) (Farooqui et al. 2012), but also produce alterations in the hypothalamic–pituitary adrenal axis (Kodl and Seaquist 2008). Furthermore, cognitive problems associated with diabetes and AD can also be attributed to atherosclerosis (Reaven 2004). Chronic hyperglycemia produces biological toxicity not only through the generation of glycated proteins and altered redox potential, but also through altered signal transduction and generation of reactive oxygen species (McCall 2005). The resulting oxidative stress contributes to vascular/endothelial dysfunction and this microvascular damage may be an important determinant of the observed cognitive deficits. This suggests that hyperglycemia or altered insulin signaling or both may be closely associated with changes in cognitive function, which not only refers to the acquisition, processing, and integration but also to storage and retrieval of information. In addition to hyperglycemia and impaired insulin action, type II diabetes is also accompanied by hypercholesterolemia, hyperlipidemia, and hypertension, which may provide additional risk factors. The clustering of above mentioned pathogenetic factors, which interact with each other at the molecular level in type II diabetes, may cause impaired cognition and produce structural substrates of AD (Ott et al. 1999; Arvanitakis et al. 2004).

Although initially it was thought that brain is an insulin-insensitive organ, now it is well known that circulating insulin can cross the BBB via an active transport mechanism and interacts with IRs, which are widely distributed in the brain with variable densities in different brain areas. The highest concentration of IRs is found in the olfactory bulb, cerebral cortex, hippocampus, cerebellum, and hypothalamus (Unger et al. 1989). Two types of IRs have been reported in the adult mammalian brain: a peripheral type on glial cells and brain-specific type with high concentrations on neurons (Adamo et al. 1989). In addition, the brain also contains insulin-like growth factors I (IGF-1) and II (IGF-2) receptors which can be detected throughout neurons and glia (Garcia-Segura et al. 1997), particularly in hippocampus, amygdala, parahippocampal gyrus, cerebellum, cerebral cortex, and caudate nucleus, being less abundant in substantia nigra, red nucleus, white matter, and cerebral peduncles (Zhao et al. 1999). Due to structural and functional homology, insulin and IGF-1 can bind to (and activate) both IR and IGF-1R, with insulin binding to the IR with higher affinity (<1 nM) than IGF-1 (100–500-fold lower affinity), whereas IGF-1R preferentially binds IGF-1 (<1 nM) as compared to insulin (100–500-fold lower affinity) (Conejo and Lorenzo 2001). In AD brain, deficits in insulin and IGF signaling progress with severity of the disease (Rivera et al. 2005) and involve pathways associated with survival of neurons, energy production, gene expression, and neural cell plasticity (Frolich et al. 1998). It is interesting to note that nearly all of the critical features of AD, including the increase in (a) activation of kinases involved in tau phosphorylation that leads to the accumulation of neurofibrillary tangles; (b) expression of APP and accumulation of Aβ peptides that are involved in senile plaque formation; (c) oxidative and ER stresses, which propagate cell death cascades; (d) induction of mitochondrial dysfunction which causes energy deficits; (e) increase in levels of lipid mediators, which are generated through the activation of phospholipases $A_2$, sphingomyelinases, and cholesterol hydroxylases; and (f) disruption of cholinergic homeostasis needed for neuronal plasticity, memory, and cognition, can be explained through the involvement of brain insulin/IGF resistance and signaling.

In the brain, insulin binds with the extracellular domain of the IR and promotes the autophosphorylation by triggering intrinsic tyrosine kinase activity. Activated IR interacts not only with insulin receptor substrates (IRS) but also with the Src homology 2-containing protein (Saltiel and Pessin 2002). The phosphorylation of intracellular substrates leads to the recruitment and activation of multiple proteins and the initiation of several signaling cascades, including PtdIns 3K/Akt and the mitogen-activated protein kinase (MAPK) signaling pathways (Johnston et al. 2003; van der Heide et al. 2006). PtdIns 3K/Akt signaling cascade triggers the translocation of the insulin-sensitive glucose transporter 4 (GLUT4) to the membrane surface, which consequently enhances cellular glucose uptake (Johnston et al. 2003). In the brain, activated PtdIns 3K/Akt also phosphorylates and inhibits both α- and β-cytosolic forms of glycogen synthase kinase 3 (GSK3) (Cross et al. 1995). This enzyme is involved in APP processing and regulates the formation of Aβ peptides (Phiel et al. 2003). Relationship between AD and MetS is also supported by several other observations. Thus, Aβ plaques in AD and atherosclerotic plaques in MetS have many similar characteristics (90 % structural similarity) (Janson et al. 2004). Both types of plaques are derived from APP and are damage control responses of the body to high cholesterol in a proinflammatory environment (Umeda et al. 2010). In AD and MetS, Aβ plays an apolipoprotein-like role in helping to maintain cerebral homeostasis. Atherosclerotic plaques in MetS also play a similar role in arterial cholesterol homeostasis (Umeda et al. 2010). Thus, cholesterol contributes to the pathogenesis of AD not only through stimulation of APP processing by β-secretase and Aβ deposition but also through the involvement of ApoE4 (Vassar 2004; Cui et al. 2011). Insulin modulates cholesterol synthesis by stimulating activity of 3-hydroxy-3-methylglutaryl-CoA reductase (HMG-CoA reductase), a rate-limiting enzyme, which catalyzes the conversion of HMG-CoA into mevalonate in cholesterol biosynthesis. In addition, insulin not only modulates levels of classic neurotransmitters such as acetylcholine, norepinephrine, and dopamine that play important roles in cognition (Craft 2007) but also affects membrane potentials, neuronal physiology, and long-term potentiation (LTP), all of which influence the synaptic remodeling processes thought to underlie memory formation (Craft 2007).

Although the relation between brain and plasma Aβ levels in control subjects and AD patients is complex, studies on AD transgenic mice indicate that within a few months of birth, APP transgenic mice secrete more Aβ42 and Aβ40 than their wild-type littermates (Kawarabayashi et al. 2001; Kuo et al. 2001), and Aβ secretion is increased throughout the life of APP transgenic mice. Both plasma and cerebrospinal fluid levels of Aβ42, and to a lesser extent Aβ40, increase with age, but both precipitously decrease as Aβ42 and Aβ40 levels rise in the brain (Kawarabayashi et al. 2001; Schupf et al. 2008). After 1 year, transgenic mice not only display the deposition of Aβ plaques but also show characteristic age-related behavioral changes in memory. However, others have reported that the behavioral effects are not the result of the Aβ accumulation in the brain, but due to the increase in the protofibrillar subspecies of Aβ (Lesne et al. 2006). It is stated that increase in Aβ levels along with behavioral changes in transgenic mice is similar not only to human AD patients (Mayeux et al. 1999, 2003) but also in nondemented elderly women with mild cognitive impairment (Assini et al. 2004) and in adults with Down syndrome who subsequently develop dementia (Schupf et al. 2001, 2007, 2008). Levels of CSF Aβ42 are inversely related to brain amyloid load as imaged with Pittsburgh Compound B (Fagan et al. 2006, 2007). Based on these results, it is proposed that plasma levels of Aβ42 increase before the onset of AD and decline shortly after the onset of dementia and with progression of symptoms (Assini et al. 2004; Mayeux et al. 2003; Kanai et al. 1998; Mehta et al. 2000).

Hypercholesterolemia is another important risk factor for AD and type II diabetes (Kompoti et al. 2006). Transgenic mice models overexpressing islet amyloid polypeptide develop diabetes and generally subsequent to amyloid deposits (Janson et al. 1996). Conversely, targeted disruption of islet amyloid polypeptide leads to enhanced insulin secretion and improved glucose tolerance (Gebre-Medhin et al. 1998). Defective insulin signaling not only makes neurons energy deficient and vulnerable to other metabolic insults (oxidative stress and neuroinflammation) but also impairs synaptic plasticity (Bosco et al. 2011). In fact, oxidative stress-induced alterations in mitochondrial function are closely associated with loss of synapse in AD (Farooqui 2010). Hyperinsulinemia as well as complete lack of insulin causes an increase in τ phosphorylation, resulting in an imbalance in insulin-regulated τ kinases and phosphatase. Chronic hyperinsulinemia may also exacerbate inflammatory responses and increase markers of oxidative stress (Bosco et al. 2011). In addition, insulin appears to act as "neuromodulator," influencing release and reuptake of neurotransmitters and improving learning and memory. Thus, experimental and clinical evidences support the view that insulin not only modulates brain functions but initiates insulin resistance, which is closely associated with the pathogenesis of AD (Farooqui 2010; Bosco et al. 2011). These observations are consistent with the concept that insulin has important functions in the brain, including metabolic, neurotrophic, neuromodulatory, and neuroendocrine actions, and accumulation of β-amyloid–lipoprotein complexes may act as an amplifier of diet-induced inflammation (Takechi et al. 2010).

APP processing is modulated by the presence of docosahexaenoic acid (DHA) in diet (Farooqui 2009; Farooqui et al. 2012). DHA effects APP processing in AD brain and MetS arteries in a similar manner. Thus, it reduces amyloidogenic processing by inhibiting β- and γ-secretase activity (Grimm et al. 2011a). The beneficial effects of DHA in patients with MetS may be not only due to similar mechanism (inhibition of β- and γ-secretase activity) in the muscles, liver, platelets, endothelial cells, and macrophages but also through improvement of endothelial cell function. Furthermore, in the brain as well as arteries, soluble Aβ peptide upregulates inducible nitric oxide synthase (iNOS), leading to increased generation of ROS and reactive nitrogen species (RNS). Increase in ROS and RNS generation results in macrophage activation in the arteries and microglial cell activation in the brain, both of which facilitate β-secretase-mediated degradation of APP, increased production of Aβ, and increased expression of proinflammatory cytokines-potentially vicious cycle, which promotes apoptotic cell death in visceral circulatory system as well as in the brain (Farooqui 2010). NO produces differential regulation on β-secretase. At low concentration, it increases β-secretase activity, whereas at modest to high concentrations, it suppresses β-secretase activity either transcriptionally or posttranslationally through *S*-nitrosylation. Thus, *S*-nitrosylation may represent a basic regulatory mechanism for maintaining β-secretase at physiological levels (Kwak et al. 2011). In contrast, as mentioned above, α-secretase in the brain promotes neuroprotection and in the visceral organs promotes insulin sensitivity and optimal mitochondrial and endothelial function.

AMP-activated protein kinase (AMPK) is a heterotrimeric enzyme which acts as energy sensor and plays a pivotal role in the regulation of metabolic homeostasis by phosphorylating targets that are involved in glucose, carbohydrate, lipid, and protein metabolism and mitochondrial biogenesis (Fig. 8.6) (Lage et al. 2008; Kodiha and Stochaj 2011). AMPK may also be a crucial link between MetS and sporadic AD. Like visceral organs, in the brain, AMPK is activated by a variety of stimuli, including low-nutrient supply, ROS, hormones, cytokines, and various physiological or environmental stressors. Its activity is diminished in diabetes and AD (Lage et al. 2008; Kodiha and Stochaj 2011). Alterations in AMPK activity may be caused by (a) mitochondrial dysfunction, which is a crucial factor in both AD (Galindo et al. 2010) and MetS (Ren et al. 2010), (b) alterations in cholesterol and lipid metabolism and ApoE4 as a risk allele for both diseases (Olivieri et al. 2007; Bhat 2010), and (c) increase in endoplasmic reticulum stress and inflammatory disorders (Dong et al. 2010; Hotamisligil 2010). Furthermore, onset of AD and MetS usually occurs after midlife. Mounting evidence suggests that MetS is a risk factor for AD.

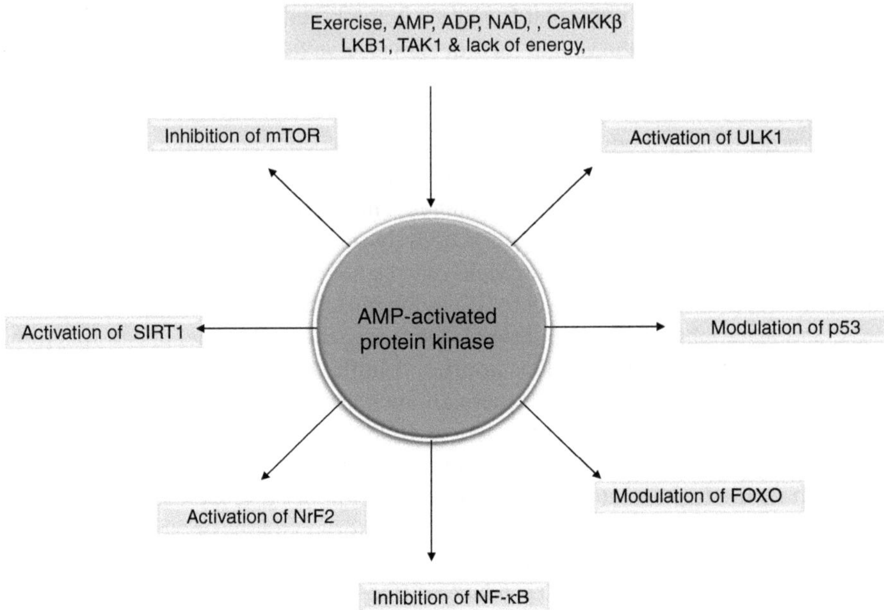


**Fig. 8.6** Effect of AMP-kinase on transcription factors

## 8.5 Ceramide-Mediated Insulin Resistance in the Brain and Visceral Tissues

Insulin resistance is defined as a condition in which insulin's target organs are resistant to its action, so that higher concentrations of this hormone are needed to obtain a normal biological effect. Therefore, hyperinsulinemia is an obvious consequence of insulin resistance, which contributes to the development of endothelial dysfunction, playing a key role in the establishment and progression of atherosclerosis (DeFronzo 1997). Insulin resistance increases with age, and the organism maintains normal glucose levels as long as it can produce enough insulin (hyperinsulinemia). Some individuals are less capable than others to mount sustained hyperinsulinemia and will develop glucose intolerance and diabetes (Fig. 8.7) (Festa et al. 2006). Other individuals with insulin resistance will maintain normal glucose levels at the expense of hyperinsulinemia, but their pancreas will eventually "burn out," will not be able to sustain hyperinsulinemia, and will develop glucose intolerance and diabetes (Festa et al. 2006).

Long-term consequences of insulin resistance include cellular energy failure, elevated plasma lipids, and high blood pressure. In addition, chronic hyperinsulinemia predicts future development of diabetes mellitus (Dankner et al. 2009). It is also stated that insulin resistance is an independent predictor of cerebrovascular and cardiovascular disease, hypertension, nonalcoholic fatty liver disease (NAFLD), metabolic syndrome, malignancy, and neurological disorders (Faulds and

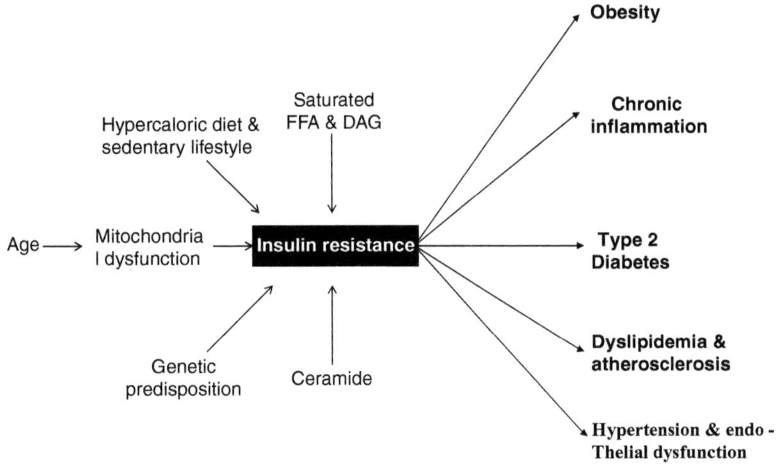


**Fig. 8.7** Effect of insulin resistance on various parameters associated with metabolic syndrome

Dahlman-Wright 2012; Farooqui et al. 2012; Colonna et al. 2012). The oversupply of lipid to peripheral tissues (skeletal muscle and liver) contributes to the development of insulin resistance. Insulin-resistant subjects display signs of abnormal lipid metabolism (including obesity, increased circulating free fatty acid (FFA), diacylglycerol (DAG), and ceramide levels) and elevated intra-myocellular lipid levels in peripheral tissues (Farooqui et al. 2012). The metabolic events associated with FFA oversupply to stress kinase activation (e.g., JNK, IKKβ) and insulin resistance is not fully understood. However, in the skeletal muscle, DAG accumulates via shuttling of FFA through the glycerolipid synthesis pathway and from hydrolysis of TAG. Accumulation of DAG has been reported to occur in the skeletal muscle of obese rodents and humans (Turinsky et al. 1990). Many studies indicate that increase in plasma FFA is related to the accumulation of DAG and insulin resistance (Yu et al. 2002). DAG promotes the phosphorylation of IRS-1 through the activation of PKCζ and IKKβ (Cai et al. 2005; Kim et al. 2004).

Ceramide is an important sphingolipid that regulates diverse signaling associated with apoptosis, cell senescence, the cell cycle, and differentiation. During signal transduction, ceramide and DAG are generated simultaneously, but they work in opposite directions in the sphingomyelin cycle (Ruvolo 2003). While ceramide stimulates signal transduction pathways that are associated with cell death, DAG activates several isoforms of the PKC family. These PKC isoforms are associated with cell growth and cell survival. Ceramide not only inhibits the phospholipase D and c-Fos-dependent signaling pathway (Cutler and Mattson 2001) but also reduces the insulin/PtdIns 3-kinase-mediated Akt/protein kinase B translocation to membranes and its activation (Stratford et al. 2004). In contrast, DAG activation of PKC stimulates other signal transduction pathways that support cell proliferation (e.g., MAPK pathways). Thus, ceramide and DAG generation may serve to monitor cellular homeostasis by inducing pro-death or pro-growth pathways, respectively.

Among abovementioned lipid mediators, ceramide has received considerable attention. Numerous cell culture and animal model studies have shown that increasing ceramide levels inhibits insulin signaling and causes insulin resistance (Liu et al. 2007). Inhibition of de novo ceramide synthesis by myriocin, an inhibitor of serine palmitoyltransferase one (SPT1), can prevent insulin resistance mediated by corticosteroids and saturated FFA (Pickersgill et al. 2007). SPT is the rate-limiting enzyme in the biosynthesis of sphingolipids. This enzyme is linked with APP processing and is affected in AD. Increase in SPT activity in AD has been observed in PS1/2- and APP/APLP2-deficient cell lines (Grimm et al. 2011b). The elevated SPT activity is caused by increased expression of the SPT subunit SPTLC2 (63 kDa subunit), which represents the catalytic subunit of the SPT heterodimer (Hojjati et al. 2005a, b). It is speculated that increase in SPTLC2 expression exerts its toxic effect by increasing Aβ generation through the amyloidogenic processing of APP in lipid raft of the membrane. More studies are needed to clarify the question whether the observed change in SPTLC2 levels in the human sporadic and familiar AD brains is the cause or consequence of AD (Grimm et al. 2011a, b).

Another choline-containing sphingolipid metabolism metabolite, sphingosylphosphorylcholine (SPC), produces suppressive effect on Aβ production in PC12 cells which stably express Swedish mutant of amyloid precursor protein (APPsw) (Yi et al. 2011). Thus, in PC12 cells, SPC (>3 μM) lowers the accumulation of Aβ40/42 and the expression of BACE1, but the transcriptions of ADAM10 and presenilin 1 are not affected by the treatment with SPC. It is proposed that SPC-mediated downregulation of BACE1 expression seems to be mediated by NF-κB which is known to suppress the transactivation of BACE1 promoter in PC12 cells. Emerging evidence suggests that SPC has the potential to ameliorate Aβ pathology in neurons by downregulating the BACE1-mediated amyloidogenic pathway (Yi et al. 2011).

Myriocin-mediated inhibition of SPT1 in human muscle cells also prevents the inhibition of insulin-stimulated glycogen synthesis induced by palmitic acid (Holland et al. 2007). Ceramide is primarily converted into more complex sphingolipids in the Golgi apparatus. Some ceramide is either metabolized to ceramide-1-phosphate (C1P) or hydrolyzed by ceramidase into sphingosine, which is transformed into sphingosine-1-phosphate (S1P) (Fig. 8.8). S1P acts through a family of G protein-coupled receptors and opposes the actions of ceramide. In sharp contrast to ceramide, S1P promotes nutrient uptake (Yea et al. 2008) and increases basal and insulin-stimulated glucose uptake (Rapizzi et al. 2009). S1P activates Akt/PKB in a wide variety of cell types (Spiegel and Milstien 2003). It has not only been implicated in insulin receptor activation (Rapizzi et al. 2009) but also in AMPK activation (Levine et al. 2007). It is necessary for adiponectin-induced AMPK activation (Holland et al. 2010). Although the addition of ceramide inhibitor retards adiponectin-mediated AMPK phosphorylation, it has no effect on S1P-mediated signal transduction processes.

Glucose increases S1P levels through the activation of sphingosine kinase 2 (SphK2) (Holland and Summers 2008; Stanford et al. 2012). Increase in S1P levels correlates with elevation in glucose-stimulated insulin secretion. Reduction in S1P

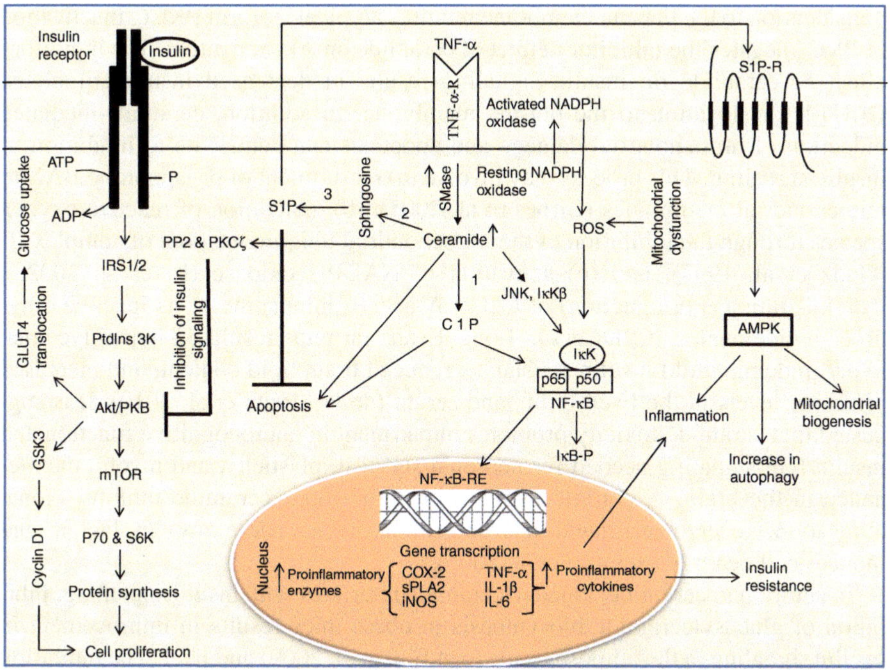


**Fig. 8.8** Effect of ceramide on insulin signaling in metabolic syndrome. Phosphatidylcholine (PtdCho), sphingomyelin (SM), cytosolic phospholipase $A_2$ ($cPLA_2$), sphingomyelinase (SMase), arachidonic acid (ARA), reactive oxygen species (ROS), cyclooxygenase-2 (COX2), ceramide-1-kinase (1), ceramidase (2), sphingosine-1-kinase (3), ceramide-1-phosphate (C1P), sphingosine-1-phosphate (S1P), mammalian target of rapamycin complex 1 (mTORC1), and plasma membrane (PM). Proinflammatory genes include TNF-α, IL-1β, nitric oxide synthase, cyclooxygenase-2 (COX-2), $sPLA_2$. Ceramide activates protein phosphatase 2A (PP2A) and protein kinase Cf (PKCζ) to inhibit Akt/PKB and activates the serine/threonine kinase c-Jun N-terminal kinase (JNK) and inhibitor of κB-kinase β (IκKβ) to activate the transcription factors Jun and NF-κB

levels by treatment of MIN6 cells with the sphingosine kinase inhibitor decreases glucose-stimulated insulin secretion (Stanford et al. 2012). Moreover, knockdown of SphK2 alone results in decreased glucose-stimulated insulin secretion, whereas knockdown of the S1P phosphatase leads to a rise in glucose-stimulated insulin secretion. Reduction of S1P levels in mice, by treatment with sphingosine kinase inhibitor, impairs glucose disposal due to decrease in plasma insulin levels. Collective evidence suggests that glucose activates sphingosine kinase 2 in pancreatic β cells leading to a rise in S1P levels, which is important for glucose-stimulated insulin secretion (Stanford et al. 2012). It is suggested that ceramide may be a putative intermediate that links both excess nutrients such as saturated FFA (a component of ceramide) and inflammatory adipocytokines such as TNF-α to MetS (Samad 2007). Ceramide blocks insulin stimulation of PKB/Atk by two independent mechanisms, i.e., translocation of Akt to the plasma membrane and stimulation of Akt dephosphorylation by protein phosphatase 2A (PPA2) (Fig. 8.8). Blocking of Akt translocation to the plasma membrane requires atypical isoform PKCζ. Inactivation of PKCζ negates the inhibitory effect of ceramide on Akt activation. The inhibitory effect of ceramide on insulin signaling results in decreased insulin-stimulated GLUT4 translocation to the plasma membrane. In addition, ceramide-mediated induction of mitochondrial damage and apoptosis can contribute to inhibition of insulin signaling. This process may involve (a) recruitment of proapoptotic BAX to mitochondrial membranes (Birbes et al. 2005), (b) generation of reactive oxygen species through the inhibition of the mitochondrial ubiquinone pool of complex III (Gudz et al. 1997), and (c) activation of NADPH oxidase (Li et al. 2002a). Administration (i.p. injection) of C2Cer:*N*-acetylsphinganine or its inactive dihydroceramide analog (C2DCer) in Long–Evans rat pups results in hyperglycemia, hyperlipidemia, mild insulin resistance, reduced brain lipid content, and increased ceramide levels in the liver, brain, and serum (de la Monte et al. 2010). It is suggested that ceramide toxicity promotes impairment in neurocognitive function and insulin/IGF signaling needed for neuronal survival, plasticity, and myelin maintenance in the brain. A chronic increase in intracellular ceramide inhibits axonal elongation, receptor-mediated internalization of the nerve growth factor, and induces cell death (Costantini et al. 2005).

In addition to ceramide, glucosylceramide interferes with insulin signaling: inhibition of glucosylceramide biosynthesis in obese mice results in improvement in insulin signaling in the adipose tissue (van Eijk et al. 2009) and liver. The inhibition of glucosylceramide synthase also corrects insulin resistance in cultured adipocytes from obese individuals (Aerts et al. 2007). Based on above information, ceramide (galactosylceramide and glucosylceramide), a metabolite of sphingolipid metabolism, can be regarded as a potential endogenous neurotoxin, whose level and composition may not only be closely associated with the pathogenesis of chronic visceral diseases (diabetes and cancer) but also be involved in neural cell death in neurotraumatic and neurodegenerative diseases (Farooqui 2011).

## 8.6 Cognitive Dysfunction in Metabolic Syndrome and Alzheimer Disease

It is well known that the brain utilizes glucose, largely through oxidative metabolism. Glucose is continually supplied to the brain either through the glucose transport proteins (GLUT), which allow glucose entry into individual cells (neurons and glial cells) (Clarke and Sokoloff 1999) or by endothelial cells that form the BBB to reach neurons and glial cells by facilitated diffusion. In the brain, glucose homeostasis is critical for energy generation, neuronal maintenance, neurogenesis, neurotransmitter regulation, cell survival, and synaptic plasticity. Glucose also plays a key role in cognitive function. Insulin neither influences basal cerebral glucose metabolism nor transport of glucose into the brain (Bingham et al. 2002), but it affects the use of glucose in specific regions of the brain, most likely by selective distribution of insulin-sensitive GLUT isoforms, which overlap with the distribution of insulin and IRs in the brain (Reagan et al. 2001). Insulin, through interactions with insulin receptors that are localized in the hypothalamus, contributes to the regulation of food intake and energy homeostasis and leads to changes in body weight by anorexigenic or orexigenic effects produced by increasing or decreasing levels of insulin in the brain, respectively. Moreover, insulin in the brain modifies peripheral glucose metabolism by increasing insulin sensitivity in the peripheral tissue (Niswender and Schwartz 2003).

In the brain, insulin produces cognition-enhancing effect, which is independent of changes in peripheral glucose. Insulin also modulates CNS concentration of neurotransmitters associated with important roles in cognition such as acetylcholine, norepinephrine, and dopamine (Kopf and Baratti 1999). Patients with type I and II diabetes display impairments in learning, memory, and problem-solving skills and delayed development of mental speed and flexibility and patients with type I or type II diabetes have at least a two- to threefold higher risk for developing significant worsening in cognition as compared to normal subjects. These changes have been attributed to episodic hypoglycemia resulting from glycemic control (Strachan et al. 1999). In addition, other factors, such as hypertension, dyslipidemia, hypothalamic pituitary adrenocortical (HPA) axis abnormalities and inflammation, microvascular dysfunction, and duration of diabetes can also contribute to the severity of cognitive decline (Brands et al. 2004). It is suggested that prolonged hyperglycemia plays a major role in the pathogenesis of cognitive impairment by altering both metabolic and vascular function in the brains of patients with diabetes and AD (Fig. 8.9) (Pasquier et al. 2006). Hyperglycemia may produce increase in levels of glucose in the brain. Excess of glucose is converted into sorbitol and fructose, which not only induces the formation of toxic advanced glycation end products (AGE) and ROS but also causes hyperhomocysteinemia.

AGEs are linked to the cytoskeletal changes that occur in AD (Kuhla et al. 2007). In fact, glycation of Aβ is reported to enhance its aggregation and subsequent formation of senile plaques (Sasaki et al. 1998; Smith et al. 1994). Hyperglycemia is also associated with both structural and functional alterations in the cerebral vascular system.

Positron emission tomography (PET) has indicated that metabolic demands of the brain are efficiently met when cerebral blood flow is maintained to approximately 750 ml per minute or about 50 ml/100 ml of brain tissue per minute (Walters 1998; Ito et al. 2005; Roher et al. 2012). Cerebral blood flow has been reported to be decreased in patients with diabetes and AD (30–40 %) compared to age-matched control subjects (Erol 2008). This decrease in brain–blood flow may cause a neuronal energy crisis through reduction in ATP synthesis. The ensuing metabolic energy crisis initially carves up ischemic-sensitive neurons in the hippocampus and posterior parietal cortex setting up cognitive meltdown and progressive neurodegenerative and atrophic changes in the brain. Furthermore, neuronal energy compromise enhances oxidative stress, aberrant protein synthesis, ionic membrane pump dysfunction, signal transduction impairment, neurotransmitter failure, abnormal processing of amyloid precursor protein resulting in β-amyloid deposition, and axonal

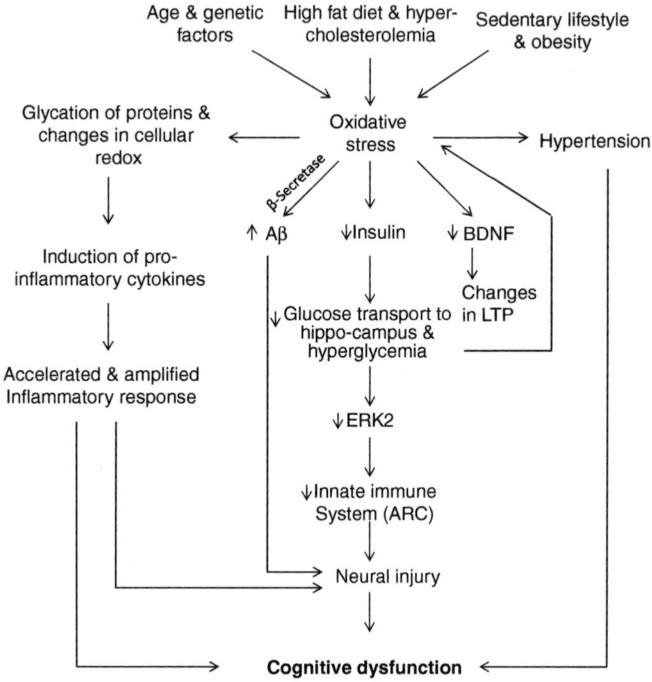


**Fig. 8.9** Factors modulating cognitive dysfunction

microtubule disruption from tau hyperphosphorylation. In addition, decrease in cerebral blood flow negatively affects the synthesis of proteins required for the formation of memory and learning and may eventually lead to neuritic injury and neuronal death (Erol 2008; Farooqui 2010; Austin et al. 2011).

Alterations in insulin secretion and insulin resistance, both systemically and centrally, have an important role in cognitive function and neurodegeneration (Celik et al. 2007) (Fig. 8.9). Patients with diabetes have various complications of microvascular (retinopathy, nephropathy, and peripheral neuropathy) and macrovascular (coronary heart disease and stroke) diseases (Vinereanu 2006). These vascular disorders in the brain are considered as one of the major mechanism in AD- and diabetes-induced cognitive impairment. In addition, intracerebroventricular STZ administration results in many of the behavioral and pathological symptoms that are shared by sporadic AD and brain insulin resistance. Although mechanisms that trigger the onset of sporadic AD remain elusive, recent studies have indicated that synaptic transmission and plasticity as well as the integrin-linked kinase (ILK)–glycogen synthase 3β (GSK3β) signaling pathway are affected in the streptozotocin model of sporadic AD (Shonesy et al. 2012). In addition, there is a decrease not only in expression of brain-derived neurotropic factor (BDNF) but also in levels of stargazing (an AMPAR auxiliary subunit), which are needed for driving AMPA receptors to the surface of the postsynaptic membrane. Collective evidence suggests that central insulin resistance produces hippocampal glutamatergic dysfunction, probably not only due to alterations in BDNF and ILK–GSK3β signaling but also due to synaptic deficits, which is closely associated with memory loss (Shonesy et al. 2012).

The involvement of BBB in cognitive dysfunction is controversial. However, disruption of the BBB, occurrence white matter lesions, decrease in the level of tight junction proteins, and increase in matrix metalloproteinase activity have been reported to occur in many diabetic patients and experimental models of diabetes (Mogi and Horiuchi 2011). Diabetes-induced cognitive decline may be induced via disruption of neurovascular coupling, with not only vascular disorder but also impairment of astrocytic trafficking (Akisaki et al. 2006; Mogi and Horiuchi 2011). As stated above, insulin resistance increases the risk of age-related memory impairment and AD. Possible mechanisms through which these risks are increased include the effects of peripheral hyperinsulinemia on memory, neuroinflammation, and regulation of the Aβ production.

For many years, it was thought that insoluble deposits of Aβ may be the leading cause of memory loss in AD. More recent studies indicate that soluble Aβ oligomers may be the cause of memory loss, especially in the early stages of AD, because Aβ oligomers inhibit long-term potentiation in neurons (Walsh et al. 2002). In addition, there are pathological connections between diabetes and AD. Thus, elderly people with diabetes and AD have increased cortical and hippocampal atrophy (Biessels et al. 2006) and an increased accumulation of advanced glycation end products (AGEs) (Sato et al. 2006) as compared to age-matched nondiabetic elderly patients. Collective evidence suggests that type II diabetes mellitus and AD may increase the risk of cognitive impairment, dementia, and neurodegeneration through many mechanisms, including the induction of insulin resistance, increase in oxidative stress, formation of AGEs, upregulation of inflammatory cytokines, and initiation of microvascular and macrovascular disease (Biessels et al. 2006).

## 8.7 Link Between Metabolic Syndrome and Alzheimer Disease

Insulin accelerates glucose entry into the skeletal muscle and adipose tissue by promoting the translocation of GLUT4 glucose transporters from intracellular stores to the plasma membrane. In addition, insulin also regulates GSK3β and pp70 S6-kinase to promote storage of the incoming glucose as glycogen, triglyceride, or protein (Jope et al. 2007). The maintenance of glucose levels in normal range is achieved by the balance between the ability of peripheral tissues (muscle, adipose tissue, liver) to take glucose into cells and the ability of pancreas to produce and secrete insulin. Insulin resistance increases with age, and individuals can maintain normal glucose levels as long as their pancreas can produce enough insulin.

Prolonged peripheral hyperinsulinemia produces insulin resistance and reduces insulin transport across the BBB, subsequently lowering insulin levels and activity in the brain. These effects may be responsible not only for reduction in CSF insulin and decrease in brain insulin receptor activity but also in progressive reduction of brain glucose metabolism in AD (Craft et al. 1998), supporting the view that impaired brain insulin signaling plays a critical role in the etiology of AD. Reduction in brain insulin signaling leads to increase in tau phosphorylation and increase in Aβ levels in a streptozotocin mouse model of diabetes (Jolivalt et al. 2008). Insulin not only promotes release of intracellular Aβ in neuronal cultures but also stimulates Aβ trafficking to the plasma membrane (Gasparini et al. 2001). Furthermore, intravenous injections of insulin increase plasma levels of Aβ42 in AD patients, but not in normal subjects, an effect that can be exaggerated in AD patients with higher body mass index (Kulstad et al. 2006). 4-HNE, which accumulates in the visceral tissues in MetS and in the brains from AD patients, also stimulates the production of Aβ through the upregulation of β-secretases (Mattson 2009).

Aβ oligomers have been shown to induce neuronal insulin resistance in the AD brain not only by inhibiting the insulin network by targeting the insulin/Akt pathway (Townsend et al. 2007) but also by removing insulin receptors after binding at the dendrites of synaptic sites (Zhao et al. 2008). Impaired insulin signaling cannot efficiently block GSK3β, and therefore, the activated GSK3β not only triggers APP γ-secretase activity but also increases tau phosphorylation (Hooper et al. 2008), simultaneously aggravating the two major pathological substrates of AD. Aβ oligomers are small, diffusible aggregates that can specifically attach with synapses in hippocampal and cortical neurons and act as pathogenic ligands for the loss of synapse in AD (De Felice et al. 2009). The oligomer binding induces AD-like pathology, oxidative stress (Farooqui 2010), synapse deterioration and loss (De Felice et al. 2009; Lacor et al. 2007), and inhibition of synaptic plasticity (Lambert et al. 1998). Collective evidence suggests that insulin not only modulates levels of Aβ aggregates but also inhibits the activity of GSK3β, preventing the phosphorylation of tau protein and therefore inhibiting the formation of neurofibrillary tangles. Stimulation of insulin signaling also protects neurons from Aβ oligomer-induced impairment of LTP (Townsend et al. 2007) and accumulation of hyperphosphorylated tau (Escribano et al. 2010). It is suggested that insulin signaling provides a physiological defense mechanism against oligomer-induced synapse loss (De Felice et al. 2009). Thus, insulin not only downregulates oligomer binding sites in neurons through a mechanism requiring insulin receptor tyrosine kinase activity (De Felice et al. 2009) but also protects neurons from oligomer-induced impairment of LTP and accumulation of hyperphosphorylated tau (Escribano et al. 2010).

Insulin resistance in diabetes and AD also contributes to the development of endothelial dysfunction and plays a key role in the establishment and progression of chronic inflammation, diabetes, hypertension, and obesity (Fig. 8.7) (Madonna and De Caterina 2011). These parameters are also related to obesity (Shoelson and Goldfine 2009) and hypertension (Savoia and Schiffrin 2006) and are common features of MetS. The molecular pathways that mediate obesity-associated inflammation involve IKKβ/NF-κB signaling (Shoelson and Goldfine 2009; Shoelson et al. 2006;

Cai 2009). IKKβ/NF-κB signaling in the hypothalamus is activated under obesogenic condition to promote energy, body weight, and glucose imbalance (Kleinridders et al. 2009; Posey et al. 2009). Obesity and hypertension are mechanistically coupled with each other by IKKβ/NF-κB signaling in the hypothalamus (Purkayastha et al. 2011). This form of hypothalamic inflammation-induced hypertension involves the sympathetic upregulation of hemodynamics and is reversed by sympathetic suppression. Inhibition of NF-κB in the mediobasal hypothalamus counteracts obesity-related hypertension in a manner that is dissociable from changes in body weight (Purkayastha et al. 2011; Ewing 2010).

Type II diabetes (an important component of MetS) increases the risk of AD. Although the nature of link between type II diabetes mellitus and AD risk remains unclear, several mechanisms including aging, insulin resistance, inflammatory cytokines, and oxidative stress may contribute to this link (Haan 2006; de la Monte et al. 2010). Insulin provokes Aβ accumulation by limiting Aβ degradation via direct competition for the IDE, a 110 kDa thiol zinc metalloendopeptidase that cleaves small proteins of diverse sequence, many of which share a propensity to form β-pleated sheet-rich amyloid fibrils such as Aβ, insulin, glucagon, amylin, atrial natriuretic factor, and calcitonin (Bennett et al. 2000; Kurochkin 2001). Moreover, insulin itself promotes Aβ accumulation by accelerating amyloid precursor protein (APP)/Aβ trafficking from the *trans*-Golgi network, a major cellular site for Aβ generation, to the plasma membrane (Gasparini et al. 2001). As stated above, Aβ peptide is a substrate for IDE, which rapidly degrades monomeric but not the oligomeric Aβ peptide (Qiu et al. 1997). The relationship of IDE with Aβ is supported by recent evidence indicating that IDE activity in the brain is negatively correlated with Aβ content and that IDE expression is decreased in the AD brain (Cook et al. 2003). Insulin regulates the levels of IDE. Hyperinsulinemia reduces Aβ degradation by reduction of IDE levels and binding competition (Farris et al. 2003). It is also reported that the expression of sortilin-related receptor (SORL1), a neuronal sorting protein that regulates APP processing to decrease Aβ production in a dose-dependent manner, is decreased in the brains of patients suffering from AD and that under expression of SORL1 leads to overproduction of Aβ (Ma et al. 2007; Dodson et al. 2008). This protein is also genetically associated with types I and II diabetes, supporting the view that SORL1 deficiency might be a common genetic risk factor underlying the predisposition to AD as well as type I and II diabetes. It is proposed that probing the molecular consequences of SORL1 dysfunction may lead to pathways that elucidate the link between AD and diabetes (Lane et al. 2010).

Accumulating evidence thus suggests that under diabetic conditions, impairment in certain insulin receptor-responsive cellular signaling pathways may mechanistically promote AD-related neuropathology and cognitive deterioration (Craft 2007; Talbot et al. 2012). In addition, cerebrovascular alterations are common pathological features of diabetes and AD. Chronic hyperglycemia produces advanced glycation end products (AGEs) via Maillard reactions, during which reducing sugars can react with the amino groups of proteins to produce cross-linked complexes and unstable compounds (Brownlee 1995). AGEs not only produce complications of diabetes and promote the pathogenesis of AD by interacting with the receptors for

AGE (RAGE) that are located on the vascular cells and microglia. RAGE is a multiligand receptor, which interacts with a diverse range of molecules (AGEs, Aβ, amphoterin, and S100B protein), with a preference for ligands that are prone to aggregation and posttranslational modifications. The engagement of RAGE by these ligands promotes oxidative stress, as well as inflammatory and thrombogenic responses, which play a crucial role in cardiovascular, neurodegenerative, inflammatory, and autoimmune disorders (Farooqui 2010). In addition, RAGE acts as a putative receptor for Aβ (Deane et al. 2003; Nogueira-Machado and Chaves 2008) and transduces an inflammatory response in the brain. Both AD and type II diabetes upregulate RAGE expression in neural cells (neurons, astrocytes, and microglia) and the cerebral vasculature (endothelial cells, pericytes, and smooth muscle cells) (Deane et al. 2003; Liu et al. 2009), suggesting the upregulation of RAGE in the cerebral vasculature combined with the presence of Aβ in the brain of $APP^{+}$-ob/ob mice may also cause cerebrovascular dysfunction, resulting in early onset of severe cognitive impairment. Vascular dysregulation produced by the RAGE–Aβ interaction may be another important factor shared by type II diabetes and AD. These interactions may play an important role at multiple points in the AD pathway. High-affinity binding of RAGE with Aβ may produce adverse effects, which are potentially relevant to AD pathogenesis. These effects include potentiation of Aβ aggregation, perturbation of neuronal function, elevation of oxidative stress, amyloidosis, augmentation of microglial inflammatory responses, vascular dysfunction, and stimulation of autoimmunity to both Aβ and RAGE (Mruthinti et al. 2006; Fang et al. 2010). It is suggested that ligand-bound RAGE (RAGE/Aβ) is known to enhance neural cell-damaging ROS, which in turn promote AGE generation and neuroinflammation (Mruthinti et al. 2006). This cycle may be closely associated with eventual tissue dysfunction and irreparable neuronal damage in AD (Ramasamy et al. 2005).

There are many other similarities between type II diabetes and AD. Both pathological conditions involve abnormal cholesterol levels, elevation in levels of lipid mediators, increase in expression of cytoadipokines/cytokines, aggregation of Aβ, stimulation in GSK3β activity, and deregulation of protein physiological and abnormal processes. These processes not only include age-related changes in metabolism, increase in phosphorylation (Ristow 2004) along with blood vessel abnormalities (Brands et al. 2004), and increased production of AGEs (Farooqui 2011). Based on the above discussion, it is now widely accepted that type II diabetes and AD share several common abnormalities including impairment in glucose metabolism, elevation in oxidative stress, increase in insulin resistance, alterations in amyloidogenesis, as well as peripheral oxidative and inflammatory stress, neural atrophy and/or degeneration, and cognitive decline, supporting the view that type II diabetes and AD disrupt common molecular pathways and each disease compounds the progression of the other. For example, accumulation of Aβ in AD is accompanied by the induction of neuronal insulin resistance. Conversely, disruption of normal glucose metabolism in transgenic animal models of AD that overexpress the human amyloid precursor protein (hAPP) promotes amyloid-peptide aggregation and accelerates the disease progression (Zhao and Townsend 2009). At the cellular level, hallmark of both diseases (insulin resistance and Aβ aggregation) is promoted and supported by excitotoxicity, aberrant $Ca^{2+}$ signals, and secretion of TNF-α. At the molecular level, insulin resistance and Aβ disrupt common signal transduction cascades including the insulin receptor family/PtdIns 3-kinase/Akt/GSK3β pathway (Zhao and Townsend 2009). In humans, aging is accompanied by decrease in metabolic turnover, reduction in glucose utilization, as well as decline in insulin and insulin-like growth factor (IGF-1) signal transduction due to receptor desensitization (Lindeboom and Weinstein 2004; Watson and Craft 2004). These abnormalities are magnified in AD not only because of high oxidative stress but also due to the degradation of both the insulin and IGF-1 receptor and their consequent effects on glucose metabolism (Hoyers 2000; Gasparini et al. 2002). In addition, other factors, such as cortisol, catecholamines, and cholesterol, may indirectly contribute to impairment in insulin signaling (Farooqui et al. 2012). Deficiencies in insulin, IGF-1, and NGF converge in their signal transduction, diminishing tyrosine kinase signaling and impairing PtdIns 3K activity, thereby activating JNK and p38 and leading to apoptotic cell death through the activation of caspase-3 (Fig. 8.4). In addition, insulin deficiency also activates FAS including NGFR-p75 with activation of caspase-8, which is involved in the splicing of APP into C-terminal fragments and into intracellular processing into Aβ42 (Galvan et al. 2002). Oxidative stress and mitochondrial dysfunction mediate abnormal APP processing through the involvement of caspase-9 (Veinbergs et al. 2002). Therefore, several apoptotic stressors are likely to contribute to abnormal intracellular processing of APP.

As stated earlier, in the brain, insulin receptors are found not only in the hippocampus and parahippocampal areas but also in the hypothalamus, amygdala, cerebellum, and cortex. Presence of insulin receptors in hippocampus and medial temporal cortex is consistent with the view that insulin is closely associated not only with the modulation of long-term potentiation and memory formation but also with the release of neurotransmitters such as acetylcholine and norepinephrine, which are known to influence cognitive function (Zhao and Townsend 2009). In the brain, insulin and insulin receptor signaling are also involved in the induction/expression of hippocampal synaptic plasticity changes, such as LTP and memory formation (English and Sweatt 1997; Man et al. 2003; Opazo et al. 2003). Insulin improves cognitive performance in humans and animals in a wide variety of settings, including healthy subjects, aged subjects, AD patients, and in experimental models of insulin resistance (Zhao and Townsend 2009; Man et al. 2003; Opazo et al. 2003). Animal studies have shown that high-calorie diets impair the structure and function of the hippocampus, a brain region critical for learning and memory (Farr et al. 2008; Wu et al. 2004). It is also shown that adverse effects of high-calorie diets on learning and memory are associated with impaired hippocampal synaptic plasticity and neurogenesis (Farr et al. 2008; Lindqvist et al. 2006), supporting the view that the hippocampus may be particularly sensitive to changes in dietary energy intake.

As stated above, insulin receptor is a transmembrane protein tyrosine kinase (PTK) that upon binding with insulin phosphorylates itself as well as target substrates, such as the IR substrate 1 (IRS-1) (Bevan 2001; White 2003), leading to phosphorylation of receptor substrates and adapter proteins, which form docking sites for other downstream effects within the cell (Gallagher et al. 2008). These events allow for the recruitment and activation of signaling pathways, including the Ras/mitogen-activated protein kinase (MAPK) and PtdIns 3K/Akt pathways that mediate the metabolic, transcriptional, and mitogenic actions of insulin (White 2003; de la Monte et al. 2010). Like visceral organs in the brain, activation of PtdIns 3K results in the synthesis of phosphatidylinositol 3,4,5-trisphosphate (PtdIns 3,4,5-P3), which is responsible for the activation of PDK (phosphoinositide-dependent kinase) and Akt (PKB). Akt plays an important role in glucose transport and storage, protein synthesis, and stopping lipid degradation (Chang et al. 2004). These pathways are compromised in the brain and visceral tissues of experimental models of diabetes (Biessels et al. 1996; Stranahan et al. 2008a, b). PtdIns 3K activity mediates activation of Akt, which phosphorylates and thereby deactivates forkhead transcription factors (FOXOs). These transcription factors regulate the transcription of many genes involved in glucose and lipid metabolism, growth, stress response, and aging process. Thus, insulin-like signaling controls many processes through FOXO regulation and other signaling cascades (Huang and Tindall 2007; Belgardt et al. 2010). In neurons, FOXO factors are mainly involved in the initiation of apoptosis through the induction of proapoptotic proteins rather than driving cells into a state of quiescence (van der Heide et al. 2006).

Activation of PtdIns 3K/Akt signaling also modulates GSK3β, a multitasking serine/threonine kinase, which is involved in many cellular functions including the regulation of apoptosis, cell cycle, gene expression, and many other functions (Jope et al. 2007). GSK3β is closely associated with key steps of the pathogenesis of AD (Figs. 8.5 and 8.8). First, generation of Aβ is promoted by GSK3β (Phiel et al. 2003). Second, GSK3β contributes to the abnormal hyperphosphorylation of tau protein, which promotes tau protein aggregation and neurotoxicity (Cho and Johnson 2004), the characteristic features of AD. Third, the increased GSK3β activity promotes p53-induced gene expression and apoptosis (Watcharasit et al. 2003). Oxidative stress also induces insulin resistance through activation of JNK that results in the attenuation of intracellular insulin signal transduction (Ozcan et al. 2004). JNK activation may promote pathogenesis of AD by promoting tau phosphorylation at Thr205 and Ser422 and Aβ-mediated Fas ligand-dependent neuronal apoptosis (Troy et al. 2001). Stimulation of JNK promotes FOXO activation and sustained activation of FOXO proteins increases the transcription of gluconeogenic enzymes in the liver leading to hyperglycemia and subsequent hyperinsulinemia—both are closely associated to the pathophysiology of AD (Neumann et al. 2008). Interestingly, it is reported that hyperinsulinemia coincides with Aβ accumulation in mice lacking insulin-degrading enzyme, a protease involved in the clearance of both insulin and Aβ (Farris et al. 2003). Diabetes and obesity are common consequences of excessive energy intake, which compromises hippocampal neurogenesis, LTP, and memory formation (Stranahan et al. 2008a). Excessive caloric intake may also increase the risk for developing AD through alterations in permeability and integrity of the BBB (Kanoski et al. 2010), resulting in deficient neurovascular coupling. Through these mechanisms and others, conditions arising from excessive energy intake have an impact on brain regions implicated in various facets of cognition (Stranahan et al. 2008b). Chronic intake of high-fat diet leads to hyperlipidemia and hyperglycemia, which through several mechanisms (including JNK1 activation) reduce cellular insulin sensitivity, thereby disrupting metabolic homeostasis and inducing obesity (another component of MetS), which may influence the onset of AD through their effects on oxidative stress, hypertension, insulin resistance, and neuroinflammation (Haan 2006). JNK utilizes Akt, p53, and microtubule-associated protein tau as substrates (Bogoyevitch and Kobe 2006). Excessive ROS production stimulates JNK activity in susceptible neurons that surround the amyloid plaques. These cells in turn produce more Aβ peptide through JNK-dependent, γ-secretase-mediated APP cleavage (Shen et al. 2008). The phosphorylation of tau reduces its ability to promote microtubule assembly, and hyperphosphorylation inhibits binding to microtubules (Yoshida et al. 2004). Hyperphosphorylation of tau protein results in neurodegeneration through deleterious effects on axonal transport mechanisms and cell shape. The overexpression of tau also induces metabolic stress through the stimulation of JNK. These processes are supported by caspase activation and are required for NFT formation and neuronal death in neurodegenerative diseases, such as AD (Fig. 8.5) (Sahara et al. 2008). The phosphorylation of IRS-1serine (IRS-1pSer) is common to diabetes and AD (Bomfim et al. 2012). Brain tissues from AD patients and peripheral tissues from diabetes patients have elevated levels of IRS-1pSer and activated JNK activity compared to age-matched controls.

Insulin not only functions by controlling neurotransmitter release at the synapses but also acts by activating signaling pathways associated with learning and long-term memory. Insulin regulates processes such as neuronal survival, energy metabolism, and plasticity (de la Monte 2009). Two main pathophysiological mechanisms may be associated with brain insulin resistance in AD: (a) progressive loss of insulin/IGF responsive neurons induced by trophic factor withdrawal and (b) impairment in insulin/IGF ligand–receptor binding due to pathological alterations in membrane lipid composition along with reduction in membrane receptor expression. It is realized that AD may be initiated by decrease in the blood flow. However, insulin/IGF deficiency and insulin/IGF resistance in brain may also contribute to the pathogenesis of AD. Based on animal and cell culture experiments, it is proposed that AD represents a form of diabetes mellitus that selectively afflicts the brain (de la Monte 2009). It is called as type 3 diabetes (de la Monte and Wands 2008). It is well known that diabetes is characterized by hyperglycemia in peripheral tissues (Sacks et al. 2011), but at present, there is no evidence that the brain from AD patient is hyperglycemic (Talbot et al. 2012). Moreover, the insulin resistance in AD brains occurs in the absence of diabetes and does not involve glucose uptake in neurons the way peripheral insulin resistance does in the muscle, fat, and liver (Talbot et al. 2012; Liu et al. 2011). The role of impaired brain insulin signaling in the pathogenesis of AD is supported by several recent studies demonstrating improvements in cognition and memory by treatment with insulin or insulin sensitizers in AD patients (Reger et al. 2006, 2008) and in rodent models of AD and diabetes (Pedersen et al. 2006; Francis et al. 2008). It is interesting to note that rosiglitazone, an antidiabetic drug that increases insulin sensitivity, improves attention and memory in a subgroup of AD patients who do not carry the ApoE4 allele (Risner et al. 2006).

Based on these observations, the use of the term type III diabetes has been challenged and a more accurate term, insulin-resistant brain state (Correia et al. 2011), a neural variant of the insulin resistance syndrome (Reaven 2004) found in AD patients, has been recommended (Talbot et al. 2012). This syndrome in AD has been reported to be independent of APOE ε4 (Kuusisto et al. 1997).

In AD and MetS, insulin and insulin signaling are decreased compared to age-matched healthy control subjects (Craft et al. 1998; Rivera et al. 2005; Steen et al. 2005). Insulin/insulin receptor (IGF-R) signaling is closely associated with the regulation of peripheral glucose metabolism and energy homeostasis (Fig. 8.8). In the brain, the insulin/IGF-R complex is abundantly distributed in synaptic membranes of the cerebral cortex and hippocampus (Heidenreich et al. 1988; Zhao et al. 1999) and functions to regulate synaptic activities that are required for learning and memory (Zhao and Alkon 2001). Neurochemical changes in transgenic mice model of AD (Tg2576) are accompanied by the development of insulin resistance, suggesting a link between Aβ toxicity and insulin signaling (Pedersen and Flynn 2004). As stated above, during physiological insulin signaling, various ligands (trophic factors) initiate the autophosphorylation of tyrosine kinases. This leads to the recognition by insulin receptor substrate (IRS-1 and IRS-2) adaptors (White 2003). IRS is then phosphorylated at tyrosines, permitting recruitment of SH2 (Src homology 2) domain-containing proteins, including PtdIns 3K (Sun et al. 1995). This activates signaling pathways responsible for the pleiotropic actions of insulin. In contrast, during insulin resistance in diabetics, IRS-1 is phosphorylated at Ser312 by insulin-stimulated or stress-activated kinases, including c-Jun N-terminal kinase (JNK), which uncouples IRS-1 (Aguirre et al. 2002) and triggers rapid IRS-1 degradation, yielding a deficient signal transduction response (Sun et al. 1999). IRS deficits contribute to insulin resistance in animal models and diabetics (Saad et al. 1992; Rondinone et al. 1997). Significantly reduced IRS-1 and IRS-2 occur in AD brain, accompanied by elevated cytosolic phospho-IRS-1 (Ser312 and Ser616). These phosphoserine epitopes colocalize with neurofibrillary tangles, suggesting a possible link between "insulin resistance" and tau pathology (Moloney et al. 2008).

Studies on human postmortem brain tissue link many of the characteristic molecular and pathological features of AD to reduce expression of the insulin and insulin-like growth factor (IGF) genes and their corresponding receptors (Lester-Coll et al. 2006). Furthermore, intracerebral streptozotocin (STZ) produces chemical depletion of insulin and IGF signaling mechanisms combined with oxidative injury. These changes produce neurodegeneration similar to AD (de la Monte et al. 2010). Alterations in insulin sensitivity and signaling may contribute to the unstability of synapses in AD (Kroner 2009). Insulin resistance and obesity also result in changes in levels of a number of different energy-regulating signaling peptides and small proteins. For example, leptin enhances hippocampal synaptic plasticity and improves performance of rodents in learning and memory tasks (Harvey 2007). Studies on transgenic mice indicate that excess accumulation of pancreatic amyloid leads to β-cell dysfunction, disruption in glucose homeostasis, and diabetes

(Fig. 8.9). Similarly, in humans, on autopsy islet, amyloid polypeptide and hyperphosphorylated tau are found in pancreatic islet cells of patients with type II diabetes. Furthermore, an increased amount of neurofibrillary tangles and amyloid plaques in the hippocampus have been found on autopsy in patients with diabetes (Miklossy et al. 2010; Peila et al. 2002), supporting the view that deposition of amyloid in the brain and pancreatic islet cells represent a pathogenic similarity between AD and type II diabetes. Mounting evidence suggests that Aβ accumulation, insulin signaling, and tau phosphorylation may play crucial roles in the pathological interaction between AD and diabetes. Aβ accumulation and diabetes have been shown to impair insulin signaling, thus accelerating tau phosphorylation and neurodegeneration.

## 8.8 Leptin in Diabetes

Diabetes is a chronic disease that is characterized by the failure of pancreatic islets to produce insulin and/or the sensitivity of glucose-metabolizing tissues to insulin decreases. In type I diabetes, pancreatic islet β cells are mainly destroyed by autoimmune-induced apoptosis, leading to the loss of insulin production (Kawasaki et al. 2004). In contrast, type II diabetes is caused by the failure of β-cell function and β-cell mass reduction along with the increase in circulating adipocytokines, elevation in FFAs, and with persistent hyperglycemia (Stumvoll et al. 2005). Chronic exposure of β cell to adipocytokines, FFAs, and high glucose induces excessive ROS production and activation of caspases, which inhibit insulin secretion and promote apoptosis of pancreatic β cells. Diabetes is accompanied by increase in adiposity that causes insulin resistance. Insulin resistance results in hyperinsulinemia to maintain normal glucose levels.

Inability of pancreas to overcome insulin resistance results in type II diabetes. These processes are components of MetS, which includes hypertension, dyslipidemia, and elevated systemic inflammation.

Leptin (16 kDa) is an adipocyte hormone that controls energy balance and food intake acting in the brain but can also modulate the function of several peripheral organs and tissues. Leptin levels are significantly elevated in obese individuals and are higher in women than in man (Schwartz et al. 2002). Leptin modulates the control on food intake and energy expenditure by acting on brain centers within the hypothalamus that control satiety and body weight. Thus, leptin mediates the transfer information from adipocytes to the brain. This endocrine function of leptin couples feeding behavior, metabolism, and endocrine physiology to the nutritional state of the organism. Leptin induces its action by interacting with specific receptors (Ob-R). Alternate splicing of Ob-R gives rise to six isoforms (Ob-Ra, Ob-Rb, Ob-Rc, Ob-Rd, Ob-Re, and Ob-Rf), all of which possess the extracellular leptin binding domain, but only Ob-Rb mediates signaling, due to the presence of full intracellular domain (Elmquist et al. 1998). Thus, ObRs show the greatest homology with the class I cytokine receptor superfamily (Tartaglia et al. 1995): receptors that lack intrinsic tyrosine kinase activity but signal via association with Janus tyrosine kinases (JAKs). Activation of Ob-R facilitates the phosphorylation of JAK2, which subsequently phosphorylates STAT3 to mediate transcription of a number of genes. Leptin receptor activation also stimulates PtdIns 3K and Ras/mitogen-activated protein kinase (MAPK), NF-κB, and AMPK signaling pathways, all of which are downstream of JAK. Leptin deficiency can be induced either by a defect in leptin transport across the BBB or by suppression of leptin signaling (Flier 2004), which is inhibited not only by suppressors of cytokine signaling-3 (SOCS3) but also by dephosphorylation via protein tyrosine phosphatase 1B. SOCS3- and PTP-1B-deficient mice exhibit increased sensitivity to leptin and resistance to obesity (Flier 2004; Doherty 2011). The deficiency of leptin or its receptor in rodents and humans causes severe hyperphagia, obesity, insulin resistance, and neuroendocrine and reproductive dysfunction (Gautron and Elmquest 2011). Insulin resistance caused by leptin deficiency can be reversed by administration of exogenous leptin (Halaas et al. 1997). Leptin administration lowers circulating glucagon levels and decreases gluconeogenic gene expression in animals with type I diabetes (Yu et al. 2008). In the skeletal muscle, leptin increases fatty acid oxidation and lowers triacylglycerol (TAG) levels (Havel 2004). Leptin administration also improves insulin sensitivity in rodent models of obesity and type I diabetes (Halaas et al. 1997; Chinookoswong et al. 1999). Although mechanisms by which leptin improves insulin sensitivity remain undefined, it is becoming increasingly evident that leptin either may act on peripheral tissues, such as liver and skeletal muscle, or indirectly through its effects in the brain to activate neural or endocrine signals to the periphery (Rossetti et al. 1997; Havel et al. 1998; Myers et al. 2008). Collective evidence suggests that diabetes is caused by relative insulin deficiency, which may be the result of insufficient insulin supply due to defective insulin secretion and/or reduced insulin-secreting β-cell mass and/or impaired insulin sensitivity in peripheral metabolic organs such as liver and muscle along with alterations in leptin signaling cascade.

## 8.9 Leptin in Alzheimer Disease

Leptin enters the brain through two distinct mechanisms: One mechanism that involves a saturable transport system promotes leptin entry through the blood–brain barrier via receptor-mediated transcytosis (Banks et al. 1996), and another mechanism promotes the transport of leptin in the brain through the cerebrospinal fluid (CSF; Schwartz et al. 1996). This process may involve the choroid plexus, the key site for production of CSF. Choroid plexus is enriched in Ob-Ra, which may mediate transport of leptin from the blood to the CSF (Bjorbaek et al. 1998). In mice, impairment in leptin transport across the blood–brain barrier promotes obesity, which can be reversed by modest weight reduction (Banks and Farrell 2003). Moreover, recent studies have also shown that leptin transport is regulated by epinephrine and triacylglycerol (Banks 2001; Banks et al. 2004). Although in the hypothalamus, leptin plays an important role in energy homeostasis, in the hippocampus, midbrain, and hindbrain, leptin contributes to neurogenesis, axonal growth, and synaptogenesis (Bouret 2010). Leptin-deficient patients show reduction in brain weight, myelination, and synaptogenesis (Frank et al. 2011). In addition, in the cortex, leptin acts as a potent neurogenic factor, not only inducing structural changes but also modifying neuronal excitability (Paz-Filho et al. 2010). This observation suggests a cognitive role of leptin in this area. In hippocampus, leptin modulates synaptic plasticity. In rodents (db/db mice or fa/fa rats), mutation in leptin receptor is associated with impairments in the hippocampal LTP and long-term depression as well as deficits in the hippocampal-specific memory tasks (Li et al. 2002b). Direct administration of leptin into the hippocampus improves learning and memory performance (Farr et al. 2006; Oomura et al. 2006) and facilitates hippocampal LTP (Wayner et al. 2004). Furthermore, leptin facilitates the conversion of short-term potentiation (STP) into LTP (Shanley et al. 2001) and enhances LTP at hippocampal CA1 synapses (Oomura et al. 2006). Leptin modulates synaptic plasticity by enhancing *N*-methyl-D-aspartate (NMDA) receptor function (Shanley et al. 2001; Oomura et al. 2006). Since, synaptic plasticity is the cellular basis of memory formation, it is suggested that leptin may produce beneficial effects on the memory loss in animal models of AD. Thus, rodents that have mutations in Ob-R show memory impairment (Li et al. 2002b). Furthermore, treatment of transgenic mice with exogenous leptin not only reduces AD-linked pathological changes but also enhances memory in these mice (Greco et al. 2010). In addition, studies on the effect of leptin on NMDA receptor indicate that leptin facilitates both NMDA receptor-dependent synaptic currents in hippocampal slices from 2-week-old to 3-week-old rats and NMDA-induced $Ca^{2+}$ influx in hippocampal cultures via activation of PtdIns 3K and MAPK-dependent signaling cascades (Shanley et al. 2001). These studies support the view that leptin plays an important role in hippocampal synaptic plasticity as well as in the induction and maintenance of hippocampal LTP. Moreover, leptin inhibits Aβ(1–42)-driven facilitation of long-term depression (LTD) and internalization of the 2-amino-3-(5-methyl-3-oxo-1,2-oxazol-4-yl)propanoic acid (AMPA) receptor subunit, GluR1, via activation of PtdIns 3K (Doherty et al. 2013).

Epidemiological studies have indicated that dysregulation of plasma leptin levels is also associated with the development of AD. Lower circulating levels of leptin have been reported in patients with AD (Power et al. 2001; Lieb et al. 2009). Leptin modulates Aβ production and tau hyperphosphorylation in vivo and in vitro (Fig. 8.10) (Tezapsidis et al. 2009). Chronic leptin administration has been reported to reduce Aβ levels in Tg2576 (Fewlass et al. 2004) and improve cognitive performance in CRND8 transgenic mice models for AD (Greco et al. 2010; Beccano-Kelly and Harvey 2012). Although the molecular mechanism associated with the effect of leptin is not fully understood, it is suggested that leptin decreases the activity of β-secretase (BACE-1) in the SH-SY5Y cells (Fewlass et al. 2004). It is also reported that Aβ and leptin interact with the Akt/mammalian target of rapamycin complex1 (mTORC1) signaling pathway (Marwarha et al. 2010). Akt/mTORC1 activation retards tau phosphorylation by blocking the downstream

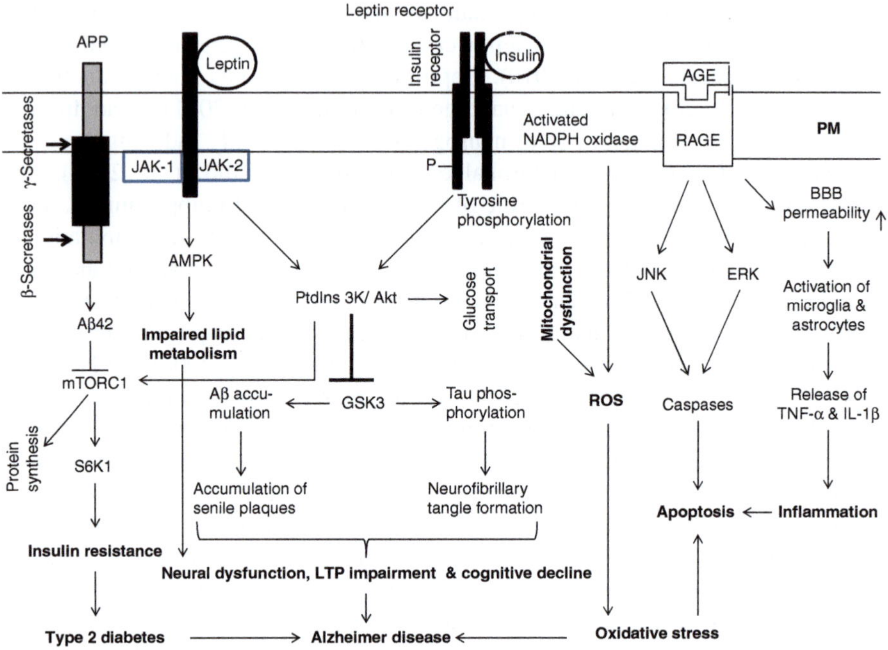


**Fig. 8.10** Interactions between leptin and insulin signaling and their relationship with development of Alzheimer disease. Phosphatidylcholine (PtdCho), sphingomyelin (SM), cytosolic phospholipase $A_2$ ($cPLA_2$), sphingomyelinase (SMase), arachidonic acid (ARA), reactive oxygen species (ROS), cyclooxygenase-2 (COX-2), reactive oxygen species (ROS), platelet activating factor (PAF), phosphatidylinositol 3-kinase (PtdIns 3K), glycogen synthase kinase 3 (GSK3), amyloid precursor protein (APP), β-amyloid (Aβ), and Janus kinase (JAK)

enzyme GSK3β. mTORC1 also regulates translation of many proteins including leptin. Aβ inactivates Akt, inhibits mTORC1, and facilitates the phosphorylation of tau. In contrast, leptin activates both Akt and mTORC1 and reduces tau phosphorylation. Incubation of organotypic slices from rabbit hippocampus with Aβ results in downregulation of leptin expression, inhibition of Akt, activation of GSK3β, elevation in tau phosphorylation, and inactivation of mTORC1 (Marwarha et al. 2010). Treatment with leptin reverses Aβ effects by alleviating Akt inhibition, blocking GSK3β activation, inhibiting tau phosphorylation, and stimulating mTORC1. On the other hand, Rapamycin, an allosteric inhibitor of mTORC1, downregulates leptin expression, increases tau phosphorylation, and does not affect Akt and GSK3β (Marwarha et al. 2010). Collective evidence suggests that leptin downregulation precedes and triggers accumulation of Aβ and phosphorylation of tau. In addition, Aβ regulates leptin expression and tau phosphorylation through mTORC1. Impaired leptin signaling contributes to the diet-induced deficits in hippocampal plasticity in AD brain (Marwarha et al. 2010). Leptin also protects cortical neurons from Aβ-mediated cell death by a signal transducer and activator of transcription 3 (STAT3)-dependent mechanism (Doherty et al. 2013). Based on the importance of leptin signaling in the brain, it is suggested that leptin signaling is associated with modulation of Aβ and its involvement in the pathogenesis of AD. It is recently shown that individuals with higher leptin levels have a much lower risk of developing AD (Holden et al. 2009; Lieb et al. 2009). Future studies on cross talk between defective leptin signaling and perturbed insulin signaling may shed more light not only on the significance of leptin/insulin signaling in regulation of Aβ and its oligomerization in the pathogenesis of AD but also on the link between obesity, diabetes, and AD. In addition, these studies may provide molecular mechanism of accelerated AD progression in insulin deficiency. Patients with type II diabetes are characterized by insulin resistance. They are at an increased risk of getting AD not only due to aging but also because peripheral insulin resistance induces brain insulin resistance, which accelerates the assembly of Aβ by increasing and clustering of GM1 in detergent-resistant membrane microdomain of neuronal membranes (Yamamoto et al. 2012). This suggestion is supported by reduction in insulin receptor mRNA, insulin protein levels, and abnormal insulin receptor downstream signaling in autopsied AD brains (Craft et al. 1998; Steen et al. 2005; Moloney et al. 2010).

## 8.10 Metal Ions in Alzheimer Disease and Metabolic Syndrome

Iron is essential for normal neuronal metabolism, including mitochondrial energy generation, transport of oxygen, electron transfer, synthesis of neurotransmitters, and myelination (Kell 2010). However, excessive levels of brain iron may exert iron-induced oxidative stress and thus lead to neurodegeneration (Zecca et al. 2004). As stated above, iron accumulation has been reported to occur in the hippocampus and the cortex, brain regions that are affected in AD patients (Crichton et al. 2011; Smith et al. 2010). In these regions, iron and proteins that bind iron accumulate in the amyloid plaques (Silvestri and Camaschella 2008) and appear to play a role in the degree of tau phosphorylation (Egana et al. 2003). It is suggested that dysregulation of the level of iron in the brain may enhance the amyloidogenic pathway and potentiate the aggregation and toxicity of Aβ. Similarly, levels of iron are markedly increased in type II diabetes and MetS (Swaminathan et al. 2007; Rajpathak et al. 2009). In addition, iron also impedes insulin extraction in the liver, impair pancreatic insulin secretion, and interfere with insulin action and glucose uptake in adipocytes. Collectively, these studies indicate that AD and MetS are accompanied by an imbalance of transition metal (copper and iron) levels in the brain tissue. These metals promote oxygen transport in nervous tissue. Copper and iron imbalance can induce the generation of ROS which are closely associated with the pathogenesis of

AD, diabetes, and MetS (Chadwick et al. 2010; Leiva et al. 2013; Le Blanc et al. 2012; Jiang et al. 2009; Barnham et al. 2004). ROS have been reported to play a substantive role not only in insulin resistance (West 2000; Evans et al. 2005) and in a variety of diabetic sequelae, but also in mitochondrial dysfunction, which may be an early step in this process (Kelley et al. 2002). It should be noted that clioquinol–Cu metal complexes, which are neural membrane permeable, have been used to ameliorate AD pathology via their ability to stimulate epidermal growth factor receptor in neuronal-derived cell lines producing a reduction in β-amyloid levels (Price et al. 2008). Imbalance of copper may also be involved in the activation of trabecular meshwork matrix metalloproteinases (MMPs) as well as the nonreceptor tyrosine kinase c-Src (Price et al. 2008). These studies support the view that dysregulation of iron and copper in the brain and visceral tissues may be a common link that contributes to the pathogenesis of AD and MetS.

## 8.11 Conclusion

AD is a neurodegenerative disease, which is characterized by the accumulation of intracellular neurofibrillary tangles (NFTs), containing an abnormal hyperphosphorylated form of τ protein and extracellular senile plaques (SPs), mainly composed of aggregated Aβ. Under normal conditions, mature APP is proteolytically degraded by nonamyloidogenic α-secretase pathways. This enzyme cleaves APP within the Aβ domain, yielding soluble N-terminal APPα and a 10 kDa C-terminal fragment (α-CTF) that can be further processed by γ-secretase to generate Aβ17–40/Aβ17–42 fragments. In AD, APP processing is performed by β-secretase, which cleaves APP to generate soluble APPβ and an Aβ-containing C-terminal fragment (β-CTF) (Kodam et al. 2011). This fragment is subsequently processed via γ-secretase to yield full-length Aβ1–40/Aβ1–42 peptides. Deposition of hyperphosphorylated form of τ protein containing NFTs and hyperphosphorylated form of τ protein containing SPs is supported by (a) damage of the brain cerebrovascular system over the years leading to metabolic disturbances in the brain and (b) a disbalance in the clearance of Aβ from brain to blood or vice versa which may increase brain Aβ1–40/Aβ1–42 levels. A cascade of downstream events may lead to Aβ and tau pathology.

MetS is a pathological condition, which is characterized by a group of biochemical abnormalities, such as disturbed glucose and insulin metabolism resulting in hyperglycemia and hyperinsulinemia, dyslipidemia, insulin resistance, visceral obesity, elevated triacylglycerol, low-grade inflammation, hypertension, and type II diabetes. MetS and AD are more prevalent in the aging population. Glucose homeostasis plays a key role in cognitive function. During insulin resistance, there is a reduction in sensitivity to insulin resulting in hyperinsulinemia. This condition persists for several years before becoming full-blown type II diabetes, which rapidly leads to MetS. Toxic levels of insulin negatively modulate neuronal function and survival. Peripheral hyperinsulinemia correlates with an abnormal removal of the Aβ peptide and an increase in tau hyperphosphorylation due to alterations in cdk5 and GSK3β activities. Some uncertainty remains into the exact pathogenesis of type II diabetes and MetS. However, it is becoming increasingly evident that several mechanisms, including hyperglycemia, insulin resistance, microvascular disease, glucose toxicity, oxidative stress, inflammation, and atherosclerosis may modulate the brain function and contribute to the pathogenesis of type II diabetes and MetS. At the molecular level, the modification of neuronal insulin and leptin signaling by diabetic conditions may contribute to AD progression. In addition, cerebrovascular alteration may also be common to type II diabetes and AD.

Accumulating evidence suggests the importance of Aβ-induced cerebrovascular dysfunction in AD and indicates that pathological interactions between the receptor for advanced glycation end products (RAGE) and Aβ peptides may play a role in this dysfunction. Long-term presence of above processes may result in cerebrovascular disease, which may accelerate cognitive decline and promote dementia. Animal models of type II diabetes show reduced insulin transport to the brain, reduced insulin uptake, and reduced neuronal insulin. This is similar to reported reduced insulin levels, insulin receptor expression, and insulin resistance in the brains of AD patients. There are several commonalities between AD and MetS. These commonalities include not only occurrence of NFTs and SP, cell loss, alterations in leptin levels, and degenerative changes but also alterations in activities of several enzymes (glutamic acid decarboxylase, DOPA decarboxylase) and growth factor receptors (p75 receptors, neuronal growth factor receptors, and thyrotrophin-releasing hormone). Furthermore, mitochondrial dysfunction, which is caused by increased production of ROS and AGEs, is also common in MetS and AD. AD is accompanied with neurodegeneration in the nucleus basalis and hippocampus. Similarly, type II diabetes results in the selective destruction of pancreatic β cells. Common risk factors for the onset MetS and AD include higher cholesterol, deposition of Aβ, involvement of β, and tau protein phosphorylation; induction of oxidative stress, increase in phospholipid-, sphingolipid-, and cholesterol-derived lipid mediators; induction of inflammation, alterations in leptin levels, involvement of ApoE4, and apoptosis. Studies in animal and cell culture models have shown that there are synergistic interactions among risk factors of AD, diabetes, and MetS along with elevation in Aβ peptide levels. Some of these parameters can be modified by diet. However, due to the multifactorial nature of AD, diabetes, MetS, and multiplicity of physiological pathways involved, it may not be possible to treat AD or complication of diabetes and MetS with a single drug. It is becoming increasingly evident that leptin prevents the toxic accumulation of Aβ and phosphorylated tau in neurons, and it has the ability to improve performance in various memory tasks in murine AD models. These findings, coupled with the already established safety of leptin in humans, make this hormone or related leptin mimetics in concert with changes in diet and lifestyle novel therapeutic candidates for the treatment of AD as well as MetS.

## References


Adamo M, Raizada MK, LeRoith D (1989) Insulin and insulin-like growth factor receptors in the nervous system. Mol Neurobiol 3:71–100

Aerts JM, Ottenhoff R, Powlson AS, Grefhorst van Ejik AM, Dubbelhuis PF, Aten J, Kuipers F, Serlie MJ, Wennekes T, Sethi JK, O'Rahilly S, Overkleeft HS (2007) Pharmacological inhibition of glucosylceramide synthase enhances insulin sensitivity. Diabetes 56:1341–1349

Aguirre V, Werner ED, Giraud J, Lee YH, Shoelson SE, White MF (2002) Phosphorylation of Ser307 in insulin receptor substrate-1 blocks interactions with the insulin receptor and inhibits insulin action. J Biol Chem 277:1531–1537

Akisaki T, Sakurai T, Takata T, Umegaki H, Araki A, Mizuno S, Tanaka S, Ohashi Y, Iguchi A, Yokono K, Ito H (2006) Cognitive dysfunction associates with white matter hyperintensities and subcortical atrophy on magnetic resonance imaging of the elderly diabetes mellitus Japanese elderly diabetes intervention trial (J-EDIT). Diabetes Metab Res Rev 22:376–384

Alzheimer's Association (2010) Alzheimer's disease facts and figures. Alzheimers Dement 6:158–194

Alzheimer's Association (2012) 2011 Alzheimer's diseases facts and figures: prevalence. Alzheimers Dement 7:12–13

Amaducci L, Tesco G (1994) Aging as a major risk for degenerative diseases of the central nervous system. Curr Opin Neurol 7:283–286

Arvanitakis Z, Wilson RS, Bievias JL, Evans DA, Bennett DA (2004) Diabetes mellitus and risk of Alzheimer's disease and decline in cognitive function. Arch Neurol 61:661–666

Assini A, Cammarata S, Vitali A, Colucci M, Giliberto L, Borghi R, Inglese ML, Volpe S, Ratto S, Dagna-Bricarelli F, Baldo C, Argusti A, Odetti P, Piccini A, Tabaton M (2004) Plasma levels of amyloid β-protein 42 are increased in women with mild cognitive impairment. Neurology 63:828–831

Austin BP, Nair VA, Mejer TB, Xu G, Rowley HA, Carlsson CM, Johnson SC, Prabhakaran V (2011) Effects of hypoperfusion in Alzheimer's disease. J Alzheimers Dis 26:123–133

Bale TL, Vale WW (2004) CRF and CRF receptors: role in stress responsivity and other behaviors. Annu Rev Pharmacol Toxicol 44:525–557

Banks WA (2001) Enhanced leptin transport across the blood brain barrier by alpha 1-adrenergic agents. Brain Res 899:209–217

Banks WA, Farrell CL (2003) Impaired transport of leptin across the blood brain barrier in obesity is acquired and reversible. Am J Physiol Endocrinol Metab 285:E10–E15

Banks WA, Kastin AJ, Huang W, Jaspan JB, Maness LM (1996) Leptin enters the brain by a saturable system independent of insulin. Peptides 17:305–311

Banks WA, Coon AB, Robinson SM, Moinuddin A, Shultz JM, Nakaoke R, Morley JE (2004) Triglycerides induce leptin resistance at the blood brain barrier. Diabetes 53:1253–1260

Barnham KJ, Masters CL, Bush AI (2004) Neurodegenerative diseases and oxidatives stress. Nat Rev Drug Discov 3:205–214

Barrier L, Ingrand S, Damjanac M, Rioux Bilan A, Hugon J, Page G (2007) Genotype-related changes of ganglioside composition in brain regions of transgenic mouse models of Alzheimer's disease. Neurobiol Aging 28:1863–1872

Bayatti N, Behl C (2005) The neuroprotective actions of corticotropin releasing hormone. Ageing Res Rev 4:258–270

Beccano-Kelly D, Harvey J (2012) Leptin: a novel therapeutic target in Alzheimer's disease? Int J Alzheimers Dis 2012:594137

Behan DP, Khongsaly O, Owens MJ, Chung HD, Nemeroff CB, De Souza EB (1997) Corticotropin-releasing factor (CRF), CRF-binding protein (CRF-BP), and CRF/CRF-BP complex in Alzheimer's disease and control postmortem human brain. J Neurochem 68:2053–2060

Belgardt BF, Mauer J, Brüning JC (2010) Novel roles for JNK1 in metabolism. Aging (Albany NY) 2(9):621–626

Bell RD, Zlokovic BV (2009) Neurovascular mechanisms and blood–brain barrier disorder in Alzheimer's disease. Acta Neuropathol 118:103–113

Bellingham SA, Ciccotosto GD, Needham BE, Fodero LR, White AR, Masters CL, Cappai R, Camakaris J (2004) Gene knockout of amyloid precursor protein and amyloid precursor-like protein-2 increases cellular copper levels in primary mouse cortical neurons and embryonic fibroblasts. J Neurochem 91:423–428

Bennett RG, Duckworth WC, Hamel FG (2000) Degradation of amylin by insulin-degrading enzyme. J Biol Chem 275:36621–36625

Bevan P (2001) Insulin signalling. J Cell Sci 114:1429–1430

Bhat NR (2010) Linking cardiometabolic disorders to sporadic Alzheimer's disease: a perspective on potential mechanisms and mediators. J Neurochem 115:551–562

Biessels GJ, Kamal A, Ramakers GM, Urban IJ, Spruijt BM, Erkelens DW, Gispen WH (1996) Place learning and hippocampal synaptic plasticity in streptozotocin-induced diabetic rats. Diabetes 45:1259–1266

Biessels GJ, De Leeuw FE, Lindeboom J, Barkhof F, Scheltens P (2006) Increased cortical atrophy in patients with Alzheimer's disease and type 2 diabetes mellitus. J Neurol Neurosurg Psychiatry 77:304–307

Bingham EM, Hopkins D, Smith D (2002) The role of insulin in human brain glucose metabolism: an 18-fluoro-deoxyglucose positron emission tomography study. Diabetes 51:3384–3390

Birbes H, Luberto C, Hsu YT, El Bawab S, Hannun YA, Obeid LM (2005) A mitochondrial pool of sphingomyelin is involved in TNFalpha-induced Bax translocation to mitochondria. Biochem J 386:445–451

Bjorbaek C, Elmquist JK, Michl P, Ahima RS, van Bueren A, McCall AL, Flier JS (1998) Expression of leptin receptor isoforms in rat brain microvessels. Endocrinology 139:3485–3491

Bogoyevitch MA, Kobe B (2006) Uses for JNK: the many and varied substrates of the c-Jun N-terminal kinases. Microbiol Mol Biol Rev 70:1061–1095

Boldyrev AA, Johnson P (2007) Homocysteine and its derivatives as possible modulators of neuronal and non-neuronal cell glutamate receptors in Alzheimer's disease. J Alzheimers Dis 11:219–228

Bomfim TR, Forny-Germano L, Sathler LB, Brito-Moreira J, Houzel JC, Decker H, Silverman MA, Kazi H, Melo HM, McClean PL, Holscher C, Arnold SE, Talbot K, Klein WL, Munoz DP, Ferreira ST, De Felice FG (2012) An anti-diabetes agent protects the mouse brain from defective insulin signaling caused by Alzheimer's disease-associated Aβ oligomers. J Clin Invest 122:1339–1353

Bosco D, Fava A, Plastino M, Montaecini T, Puja A (2011) Possible implications of insulin resistance and glucose metabolism in Alzheimer's disease pathogenesis. J Cell Mol Med 15:1807–1821

Bottiglieri T, Laundy M, Crellin R, Toone BK, Carney MW, Reynolds EH (2000) Homocysteine, folate, methylation, and monoamine metabolism in depression. J Neurol Neurosurg Psychiatry 69:228–232

Bouret SG (2010) Neurodevelopmental actions of leptin. Brain Res 1350:2–9

Brands AM, Kessels RP, de Haan EH, Kappelle LJ, Biessels GJ (2004) Cerebral dysfunction in type 1 diabetes: effects of insulin, vascular risk factors and blood-glucose levels. Eur J Pharmacol 490:159–168

Brownlee M (1995) Advanced protein glycosylation in diabetes and aging. Annu Rev Med 46:223–234

Buchsteiner A, Hauß T, Dante S, Dencher NA (2010) Alzheimer's disease amyloid-β peptide analogue alters the ps-dynamics of phospholipid membranes. Biochim Biophys Acta 1798:1969–1976

Butterfield DA, Bader Lange ML, Sultana R (2010) Involvements of the lipid peroxidation product, HNE, in the pathogenesis and progression of Alzheimer's disease. Biochim Biophys Acta 1801:924–929

Buxbaum JD, Cullen EI, Friedhoff LT (2002) Pharmacological concentrations of the HMG-CoA reductase inhibitor lovastatin decrease the formation of the Alzheimer beta-amyloid peptide in vitro and in patients. Front Biosci 7:a50–a59

Cai D (2009) NFkappaB-mediated metabolic inflammation in peripheral tissues versus central nervous system. Cell Cycle 8:2542–2548

Cai D, Yuan M, Frantz DF, Melendez PA, Hansen L, Lee J, Shoelson SE (2005) Local and systemic insulin resistance resulting from hepatic activation of IKK-b and NF-kappaB. Nat Med 11:183–190

Castillo MJ, Scheen AJ, Letiexhe MR, Lefebvre PJ (1994) How to measure insulin clearance. Diabetes Metab Rev 10:119–150

Celik T, Iyisoy A, Kardesoglu E, Isik E (2007) The clinical significance of microvascular impairment in patients with pure uncomplicated diabetes mellitus. Int J Cardiol 133:394–396

Chadwick W, Zhou Y, Park SS, Wang L, Mitchell N, Stone MD, Becker KG, Martin B, Maudsley S (2010) Minimal peroxide exposure of neuronal cells induces multifaceted adaptive responses. PLoS One 5:e14352

Chang L, Chian S, Saltiel AR (2004) Insulin signaling and regulation of glucose transport. Mol Med 10:65–71

Chinookoswong N, Wang JL, Shi ZQ (1999) Leptin restores euglycemia and normalizes glucose turnover in insulin-deficient diabetes in the rat. Diabetes 48:1487–1492

Chiti F, Dobson CM (2006) Protein misfolding, functional amyloid, and human disease. Annu Rev Biochem 75:333–366

Cho JH, Johnson GVW (2004) Primed phosphorylation of tau at Thr 231 by glycogen synthase kinase 3β (GSK3β) plays a critical role in regulating tau's ability to bind and stabilize microtubules. J Neurochem 88:349–358

Clarke DD, Sokoloff L (1999) Circulation and energy metabolism of the brain. In: Siegel GJ, Agranoff BW, Albers RW, Molinoff SK, Fisher PB, Uhler MD (eds) Basic neurochemistry. Lippincott-Ratven, Philadelphia, PA, pp 637–669

Clippingdale AB, Wade JD, Barrow CJ (2001) The amyloid-β peptide and its role in Alzheimer's disease. J Peptide Sci 7:227–249

Cole SL, Vassar R (2007) The Alzheimer's disease β-secretase enzyme, BACE1. Mol Neurodegener 2:22

Colonna SV, Douglas Case L, Lawrence JA (2012) A retrospective review of the metabolic syndrome in women diagnosed with breast cancer and correlation with estrogen receptor. Breast Cancer Res Treat 131:325–331

Conejo R, Lorenzo M (2001) Insulin signaling leading to proliferation, survival, and membrane ruffling in C2C12 myoblasts. J Cell Physiol 187:96–108

Connor JR, Snyder BS, Beard JL, Fine RE, Mufson EJ (1992) Regional distribution of iron and iron-regulatory proteins in the brain in aging and Alzheimer's disease. J Neurosci Res 31:327–335

Cook DG, Leverenz JB, McMillan PJ, Kulstad JJ, Ericksen S, Roth RA, Schellenberg GD, Jin LW, Kovacina KS, Craft S (2003) Reduced hippocampal insulin-degrading enzyme in late-onset alzheimer's disease is associated with the apolipoprotein Eepsilon4 allele. Am J Pathol 162:313–319

Correia SC, Santos RX, Perry G, Zhu X, Moreira PI, Smith MA (2011) Insulin-resistant brain state: the culprit in sporadic Alzheimer's disease? Ageing Res Rev 10:264–273

Costa RM, Drew J, Silva AJ (2005) Notch to remember. Trends Neurosci 28:429–435

Costantini C, Kolasani RMK, Puglielli L (2005) Ceramide and cholesterol: possible connections between normal aging of the brain and Alzheimer's disease. Just hypotheses or molecular pathways to be identified? Alzheimers Dement 1:43–50

Craft S (2007) Insulin resistance and Alzheimer's disease pathogenesis: potential mechanisms and implications for treatment. Curr Alzheimer Res 4:147–152

Craft S, Newcomer J, Kanne S, Dagogo-Jack S, Cryer P, Sheline Y, Luby J, Dagogo-Jack A, Alderson A (1996) Memory improvement following induced hyperinsulinemia in Alzheimer's disease. Neurobiol Aging 17:123–130

Craft S, Peskind E, Schwartz MW, Schellenberg GD, Raskind M, Porte D Jr (1998) Cerebrospinal fluid and plasma insulin levels in Alzheimer's disease: relationship to severity of dementia and apolipoprotein E genotype. Neurology 50:164–168

Crichton RR, Dexter DT, Ward RJ (2011) Brain iron metabolism and its perturbation in neurological diseases. J Neural Transm 118:301–314

Cross DA, Alessi DR, Cohen P, Andjelkovich M, Hemmings BA (1995) Inhibition of glycogen synthase kinase-3 by insulin mediated by protein kinase B. Nature 378:785–789

Cui W, Sun Y, Wang Z, Xu C, Xu L, Wang F, Chen Z, Peng Y, Li R (2011) Activation of liver X receptor decreases BACE1 expression and activity by reducing membrane cholesterol levels. Neurochem Res 36:1910–1921

Cutler RG, Mattson MP (2001) Sphingomyelin and ceramide as regulators of development and lifespan. Mech Ageing Dev 122:895–908

Cutler RG, Kelly J, Storie K, Pedersen WA, Tammara A, Hatanpaa K, Troncoso JC, Mattson MP (2004) Involvement of oxidative stress-induced abnormalities in ceramide and cholesterol metabolism in brain aging and Alzheimer's disease. Proc Natl Acad Sci U S A 17:2070–2075

D'Ercole AJ, Ye P (2008) Expanding the mind: insulin-like growth factor I and brain development. Endocrinology 149:5958–5962

Dankner R, Chetrit A, Shanik MH, Raz I, Roth J (2009) Basal-state hyperinsulinemia in healthy normoglycemic adults is predictive of type 2 diabetes over a 24-year follow-up: a preliminary report. Diabetes Care 32:1464–1466

Deane R, Du Yan S, Submamaryan RK, LaRue B, Jovanovic S, Hogg E, Welch D, Manness L, Lin C, Yu J, Zhu H, Ghiso J, Frangione B, Stern A, Schmidt AM, Armstrong DL, Arnold B, Liliensiek B, Nawroth P, Hofman F, Kindy M, Stern D, Zlokovic B (2003) RAGE mediates amyloid-beta peptide transport across the blood-brain barrier and accumulation in brain. Nat Med 9:907–913

Deane R, Wu Z, Zlokovic BV (2004) RAGE (yin) versus LRP (yang) balance regulates Alzheimer amyloid β-peptide clearance through transport across the blood–brain barrier. Stroke 35:2628–2631

De Felice FG, Velasco PT, Lambert MP, Viola K, Fernandez SJ, Ferreira ST, Klein WL (2007) A beta oligomers induce neuronal oxidative stress through an N-methyl-D-aspartate receptor-dependent mechanism that is blocked by the Alzheimer drug memantine. J Biol Chem 282:11590–11601

De Felice FG, Wu D, Lambert MP, Fernandez SJ, Velasco PT, Lacor PN, Bigio EH, Jerecic J, Acton PJ, Shughrue PJ, Chen-Dodson E, Kinney GG, Klein WL (2008) Alzheimer's disease-type neuronal tau hyperphosphorylation induced by A beta oligomers. Neurobiol Aging 29:1334–1347

De Felice FG, Vieira MN, Bomfim TR, Decker H, Velasco PT, Lambert MP, Viola KL, Zhao WQ, Ferreira ST, Klein WL (2009) Protection of synapses against Alzheimer's-linked toxins: insulin signaling prevents the pathogenic binding of Abeta oligomers. Proc Natl Acad Sci U S A 106:1971–1976

de la Monte SM (2009) Insulin resistance and Alzheimer's disease. BMB Rep 42:475–481

de la Monte SM, Wands JR (2005) Review of insulin and insulin-like growth factor expression, signaling, and malfunction in the central nervous system: relevance to Alzheimer's disease. J Alzheimers Dis 7:45–61

de la Monte SM, Wands JR Jr (2008) Alzheimer's disease is type 3 diabetes-evidence reviewed. J Diabetes Sci Technol 2:1101–1113

de la Monte SM, Tong M, Nguyen V, Setshedi M, Longato L, Wands JR (2010) Ceramide-mediated insulin resistance and impairment of cognitive-motor functions. J Alzheimers Dis 21:967–984

DeFronzo RA (1997) Pathogenesis of type 2 diabetes: metabolic and molecular implications for identifying diabetes. Diabetes Rev 5:177–269

Dodson SE, Andersen OM, Karmali V, Fritz JJ, Cheng D, Peng J, Levey AI, Willnow TE, Lah JJ (2008) Loss of LR11/SORLA enhances early pathology in a mouse model of amyloidosis: evidence for a proximal role in Alzheimer's disease. J Neurosci 28:12877–12886

Doherty GH (2011) Obesity and the ageing brain: could leptin play a role in neurodegeneration? Curr Gerontol Geriatr Res 2011:708154

Doherty GH, Beccano-Kelly D, Yan SD, Gunn-Moore FJ, Harvey J (2013) Leptin prevents hippocampal synaptic disruption and neuronal cell death induced by amyloid β. Neurobiol Aging 34(1):226–237

Dohm GL, Elton CW, Friedman JE, Pilch PF, Pories WJ, Atkinson SM Jr, Caro JF (1991) Decreased expression of glucose transporter in muscle from insulin-resistant patients. Am J Physiol 260:E459–E463

Dong Y, Zhang M, Liang B, Xie Z, Zhao Z, Asfa S, Choi HC, Zou MH (2010) Reduction of AMP-activated protein kinase A2 increases endoplasmic reticulum stress and atherosclerosis in vivo. Circulation 121:792–803

Donmez G, Wang D, Cohen DE, Guarente L (2010) SIRT1 suppresses beta-amyloid production by activating the alpha-secretase gene ADAM10. Cell 142:320–332

Drake J, Petroze R, Castegna A, Ding Q, Keller JN, Markebery WR, Lovell MA, Butterfield DA (2004) 4-Hydroxynonenal oxidatively modifies histones: implications for Alzheimer's disease. Neurosci Lett 356:155–158

Duce JA, Tsatsanis A, Cater MA, James SA, Robb E, Wikhe K, Leong SL, Perez K, Johanssen T, Greenough MA, Cho HH, Galatis D, Moir RD, Masters CL, McLean C, Tanzi RE, Cappai R, Barnham KJ, Ciccotosto GD, Rogers JT, Bush AI (2010) Iron-export ferroxidase activity of beta-amyloid precursor protein is inhibited by zinc in Alzheimer's disease. Cell 142:857–867

Duckworth WC, Kitabchi AE (1981) Insulin metabolism and degradation. Endocr Rev 2:210–233

Duckworth WC, Bennett RG, Hamel FG (1998) Insulin degradation: progress and potential. Endocr Rev 19:608–624

Eckert GP, Wood WG, Müller WE (2010) Lipid membranes and β-amyloid: a harmful connection. Curr Protein Pept Sci 11:319–325

Egana JT, Zambrano C, Nunez MT, Gonzalez-Billault C, Maccioni RB (2003) Iron-induced oxidative stress modify tau phosphorylation patterns in hippocampal cell cultures. Biometals 16:215–223

Elmquist JK, Bjørbæk C, Ahima RS, Flier JS, Saper CB (1998) Distributions of leptin receptor mRNA isoforms in the rat brain. J Comp Neurol 395:535–547

English JD, Sweatt JD (1997) A requirement for the mitogen-activated protein kinase cascade in hippocampal long term potentiation. J Biol Chem 272:19103–19106

Erol A (2008) An integrated and unifying hypothesis for the metabolic basis of sporadic Alzheimer's disease. J Alzheimers Dis 13:241–253

Escribano L, Simón AM, Gimeno E, Cuadrado-Tejedor M, López de Maturana R, García-Osta A, Ricobaraza A, Pérez-Mediavilla A, Del Río J, Frechilla D (2010) Rosiglitazone rescues memory impairment in Alzheimer's transgenic mice: mechanisms involving a reduced amyloid and tau pathology. Neuropsychopharmacology 35:1593–1604

Evans JL, Maddux BA, Goldfine ID (2005) The molecular basis for oxidative stress-induced insulin resistance. Antioxid Redox Signal 7:1040–1052

Ewing GW (2010) Mathematical modeling the neuroregulation of blood pressure using a cognitive top-down approach. N Am J Med Sci 2:341–352

Fagan AM, Mintun MA, Mach RH, Lee SY, Dence CS, Shah AR, LaRossa GN, Spinner ML, Klunk WE, Mathis CA, DeKosky ST, Morris JC, Holtzman DM (2006) Inverse relation between in vivo amyloid imaging load and cerebrospinal fluid Abeta42 in humans. Ann Neurol 59:512–519

Fagan AM, Roe CM, Xiong C, Mintun MA, Morris JC, Holtzman DM (2007) Cerebrospinal fluid tau/beta-amyloid(42) ratio as a prediction of cognitive decline in nondemented older adults. Arch Neurol 64:343–349

Fahrenholz F (2007) Alpha-secretase as a therapeutic target. Curr Alzheimer Res 4:412–417

Fang F, Lue LF, Yan S, Xu H, Luddy JS, Chen D, Walker DG, Stern DM, Yan S, Schmidt AM, Chen JX, Yan S (2010) RAGE-dependent signaling in microglia contributes to neuroinflammation, Abeta accumulation, and impaired learning/memory in a mouse model of Alzheimer's disease. FASEB J 24:1043–1055

Farooqui AA (2009) Hot topics in neural membrane lipidology. Springer, New York

Farooqui AA (2010) Neurochemical aspects of neurotraumatic and neurodegenerative diseases. Springer, New York

Farooqui AA (2011) Lipid mediators and their metabolism in the brain. Springer, New York

Farooqui AA (2012) Lipid mediators and their metabolism in the nucleus: implications for Alzheimer's disease. J Alzheimers Dis 30(Suppl 2):S163–S178

Farooqui AA, Ong WY, Horrocks LA (2003) Plasmalogens, docosahexaenoic acid and neurological disorders. Adv Exp Med Biol 544:335–354

Farooqui AA, Ong WY, Horrocks LA (2006) Inhibitors of brain phospholipase $A_2$ activity: their neuropharmacological effects and therapeutic importance for the treatment of neurologic disorders. Pharmacol Rev 58:591–620

Farooqui AA, Ong WY, Farooqui T (2010) Lipid mediators in the nucleus: their potential contribution to Alzheimer's disease. Biochim Biophys Acta 1801:906–916

Farooqui AA, Farooqui T, Panza F, Frisardi V (2012) Metabolic syndrome as a risk factor for neurological disorders. Cell Mol Life Sci 69:741–762

Farr SA, Banks WA, Morley JE (2006) Effects of leptin on memory processing. Peptides 27:1420–1425

Farr SA, Yamada KA, Butterfield DA, Abdul HM, Xu L, Miller NE, Banks WA, Morley JE (2008) Obesity and hypertriglyceridemia produce cognitive impairment. Endocrinology 149:2628–2636

Farris W, Mansourian S, Chang Y, Lindsley L, Eckman EA, Frosch MP, Eckman CB, Tanzi RE, Selkoe DJ, Guenette S (2003) Insulin-degrading enzyme regulates the levels of insulin, amyloid beta-protein, and the beta-amyloid precursor protein intracellular domain in vivo. Proc Natl Acad Sci U S A 100:4162–4167

Faulds MH, Dahlman-Wright K (2012) Metabolic diseases and cancer risk. Curr Opin Oncol 24:58–61

Festa A, Williams K, D'Agostino R Jr, Wagenknecht L, Hafner S (2006) The natural course of β-cell function in nondiabetic and diabetic individuals: the Insulin Resistance Atherosclerosis Study. Diabetes 55:1114–1120

Fewlass DC, Noboa K, Pi-Sunyer FX, Johnston JM, Yan SD, Tezapsidis N (2004) Obesity-related leptin regulates Alzheimer's Abeta. FASEB J 18:1870–1878

Flier JS (2004) Obesity wars: molecular progress confronts an expanding epidemic. Cell 116:337–350

Folstein M, Liu T, Peter I, Buell J, Arsenault L, Scott T, Qiu WW (2007) The homocysteine hypothesis of depression. Am J Psychiatry 164:861–867

Francis GJ, Martinez JA, Liu WQ, Xu K, Ayer A, Fine J, Tuor UI, Glazner G, Hanson LR, Frey WH II, Toth C (2008) Intranasal insulin prevents cognitive decline, cerebral atrophy and white matter changes in murine type I diabetic encephalopathy. Brain 131:3311–3334

Frank S, Heni M, Moss A, von Schnurbein J, Fritsche A, Häring HU, Farooqi S, Preissl H, Wabitsch M (2011) Leptin therapy in a congenital leptin-deficient patient leads to acute and long-term changes in homeostatic, reward, and food-related brain areas. J Clin Endocrinol Metab 96:E1283–E1287

Frasca F, Pandini G, Scalia P, Sciacca L, Mineo R, Costantino A, Goldfine ID, Belfiore A, Vigneri R (1999) Insulin receptor isoform A, a newly recognized, high-affinity insulin-like growth factor II receptor in fetal and cancer cells. Mol Cell Biol 5:3278–3288

Friedman R, Pellarin R, Caflisch A (2009) Amyloid aggregation on lipid bilayers and its impact on membrane permeability. J Mol Biol 387:407–415

Frolich L, Dirr A, Götz ME, Gsell W, Reichmann H, Riederer P, Maurer K (1998) Brain insulin and insulin receptors in aging and sporadic Alzheimer's disease. J Neural Transm 105:423–438

Funaki M (2009) Saturated fatty acids and insulin resistance. J Med Invest 56:88–92

Galindo MF, Ikuta I, Zhu X, Casadesus G, Jordan J (2010) Mitochondrial biology in Alzheimer's disease pathogenesis. J Neurochem 114:933–945

Gallagher EJ, LeRoith D, Karnieli E (2008) The metabolic syndrome—from insulin resistance to obesity and diabetes. Endocrinol Metab Clin North Am 37:559–579

Galvan V, Chen S, Lu D, Logvinova A, Goldsmith P, Koo EH, Bredesen DE (2002) Caspase cleavage of members of the amyloid precursor family of proteins. J Neurochem 82:283–294
Garcia A, Zanibbi K (2004) Homocysteine and cognitive function in elderly people. CMAJ 171:897–904
Garcia-Segura LM, Rodriguez JR, Torres-Aleman I (1997) Localization of the insulin-like growth factor I receptor in the cerebellum and hypothalamus of adult rats: an electron microscopic study. J Neurocytol 26:479–490
Gasparini L, Gouras GK, Wang R, Gross RS, Beal MF, Greengard P, Xu H (2001) Stimulation of beta-amyloid precursor protein trafficking by insulin reduces intraneuronal beta-amyloid and requires mitogen-activated protein kinase signaling. J Neurosci 21:2561–2570
Gasparini L, Netzer WJ, Greengard P, Xu H (2002) Does insulin dysfunction play a role in Alzheimer's disease? Trends Pharmacol Sci 23:288–293
Gautron L, Elmquest JK (2011) Sixteen years and counting: an update on leptin in energy balance. J Clin Invest 121:2087–2093
Gebre-Medhin S, Mulder H, Pekny M, Westermark G, Tornell J, Westermark P, Sundler F, Ahrén B, Betsholtz C (1998) Increased insulin secretion and glucose tolerance in mice lacking islet amyloid polypeptide (amylin). Biochem Biophys Res Commun 250:271–277
Gralle M, Ferreira ST (2007) Structure and functions of the human amyloid precursor protein: the whole is more than the sum of its parts. Prog Neurobiol 82:11–32
Grant WB, Campbell A, Itzhaki RF, Savory J (2002) The significance of environmental factors in the etiology of Alzheimer's disease. J Alzheimers Dis 4:179–189
Greco SJ, Bryan KJ, Sarkar S, Zhu X, Smith MA, Ashford JW, Johnston JM, Tezapsidis N, Casadesus G (2010) Chronic leptin supplementation ameliorates pathology and improves cognitive performance in a transgenic mouse model of Alzheimer's disease. J Alzheimers Dis 19:155–1167
Green KN, Billings LM, Roozendaal B, McGaugh JL, LaFerla FM (2006) Glucocorticoids increase amyloid-β and tau pathology in a mouse model of Alzheimer's disease. J Neurosci 26:9047–9056
Grimm MO, Grimm HS, Patzold AJ, Zinser EG, Halonen R, Duering M, Tschäpe JA, De Strooper B, Müller U, Shen J, Hartmann T (2005) Regulation of cholesterol and sphingomyelin metabolism by amyloid-β and presenilin. Nat Cell Biol 27:1118–1123
Grimm MO, Grimm HS, Hartmann T (2007) Amyloid beta as a regulator of lipid homeostasis. Trends Mol Med 13:337–344
Grimm MO, Rothhaar TL, Grösgen S, Burg VK, Hundsdörfer B, Haupenthal VJ, Friess P, Kins S, Grimm HS, Hartmann T (2011a) Trans fatty acids enhance amyloidogenic processing of the Alzheimer amyloid precursor protein (APP). J Nutr Biochem 23:1214–1223
Grimm MO, Kuchenbecker J, Grösgen S, Burg VK, Hundsdörfer B, Rothhaar TL, Friess P, de Wilde MC, Broersen LM, Penke B, Péter M, Vígh L, Grimm HS, Hartmann T (2011b) Docosahexaenoic acid reduces amyloid beta production via multiple pleiotropic mechanisms. J Biol Chem 286:14028–14039
Grimm MO, Zinser EG, Grösgen S, Hundsdörfer B, Rothhaar TL, Burg VK, Kaestner L, Bayer TA, Lipp P, Müller U, Grimm HS, Hartmann T (2012) Amyloid precursor protein (APP) mediated regulation of ganglioside homeostasis linking Alzheimer's disease pathology with ganglioside metabolism. PLoS One 7:e34095
Grünblatt E, Bartl J, Riederer P (2011) The link between iron, metabolic syndrome, and Alzheimer's disease. J Neural Transm 118:371–379
Gudz TI, Tserng KY, Hoppel CL (1997) Direct inhibition of mitochondrial respiratory chain complex III by cell-permeable ceramide. J Biol Chem 272:24154–24158
Haan MN (2006) Therapy Insight: type 2 diabetes mellitus and the risk of late-onset alzheimer's disease. Nat Clin Pract Neurol 2:159–166
Halaas JL, Boozer C, Blair-West J, Fidahusein N, Denton DA, Friedman JM (1997) Physiological response to long-term peripheral and central leptin infusion in lean and obese mice. Proc Natl Acad Sci U S A 94:8878–8883

Han X, M Holtzman D, McKeel DW Jr, Kelley J, Morris JC (2002) Substantial sulfatide deficiency and ceramide elevation in very early Alzheimer's disease: potential role in disease pathogenesis. J Neurochem 82:809–818

Hannun YA, Obeid LM (2008) Principles of bioactive lipid signalling: lessons from sphingolipids. Nat Rev Mol Cell Biol 9:139–150

Hartmann D, Tournoy J, Saftig P, Annaert W, De Strooper B (2001) Implication of APP secretases in notch signaling. J Mol Neurosci 17:171–181

Harvey J (2007) Leptin regulation of neuronal excitability and cognitive function. Curr Opin Pharmacol 7:643–667

Havel PJ (2004) Update on adipocyte hormones: regulation of energy balance and carbohydrate/ lipid metabolism. Diabetes 53(Suppl 1):S143–S151

Havel PJ, Uriu-Hare JY, Liu T, Stanhope KL, Stern JS, Keen CL, Ahrén B (1998) Marked and rapid decreases of circulating leptin in streptozotocin diabetic rats: reversal by insulin. Am J Physiol 274:R1482–R1491

He X, Huang Y, Li B, Gong CX, Schuchman EH (2010) Deregulation of sphingolipid metabolism in Alzheimer's disease. Neurobiol Aging 31:398–408

Hebert LE, Scherr PA, Bienias JL, Bennett DA, Evans DA (2003) Alzheimer's disease in the U.S. population: prevalence estimates using the 2000 Census. Arch Neurol 60:1119–1122

Heidenreich KA, de Vellis G, Gilmore PR (1988) Functional properties of the subtype of insulin receptor found on neurons. J Neurochem 51:878–887

Henriques AG, Vieira SI, Rebelo S, Domingues SC, da Cruz e Silva EF, da Cruz e Silva OA (2007) Isoform specific amyloid-β protein precursor metabolism. J Alzheimers Dis 11:85–95

Ho L, Qin W, Pompl PN, Xiang Z, Wang J, Zhao Z, Peng Y, Cambareri G, Rocher A, Mobbs CV, Hof PR, Pasinetti GM (2004) Diet-induced insulin resistance promotes amyloidosis in a transgenic mouse model of Alzheimer's disease. FASEB J 18:902–904

Hoe HS, Lee KJ, Carney RS, Lee J, Markova A, Lee JY, Howell BW, Hyman BT, Pak DT, Bu G, Rebeck GW (2009) Interaction of reelin with amyloid precursor protein promotes neurite outgrowth. J Neurosci 29:7459–7473

Hojjati MR, Li Z, Jiang XC (2005a) Serine palmitoyl-CoA transferase (SPT) deficiency and sphingolipid levels in mice. Biochim Biophys Acta 1737(1):44–51

Hojjati MR, Li Z, Zhou H, Tang S, Huan C, Ooi E, Lu S, Jiang XC (2005b) Effect of myriocin on plasma sphingolipid metabolism and atherosclerosis in apoE-deficient mice. J Biol Chem 280:10284–10289

Holden KF, Lindquist K, Tylavsky FA, Rosano C, Harris TB, Yaffe K (2009) Serum leptin level and cognition in the elderly: findings from the Health ABC Study. Neurobiol Aging 30:1483–1489

Holland WL, Summers SA (2008) Sphingolipids, insulin resistance, and metabolic disease: new insights from in vivo manipulation of sphingolipid metabolism. Endocr Rev 29:381–402

Holland WL, Brozinick JT, Wang LP, Hawkins ED, Sargent KM, Liu Y, Narra K, Hoehn KL, Knotts TA, Siesky A, Nelson DH, Karathanasis SK, Fontenot GK, Birnbaum MJ, Summers SA (2007) Inhibition of ceramide synthesis ameliorates glucocorticoid-, saturated-fat-, and obesity-induced insulin resistance. Cell Metab 5:167–179

Holland WL, Miller RA, Wang ZV, Sun K, Barth BM, Bui HH, Davis KE, Bikman BT, Halberg N, Rutkowski JM, Wade MR, Tenorio VM, Kuo MS, Brozinick JT, Zhang BB, Birnbaum MJ, Summers SA, Scherer PE (2010) Receptor-mediated activation of ceramidase activity initiates the pleiotropic actions of adiponectin. Nat Med 17:55–63

Hooper C, Killick R, Lovestone S (2008) The GSK3 hypothesis of Alzheimer's disease. J Neurochem 104:1433–1439

Hoozemans JJ, van Haastert ES, Nijholt DA, Rozemuller AJ, Eikelenboom P, Scheper W (2009) The unfolded protein response is activated in pretangle neurons in Alzheimer's disease hippocampus. Am J Pathol 174:1241–1251

Hosoi T, Ozawa K (2012) Molecular approaches to the treatment, prophylaxis, and diagnosis of Alzheimer's disease: endoplasmic reticulum stress and immunological stress in pathogenesis of Alzheimer's disease. J Pharmacol Sci 118(3):319–324

Hotamisligil GS (2010) Endoplasmic reticulum stress and the inflammatory basis of metabolic disease. Cell 140:900–917

Hoyer S (2000) The aging brain. Changes in the neuronal insulin/insulin receptor signal transduction cascade trigger late-onset sporadic Alzheimer disease (SAD). A mini review. J Neural Transm 109:991–1002

Huang H, Tindall DJ (2007) Dynamic FoxO transcription factors. J Cell Sci 120:2479–2487

Huang HJ, Liang KC, Ke HC, Chang YY, Hsieh-Li HM (2011) Long-term social isolation exacerbates the impairment of spatial working memory in APP/PS1 transgenic mice. Brain Res 1371:150–160

Humpel C, Marksteiner J (2005) Cerebrovascular damage as a cause for Alzheimer's disease? Curr Neurovasc Res 2:341–347

Iadecola C (2004) Neurovascular regulation in the normal brain and in Alzheimer's disease. Nat Rev Neurosci 5:347–360

Ito H, Kanno I, Fukuda H (2005) Human cerebral circulation: positron emission tomography studies. Ann Nucl Med 19:65–74

Janson J, Soeller WC, Roche PC, Nelson RT, Torchia AJ, Kreutter DK, Butler PC (1996) Spontaneous diabetes mellitus in transgenic mice expressing human islet amyloid polypeptide. Proc Natl Acad Sci U S A 93:7283–7288

Janson J, Laedtke T, Parisi JE, O'Brien P, Petersen RC, Butler PC (2004) Increased risk of type 2 diabetes in Alzheimer disease. Diabetes 53:474–481

Jeong S, Yoon M (2011) 17beta-Estradiol inhibition of PPARgamma-induced adipogenesis and adipocyte-specific gene expression. Acta Pharmacol Sin 32:230–238

Jiang D, Li X, Williams R, Patel S, Men L, Wang Y, Zhou F (2009) Ternary complexes of iron, amyloid-β, and nitrilotriacetic acid: binding affinities, redox properties, and relevance to iron-induced oxidative stress in Alzheimer's disease. Biochemistry 48:7939–7947

Johnston AM, Pirola L, Van Obberghen E (2003) Molecular mechanisms of insulin receptor substrate protein-mediated modulation of insulin signalling. FEBS Lett 546:32–36

Jolivalt CG, Lee CA, Beiswenger KK, Smith JL, Orlov M, Torrance MA, Masliah E (2008) Defective insulin signaling pathway and increased glycogen synthase kinase-3 activity in the brain of diabetic mice: parallels with Alzheimer's disease and correction by insulin. J Neurosci Res 86:3265–3274

Jope RS, Yuskatis CJ, Beurel E (2007) Glycogen synthase kinase-3 (GSK3): inflammation, diseases, and therapeutics. Neurochem Res 32:577–595

Kahn BB (1996) Lilly lecture 1995. Glucose transport: pivotal step in insulin action. Diabetes 45:1644–1654

Kalaria RN (2010) Vascular basis for brain degeneration: faltering controls and risk factors for dementia. Nutr Rev 68:S74–S87

Kalmijn S (2000) Fatty acid intake and the risk of dementia and cognitive decline: a review of clinical and epidemiological studies. J Nutr Health Aging 4:202–207

Kalvodova L, Kahya N, Schwille P, Ehehalt R, Verkade P, Drechsel D, Simons K (2005) Lipids as modulators of proteolytic activity of BACE: involvement of cholesterol, glycosphingolipids, and anionic phospholipids in vitro. J Biol Chem 280:36815–36823

Kanai M, Matsubara E, Isoe K, Urakami K, Nakashima K, Arai H, Sasaki H, Abe K, Iwatsubo T, Kosaka T, Watanabe M, Tomidokoro Y, Shizuka M, Mizushima K, Nakamura T, Igeta Y, Ikeda Y, Amari M, Kawarabayashi T, Ishiguro K, Harigaya Y, Wakabayashi K, Okamoto K, Hirai S, Shoji M (1998) Longitudinal study of cerebrospinal fluid levels of τ, Aβ1–40, and Aβ1–42(43) in Alzheimer's disease: a study in Japan. Ann Neurol 44:17–26

Kanfer JN, Sorrentino G, Sitar DS (1998) Phospholipases as mediators of amyloid beta peptide neurotoxicity: an early event contributing to neurodegeneration characteristic of Alzheimer's disease. Neurosci Lett 257:93–96

Kang JE, Cirrito JR, Dong H, Csernansky JG, Holtzman DM (2007) Acute stress increases interstitial fluid amyloid-beta via corticotropin-releasing factor and neuronal activity. Proc Natl Acad Sci U S A 104:10673–10678

Kanoski SE, Zhang Y, Zheng W, Davidson TL (2010) The effects of a high-energy diet on hippocampal function and blood–brain barrier integrity in the rat. J Alzheimers Dis 21:207–219

Katayama T, Imaizumi K, Sato N, Miyoshi K, Kudo T, Hitomi J, Morihara T, Yoneda T, Gomi F, Mori Y, Nakano Y, Takeda J, Tsuda T, Itoyama Y, Murayama O, Takashima A, St George-Hyslop P, Takeda M, Tohyama M (1999) Presenilin-1 mutations downregulate the signalling pathway of the unfolded-protein response. Nat Cell Biol 1:479–485

Katayama T, Imaizumi K, Honda A, Yoneda T, Kudo T, Takeda M, Mori K, Rozmahel R, Fraser P, George-Hyslop PS, Tohyama M (2001) Disturbed activation of endoplasmic reticulum stress transducers by familial Alzheimer's disease-linked presenilin-1 mutations. J Biol Chem 276:43446–43454

Kawarabayashi T, Younkin LH, Saido TC, Shoji M, Ashe KH, Younkin SG (2001) Age-dependent changes in brain, CSF, and plasma amyloid (beta) protein in the Tg2576 transgenic mouse model of Alzheimer's disease. J Neurosci 21:372–381

Kawasaki E, Abiru N, Eguchi K (2004) Prevention of type 1 diabetes: from the view point of β cell damage. Diabetes Res Clin Pract 66(Suppl 1):S27–S32

Kell DB (2010) Towards a unifying, systems biology understanding of large-scale cellular death and destruction caused by poorly liganded iron: Parkinson's, Huntington's, Alzheimer's, prions, bactericides, chemical toxicology and others as examples. Arch Toxicol 84:825–889

Kelley DE, He J, Menshikova EV, Ritov VB (2002) Dysfunction of mitochondria in human skeletal muscle in type 2 diabetes. Diabetes 51:2944–2950

Kim YB, Nikoulina SE, Ciaraldi TP, Henry RR, Kahn BB (1999) Normal insulin-dependent activation of Akt/protein kinase B, with diminished activation of phosphoinositide 3-kinase, in muscle in type 2 diabetes. J Clin Invest 104:733–741

Kim JK, Fillmore JJ, Sunshine MJ, Albrecht B, Higashimori T, Kim DW, Liu ZX, Soos TJ, Cline GW, O'Brien WR, Littman DR, Shulman GI (2004) PKC-θ knockout mice are protected from fat-induced insulin resistance. J Clin Invest 114:823–827

Kitaguchi N, Takahashi Y, Tokushima Y, Shiojiri S, Ito H (1988) Novel precursor of Alzheimer's disease amyloid protein shows protease inhibitory activity. Nature 331:530–532

Kleinridders A, Schenten D, Könner AC, Belgardt BF, Mauer J, Okamura T, Wunderlich FT, Medzhitov R, Brüning JC (2009) MyD88 signaling in the CNS is required for development of fatty acid-induced leptin resistance and diet-induced obesity. Cell Metab 10:249–259

Kodam A, Maulik M, Peake K, Amritraj A, Vetrivel KS, Thinakaran G, Vance JE, Kar S (2011) Altered levels and distribution of amyloid precursor protein and its processing enzymes in Niemann-Pick type C1-deficient mouse brains. Glia 58:1267–1281

Kodiha M, Stochaj U (2011) AMP kinase: the missing link between type 2 diabetes and neurodegenerative diseases? Trends Mol Med 17:613–614

Kodl CT, Seaquist ER (2008) Cognitive dysfunction and diabetes mellitus. Endocr Rev 29:494–511

Kompoti M, Mariolis A, Alevizos A, Kyrazis I, Protopsaltis I, Dimou E, Lentzas I, Levisianou D, Gova A, Melidonis A (2006) Elevated serum triglycerides are the strongest single indicator for the presence of metabolic syndrome in patients with type 2 diabetes. Cardiovasc Diabetol 5:21

Kopf S, Baratti C (1999) Effects of posttraining administration of insulin on retention of a habituation response in mice: participation of a central cholinergic mechanism. Neurobiol Learn Mem 71:50–61

Kroner Z (2009) The relationship between Alzheimer's disease and diabetes: type 3 diabetes? Altern Med Rev 14:373–379

Kruman II, Culmsee C, Chan SL, Kruman Y, Guo Z, Penix L, Mattson MP (2000) Homocysteine elicits a DNA damage response in neurons that promotes apoptosis and hyper sensitivity to exitotoxicity. J Neurosci Res 20:6920–6926

Kuemmerle JF (2012) Insulin-like growth factors in the gastrointestinal tract and liver. Endocrinol Metab Clin North Am 41:409–423

Kuhla B, Haase C, Flach K, Lüth HJ, Arendt T, Münch G (2007) Effect of pseudophosphorylation and cross-linking by lipid peroxidation and advanced glycation end product precursors on tau aggregation and filament formation. J Biol Chem 282:6984–6991

Kulstad JJ, Green PS, Cook DG, Watson GS, Reger MA, Baker LD, Plymate SR, Asthana S, Rhoads K, Mehta PD, Craft S (2006) Differential modulation of plasma beta-amyloid by insulin in patients with Alzheimer disease. Neurology 66:1506–1510

Kumar S, Walter J (2011) Phosphorylation of amyloid beta (Aβ) peptides—a trigger for formation of toxic aggregates in Alzheimer's disease. Aging (Albany NY) 3:PMC3184981

Kuo YM, Beach TG, Sue LI, Scott S, Layne KJ, Kokjohn TA, Kalback WM, Luehrs DC, Vishnivetskaya TA, Abramowski D, Sturchler-Pierrat C, Staufenbiel M, Weller RO, Roher AE (2001) The evolution of a beta peptide burden in the APP23 transgenic mice: implications for A beta deposition in Alzheimer disease. Mol Med 7:609–618

Kurochkin IV (2001) Insulin-degrading enzyme: embarking on amyloid destruction. Trends Biochem Sci 26:421–425

Kuusisto J, Koivisto K, Mykkanen L, Helkala EL, Vanhanen M, Hanninen T, Kervinen K, Kesaniemi YA, Riekkinen PJ, Laakso M (1997) Association between features of the insulin resistance syndrome and Alzheimer's disease independently of apolipoprotein E4 phenotype: cross sectional population based study. Br Med J 315:1045–1049

Kwak Y-D, Wang R, Li JJ, Zhang Y-W, Xu H, Liao F-F (2011) Differential regulation of BACE1 expression by oxidative and nitrosative signals. Mol Neurodegener 6:17

Lacor PN, Buniel MC, Chang L, Fernandez SJ, Gong Y, Viola KL, Lambert MP, Velasco PT, Bigio EH, Finch CE, Krafft GA, Klein WL (2004) Synaptic targeting by Alzheimer's-related amyloid beta oligomers. J Neurosci 24:10191–10200

Lacor PN, Buniel MC, Furlow PW, Clemente AS, Velasco PT, Wood M, Viola KL, Klein WL (2007) Abeta oligomer-induced aberrations in synapse composition, shape, and density provide a molecular basis for loss of connectivity in Alzheimer's disease. J Neurosci 27:796–807

Lage R, Dieguez C, Vidal-Puig A, Lopez M (2008) AMPK: a metabolic gauge regulating whole-body energy homeostasis. Trends Mol Med 14:539–549

Lambert MP, Barlow AK, Chromy BA, Edwards C, Freed R, Liosatos M, Morgan TE, Rozovsky I, Trommer B, Viola KL, Wals P, Zhang C, Finch CE, Krafft GA, Klein WL (1998) Diffusible, nonfibrillar ligands derived from Abeta1-42 are potent central nervous system neurotoxins. Proc Natl Acad Sci U S A 95:6448–6453

Landfield PW, Blalock EM, Chen KC, Porter NM (2007) A new glucocorticoid hypothesis of brain aging: implications for Alzheimer's disease. Curr Alzheimer Res 4:205–212

Lane RF, Raines SM, Steele JW, Ehlich ME, Lah JA, Small SA, Tanzi RE, Attie AD, Gandy S (2010) Diabetes-associated SorCS1 regulates Alzheimer's amyloid-β metabolism: evidence for involvement of SorL1 and the retromer complex. J Neurosci 30:13110–13115

Lau TL, Gehman JD, Wade JD, Masters CL, Barnham KJ, Separovic F (2007) Membrane interactions and the effect of metal ions of the amyloidogenic fragment Aβ(25–35) in comparison to Aβ(1–42). Biochim Biophys Acta 1768:2400–2408

Le Blanc S, Villarroel P, Candia V, Gavilán N, Soto N, Pérez-Bravo F, Arredondo M (2012) Type 2 diabetic patients and their offspring show altered parameters of iron status, oxidative stress and genes related to mitochondrial activity. Biometals 25:725–735

Lee KW, Kim JB, Seo JS, Kim TK, Im JY, Baek IS, Kim KS, Lee JK, Han PL (2009) Behavioral stress accelerates plaque pathogenesis in the brain of Tg2576 mice via generation of metabolic oxidative stress. J Neurochem 108:165–175

Leiva E, Mujica V, Sepúlveda P, Guzmán L, Núñez S, Orrego R, Palomo I, Andrews M, Arredondo MA (2013) High levels of iron status and oxidative stress in patients with metabolic syndrome. Biol Trace Elem Res 151(1):1–8

Lesne S, Koh MT, Kotilinek L, Kayed R, Glabe CG, Yang A, Gallagher M, Ashe KH (2006) A specific amyloid-beta protein assembly in the brain impairs memory. Nature 440:352–357

Lester-Coll N, Rivera EJ, Soscia SJ, Doiron K, Wands JR, de la Monte SM (2006) Intracerebral streptozotocin model of type 3 diabetes: relevance to sporadic Alzheimer's disease. J Alzheimers Dis 9:13–33

Levine YC, Li GK, Michel T (2007) Agonist-modulated regulation of AMP-activated protein kinase (AMPK) in endothelial cells. Evidence for an AMPK ->Rac1 ->Akt ->endothelial nitric-oxide synthase pathway. J Biol Chem 282:20351–20364

Li H, Junk P, Huwiler A, Burkhardt C, Wallerath T, Pfeilschifter J, Forstermann U (2002a) Dual effect of ceramide on human endothelial cells: induction of oxidative stress and transcriptional upregulation of endothelial nitric oxide synthase. Circulation 106:2250–2256

Li XL, Aou S, Oomura Y, Hori N, Fukunaga K, Hori T (2002b) Impairment of long-term potentiation and spatial memory in leptin receptor-deficient rodents. Neuroscience 113:607–615

Li R, Lindholm K, Yang LB, Yue X, Citron M, Yan R, Beach T, Sue L, Sabbagh M, Cai H, Wong P, Price D, Shen Y (2004) Amyloid β peptide load is correlated with increased β-secretase activity in sporadic Alzheimer's disease patients. Proc Natl Acad Sci U S A 101:3632–3637

Li S, Wang C, Wang W, Dong H, Hou P, Tang Y (2008) Chronic mild stress impairs cognition in mice: from brain homeostasis to behavior. Life Sci 82:934–942

Lieb W, Beiser AS, Vasan RS, Tan ZS, Au R, Harris TB, Roubenoff R, Auerbach S, DeCarli C, Wolf PA, Seshadri S (2009) Association of plasma leptin levels with incident Alzheimer disease and MRI measures of brain aging. JAMA 302:2565–2572

Lim JH, Lee HJ, Ho Jung M, Song J (2009) Coupling mitochondrial dysfunction to endoplasmic reticulum stress response: a molecular mechanism leading to hepatic insulin resistance. Cell Signal 21:169–177

Lindeboom J, Weinstein H (2004) Neuropsychology of cognitive aging, minimal cognitive impairment, Alzheimer's disease, and vascular cognitive impairment. Eur J Pharmacol 490:83–86

Lindqvist A, Mohapel P, Bouter B, Frielingsdorf H, Pizzo D, Brundin P, Erlanson-Albertsson C (2006) High-fat diet impairs hippocampal neurogenesis in male rats. Eur J Neurol 13:1385–1388

Lipton SA, Kim WK, Choi YB, Kumar S, D'Emilia DM, Rayudu PV, Arnelle DR, Stamler JS (1997) Neurotoxicity associated with dual actions of homocysteine at the N-methyl-D-aspartate receptor. Proc Natl Acad Sci U S A 94:5923–5928

Liu L, Zhang Y, Chen N, Shi X, Tsang B, Yu YH (2007) Upregulation of myocellular DGAT1 augments triglyceride synthesis in skeletal muscle and protects against fat-induced insulin resistance. J Clin Invest 117:1679–1689

Liu LP, Hong H, Liao JM, Wang TS, Wu J, Chen SS, Li YQ, Long Y, Xia YZ (2009) Upregulation of RAGE at the blood–brain barrier in streptozotocin induced diabetic mice. Synapse 63:636–642

Liu Y, Liu F, Grundke-Iqbal I, Iqbal K, Gong CX (2011) Deficient brain insulin signalling pathway in Alzheimer's disease and diabetes. J Pathol 225:54–62

Luo Y, Bolon B, Kahn S, Bennett BD, Babu-Khan S, Denis P, Fan W, Kha H, Zhang J, Gong Y, Martin L, Louis JC, Yan Q, Richards WG, Citron M, Vassar R (2001) Mice deficient in BACE1, the Alzheimer's β-secretase, have normal phenotype and abolished β-amyloid generation. Nat Neurosci 4:231–232

Ma QL, Teter B, Ubeda OJ, Morihara T, Dhoot D, Nyby MD, Tuck ML, Frautschy SA, Cole GM (2007) Omega-3 fatty acid docosahexaenoic acid increases SorLA/LR11, a sorting protein with reduced expression in sporadic Alzheimer's disease (AD): relevance to AD prevention. J Neurosci 27:14299–14307

Macotela Y, Boucher J, Tran TT, Kahn CR (2009) Sex and depot differences in adipocyte insulin sensitivity and glucose metabolism. Diabetes 58:803–812

Madonna R, De Caterina R (2011) Cellular and molecular mechanisms of vascular injury in diabetes—part I: pathways of vascular disease in diabetes. Vascul Pharmacol 54:68–74

Man HY, Wang Q, Lu WY, Ju W, Ahmadian G, Liu L, D'Souza S, Wong TP, Taghibiglou C, Lu J, Becker LE, Pei L, Liu F, Wymann MP, MacDonald JF, Wang YT (2003) Activation of PI3-kinase is required for AMPA receptor insertion during LTP of mEPSCs in cultured hippocampal neurons. Neuron 38:611–624

Marwarha G, Dasari B, Prabhakara JP, Schommer J, Ghribi O (2010) β-Amyloid regulates leptin expression and tau phosphorylation through the mTORC1 signaling pathway. J Neurochem 115:373–384

Masliah E, Mallory M, Alford M, DeTeresa R, Hansen LA, McKeel DW Jr, Morris JC (2001) Altered expression of synaptic proteins occurs early during progression of Alzheimer's disease. Neurology 56:127–129

Masters CL, Simms G, Weinman NA, Multhaup G, McDonald BL, Beyreuther K (1985) Amyloid plaque core protein in Alzheimer disease and Down syndrome. Proc Natl Acad Sci U S A 82:4245–4249

Matheny RW Jr, Nindl BC, Adamo ML (2010) Minireview: mechano-growth factor: a putative product of IGF-I gene expression involved in tissue repair and regeneration. Endocrinology 151:865–875

Mattson MP (2009) Roles of the lipid peroxidation product 4-hydroxynonenal in obesity, the metabolic syndrome, and associated vascular and neurodegenerative disorders. Exp Gerontol 44:625–633

Mayeux R, Tang MX, Jacobs DM, Manly J, Bell K, Merchant C, Small SA, Stern Y, Wisniewski HM, Mehta PD (1999) Plasma amyloid β-peptide 1–42 and incipient Alzheimer's disease. Ann Neurol 46:412–416

Mayeux R, Honig LS, Tang MX, Manly J, Stern Y, Schupf N, Mehta PD (2003) Plasma Aβ40 and Aβ42 and Alzheimer's disease: relation to age, mortality, and risk. Neurology 61:1185–1190

McCall AL (2005) Altered glycemia and brain-update and potential relevance to the aging brain. Neurobiol Aging 26:70–75

McClain DA (1991) Different ligand affinities of the two human insulin receptor splice variants are reflected in parallel changes in sensitivity for insulin action. Mol Endocrinol 5:734–739

McEwen BS, Magariños AM, Reagan LP (2002) Studies of hormone action in the hippocampal formation: possible relevance to depression and diabetes. J Psychosom Res 53:883–890

McLean CA, Cherny RA, Fraser FW, Fuller SJ, Smith MJ, Beyreuther K, Bush AI, Masters CL (1999) Soluble pool of Abeta amyloid as a determinant of severity of neurodegeneration in Alzheimer's disease. Ann Neurol 46:860–866

Mehta PD, Pirttilä T, Mehta SP, Sersen EA, Aisen PS, Wisniewski HM (2000) Plasma and cerebrospinal fluid levels of amyloid beta proteins 1–40 and 1–42 in Alzheimer disease. Arch Neurol 57:100–105

Mejia S, Giraldo M, Pineda D, Ardila A, Lopera F (2003) Nongenetic factors as modifiers of the age of onset of familial Alzheimer's disease. Int Psychogeriatr 15:337–349

Melnik BC (2012) Leucine signaling in the pathogenesis of type 2 diabetes and obesity. World J Diabetes 3:38–53

Miklossy J, Qing H, Radenovic A, Kis A, Vileno B, Laszlo F, Miller L, Martins RN, Waeber G, Mooser V, Bosman F, Khalili K, Darbinian N, McGeer PL (2010) Beta amyloid and hyperphosphorylated tau deposits in the pancreas in type 2 diabetes. Neurobiol Aging 31:1503–1515

Mogi M, Horiuchi M (2011) Neurovascular coupling in cognitive impairment associated with diabetes mellitus. Circ J 75:1042–1048

Molander-Melin M, Blennow K, Bogdanovic N, Dellheden B, Mansson JE, Freedman P (2005) Structural membrane alterations in Alzheimer brains found to be associated with regional disease development; increased density of gangliosides GM1 and GM2 and loss of cholesterol in detergent-resistant membrane domains. J Neurochem 92:171–182

Moller DE, Yokota A, Caro JF, Flier JS (1989) Tissue-specific expression of two alternatively spliced insulin receptor mRNAs in man. Mol Endocrinol 8:1263–1269

Moller DE, Benecke H, Flier JS (1991) Biologic activities of naturally occurring human insulin receptor mutations. Evidence that metabolic effects of insulin can be mediated by a kinase-deficient insulin receptor mutant. J Biol Chem 266:10995–11001

Moloney AM, Griffin RJ, Timmons S, O'Connor R, Ravid R, O'Neill C (2008) Defects in IGF-1 receptor, insulin receptor and IRS-1/2 in Alzheimer's disease indicate possible resistance to IGF-1 and insulin signalling. Neurobiol Aging 31:224–243

Moloney AM, Griffin RJ, Timmons S, O'Connor R, Ravid R, O'Neill C (2010) Defects in IGF-1 receptor, insulin receptor and IRS-1/2 in Alzheimer's disease indicate possible resistance to IGF-1 and insulin signalling. Neurobiol Aging 31:224–243

Mruthinti S, Schade RF, Harrell DU, Gulati NK, Swamy-Mruthinti S, Lee GP, Buccafusco JJ (2006) Autoimmunity in Alzheimer's disease as evidenced by plasma immunoreactivity against RAGE and Abeta42: complication of diabetes. Curr Alzheimer Res 3:229–235

Myers MG, Cowley MA, Munzberg H (2008) Mechanisms of leptin action and leptin resistance. Annu Rev Physiol 70:537–556

Nakamura M, Shishido N, Nunomura A, Smith MA, Perry G, Hayashi Y, Nakayama K, Hayashi T (2007) Three histidine residues of amyloid-beta peptide control the redox activity of copper and iron. Biochemistry 46:12737–12743

Neumann KF, Rojo L, Navarrete LP, Farias G, Reyes P, Maccioni RB (2008) Insulin resistance and Alzheimer's disease: molecular links & clinical implications. Curr Alzheimer Res 5:438–447

Niswender KD, Schwartz MW (2003) Insulin and leptin revisited: adiposity signals with overlapping physiological and intracellular signaling capabilities. Front Neuroendocrinol 24:1–10

Nogueira-Machado JA, Chaves MM (2008) From hyperglycemia to AGE-RAGE interaction on the cell surface: a dangerous metabolic route for diabetic patients. Expert Opin Ther Targets 12:871–882

O'Brien RJ, Wong PC (2011) Amyloid precursor protein processing and Alzheimer's disease. Annu Rev Neurosci 34:185–204

O'Connor T, Sadleir KR, Maus E, Velliquette RA, Zhao J, Cole SL et al (2008) Phosphorylation of the translation initiation factor eIF2alpha increases BACE1 levels and promotes amyloidogenesis. Neuron 60:988–1009

O'Neill LA, Kaltschmidt C (1997) NF-kappa B: a crucial transcription factor for glial and neuronal cell function. Trends Neurosci 20:252–258

Okada T, Wakabayashi M, Ikeda K, Matsuzaki K (2007) Formation of toxic fibrils of Alzheimer's amyloid beta-protein-(1–40) by monosialoganglioside GM1, a neuronal membrane component. J Mol Biol 371:481–489

Olivieri O, Martinelli N, Bassi A, Trabetti E, Girelli D, Pizzolo F, Friso S, Pignatti PF, Corrocher R (2007) ApoEe2/e3/e4 polymorphism, ApoC-III/ApoE ratio and metabolic syndrome. Clin Exp Med 7:164–172

Oomura Y, Hori N, Shiraishi T, Fukunaga K, Takeda H, Tsuji M, Matsumiya T, Ishibashi M, Aou S, Li XL (2006) Leptin facilitates learning and memory performance and enhances hippocampal CA1 long-term potentiation and CaMK II phosphorylation in rats. Peptides 27:2738–2749

Opazo P, Watabe AM, Grant SG, O'Dell TJ (2003) Phosphatidylinositol 3-kinase regulates the induction of long-term potentiation through extracellular signal-related kinase-independent mechanisms. J Neurosci 23:3679–3688

Ott A, Stolk RP, van Harskamp F, Pols HA, Hofman A, Breteler MM (1999) Diabetes mellitus and the risk of dementia: the Rotterdam study. Neurology 58:1937–1941

Ozcan U, Cao Q, Yilmaz E, Lee AH, Iwakoshi NN, Ozdelen E et al (2004) Endoplasmic reticulum stress links obesity, insulin action, and type 2 diabetes. Science 306:457–461

Pardon MC, Rattray I (2008) What do we know about the long-term consequences of stress on ageing and the progression of age-related neurodegenerative disorders? Neurosci Biobehav Rev 32:1103–1120

Pasquier F, Boulogne A, Leys D, Fontaine P (2006) Diabetes mellitus and dementia. Diabetes Metab 32:403–414

Paz-Filho G, Wong ML, Licinio J (2010) The procognitive effects of leptin in the brain and their clinical implications. Int J Clin Pract 64:1808–1812

Pedersen WA, Flynn ER (2004) Insulin resistance contributes to aberrant stress responses in the Tg2576 mouse model of Alzheimer's disease. Neurobiol Dis 17:500–506

Pedersen WA, McMillan PJ, Kulstad JJ, Leverenz JB, Craft S, Haynatzki GR (2006) Rosiglitazone attenuates learning and memory deficits in Tg2576 Alzheimer mice. Exp Neurol 199:265–273

Peila R, Rodriguez BL, Launer LJ (2002) Type 2 diabetes, APOE gene, and the risk for dementia and related pathologies: the Honolulu-Asia Aging Study. Diabetes 51:1256–1262

Perluigi M, Sultana R, Cenini G, Di Domenico F, Memo M, Pierce WM, Coccia R, Butterfield DA (2009) Redox proteomics identification of 4-hydroxynonenal-modified brain proteins in Alzheimer's disease: role of lipid peroxidation in Alzheimer's disease pathogenesis. Proteomics Clin Appl 3:682–693

Pervanidou P, Chrousos GP (2010) Neuroendocrinology of post-traumatic stress disorder. Prog Brain Res 182:149–160
Peters I, Igbavboa U, Schütt T, Haidari S, Hartig U, Rosello X, Bottner S, Copanaki E, Deller T, Kogel D, Wood WG, Müller WE, Eckert GP (2009) The interaction of beta-amyloid protein with cellular membranes stimulates its own production. Biochim Biophys Acta 1788:964–972
Petersen KF, Shulman GI (2002) Pathogenesis of skeletal muscle insulin resistance in type 2 diabetes mellitus. Am J Cardiol 90(5A):11G–18G
Phiel CJ, Wilson CA, Lee VM, Klein PS (2003) GSK-3β regulate production of Alzheimer's disease amyloid-β peptides. Nature 423:435–439
Pickersgill L, Litherland GJ, Greenberg AS, Walker M, Yeaman SJ (2007) Key role for ceramides in mediating insulin resistance in human muscle cells. J Biol Chem 282:12583–12589
Posey KA, Clegg DJ, Printz DJ, Byun J, Morton GJ, Vivekanandan-Giri A, Pennathur S, Baskin DG, Heinecke JW, Woods SC, Schwartz MW, Niswender KD (2009) Hypothalamic proinflammatory lipid accumulation, inflammation, and insulin resistance in rats fed a high-fat diet. Am J Physiol Endocrinol Metab 296:E1003–E1012
Postina R (2008) A closer look at alpha-secretase. Curr Alzheimer Res 5:179–186
Power DA, Noel J, Collins R, O'Neill D (2001) Circulating leptin levels and weight loss in Alzheimer's disease patients. Dement Geriatr Cogn Disord 12:167–170
Powers ET, Morimoto RI, Dillin A, Kelly JW, Balch WE (2009) Biological and chemical approaches to diseases of proteostasis deficiency. Annu Rev Biochem 78:959–991
Price KA, Filiz G, Caragounis A, Du T, Laughton KM, Masters CL, Sharples RA, Hill AF, Li QX, Donnelly PS, Barnham KJ, Crouch PJ, White AR (2008) Activation of epidermal growth factor receptor by metal-ligand complexes decreases levels of extracellular amyloid beta peptide. Int J Biochem Cell Biol 40:1901–1917
Price JL, McKeel DW Jr, Buckles VD, Roe CM, Xiong C, Grundman M, Hansen LA, Petersen RC, Parisi JE, Dickson DW, Smith CD, Davis DG, Schmitt FA, Markesbery WR, Kaye J, Kurlan R, Hulette C, Kurland BF, Higdon R, Kukull W, Morris JC (2009) Neuropathology of nondemented aging: presumptive evidence for preclinical Alzheimer disease. Neurobiol Aging 30:1026–1036
Priller C, Bauer T, Mitteregger G, Krebs B, Kretzschmar HA, Herms J (2006) Synapse formation and function is modulated by the amyloid precursor protein. J Neurosci 26:7212–7221
Purkayastha S, Zhang G, Cai D (2011) Uncoupling the mechanisms of obesity and hypertension by targeting hypothalamic IKK-β and NF-κB. Nat Med 17:883–887
Qiu WQ, Ye Z, Kholodenko D, Seubert P, Selkoe DJ (1997) Degradation of amyloid beta-protein by a metalloprotease secreted by microglia and other neural and non-neural cells. J Biol Chem 272:6641–6646
Querfurth HW, LaFerla FM (2010) Alzheimer's disease. N Engl J Med 362:329–344
Rajpathak SN, Crandall JP, Wylie-Rosett J, Kabat GC, Rohan TE, Hu FB (2009) The role of iron in type 2 diabetes in humans. Biochim Biophys Acta 1790:671–681
Ramasamy R, Vannucci SJ, Yan SS, Herold K, Yan SF, Schmidt AM (2005) Advanced glycation end products and RAGE: a common thread in aging, diabetes, neurodegeneration, and inflammation. Glycobiology 15:16R
Rapizzi E, Taddei ML, Fiaschi T, Donati C, Bruni P, Chiarugi P (2009) Sphingosine 1-phosphate increases glucose uptake through trans-activation of insulin receptor. Cell Mol Life Sci 66:3207–3218
Reagan L, Gorovits N, Hoskin E (2001) Localization and regulation of GLUTx1 glucose transporter in the hippocampus of streptozotocin diabetic rats. Proc Natl Acad Sci U S A 98:2820–2825
Reaven G (2004) The metabolic syndrome or the insulin resistance syndrome? Different names, different concepts, and different goals. Endocrinol Metab Clin North Am 33:283–303
Reger MA, Watson GS, Frey WH II, Baker LD, Cholerton B, Keeling ML, Belongia DA, Fishel MA, Plymate SR, Schellenberg GD, Cherrier MM, Craft S (2006) Effects of intranasal insulin

on cognition in memory-impaired older adults: modulation by APOE genotype. Neurobiol Aging 27:451–458
Reger MA, Watson GS, Green PS, Baker LD, Cholerton B, Fishel MA, Plymate SR, Cherrier MM, Schellenberg GD, Frey WH II, Craft S (2008) Intranasal insulin administration dose-dependently modulates verbal memory and plasma amyloid-β in memory-impaired older adults. J Alzheimers Dis 13:323–331
Ren J, Pulakat L, Whaley-Connell A, Sowers JR (2010) Mitochondrial biogenesis in the metabolic syndrome and cardiovascular disease. J Mol Med 88:993–1001
Risner ME, Saunders AM, Altman JF, Ormandy GC, Craft S, Foley IM, Zvartau-Hind ME, Hosford DA, Roses AD, Rosiglitazone in Alzheimer's Disease Study Group (2006) Efficacy of rosiglitazone in a genetically defined population with mild-to-moderate Alzheimer's disease. Pharmacogenomics J 6:246–254
Ristow M (2004) Neurodegenerative disorders associated with diabetes mellitus. J Mol Med 82:510–529
Rivera EJ, Goldin A, Fulmer N, Tavares R, Wands JR, de la Monte SM (2005) Insulin and insulin-like growth factor expression and function deteriorate with progression of Alzheimer's disease: link to brain reductions in acetylcholine. J Alzheimers Dis 8:247–268
Roher AE, Debbins JP, Malek-Ahmadi M, Chen K, Pipe JG, Maze S, Belden C, Maarouf CL, Thiyyagura P, Mo H, Hunter JM, Kokjohn TA, Walker DG, Kruchowsky JC, Belohlavek M, Sabbagh MN, Beach TG (2012) Cerebral blood flow in Alzheimer's disease. Vasc Health Risk Manag 8:599–611
Rondinone CM, Wang LM, Lonnroth P, Wesslau C, Pierce JH, Smith U (1997) Insulin receptor substrate (IRS) 1 is reduced and IRS-2 is the main docking protein for phosphatidylinositol 3-kinase in adipocytes from subjects with non-insulin-dependent diabetes mellitus. Proc Natl Acad Sci U S A 94(8):4171–4175
Roselli F, Tirard M, Lu J, Hutzier P, Lamberti P, Livrea P, Morabito MK, Almeida OF (2005) Soluble beta-amyloid1–40 induces NMDA-dependent degradation of postsynaptic density-95 at glutamatergic synapses. J Neurosci 25:11061–11070
Rossetti L, Massillon D, Barzilai N, Vuguin P, Chen W, Hawkins M, Wu J, Wang J (1997) Short term effects of leptin on hepatic gluconeogenesis and in vivo insulin action. J Biol Chem 272:27758–27763
Ruvolo PP (2003) Intracellular signal transduction pathways activated by ceramide and its metabolites. Pharmacol Res 47:383–392
Saad MJ, Araki E, Miralpeix M, Rothenberg PL, White MF, Kahn CR (1992) Regulation of insulin receptor substrate-1 in liver and muscle of animal models of insulin resistance. J Clin Invest 90:1839–1849
Sacks DB, Arnold M, Bakris GL, Bruns DE, Hovath AR, Kirkman MS, Lernmark A, Metzger BE, Nathan DM (2011) Guidelines and recommendations for laboratory analysis in the diagnosis and management of diabetes mellitus. Clin Chem 57(6):e1–e47
Sadowski M, Pankiewicz J, Scholtzova H, Li Y-S, Quartermain D, Duff K, Wisniewski T (2004) Links between the pathology of Alzheimer's disease and vascular dementia. Neurochem Res 29:1257–1266
Sahara N, Murayama M, Lee B, Park JM, Lagalwar S, Binder LI, Takashima A (2008) Active c-jun N-terminal kinase induces caspase cleavage of tau and additional phosphorylation by GSK-3beta is required for tau aggregation. Eur J Neurosci 27:2897–2906
Saltiel AR, Pessin JE (2002) Insulin signaling pathways in time and space. Trends Cell Biol 12:65–71
Samad F (2007) Contribution of sphingolipids to the pathogenesis of obesity. Future Lipidol 2:625–639
Sandbrink R, Masters CL, Beyreuther K (1996) APP gene family. Alternative splicing generates functionally related isoforms. Ann N Y Acad Sci 777:281–287
Sasaki N, Fukatsu R, Tsuzuki K, Hayashi Y, Yoshida T, Fujii N, Koike T, Wakayama I, Yanagihara R, Garruto R, Amano N, Makita Z (1998) Advanced glycation end products in Alzheimer's disease and other neurodegenerative diseases. Am J Pathol 153:1149–1155

Sato T, Shimogaito N, Wu X, Kikuchi S, Yamagishi S, Takeuchi M (2006) Toxic advanced glycation end products (TAGE) theory in Alzheimer's disease. Am J Alzheimers Dis Other Demen 21:197–208

Savoia C, Schiffrin EL (2006) Inflammation in hypertension. Curr Opin Nephrol Hypertens 15:152–158

Schupf N, Kapell D, Nightingale B, Lee JH, Mohlenhoff J, Bewley S, Ottman R, Mayeux R (2001) Specificity of the fivefold increase in AD in mothers of adults with Down syndrome. Neurology 57:979–984

Schupf N, Patel B, Pang D, Zigman WB, Silverman W, Mehta PD, Mayeux R (2007) Elevated plasma beta-amyloid peptide Abeta(42) levels, incident dementia, and mortality in Down syndrome. Arch Neurol 64:1007–1013

Schupf N, Tang MX, Fukuyama H, Manly J, Andrews H, Mehta P, Ravetch J, Mayeux R (2008) Peripheral Abeta subspecies as risk biomarkers of Alzheimer's disease. Proc Natl Acad Sci U S A 105:14052–14057

Schwartz MW, Peskind E, Raskind M, Boyko EJ, Porte D Jr (1996) Cerebrospinal fluid leptin levels: relationship to plasma levels and to adiposity in humans. Nat Med 2:589–593

Schwartz MW, Woods SC, Porte D Jr, Seeley RJ, Baskin DG (2002) Central nervous system control of food intake. Nature 404:661–671

Selkoe DJ (2008) Soluble oligomers of the amyloid beta-protein impair synaptic plasticity and behavior. Behav Brain Res 192:106–113

Shanley LJ, Irving AJ, Harvey J (2001) Leptin enhances NMDA receptor function and modulates hippocampal synaptic plasticity. J Neurosci 21:RC186

Shen C, Chen Y, Liu H, Zhang K, Zhang T, Lin A, Jing N (2008) Hydrogen peroxide promotes Abeta production through JNK-dependent activation of gamma-secretase. J Biol Chem 283:7721–17730

Shoelson SE, Goldfine AB (2009) Getting away from glucose: fanning the flames of obesity-induced inflammation. Nat Med 15:373–374

Shoelson SE, Lee J, Goldfine AB (2006) Inflammation and insulin resistance. J Clin Invest 116:1793–1801

Shonesy BC, Thiruchelvam K, Parameshwaran K, Rahman EA, Karuppagounder SS, Huggins KW, Pinkert CA, Amin R, Dhanasekaran M, Suppiramaniam V (2012) Central insulin resistance and synaptic dysfunction in intracerebroventricular-streptozotocin injected rodents. Neurobiol Aging 33:430.e5–e18

Silvestri L, Camaschella C (2008) A potential pathogenetic role of iron in Alzheimer's disease. J Cell Mol Med 12:1548–1550

Simons K, Toomre D (2000) Lipid rafts and signal transduction. Nat Rev Mol Cell Biol 1:31–39

Smith MA, Taneda S, Richey PL, Miyata S, Yan SD, Stern D, Sayre LM, Monnier VM, Perry G (1994) Advanced Maillard reaction end products are associated with Alzheimer disease pathology. Proc Natl Acad Sci U S A 91:5710–5714

Smith MA, Zhu X, Tabaton M, Liu G, McKeel DW Jr, Cohen ML, Wang X, Siedlak SL, Dwyer BE, Hayashi T, Nakamura M, Nunomura A, Perry G (2010) Increased iron and free radical generation in preclinical Alzheimer disease and mild cognitive impairment. J Alzheimers Dis 19:363–372

Spiegel S, Milstien S (2003) Exogenous and intracellularly generated sphingosine 1-phosphate can regulate cellular processes by divergent pathways. Biochem Soc Trans 31:1216–1219

Stanford JC, Morris AJ, Sunkara M, Popa GJ, Larson KL, Ozcan S (2012) Sphingosine-1 phosphate (S1P) regulates glucose-stimulated insulin secretion in pancreatic beta cells. J Biol Chem 287(16):13457–13464

Steen E, Terry BM, Rivera EJ, Cannon JL, Neely TR, Tavares R, Xu XJ, Wands JR, de la Monte SM (2005) Impaired insulin and insulin-like growth factor expression and signaling mechanisms in Alzheimer's disease—is this type 3 diabetes? J Alzheimers Dis 7:63–80

Stengel A, Taché Y (2010) Corticotropin-releasing factor signaling and visceral response to stress. Exp Biol Med (Maywood) 235:1168–1178

Stone WS, Cottrill KL, Walker DL, Gold PE (1988a) Blood glucose and brain function: interactions with CNS cholinergic systems. Behav Neural Biol 50:325–334

Stone WS, Croul CE, Gold PE (1988b) Attenuation of scopolamine-induced amnesia in mice. Psychopharmacology (Berl) 96:417–442

Strachan MW, Abraha HD, Sherwood RA, Lammie GA, Deary IJ, Ewing FM, Perros P, Frier BM (1999) Evaluation of serum markers of neuronal damage following severe hypoglycaemia in adults with insulin-treated diabetes mellitus. Diabetes Metab Res Rev 15:5–12

Stranahan AM, Norman ED, Lee K, Cutler RG, Telljohann RS, Egan JM, Mattson MP (2008a) Diet-induced insulin resistance impairs hippocampal synaptic plasticity and cognition in middle-aged rats. Hippocampus 18:1085–1088

Stranahan AM, Arumugam TV, Cutler RG, Lee K, Egan JM, Mattson MP (2008b) Diabetes impairs hippocampal function through glucocorticoid-mediated effects on new and mature neurons. Nat Neurosci 11:309–317

Stratford S, Hoehn KL, Liu F, Summers SA (2004) Regulation of insulin action by ceramide: dual mechanisms linking ceramide accumulation to the inhibition of Akt/protein kinase B. J Biol Chem 279:36608–36615

Stumvoll M, Goldstein BJ, Van Haeften TW (2005) Type 2 diabetes: principles of pathogenesis and therapy. Lancet 365:1333–1346

Sun XJ, Wang LM, Zhang Y, Yenush L, Myers MG Jr, Glasheen E, Lane WS, Pierce JH, White MF (1995) Role of IRS-2 in insulin and cytokine signalling. Nature 377:173–177

Sun XJ, Goldberg JL, Qiao LY, Mitchell JJ (1999) Insulin-induced insulin receptor substrate-1 degradation is mediated by the proteasome degradation pathway. Diabetes 48:1359–1364

Swaminathan S, Fonseca VA, Alam MG, Shah SV (2007) The role of iron in diabetes and its complications. Diabetes Care 30:1926–1933

Takechi R, Galloway S, Pallebage-Gamarallage MM, Lam V, Mamo JC (2010) Dietary fats, cerebrovasculature integrity and Alzheimers disease risk. Prog Lipid Res 49:159–170

Talbot K, Wang HY, Kazi H, Han LY, Bakshi KP, Stucky A, Fuino RL, Kawaguchi KR, Samoyedny AJ, Wilson RS, Arvanitakis Z, Schneider JA, Wolf BA, Bennett DA, Trojanowski JQ, Arnold SE (2012) Demonstrated brain insulin resistance in Alzheimer's disease patients is associated with IGF-1 resistance, IRS-1 dysregulation, and cognitive decline. J Clin Invest 122:1316–1338

Tartaglia LA, Dembski M, Weng X, Deng N, Culpepper J, Devos R, Richards GJ, Campfield LA, Clark FT, Deeds J, Muir C, Sanker S, Moriarty A, Moore KJ, Smutko JS, Mays GG, Wool EA, Monroe CA, Tepper RI (1995) Identification and expression cloning of a leptin receptor OB-R. Cell 83:1263–1271

Tezapsidis N, Johnston JM, Smith MA, Ashford JW, Casadesus G, Robakis NK, Wolozin B, Perry G, Zhu X, Greco SJ, Sarkar S (2009) Leptin: a novel therapeutic strategy for Alzheimer's disease. J Alzheimers Dis 16:731–740

Thinakaran G, Koo EH (2008) Amyloid precursor protein trafficking, processing and function. J Biol Chem 283:29615–29619

Townsend M, Mehta T, Selkoe DJ (2007) Soluble Abeta inhibits specific signal transduction cascades common to the insulin receptor pathway. J Biol Chem 282:33305–33312

Troen A, Rosenberg I (2005) Homocysteine and cognitive function. Semin Vasc Med 5:209–214

Troy CM, Rabacchi SA, Xu Z, Maroney AC, Connors TJ, Shelanski ML, Greene LA (2001) Beta-amyloid-induced neuronal apoptosis requires c-Jun N-terminal kinase activation. J Neurochem 77:157–164

Turinsky J, O'Sullivan DM, Bayly BP (1990) 1,2-Diacylglycerol and ceramide levels in insulin-resistant tissues of the rat *in vivo*. J Biol Chem 265:16880–16885

Umeda T, Mori H, Zheng H, Tomiyama T (2010) Regulation of cholesterol efflux by amyloid beta secretion. J Neurosci Res 88:1985–1994

Unger J, McNeill TH, Moxley RT III, White M, Moss A, Livingston JN (1989) Distribution of insulin receptor-like immunoreactivity in the rat forebrain. Neuroscience 31:143–157

van der Heide LP, Ramakers GMJ, Smidt MP (2006) Insulin signaling in the central nervous system: learning to survive. Prog Neurobiol 79:205–221
van Eijk M, Aten J, Bijl N, Ottenhoff R, van Roomen CP, Dubbelhuis PF, Seeman I, Ghauharali-van der Vlugt K, Overkleeft HS, Arbeeny C, Groen AK, Aerts JM (2009) Reducing glycosphingolipid content in adipose tissue of obese mice restores insulin sensitivity, adipogenesis and reduces inflammation. PLoS One 4:e4723
Vander Haar E, Lee SI, Bandhakavi S, Griffin TJ, Kim DH (2007) Insulin signalling to mTOR mediated by the Akt/PKB substrate PRAS40. Nat Cell Biol 9:316–323
Vassar R (2004) BACE1: the beta-secretase enzyme in Alzheimer's disease. J Mol Neurosci 23:105–114
Veinbergs I, Everson A, Sagara Y, Masliah E (2002) Neurotoxic effects of apolipoprotein E4 are mediated via dysregulation of calcium homeostatis. J Neurosci Res 67:379–387
Vinereanu D (2006) Risk factors for atherosclerotic disease: present and future. Herz 31(Suppl 3):5–24
Wakabayashi M, Okada T, Kozutsumi Y, Matsuzaki K (2005) GM1 ganglioside-mediated accumulation of amyloid beta-protein on cell membranes. Biochem Biophys Res Commun 328:1019–1023
Walsh DM, Selkoe DJ (2007) A beta oligomers—a decade of discovery. J Neurochem 101:1172–1184
Walsh DM, Klyubin I, Fadeeva JV, Cullen WK, Anwyl R, Wolfe MS, Rowan MJ, Selkoe DJ (2002) Naturally secreted oligomers of amyloid beta protein potently inhibit hippocampal long-term potentiation in vivo. Nature 416:535–539
Walters FJM (1998) Intracranial pressure and cerebral blood flow. Physiology 8:1–4
Wang Z, Wang B, Yang L, Guo Q, Aithmitti N, Songyang Z, Zheng H (2009) Presynaptic and postsynaptic interaction of the amyloid precursor protein promotes peripheral and central synaptogenesis. J Neurosci 29:10788–10801
Watcharasit P, Bijur GN, Song L, Zhu J, Chen X, Jope RS (2003) Glycogen synthase-3β (GSK3β) binds to and promotes the actions of p53. J Biol Chem 278:48872–48879
Watson GS, Craft S (2004) Modulation of memory by insulin and glucose: neuropsychological observations in Alzheimer's disease. Eur J Pharmacol 490:97–113
Wayner MJ, Armstrong DL, Phelix CF, Oomura Y (2004) Orexin-A (Hypocretin-1) and leptin enhance LTP in the dentate gyrus of rats in vivo. Peptides 25:991–996
Welsh B, Wecker L (1991) Effects of streptozotocin-induced diabetes on acetylcholine metabolism in rat brain. Neurochem Res 16:453–460
Wenk GL (2006) Neuropathologic changes in Alzheimer's disease: potential targets for treatment. J Clin Psychiatry 67(Suppl 3):3–7
West IC (2000) Radicals and oxidative stress in diabetes. Diabet Med 17:171–180
White MF (2003) Insulin signaling in health and disease. Science 302:1710–1711
Wilson RS, Arnold SE, Schneider JA, Kelly JF, Tang Y, Bennett DA (2006) Chronic psychological distress and risk of Alzheimer's disease in old age. Neuroepidemiology 27:143–153
Wolf OT (2009) Stress and memory in humans: twelve years of progress. Brain Res 1293:142–154
Wolozin B (2004) Cholesterol and the biology of Alzheimer's disease. Neuron 41(1):7–10
Wu A, Ying Z, Gomez-Pinilla F (2004) The interplay between oxidative stress and brain-derived neurotrophic factor modulates the outcome of a saturated fat diet on synaptic plasticity and cognition. Eur J Neurosci 19:1699–1707
Xu WL, Atti AR, Gatz M, Pedersen NL, Johansson B, Fratigleoni L (2011) Midlife overweight and obesity increase late-life dementia risk. Neurology 76:1568–1574
Yamaguchi Y, Flier JS, Yokota A, Benecke H, Backer JM, Moller DE (1991) Functional properties of two naturally occurring isoforms of the human insulin receptor in Chinese hamster ovary cells. Endocrinology 4:2058–2066
Yamamoto A, Shin RW, Hasegawa K, Naiki H, Sato H, Yoshimasu F, Kitamoto T (2002) Iron (III) induces aggregation of hyperphosphorylated tau and its reduction to iron (II) reverses the

aggregation: implications in the formation of neurofibrillary tangles of Alzheimer's disease. J Neurochem 82:1137–1147
Yamamoto N, Matsubara T, Sato T, Yanagisawa K (2008) Age-dependent high-density clustering of GM1 ganglioside at presynaptic neuritic terminals promotes amyloid beta-protein fibrillogenesis. Biochim Biophys Acta 1778:2717–2726
Yamamoto N, Matsubara T, Sobue K, Tanida M, Kasahara R, Naruse K, Taniura H, Sato T, Suzuki K (2012) Brain insulin resistance accelerates Aβ fibrillogenesis by inducing GM1 ganglioside clustering in the presynaptic membranes. J Neurochem 121:619–628
Yang L, Wang B, Long C, Wu G, Zheng H (2007) Increased asynchronous release and aberrant calcium channel activation in amyloid precursor protein deficient neuromuscular synapses. Neuroscience 149:768–778
Ye R, Jung DY, Jun JY, Li J, Luo S, Ko HJ, Kim JK, Lee AS (2010) Grp78 heterozygosity promotes adaptive unfolded protein response and attenuates diet-induced obesity and insulin resistance. Diabetes 59:6–16
Yea K, Kim J, Lim S, Park HS, Park KS, Suh PG, Ryu SH (2008) Lysophosphatidic acid regulates blood glucose by stimulating myotube and adipocyte glucose uptake. J Mol Med 86:211–220
Yi H, Lee SJ, Lee J, Myung CS, Park WK, Lee GH, Kong JY, Cho H (2011) Sphingosylphosphorylcholine attenuated β-amyloid production by reducing BACE1 expression and catalysis in PC12 cells. Neurochem Res 36:2083–2090
Yoshida H, Hastie CJ, McLauchlan H, Cohen P, Goedert M (2004) Phosphorylation of microtubule-associated protein tau by isoforms of c-jun N-terminal kinase (JNK). J Neurochem 90:352–358
Yu C, Chen Y, Cline GW, Zhang D, Zong H, Wang Y, Bergeron R, Kim JK, Cushman SW, Cooney GJ, Atcheson B, White MF, Kraegen EW, Shulman GI (2002) Mechanism by which fatty acids inhibit insulin activation of insulin receptor substrate-1 (IRS-1)-associated phosphatidylinositol 3-kinase activity in muscle. J Biol Chem 277:50230–50236
Yu X, Park BH, Wang MY, Wang ZV, Unger RH (2008) Making insulin-deficient type 1 diabetic rodents thrive without insulin. Proc Natl Acad Sci U S A 105:14070–14075
Zecca L, Youdim MBH, Riederer P, Connor JR, Crichton RR (2004) Iron, brain ageing and neurodegenerative disorders. Nat Rev Neurosci 5:863–873
Zha Q, Ruan Y, Hartmann T, Beyreuther K, Zhang D (2004) GM1 ganglioside regulates the proteolysis of amyloid precursor protein. Mol Psychiatry 9:946–952
Zhang X, Dong F, Ren J, Driscoll MJ, Culver B (2005) High dietary fat induces NADPH oxidase-associated oxidative stress and inflammation in rat cerebral cortex. Exp Neurol 191:318–325
Zhao WQ, Alkon DL (2001) Role of insulin and insulin receptor in learning and memory. Mol Cell Endocrinol 177:125–134
Zhao WQ, Townsend M (2009) Insulin resistance and amyloidogenesis as common molecular foundation for type 2 diabetes and Alzheimer's disease. Biochim Biophys Acta 1792:482–496
Zhao W, Chen H, Xu H, Moore E, Meiri N, Quon MJ, Alkon DL (1999) Brain insulin receptors and spatial memory. Correlated changes in gene expression, tyrosine phosphorylation, and signaling molecules in the hippocampus of water maze trained rats. J Biol Chem 274:34893–34902
Zhao WQ, De Felice FG, Fernandez S, Chen H, Lambert MP, Quon MJ, Krafft GA, Klein WL (2008) Amyloid beta oligomers induce impairment of neuronal insulin receptors. FASEB J 22:246–260
Zlokovic BV (2005) Neurovascular mechanisms of Alzheimer's neurodegeneration. Trends Neurosci 28:202–208

# Chapter 9
# Metabolic Syndrome as a Risk Factor for Depression

## 9.1 Introduction

Depression is a multisystem and multifactorial mental disorder characterized by poor (low) mood, changes in weight (decrease or increase), sleep disturbances (insomnia or hypersomnia), psychomotor retardation or agitation, fatigue, feelings of worthlessness or guilt, diminished cognitive functioning, and recurrent thoughts of death. Other symptoms of depression include psychomotor change, loss of energy, concentration difficulties/indecisiveness, irritability, and low self-esteem (Davidson et al. 2002a, b). Not all individuals show all of the symptoms of depression (Anisman and Matheson 2005). Clinical features of depression include reduction in plasma levels of brain-derived neurotrophic factor (BDNF) (Sen et al. 2008), elevated blood levels of IL-6 and TNF-α (Dowlati et al. 2010), and low levels of magnesium. In addition, overactivity of hypothalamic-pituitary-adrenal (HPA) axis (Varghese and Brown 2001), increase in ventricle/brain ratio along with localized atrophy of the prefrontal cortex, cingulated gyrus, ventral striatum, amygdale, cerebellum, and hippocampus are also considered as morphological features of depression (aan het Rot et al. 2009). The molecular mechanism of depression is not fully understood. However, it is becoming increasingly evident that changes in neurotransmitters, neuropeptides (vasopressin), cytokines, and gene environmental interactions may contribute to the pathogenesis of major depression (Fig. 9.1). It is well known that glutamate (Glu), γ-aminobutyric acid (GABA), and monoaminergic systems are the principal excitatory and inhibitory neurotransmitter systems in the central nervous system (CNS), and increasing evidences suggest that alterations in these neurotransmitter systems along with alterations in biogenic amine (dopamine, norepinephrine and serotonin) metabolism along with modification in neuronal synaptic density and expression of neurotrophic factors may contribute to the pathophysiology of depression (Krystal et al. 2002; Cryan and Slattery 2010).

Depression is the fourth leading cause of disability worldwide according to the World Health Organization (Ustun et al. 2004). Prevalence rates of depression are between 5 and 10 %, whereas it is estimated to be about 12–18 % in patients with

A.A. Farooqui, *Metabolic Syndrome: An Important Risk Factor for Stroke, Alzheimer Disease, and Depression*, DOI 10.1007/978-1-4614-7318-3_9,

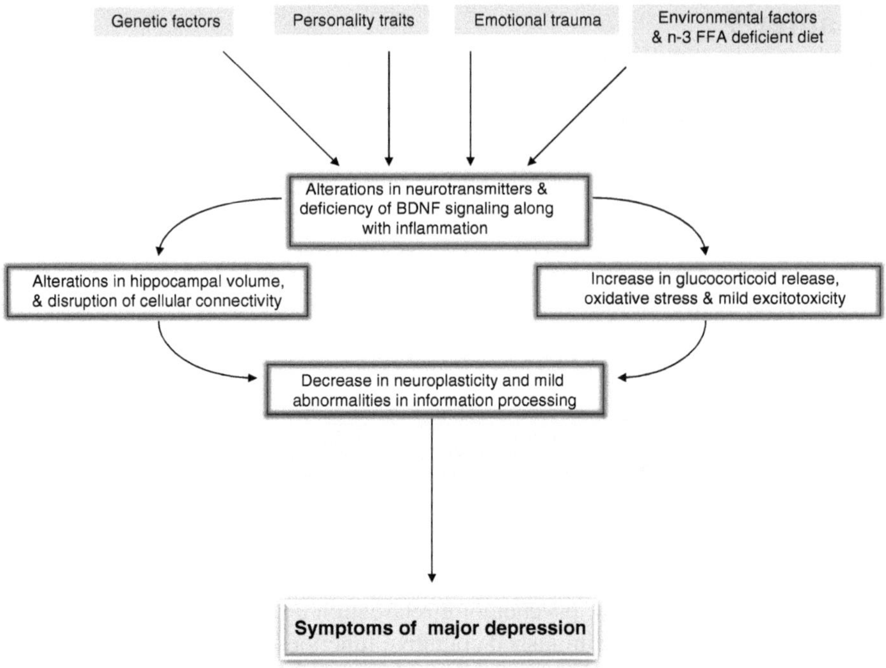


**Fig. 9.1** Factors that modulate depression

diabetes (Ali et al. 2006). Moreover, 31 % of elderly diabetic patients have been reported to have severe depressive symptoms (Black 1999). Women are at a greater risk for major depression than men (Wise et al. 2001). In most studies, the ratio of prevalence rates in women to men is between the range of 1.5 and 2.5 (Fava and Kendler 2000). A meta-analysis has indicated that genetic factors also influence the pathogenesis of depression (Sullivan et al. 2000). Genes can predispose individuals to major depressive disorder in many ways. For example, aberrant genes can predispose one to depression by diminishing production of growth factors that act and play important roles during brain development. Aberrant genes can also increase or decrease a neurotransmitter's release into synapses. The gene products of greatest interest at present are those involved in the synthesis and actions of serotonin, among them, the serotonin-uptake protein localized within the terminals and dendrites of serotonin-releasing neurons (Wurtman 2005). In addition, genes also modulate synaptic connections, the intracellular transduction of neuronal signals, and the speed with which all of these can change in response to environmental stressors (Wurtman 2005). The serotonin transporter gene is the most studied gene in major depressive disorder (Levinson 2006). However, vulnerability to depression is only partly genetic, with environmental factors playing a greater role when it comes to non-related individuals (Sullivan et al. 2000).

Nongenetic factors for depression include stress, emotional trauma, and obesity, (Krysinska and Lester 2010), adverse childhood experience, and individual personality traits (Fig. 9.1) (Anda et al. 2002). The term stress refers to an internal or external (environmental) event that disrupts normal homeostasis. Stressful events evoke an extensive multisystem and highly integrative physiological response (Goldstein and Kopin 2007). Animals respond to stress by inducing a physiological stress response to regain equilibrium lost during the stress. Stress may be induced by external stressors such as social strains, lower socioeconomic status, and job strain or by intrinsic stressors such as sleep deprivation. Although an acute short-term stress response is necessary for homeostasis recovery, chronic stress responses can be harmful and may cause several disease states, such as depression (Peeke and Chrousos 1995). Chronic stress plays an important role in the etiology of depression. The activation of the HPA axis constitutes a major component of the neuroendocrine stress response (Smith and Vale 2006). Depressed patients often show hyperactivity in the HPA axis such as hypersecretion of basal corticosterone and increased adrenal weight (Zheng et al. 2006). Collective evidence suggests that rodents exposed to chronic unpredictable stress or chronic social defeat stress develop behavioral deficits and endocrine abnormalities, mimicking the symptoms of human depression. Rats exposed to chronic unpredictable stress or chronic social defeat stress show decrease in basal levels of leptin, a 16 kDa hormone produced by white adipose tissue in plasma (Lu et al. 2006). Systemic administration of leptin reverses the hedonic-like deficit mediated by chronic unpredictable stress and improves behavioral despair dose dependently in the forced swim test (FST). The behavioral effects of leptin in the FST are accompanied by an increase in neuronal activation in the hippocampus. Intra-hippocampal infusion of leptin produces a similar antidepressant-like effect in the FST as its systemic administration. In contrast, infusion of leptin into the hypothalamus not only reduces body weight but has no effect on FST behavior supporting the view that leptin can be used for the treatment of depressive disorders (Lu et al. 2006).

Stress (caused by psychosocial and/or physical stressors) also inhibits adult neurogenesis by blocking one or more phases of the neurogenetic process (Leuner et al. 2006; Leuner and Gould 2010). Thus, both brief unpredictable stress exposure and chronic mild unpredictable stress potently suppress proliferation. In addition, prolonged mild stress exposure downregulates the survival of newborn cells (Lucassen et al. 2010a, b). The molecular mechanisms of stress-induced inhibition of proliferation and neurogenesis are not fully understood. However, the consequences of stress-induced blockade of neurogenesis may not only cause apoptosis of progenitor cells but also negatively affect cell cycle (Heine et al. 2004). These suggestions are supported by observations on postmortem studies on the hippocampus of rodent model of depression. Sustained exposure to chronic stress or excessive glucocorticoids produces adverse effects on the hippocampus, a brain region that is intrinsically linked with mood, memory, and cognitive function, and may cause the hippocampal atrophy (Zheng et al. 2006; Lightman et al. 2008). There is a significant correlation between the duration of the depression and the extent of hippocampal atrophy (McEwen 1992). Acute stress not only produces the activation of resident microglia but also upregulates MHC class II (de Pablos et al. 2006; Frank et al. 2007). Similarly, chronic stress also activates microglia and induces microglial proliferation (Nair and Bonneau 2006). It is hypothesized that stress-mediated changes in microglia can "prime" or sensitize the brain to respond in a heightened fashion when presented with a subsequent peripheral immune stimulus such as lipopolysaccharide (Frank et al. 2007; de Pablos et al. 2006; Johnson et al. 2003).

Besides psychological factors, dietary factors also play a role in the susceptibility to stress and depression. Thus, increase in appetite and resulting weight gain are common symptoms of depression (Carter et al. 1994). Depression may not only lead to decrease in physical activity (Cassidy et al. 2004) but increases risk for obesity. Depression may increase risk of weight gain through its effect on binge eating (French et al. 1999), especially among women (Linde et al. 2004). It is demonstrated that stress-mediated anabolic/catabolic imbalance is closely associated in part by high cortisol, glucose, and insulin and low androgens and growth hormones leading to oxidative stress and systemic inflammation (Farooqui et al. 2012). Consumption of energy-dense food and obesity also contribute to oxidative stress and systemic inflammation. Our present-day lifestyle promotes the consumption of calorically dense food during times of stress. Induction of chronic stress, overeating, and consumption of fructose-containing soft drinks can lead to insulin resistance, generation of reactive oxygen species (ROS), formation of advanced glycation end products (AGEs), and deposition of abdominal fat tissue. Thus, humans with poorly controlled diabetes show hyperglycemia accompanied by an accelerated rate of ROS and AGEs formation and accumulation. AGEs in various tissues are known to progress during normal aging and at an accelerated rate in individuals with diabetes (Yamagishi et al. 2008). Elevated AGE levels contribute to the impairment of hippocampal neurogenesis and to behavioral deficits in rodent models of insulin-deficient diabetes with depressive behaviors. These findings land additional support to the current understanding of the pathogenetic roles of AGE in the pathogenesis of diabetic complications and depression (Wang et al. 2009).

Different organs and systems respond to oxidative stress in different ways. For example, cardiovascular system responds to stress mainly through the hyperstimulation of catecholamine and is associated with increased cardiac output and vascular resistance, enrichment of lipid metabolism and lipid mobilization, and stimulation of platelet aggregation leading to the pathogenesis of hypertension, atherosclerosis, and ischemic heart disease (Esch et al. 2002). Rat brain responds to stress through the reversible atrophy of hippocampal dendrites over the course of weeks (McEwen and Sapolsky 1995) and induction of nuclear factor-kappaB (NF-κB) and increased expression of proinflammatory cytokines leading to apoptotic cell death in cortical and hippocampal areas (McEwen and Sapolsky 1995; Lucassen et al. 2001; Munhoz et al. 2006). Atrophy of the hippocampus is also observed in humans. It is described in war-related posttraumatic stress disorder and major depression, coexisting with a decrease in neuronal activity (McEwen and Sapolsky 1995). In addition to hippocampus, posttraumatic stress-induced atrophy also affects temporal lobe, orbitofrontal cortex, and whole brain volume, resulting in decrease in blood flow of patients with posttraumatic stress (McEwen and Sapolsky 1995; Bremner et al. 2002). Chronic stress also produces changes in immune system activity. The activation of the immune system can provoke neuroendocrine and neurotransmitter changes that are similar to those provoked by physical or psychological stressors (Anisman 2009).

A correlation between low intake of n-3 fatty acids and the prevalence of oxidative stress and depression has been reported in many studies (Astorg et al. 2008; De Vriese et al. 2004; Hibbeln 1998). n-3 fatty acids retard depression by promoting and maintaining synaptic plasticity and neurogenesis in the aging brain circuits. They also reduce oxidative stress and neuroinflammation through the inhibition of lipid peroxidation and generation of neuroprotectins and resolvins (Serhan 2005; Bazan 2005; Farooqui et al. 2007; Farooqui 2009, 2012). In addition, DHA and EPA also facilitate optimal membrane fluidity, promote appropriate neurotrophic support, and inhibit production of cytokines and expression of adhesion molecules (Farooqui 2009, 2012). Collective evidence suggests that depression is a multisystem disorder involving changes in endocrine, vascular, neural, and immunologic systems and the above-mentioned factors along with elevation in indoleamine 2,3-dioxygenase (IDO) activity with decrease in levels of tryptophan and serotonin may predispose one towards depression (Maes et al. 2009, 2010). Sustained increase in brain IDO can also be deleterious because of its negative impact on monoaminergic neurotransmission and neuronal survival. It is well known that tryptophan, an essential amino acid that is actively transported into the brain, is a precursor for the synthesis of serotonin, and increased degradation of tryptophan by IDO has been postulated to reduce the availability of this amino acid for serotonin synthesis. Concurrently, activation of IDO may increase the production of several neuroactive glutamatergic metabolites, including 3-hydroxykynurenine (3HK) and quinolinic acid (QA), that play a key role in neuronal death and neurodegenerative diseases (Fu et al. 2010). Furthermore, kynurenine can also be metabolized in kynurenic acid, a metabolite that displays neuroprotective properties. However, these apparently antagonistic pathways are compartmentalized in the brain, with microglia preferentially producing quinolinic acid, whereas astrocytes produce kynurenic acid. It is proposed that impaired serotonin synthesis and IDO activation may lead to functional, structural, and behavioral alterations in the brain and these changes may be closely associated with the development of depressive-like behavior in the mouse (Fu et al. 2010). Depression is also associated with an increased risk of developing atherosclerosis, heart disease, hypertension, stroke, cognitive decline, dementia (including Alzheimer disease), osteoporosis, immune impairments (immunosenescence), obesity, metabolic syndrome, insulin resistance, and type II diabetes (Schulz et al. 2000).

## 9.2 Pathogenesis of Depression

Many hypotheses have been proposed to explain the pathogenesis of depression. These hypotheses include monoamine hypothesis, neurotransmitter receptor hypothesis, neurotrophic factor hypothesis, HPA dysregulation hypothesis, oxidative stress, cytokine hypothesis, NO pathway hypothesis, GABA-deficit hypothesis,

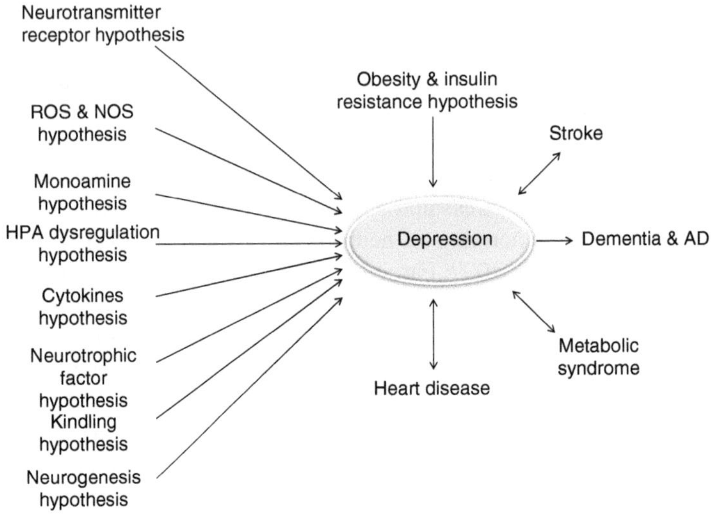


**Fig. 9.2** Hypotheses related to the pathogenesis of depression

substance withdrawal hypothesis, immune dysregulation hypothesis, and neuronal plasticity hypothesis (Fig. 9.2). Among above hypotheses, cytokine hypothesis has received considerable attention in recent years. Cytokine hypothesis of depression states that depression is caused by the actions of cytokines, which are regulatory proteins that can act in an autocrine, paracrine, and hormone-like way. Cytokines have a pleiotropic action (multiple target cells and multiple actions) and many cytokines have an overlapping spectrum of actions. They may also induce antagonistic or synergistic actions. Therefore, exposure to different cytokines simultaneously may result in qualitatively different responses in target cells. In addition, a cytokine may increase or decrease the production of another cytokines. Cytokines are rapidly cleared, and the half-life of most cytokines, which have been injected intravenously, is usually measured in minutes (Vilcek 2003). Cytokines facilitate communication not only among immune cells but also between glial cells and neurons. Although cytokines have been reported to access the brain and interact with virtually every pathophysiologic domain relevant to depression, including neurotransmitter metabolism, neuroendocrine function, and neural plasticity (Raison et al. 2006; Dantzer et al. 2008), it remains unclear whether activation of inflammatory pathways in the central nervous system (CNS) during depression originates primarily in the periphery, e.g., as a function of overt or nascent medical illness or psychological stresses (see below), and/or whether stress (e.g., vascular insults in late-life depression) induces cytokine release and inflammatory responses directly within the brain. Recent studies indicate that cytokines (TNF-α, IL-1β, and INFα), which enter the brain through humoral pathway or circumventricular organs and interact with their receptors to produce their physiological effects. They modulate virtually every pathophysiologic domain associated with depression, such as neurotransmitter metabolism, neuroendocrine function, endocrine, vascular, neural, and immunologic systems, and neural plasticity along with decrease in neurotrophic support and alterations in glutamate release/reuptake, oxidative stress, and excitotoxicity (Raison et al. 2006; Dantzer et al. 2008; Miller et al. 2009). Many inflammatory signals are mediated through NF-κB, a redox-sensitive transcription factor whose activity is affected by cellular oxidative state. NF-κB is an essential mediator at the blood–brain interface that communicates peripheral inflammatory signals to the CNS. Central blockade of NF-κB in rodents inhibits c-Fos activation in multiple brain regions following peripheral administration of IL-1β while also inhibiting IL-1β and LPS-induced behavioral changes (Nadjar et al. 2005; Godbout et al. 2005; Dantzer et al. 2008).

Stress is defined as internal or external (environmental) event that disrupts normal cell or tissue homeostasis. Stress is a state of disharmony mediated by intrinsic or extrinsic adverse forces. Stress is counteracted by intricate physiological and behavioral responses that aim to reestablish the challenged body equilibrium. The adaptive stress response depends upon an elaborate neuroendocrine, cellular, and molecular infrastructure, the stress system. Biochemically, the stress response is characterized by acute behavioral and physical adaptations, including increased cognition, analgesia, gluconeogenesis, lipolysis, and inhibition of reproduction (Kyrou et al. 2006). Crucial functions of the stress system response are mediated and modulated by the HPA axis and the central and peripheral components of the autonomic nervous system (ANS) (Bjorntorp 1993). HPA and ANS are important elements of the stress response involved in the pathogenesis of insulin resistance, obesity, and MetS. Stress is not only induced by external stressors such as social strains, lower socioeconomic status, and job strain but also by intrinsic stressors such as sleep deprivation. Although an acute short-term stress response is necessary for homeostasis recovery, chronic stress responses can be harmful and may cause several disease states, such as depression (Peeke and Chrousos 1995; Pêgo et al. 2010). As stated above, stress is perceived by sensory systems in the brain, facilitating evaluation and comparison of the existing and previous stimuli as well as the activation of hormones responsible for energy mobilization. It is hypothesized that disproportionate gains in visceral fat mass during obesity may be associated with an increase in HPA axis activity mediated by repeated exposure to stressors (Pêgo et al. 2010). The limbic system promotes the release of glucocorticoid (cortisol in humans and corticosterone in rodents), the primary stress hormones, by modulating activation of the hypothalamic paraventricular nucleus (PVN) (Pêgo et al. 2010). Cortisol is not stored in the adrenal cortex. It is synthesized on demand and immediately enters into the general circulation (blood and lymphatics). Its synthesis in the adrenal cortex is controlled along the HPA axis through one of the two negative feedback loops: one direct and the other indirect (Cowell and Buckingham 2001; Morand and Leech 2001). The amygdala plays a putative role in the evaluation of emotional events and formation of fearful memories; it is also a target of the neurochemical and hormonal mediators of stress. Clinical and experimental data show changes in the structure/function of the amygdala with emotional disorders such as anxiety. Although chronic stress is known to contribute to the pathogenesis of depression, not everyone who is exposed to stress becomes depressed. Stress

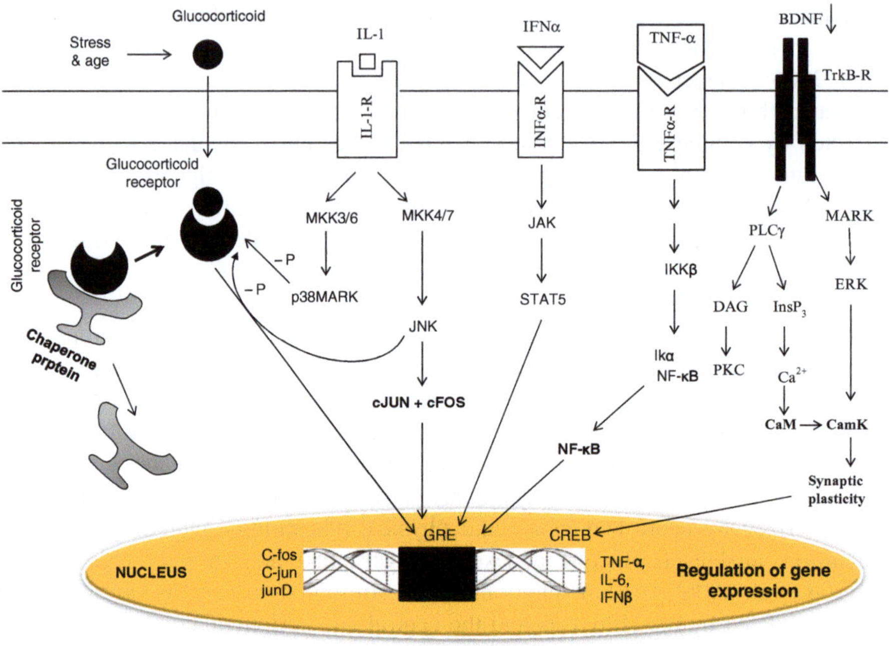


**Fig. 9.3** Secretion of cortisol and interaction between peripheral and brain cytokines

interacts with the individual's genetic makeup as well as environment factors to influence the risk for developing major depression. The molecular mechanism of stress-mediated changes in prefrontal cortex morphology and function is not fully understood. However, it is suggested that stress mediates its effect through the release of corticotropin-releasing hormone (CRH), which not only acts on the anterior pituitary to promote the secretion of adrenocorticotropic hormone (ACTH). This hormone then stimulates the adrenal cortex to release glucocorticoids (GCs) into the bloodstream (De Kloet et al. 1998). CRH also regulates BDNF expression but also modulates other central processes, such as learning and memory, synaptic plasticity, and neuroprotection during the course (Bayatti et al. 2005; Hauger et al. 2009). The stress response is principally mediated by the activation of the HPA axis via the release of glucocorticoids (cortisol), which is largely controlled by the release of corticotrophin-releasing hormone (CRH) in the parvocellular neurons of the hypothalamus. This region is known to govern various metabolic activities of the body including appetite control, energy expenditure, carbohydrate and lipid metabolism, and blood pressure homeostasis (Fig. 9.3) (Sandoval et al. 2008; Morton et al. 2006). CRH is also produced by extra-hypothalamic brain regions in response to stress. It binds to specific receptors to mediate a range of functions including stimulation of excitatory neurotransmission, synaptic plasticity, cognition, and other behavioral states relevant to depression (de Kloet et al. 2005). Cytokines, primarily IL-1β, also stimulate the release of CRH from the PVN of the hypothalamus (Fig. 9.3). CRH stimulates the release of adrenocorticotropic hormone (ACTH) from the pituitary, which travels through the systemic circulation to the adrenals where it mediates the release of cortisol (Shelton and Miller 2010). Daily cortisol dynamics are controlled by a series of factors including, but not limited to, genetic determinants, major personality dimensions, gender, negative early life events, and current negative life events.

Depressed patients have higher basal cortisol levels (Akil et al. 1993) and non-suppression of endogenous cortisol secretion after dexamethasone administration. This observation supports the view that there is impairment in feedback and consequent HPA axis hyperactivity in depression (Ehlert et al. 2001). However, HPA axis functioning in depression is dysregulated (Kasckow et al. 2001). During depression, hippocampal glucocorticoid receptors may be sensitive to elevated cortisol levels and play a crucial role in the normalization of the HPA axis (Paskitti et al. 2000; Zheng et al. 2006). Although reductions in neurogenesis and hippocampal atrophy may be characteristic features of depression, it is not known whether it is a cause, correlate, or consequence of this disorder (Kempermann and Kronenberg 2003). HPA axis dysregulation is related to many pathological conditions, such as cardiovascular disease, metabolic syndrome, cancer, and neurotraumatic and neurodegenerative diseases. Many of these pathological conditions are accompanied by oxidative stress and neuroinflammation. The risk factors for these diseases also include obesity, hypercholesterolemia, hypertriglyceridemia, increased blood pressure, elevation in production of proinflammatory cytokines, and elevated heart rate (Rosmond and Bjorntorp 2000).

Hypothalamic neurons contribute to activation of the serotonin and norepinephrine systems (Gold and Chrousos 1999). Depression is linked to the function of monoamines (serotonin, norepinephrine, and dopamine). Depletion of monoamines (serotonin and norepinephrine) induces depressive signs and symptoms in some drug-free patients (Delgado et al. 1994, 1999). Cytokines are known to influence and modulate synthesis, metabolism, and cellular transit of serotonin. Proinflammatory cytokines, which are released in response to stress, may reduce serotonin levels and subsequent platelet aggregation causing both depression and atherosclerosis. Reciprocal connections between the norepinephrine system and the hypothalamus create a feed-forward cascade in which stress progressively activates CRH and norepinephrine signaling. Collectively these studies suggest that enhancement in sympathetic activity in conjunction with increase in secretion of catecholamine, cortisol, and serotonin may contribute to oxidative stress, which may damage the arcuate nucleus as well as the hypothalamus and macrophages (Singh et al. 2012). These processes along with low levels of long-chain polyunsaturated fatty acids may damage the arcuate nucleus as well as neuropeptide-Y and pro-opiomelanocortin neurons and insulin receptors in the brain, especially from fetal life to adulthood (Singh et al. 2012).

HPA axis dysregulation and reduction in neuroplasticity in depression are consistent with the assumption that BDNF is a stress-responsive intercellular messenger modifying HPA axis activity (Schule et al. 2006). The loss of BDNF plays a major role in the pathophysiology of depression and that its restoration may represent a critical mechanism underlying antidepressant efficacy. Glucocorticoids reduce the levels of BDNF, a downstream target of antidepressants. Antidepressants promote the transcriptional activity of cyclic adenosine monophosphate (cAMP) response element-binding protein (CREB), a transcription factor responsible for increasing the expression of a number of neuroprotective and neuromodulatory proteins and BDNF important in defining memory, a process by which knowledge is encoded, stored, and later retrieved (Müller 2010). The neurotrophin hypothesis of depression has gained considerable support. The rationale for the BDNF hypothesis originates from observations that acute and chronic stress in humans decrease endogenous neurotrophin levels and can lead to significant atrophy of the hippocampus, a structure known to be involved in controlling emotionality (Duman 2004). These processes are closely associated with neurogenesis (Duman 2005). BDNF facilitates the survival of neurons in the brain by binding to tyrosine kinase receptor B (TrkB) receptors (Ibanez 1995) and the activation of the MAP kinase pathway. BDNF also promotes and regulates neuronal survival via the PtdIns 3K/Akt pathway, which phosphorylates and regulates a number of cell survival-related proteins, including IkB, the forkhead transcription factor FKHRL1, glycogen synthase kinase 3β (GSK3β), and Bad, a pro-apoptotic member of the bcl-2 family (Huang and Reichardt 2001). It is hypothesized that chronic stress may lead to a downregulation of BDNF neurotrophic support decreasing the antiapoptotic regulation of bcl-2 and thus reducing neurogenic cell survival. This has detrimental consequences for hippocampal function and ultimately leads to the development of depressive symptoms (Fig. 9.3) (Almeida et al. 2005).

Collectively, these studies suggest that glucocorticoids, which are secreted by the adrenal cortex, are powerful anti-inflammatory compounds due to their ability to inhibit all stages of the inflammatory response (Pace et al. 2007; Pace and Miller 2009) (Figs. 9.3 and 9.5). Interactions between TNF-α and its receptors trigger a signal transduction cascade that activates IκB kinase β (IKKβ or IKK2) which, in turn, phosphorylates the IκB–NF-κB dimer, releasing NF-κB. The latter then translocates to the nucleus where it interferes with glucocorticoid receptor (GR) binding to glucocorticoid receptor element segments (GRE) via protein–protein interactions (McKay and Cidlowski 1999). Similarly, IFN-α interacts with IFN-α receptors resulting in phosphorylation of Janus kinase 1 (JAK1) which activates STAT5. Like NF-κB, STAT5 blocks GR–GRE interactions through protein–protein interactions in the nucleus (Hu et al. 2009; Rogatsky and Ivashkiv 2006). Finally, binding of IL-1 to IL-1 receptor activation causes the activation of mitogen-activated protein kinase kinase (MKK)3/6 and MKK4/7. These complexes activate p38 MAPK and c-Jun N-terminal kinases (JNK), which can phosphorylate GR impeding its nuclear translocation (Fig. 9.4). JNK also phosphorylates c-Jun. Phosphorylated c-Jun binds to AP-1 complex and interferes with GR–GRE interactions (Pace et al. 2007) (Fig. 9.5).

Exposure to stress causes oxidative/nitrosative damage to the brain tissue in general and prefrontal cortex and hippocampal area in particular. These changes are mediated by catecholamines, glucocorticoids, and glutamate and are shown to be balanced by anti-inflammatory pathways in the brain. In particular, acute restraint stress is followed by the upregulation of cyclooxygenase (COX)-2 and subsequent generation of proinflammatory prostaglandin, $PGE_2$, release in the cortex.

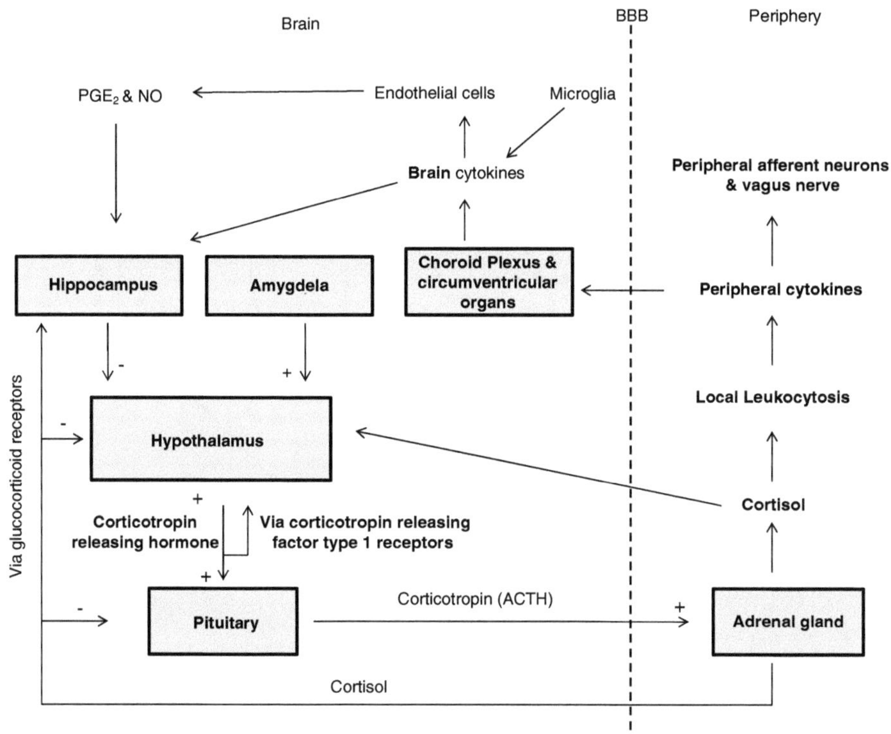


**Fig. 9.4** Inhibition of glucocorticoid signaling by cytokine signaling. Interleukin-1 (IL-1); interleukin-1 receptor (IL-1R); interferon-α (IFN-α); tumor necrosis factor-α (TNF-α); TNF-α receptor (TNF-α-R); Janus kinase 1 (JAK1); signal transducer and activator of transcription protein 5 (STAT5); nuclear factor-kappaB (NF-κB); IκB kinase α (IκBα); mitogen-activated protein kinases (MKKs); and glucocorticoid response elements (GREs)

Activation of inflammatory signaling pathways also stimulates the synthesis of nitric oxide (NO) via specific nitric oxide synthases (NOS) derived from endothelial cells (eNOS), peripheral lymphocytes (inducible nitric oxide synthase (iNOS), and neuronal tissue (nNOS) by activation of NADPH oxidase and NF-kB (Wu et al. 2008). iNOS is induced by several cytokines including IFN-γ and TNF-α secreted in response to local inflammation. NO not only modulates vasodilation and neurotransmission but is also converted to S-nitrosothiols and related derivatives. In contrast, the synthesis of anti-inflammatory prostaglandin, 15d-$PGJ_2$, and the activation of its nuclear target, the peroxisome proliferator-activated receptor γ (PPARγ), are also produced (Farooqui 2011). This constitutes a possible endogenous anti-inflammatory mechanism of defense against excessive oxidative stress and inflammation. In healthy brain, proinflammatory and anti-inflammatory pathways remain in homeostasis and without anti-inflammatory responses through the expression of anti-inflammatory transcription factors and generation of lipid mediators, such as lipoxins, resolvins, and neuroprotectins; brain tissue would be a sitting duck

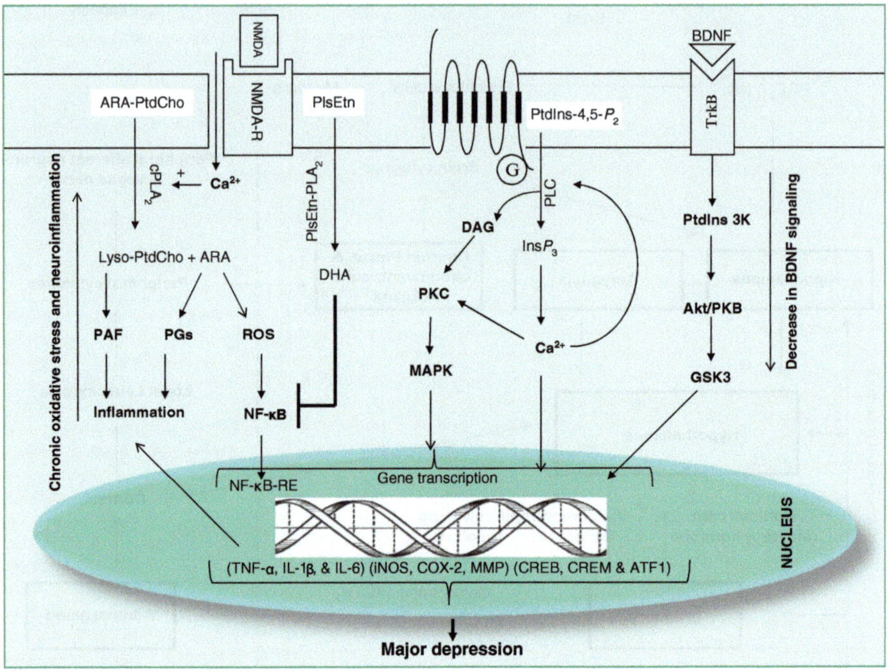


**Fig. 9.5** Hypothetical diagram showing pathogenesis of major depression. *N*-methyl-D-aspartate receptor (NMDA-R), cannabinoid (CB) receptor (CB1-R), and brain-derived neurotrophic factor (BDNF) receptors (BDNF-R); tyrosine kinase B receptor (TrkB); phospholipase C (PLC); diacylglycerol (DAG); protein kinase C (PKC); phosphatidylinositol 4,5-bisphosphate (PtdIns 4,5-$P_2$); inositol 1,4,5-trisphosphate ($InsP_3$); arachidonic acid-containing phosphatidylcholine (ARA-PtdCho); cytosolic phospholipase $A_2$ ($cPLA_2$); arachidonic acid (ARA); reactive oxygen species (ROS);prostaglandins (PGs); platelet-activating factor (PAF); phosphatidylinositol 3-kinases (PtdIns 3K); protein kinase B (Akt/PKB); glycogen synthase kinase 3 (GSK3); ); tumor necrosis factor-α (TNF-α); interleukin-1beta (IL-1β); interleukin-6 (IL-6); nuclear factor-kappaB (NF- B); inducible nitric oxide synthase (iNOS); cyclooxygenase-2 (COX-2); matrix metalloproteinase (MMP); calmodulin (CaM); calmodulin kinase (CamK); cyclic AMP response element-binding (CREB) protein; and cAMP responsive element modulator (CREM)

for various types of neuropsychiatric and neurodegenerative diseases and microbial, viral, and prion infection (Farooqui et al. 2007; Farooqui and Horrocks 2007; Farooqui 2011).

## 9.3 Neurocircuit Alterations in Depression

The hypothalamus and hippocampus have received increasing attention for their potential role in energy regulation. The hippocampus is part of a neural circuit involved with reward and energy regulation and is sensitive to satiety signals associated with learning and memory (Davidson et al. 2007; Wang et al. 2006;

Cenquizca and Swanson 2006; Tremblay 2004). Components of Western diet promote excessive food intake and weight gain and interfere with hippocampal functioning. This suggestion is supported by epidemiological studies linking diets high in saturated fat with weight gain and memory deficits (Tremblay 2004). High-fat diet-mediated maternal obesity impairs offspring hippocampal BDNF production (Tozuka et al. 2010), alters fetal hippocampal development (Niculescu and Lupu 2009), and reduces hippocampal neurogenesis during the early life of their offspring (Tozuka et al. 2009). Hippocampal BDNF signaling is involved in induction and facilitation of synaptic plasticity mechanisms associated with learning and memory. Hypothalamic BDNF expression is involved in regulation of feeding, metabolism, and stress. Selective reduction in BDNF in the hypothalamus causes hyperphagia in rodent models. Similarly, chronic administration of high levels of corticosterone also causes hyperphagia (Unger et al. 2007; Karatsoreos et al. 2010). Recent studies have also indicated that several brain regions, such as cerebral cortex, cerebellum, hippocampus, and hypothalamus, are involved in the pathogenesis of depression. Abnormalities in the above-mentioned regions involve alterations in leptin signaling and impairment in mitochondrial complexes I, III, and IV of chronically stressed rats (Rezin et al. 2008, 2009). In addition, activity of mitochondrial citrate synthase is decreased in depressed patients (Gardner et al. 2008). Based on these reports, it is proposed that mitochondrial ultrastructural alterations in cerebellum and cerebral cortex may contribute to functional alterations in depression. In addition, it is also reported that chronic stress/depression increases the vulnerability of neurons in the rat cerebral cortex by increasing caspase-3 positive neurons, supporting the view that neuronal death occurs through apoptosis (Bachis et al. 2008).

Functional neuroimaging studies examining the neural basis of depression in depressed and healthy subjects show the involvement of neurocircuits and networks in specific brain area, such as amygdala, hippocampus, limbic system structures, and prefrontal cortex. In particular, several prefrontal and limbic structures and their interconnected circuits have been reported to play a role in effective regulation associated with energy metabolism (Fig. 9.6). These areas include the ventromedial prefrontal cortex, lateral orbital prefrontal cortex, dorsolateral prefrontal cortex, anterior cingulated cortex, ventral striatum (including nucleus accumbens), amygdala, and the hippocampus. Dysregulation of the HPA axis reduces hippocampal volumes and prefrontal cortex (PFC) activity in depressed patients and disrupts homeostasis within the neurocircuit of depression. Antidepressant drugs not only increase BDNF and restore neuronal growth and activity but also modulate interactions between the neurocircuit anatomical structures (Palazidou 2012). Depressed individuals also show volumetric abnormalities in subcortical brain regions, including the amygdala and ventral striatum, and in cortical regions (anterior cingulate cortex, orbitofrontal cortex, and prefrontal cortex) compared to healthy individuals (Drevets 1998; Davidson 2003). In addition to anatomical abnormalities, neurochemical alterations have been reported to occur in these areas in depressed subjects compared with healthy control subjects

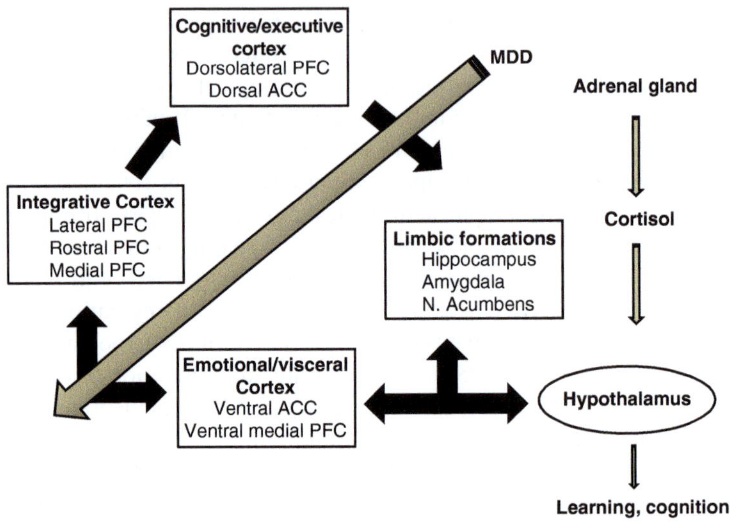


**Fig. 9.6** Hypothetical diagram showing connectivity between limbic system and cortical regions of the brain. Disruption of "connectivity" between limbic/paralimbic areas and rostral integrative prefrontal formations, leading to compromised feedback regulation of limbic activity. Prefrontal cortex (PFC) and anterior cingulated cortex (ACC) (Adapted from Maletic et al. (2007))

(Drevets 1998; Davidson 2003). Although the molecular mechanism is not fully understood, several mechanisms, including the neurotoxic effect of hypercortisolemia, and disturbances of neurogenesis have been proposed. Functional imaging studies show a common pattern of decreased activation in the dorsolateral prefrontal cortex (part of the dorsal system) and increased regional cerebral blood flow and metabolism in the subgenual cingulate, amygdala, anterior insula, and ventral striatum (parts of ventral system) during the depressive episode (Jaracz 2008). Neurophysiological alterations that occur in depression, such as prefrontal cortex and cingulate activation, may be partly responsible for reduction in attention, memory, and visuospatial capacities. Depression also affects executive processing, producing difficulties in planning strategies and mental flexibility, which, in turn, may result in impaired motivational and decision-making functions (Davidson et al. 2002a, b). Impaired cognition is thus closely associated with the severity of depression. Structural MRI studies also show volume reductions in the above-mentioned anatomical regions along with cellular aberrations, such as decreased neuronal size; increased cellular packing density suggesting a disruption in neuronal connectivity, particularly in the dorsolateral prefrontal cortex; and distortions in neuronal orientation; abnormalities of cerebral white matter, oligodendrocytes, and myelin along with alterations in signal transduction processes have been observed in immunocytochemical, ultrastructural, and neurochemical studies (Fig. 9.6) (Blitzer et al. 2005).

## 9.4 Association Between Depression and Metabolic Syndrome

As stated earlier, MetS is a cluster of common pathologies: abdominal obesity linked to an excess of visceral fat, insulin resistance, dyslipidemia, and hypertension. At the molecular level, MetS is accompanied not only by dysregulation in the expression of adipocytokines but also by alterations in levels of leptin. These changes modulate immune response and inflammation that lead to alterations in the hypothalamic "bodyweight/appetite/satiety set point," resulting in the initiation and development of metabolic syndrome. Because MetS is a multiple component condition, it remains unclear whether the relation between MetS and depression is attributable to the MetS as a whole or driven by some specific component of MetS. Studies on determination of ARA metabolism in animal model of the MetS indicate that enhancement in brain enzymic mediators of ARA and DHA metabolism along with decrease in BDNF mRNA and protein, and drebrin mRNA (Taha et al. 2012). Increases in $cPLA_2$ and $iPLA_2$ activities in streptozotocin model of diabetes support the notion that phospholipases are closely associated with the pathogenesis of neurodegeneration (Muller et al. 1998; Farooqui 2012). The decreases in BDNF and drebrin suggest increased susceptibility to synapto-dendritic injury (Muller et al. 1998).

Two hypotheses have been proposed to explain the causal link between diabetes and depression. One hypothesis proposes that depression precedes type II diabetes (a component of MetS), with onset of depression due to increase in counterregulatory hormone release and action, alterations in glucose transport function, and elevation in immuno-inflammatory activation. These physiological alterations contribute to insulin resistance and beta islet cell dysfunction inducing the development of type II diabetes and MetS (Musselman et al. 2003). The second hypothesis asserts that depression in patients with diabetes is caused by chronic psychosocial stressors (Talbot and Nouwen 2000). Insulin resistance and visceral obesity, and changes in leptin levels have been recognized as the most important pathogenic factors for depression and MetS.

Major depression and type II diabetes (an important component of MetS) share overlapping abnormalities in metabolic networks defined as glucocorticoid signaling, decrease in leptin, oxidative stress, glucose–insulin homeostasis, and inflammatory processes (McIntyre et al. 2007). Depressed patients have significantly higher basal glucose levels and greater increases in glucose concentrations over time, despite their larger cumulative insulin secretion after the oral glucose tolerance test relative to the control group. These observations indicate a functional insulin-resistant state during major depressive episodes (Winokur et al. 1988). This is consistent with evidence of increased brain activity of carnitine palmitoyltransferase (which regulates fatty acid entry from the acyl-CoA pool into mitochondria for later β-oxidation) in an animal model of the MetS (Makar et al. 1995). $^{14}C$-palmitate conversion to $^{14}C–CO_2$ also is increased in mitochondrial brain extracts of diabetic (db/db) mice (Makar et al. 1995). It should also be noted that central leptin administration fully normalizes hyperglycemia in a rodent model of uncontrolled insulin-deficient diabetes by reducing hepatic glucose production (HGP) and by increasing glucose uptake and intracerebroventricular administration of BDNF lowers blood glucose in uncontrolled insulin-deficient diabetes. Although the molecular mechanisms involved in these processes are not fully understood, it is suggested that BDNF lowers blood glucose levels by potently suppressing HGP, without affecting tissue glucose uptake, an effect associated with normalization of both plasma glucagon levels and hepatic expression of gluconeogenic genes (Meek et al. 2012). Moreover, BDNF microinjection directly into the ventromedial hypothalamic nucleus also lowers fasting blood glucose levels in uncontrolled insulin-deficient diabetes (Meek et al. 2012). Based on these results, it is proposed that in brain BDNF attenuates diabetic hyperglycemia via an insulin-independent mechanism. This action of BDNF likely involves the ventromedial hypothalamic nucleus and is associated with inhibition of glucagon secretion and a decrease in the rate of HGP (Meek et al. 2012).

Like depression, mood disorders are also characterized by inflammation, increased cellular respiration, excitotoxicity, and upregulation of glucocorticoid signaling. These processes may threaten neuronal and glial function and integrity (McIntyre et al. 2007). Insulin and insulin sensitizers exert neurotrophic, neuroprotective, anti-inflammatory, and anti-glucocorticoid effects (McIntyre et al. 2007; Rustad et al. 2011). The significant overlap between depression and diabetes is also reflected by results from neuropsychological and neuroimaging investigations (McIntyre et al. 2007). Neurocognitive deficits, which represent one measure of diminished brain function, are noted in both illnesses (McIntyre et al. 2007). In animal models, type I and type II diabetes are accompanied by impairment in hippocampal neurogenesis, which may also contribute to diabetes-related depression and cognitive decline due to decrease in BDNF (Wang et al. 2009; Krabbe et al. 2007). Thus, BDNF may be a potential common mediator of depression and type II diabetes (Krabbe et al. 2007; Rustad et al. 2011). Collective evidence suggests that depression and MetS may involve abnormalities in limbic–hypothalamic–pituitary–adrenal axis, diminution in glucocorticoid receptor function, alterations in glucose tolerance and insulin sensitivity, excitotoxicity, increases in intracellular calcium, diminished BDNF levels, and oxidative stress along with production of proinflammatory cytokines. In addition, putative mediating variables common to both conditions include insulin–glucose homeostasis, immuno-inflammatory processes, and oxidative stress mechanisms (Ceriello 2006).

## 9.5 Impaired Glucose Metabolism in Depression and Metabolic Syndrome

Pathogenesis of MetS may be closely associated with neurochemical changes that occur in depression in young adults (Kinder et al. 2004) and middle-aged populations (Heiskanen et al. 2006; Skilton et al. 2007). Thus, depression is accompanied by hyperglycemia and an increased risk for diabetic complications, and relief of depression involves improvement in glycemic control. Earlier studies have indicated that depression induces symptoms of MetS. However, recent reports indicate that depression may be caused by the MetS (Heiskanen et al. 2006; Skilton et al. 2007). To this end, it is shown that in young and middle-aged women and older men, the relationship between depression and MetS is through bidirectional association (Akbaraly et al. 2011). The mechanism of bidirectional association between depression and MetS is not fully understood. However, it is suggested that the etiology of late-onset depression may be linked to poorer lifestyle habits (such as sedentary lifestyles, high intake of saturated fats, smoking, and heavy alcohol consumption) and vascular abnormalities, including diseases of the blood vessels and circulation (Carney et al. 2002: Franko et al. 2005; Alexopoulos et al. 1997). Glucose and insulin abnormalities are most pronounced in hypercortisolemic depressed individuals (Kopf et al. 2004). This may be due to anti-insulin effects of cortisol. Hypercortisolemic depressive subjects have increased risk of developing increased abdominal (visceral) fat deposition (Weber-Hamann et al. 2002) and the MetS (Vogelzangs et al. 2007) compared with normocortisolemic subjects. Insulin resistance and decrease in cellular glucose uptake may also result in "energy crisis." In the hippocampus, "energy crisis" may promote excitotoxicity due to insufficient energy available to clear glutamate from the synapse (Sapolsky 2000). Depression is related to obesity through dysregulation of the ANS and the HPA (Bornstein et al. 2006). Stress hormones, including epinephrine, norepinephrine, and cortisol, may contribute to increased central adiposity, insulin resistance, and MetS (Goldbacher et al. 2005). In addition, changes in immune system may also contribute to the link between depressive symptoms and the MetS. Thus, C-reactive protein is a marker of systemic inflammation that is closely associated with the development of depressive symptoms (Elovainio et al. 2006) and the MetS (Haffner 2006). It is also proposed that genes, which modulate dysregulation of HPA, may also partially explain the link between depression and the MetS (Bornstein et al. 2006; Brown et al. 2004).

## 9.6 Upregulation of Oxidative Stress in Depression and Metabolic Syndrome

Oxidative and nitrosative stress is accompanied by the generation of free radicals, which are highly reactive molecules. ROS and reactive nitrogen species (RNS) consist of radicals and other reactive oxygen/nitrogen factors that can react with other substrates. Examples of ROS and RNS are superoxide, nitric oxide, peroxynitrite, and hydrogen peroxide. Under physiological conditions, these are counterbalanced by an array of defense pathways, and it needs to be emphasized that low levels of ROS and RNS have many physiological roles including signaling. In excess or in situations where defenses are compromised, ROS and RNS may react with fatty acids, proteins, and DNA, thereby causing damage to these cellular components. Oxidation of fatty acids and proteins has been described earlier. For DHA damage, it is suggested that oxidized promoter regions may potentially adopt different conformation and lose affinity for transcription factors (Lu et al. 2004). Damage on mitochondrial DNA (mtDNA), which is extremely vulnerable to oxidation due to its proximity to the site of ROS production, respiratory chain, and the absence of protective histone (Lee and Wei 2007), causes downregulation of genes related to respiratory chain and further energy metabolism impairment (Lin et al. 2002). It is also reported that several neuropsychiatric disorders including MDD, bipolar disorder, and schizophrenia not only involve mitochondrial dysfunction (Rezin et al. 2009) but are also associated with the accumulation of mtDNA damage (Trifunovic et al. 2004; Kasahara et al. 2006). These results support the view that oxidative stress plays a major role in linking aging with depression.

Recent studies indicate that levels of lipid peroxides are significantly increased plasma of depressed patients than normal controls (Maes et al. 2009, 2010). However, no significant correlations between the increased plasma peroxide values and the severity of depression have been observed. Importantly, it is reported that the peroxide levels are significantly higher in patients in a more acute phase of their depression than in patients who suffered from chronic depression (more than 2 years duration), while the latter did not differ significantly from normal controls (Maes et al. 2010). Thus, increased peroxide levels are confined to patients who are in a more (sub)acute phase of their illness. Normalization of ROS may occur when the depression tends to become more chronic. Another important finding is that increase in plasma peroxide levels is more pronounced in depressed men than in depressed women (Maes et al. 2009, 2010).

Measurements of NO obtained in individuals with depression however show mixed results. Some investigators have shown that there is significant increase in NO levels in depressed patients (Selek et al. 2008). However, other investigators have reported decrease in plasma NO and NO metabolite (NOx) in depressed patients (Ikenouchi-Sugita et al. 2009; Chrapko et al. 2004).

Thus, oxidative stress and nitrosative stress may play an important role in the brain. Anabolic/catabolic imbalance (A/C imbalance) may affect ROS and RNS production and pathogenesis of depression and MetS. Cortisol is essential for brain viability. However, when cortisol is very high for a longer time period, certain vulnerable neurons may be traumatized, in part by increases in oxidative and nitrosative stress (McIntosh and Sapolsky 1996). Dehydroepiandrosterone and estrogens have been shown to block oxidative stress-mediated damage in neurons (Behl et al. 1997). Dehydroepiandrosterone inhibits cortisol-mediated excitatory neurotoxicity, supporting the importance of the balance between cortisol and dehydroepiandrosterone in the brain (Kimonides et al. 1999).

## 9.7 Induction of Neuroinflammation in Depression and Metabolic Syndrome

As stated in Chaps. 7 and 8, adipocytes secrete a variety of hormones and proinflammatory cytokines (referred to as adipocytokines or adipokines). Examples of adipocytokines are leptin, adiponectin, resistin, and visfatin. In addition, adipocytes can also secrete IL-6 and TNF-α (Tilg and Moschen 2008). Another important mechanism for the induction of inflammation in adipose tissue is the secretion of chemokines, particularly MCP-1, which attracts leukocytes such as macrophages, T lymphocytes, and dendritic cells to adipose tissue secreting cytokines including IL1, IL6, and TNF-α (Xu et al. 1996). Thus, chemokines and cytokines produced by adipocytes may contribute to widespread immune activation, potentially causing or exacerbating diseases associated with inflammation such as type II diabetes, MetS, cardiovascular disease, cancer, and neurotraumatic and neurodegenerative diseases (Tilg and Moschen 2008; Farooqui 2011; Farooqui et al. 2012).

In MetS abnormal production and increase in levels of adipocytokines in MetS contribute to a proinflammatory state, which may also contribute to obesity, dyslipidemia, and depression. In patients with MetS and depression, adipocytokines/cytokines produce their actions through the activation NF-κB (Karmiris et al. 2006; Hallikainen et al. 2007; Norata et al. 2007; Pini et al. 2008). The activation of NF-κB not only leads to oxidative stress but also provides a link between inflammation and oxidative stress. Both processes are crucial to the development of MetS and depression (Shelton and Miller 2010) (Table 9.1).

Leptin is another important peptide produced by adipocytes that regulates dietary intake. The linkage between leptin, MetS, and depression is supported by studies on the effect of stress on depression. Chronic stress, which induces depression-like behaviors, has been shown to reduce leptin levels in rodents (Lu et al. 2006), a state that mimics leptin resistance found in obesity and MetS (Considine et al. 1996). Administration of leptin has been shown to reverse the behavioral effects of both acute and chronic stress that are thought to mimic depression (Lu et al. 2006). However, clinical studies in depressed patients indicate considerable variability in leptin levels (Rubin et al. 2002; Jow et al. 2006; Atmaca et al. 2002). This may be not only due to the clinical heterogeneity of depression but also be caused by the complexity of leptin response. Reduction in leptin signaling, whether due to leptin insufficiency or decreased receptor responsiveness, may contribute to depressive symptoms (Lu 2007). The effects of leptin are mediated by its receptors, which are mainly located in the brain, adipocytes, and endothelial cells. Leptin receptor belongs to a group of cytokine receptors, which induce their action through the activator of transcription 3 (STAT3) pathway and PtdIns 3K pathway (Buettner et al. 2006). The STAT3 pathway is essential for mediating food intake, liver glucose production, and gonadotropin secretion (Buettner et al. 2006); however, the control of adipose tissue metabolism by leptin is STAT3 independent (Buettner et al. 2008).

**Table. 9.1** Levels of lipid mediators in MetS and depression

| Lipid mediators | MetS | Depression | Reference |
|---|---|---|---|
| Prostaglandins | Increased | Increased | Farooqui et al. (2012); Yager et al. (2010) |
| Leukotrienes | Increased | Increased | Farooqui et al. (2012) |
| Thromboxanes | Increased | Increased | Farooqui et al. (2012) |
| Platelet-activating factor | Decreased | – | Farooqui et al. (2012) |
| Diacylglycerol | – | – | Farooqui et al. (2012) |
| Isoprostanes | No effect | Increased | Farooqui et al. (2012); Pomara et al. (2012) |
| Endocannabinoids | Increased | Increased | Farooqui et al. (2012); Heyman et al. (2012) |
| Ceramide | Increased | Increased | Farooqui et al. (2012); Gracia-Garcia et al. (2011) |
| Sphingosine-1-phosphate | Increased | Increased | Farooqui et al. (2012) |

There is a link between dyslipidemia, depression, and MetS. Prevalence of depression is significantly higher in the hypercholesterolemic subjects than normocholesterolemic subjects (Takeuchi et al. 2009; Tyrovolas et al. 2009). Hypercholesterolemic individuals not only have higher prevalence of depression but also have obesity, elevation in proinflammatory cytokines (TNF-α and altered levels of leptin, adiponectin, and resistin), hypertension, and diabetes, which are signs of MetS (Grundy et al. 2004). No major depression is found in hypocholesterolemic subjects (Takeuchi et al. 2009; Tyrovolas et al. 2009). Circulating leptin enters the brain where it exerts its physiological actions beyond the control of governing energy homeostasis, neuroendocrine axes, and food intake and metabolism. Systemic or intracranial administration of leptin has been reported not only to reduce food intake and body weight but also to induce antidepressant-like effects in both rats and mice (Lu et al. 2006; Yamada et al. 2011). Circulating leptin levels are reduced in animals subjected to chronic social defeat or chronic unpredictable stress, two animal models of depression (Lu et al. 2006; Finger et al. 2011). An inverse correlation between leptin levels and the severity of depression symptoms has been recently reported (Lawson et al. 2012). Based on the above studies, it is suggested that alterations in leptin signaling may be associated not only with the pathophysiology of major depressive disorders but also with modulation of the mesolimbic dopamine system in rodents. Leptin exerts its effects by activating leptin receptors, which are present in prefrontal cortex and hippocampus (Scott et al. 2009; Patterson et al. 2011; Campbell and Macqueen 2004; Koenigs and Grafman 2009). However, whether defects in leptin receptor function in prefrontal cortex and hippocampus circuits contribute to the pathogenesis of depression and whether specific neuronal populations mediate leptin's antidepressant-like activity are not yet known.

Collective evidence suggests that depression is associated with obesity, physical inactivity, and poor self-care. All these parameters are risk factors for diabetes. Furthermore, it is also possible that there may be other biologic risk factors for diabetes in depressed individuals (Golden 2007).

## 9.8 Possible Mechanisms Linking Depression with Metabolic Syndrome

Diabetes and depression share a bidirectional association—both influencing each other in multiple ways. High levels of stress hormones, which are often found in people who are depressed, can lead to problems with glucose and blood sugar metabolism, increased insulin resistance, and an accumulation of stomach fat—all risk factors for diabetes. Depression also tends to lead to unhealthy lifestyle behaviors, such as eating a poor diet and not exercising, which may further increase the risk of diabetes. Stress is triggered by numerous unexpected environmental, social, or pathological stimuli that occur during the life of animals, including humans, which induce neurochemical changes throughout the brain including increase in insulin resistance oxidative stress and neuroinflammation. In addition, depression promotes unhealthy lifestyle behaviors, such as eating a poor diet and physical inactivity, which may further increase the risk of diabetes. Although acute stress is good and essential for adaptive response-mediated neural cell survival, chronic stress can produce detrimental changes. Chronic stress is a major contributor of the development of depression due to the dysregulation of hypothalamic–pituitary–adrenal (HPA) axis. Stress triggers the release of high levels of glucocorticoid (cortisol in humans and corticosterone in rodents) and noradrenalin within minutes. The intracellular effects of glucocorticoids are mediated by the glucocorticoid receptor (GR), a 94 kDa intracellular protein, which belongs to the nuclear hormone receptor superfamily. In the absence of glucocorticoid, the GR predominantly remains in the cytoplasm as an inactive multiprotein complex that consists of the receptor, two Hsp90 molecules, one molecule each of Hsp70 and Hsp56, and an immunophilin of the FK506- and rapamycin-binding protein (Fig. 9.5). When glucocorticoid binds to GR, the receptor undergoes a change in conformation, dissociates from regulatory heat shock proteins, and is hyperphosphorylated. The activated receptor rapidly migrates to the cell nucleus, where it positively and negatively modulates gene expression through diverse mechanisms (Pariante and Lightman 2008; Lightman et al. 2008). In addition, GR is also associated with modulation of gene expression through a GRE-independent mechanism, which is mediated in part through protein–protein interactions of GR with other sequence-specific DNA-binding factors or coactivators (Pariante and Lightman 2008). The negative modulation of gene transcription operated by glucocorticoids occurs through nongenomic mechanisms (transrepression), mediated by inhibitory influences exerted by activated GR on the functions of several transcription factors (John et al. 2008). This contributes to the anti-inflammatory properties of the glucocorticoids. In humans cortisol is a powerful anti-inflammatory compound that has the ability to inhibit cytokine-mediated inflammation. Although the role of cortisol and noradrenaline has intensely been studied, the exact pathological mechanism associated with the induction of depression is still unclear. In addition, activation of the immune system can provoke neuroendocrine and neurotransmitter changes that are similar to those provoked by physical or psychological stressors (Anisman 2009). Accumulating evidence

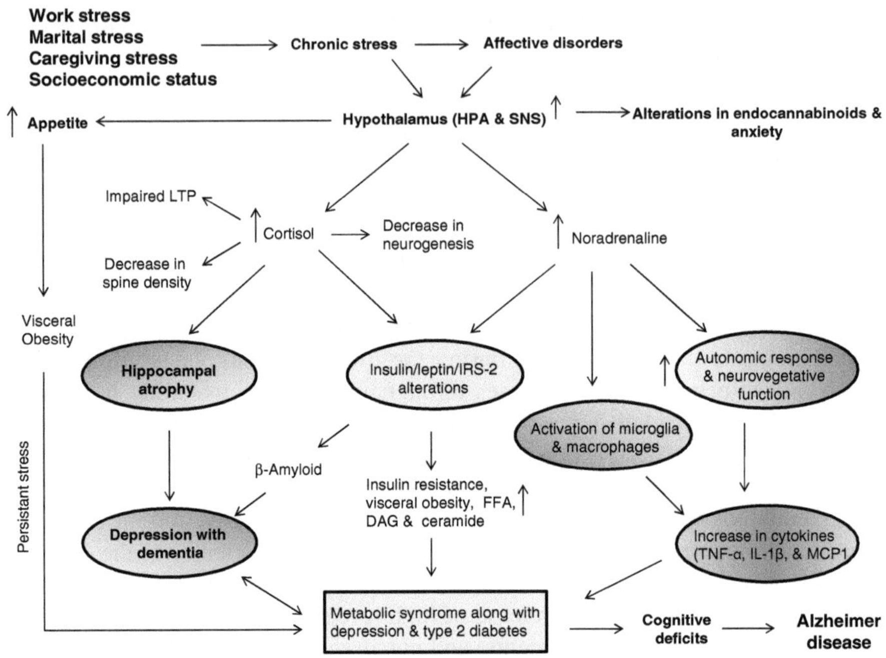


**Fig. 9.7** Stress results in development of metabolic syndrome and depression. Free fatty acids (FFA); diacylglycerol (DAG); insulin receptor substrate (IRS); tumor necrosis factor-α (TNF-α); interleukin-1beta (IL-1β); monocyte chemoattractant protein-1 (MCP-1); hypothalamic–pituitary–adrenal (HPA) axis; and the sympathetic nervous system (SNS). Upward arrow indicates increase

suggests that type II diabetes and depression are linked with each other through stress, which impairs the ability of brain to regulate corticosteroid release, which may lead to hypercortisolemia (Leonard and Myint 2009). Excessive stimulation of corticosteroid receptors in hippocampus may cause hippocampal atrophy, which may lead to depression and dementia (Fig. 9.7) (Swaab et al. 2005). Direct and indirect actions of cortisol and noradrenalin modulate lipolysis, FFA production and turnover, and very low-density lipoprotein synthesis. Accumulated FFAs are released into the circulation. This increase in FFA may not only enhance the accumulation of hepatic lipids reducing glucose uptake and activating various serine kinases which result in decreased insulin signaling and facilitate production of ROS through the oxidation of unsaturated FFA. Collective evidence suggests that stress-mediated release of cortisol and noradrenaline, generation of ROS, and membrane damage as major candidates cause an insulin-resistant state in the brain with decreased glucose/energy metabolism (Chesik et al. 2008). Noradrenergic signaling in the brain plays an essential role in circuits involving attention, mood, memory, and stress as well as providing pivotal support for autonomic function in the peripheral nervous system (Robertson et al. 2010). This may further cause a derangement in ATP-dependent cellular and molecular work, of the neural cell function in general, as well as derangements in the endoplasmic reticulum/Golgi apparatus, axon, synapses, and membranes, in particular. These studies support the view that neurochemical processes in MetS and depression involve overlapping signal transduction processes (Table 9.1).

The immune system plays an important role in the development of depression. Immune system communicates with brain through peripheral adipocytokines, which can cross the blood–brain barrier (BBB) directly via active transport mechanisms (Banks and Kastin 1991) or indirectly via activation of the afferent vagus nerve (Maier and Watkins 1998; Tracey 2002) to induce changes in brain structures. In obese subjects with MetS, adipocytes release adipocytokine (IL-6), which can penetrate BBB and interfere with long-term potentiation (Bellinger et al. 1995; Tancredi et al. 2000), neurogenesis (Monje et al. 2003), and neural plasticity (Poluektova et al. 2005). These processes impair performance on hippocampal-dependent learning and memory tests and are closely associated with the pathophysiology of depression (Arai et al. 2001). Conversely, IL-6 receptor antagonists block inflammation-mediated disruption of hippocampal LTP and ensuing cognitive sequelae (Balschun et al. 2004). Finally, IL-6 knockout mice show facilitated working memory when compared with wild-type mice (Braida et al. 2004) and are refractory to peripheral endotoxin-induced impairments of spatial memory (Sparkman et al. 2006). These findings suggest that the immune system responds to obesity and stress by enhancing the levels of proinflammatory cytokines, which modulate behavioral changes associated with depression through their effects (Wurtman 2005; Brummett et al. 2008) on neurotransmitters and neuropeptides function, synaptic plasticity, and neuroendocrine function (Wurtman 2005; Brummett et al. 2008; Miller and Raison 2008).

Another important and consistent neuropathological observation in major depressive disorder (MDD) is a decrease in the number of glial cells (Cotter et al. 2001, 2002; Uranova et al. 2004; Pittenger and Duman 2008). In animals, chronic unpredictable stress causes a reduction in the proliferation of glial and endothelial cells in the medial prefrontal cortex (mPFC) (Banasr et al. 2007) and exposure to glucocorticoids produces similar effects (Alonso 2000). It is well known that glial cells provide metabolic support to neuronal cell population; therefore, it is likely that a reduction in the numbers of glial cells may impact the function as well as morphology of mPFC pyramidal cells (Pittenger and Duman 2008). Glial cells also play an important role in both the synthesis and inactivation of glutamate, which plays an important role in the modulation of neuroplasticity. Collective evidence suggests that alteration in number or function of glial cells in MDD may therefore have impact neuroplasticity and neuronal cell survival (Pittenger and Duman 2008).

Recently, a new perspective on molecular contribution in the context of depressive disorders was drawn highlighting the importance of glutamatergic dysfunction in late-life depression and cognitive disorders (Frisardi et al. 2011). In major depression, glutamatergic hyperfunction is not only closely related to the downregulation of serotonergic and noradrenergic neurotransmission and tryptophan–kynurenine metabolism (Fig. 9.8) but also with immune system resulting in core symptoms of major depression (Muller and Schwarz 2007a, b). The activation of indoleamine

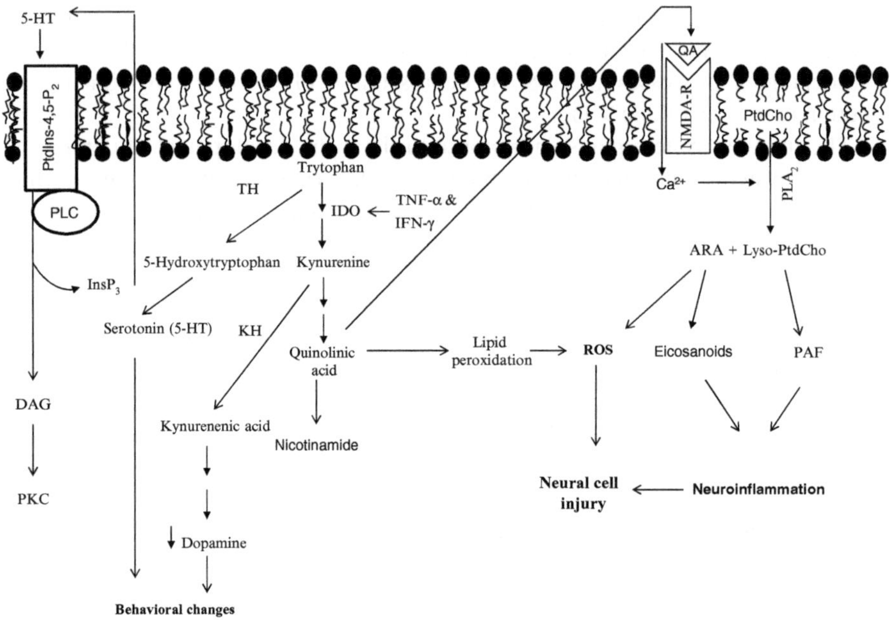


**Fig. 9.8** Metabolism of tryptophan, generation of quinolinic acid, and stimulation of NMDA receptor by quinolinic acid. *N*-methyl-D-aspartate receptor (NMDA-R); phosphatidylcholine (PtdCho); cytosolic phospholipase $A_2$ (c$PLA_2$); arachidonic acid (ARA); reactive oxygen species (ROS); platelet-activating factor (PAF); phospholipase C (PLC); indoleamine 2,3-dioxygenase (IDO); tryptophan hydroxylase (TH); kynurenine 3-monooxygenase (KMO); tumor necrosis factor-α (TNF-α); interleukin-1beta (IL-1β); interferon-gamma (INF-γ); diacylglycerol (DAG); inositol 1,4,5-$P_3$ (Ins$P_3$). Quinolinic acid not only generates reactive oxygen species but also stimulates NMDA receptor and promotes the generation of eicosanoids and platelet-activating factor. These products induce oxidative stress and facilitate neuroinflammation

2,3-dioxygenase (IDO) contributes to the symptoms of depression. This enzyme is expressed in cell types throughout the body including fibroblasts, dendritic cells, monocytes, macrophages, and microglia and can be mediated by a number of cytokines such as IFN-γ, IFN-α, and tumor necrosis factor-α (TNF-α) either alone or in combination (Robinson et al. 2006). Cytokine-mediated activation of IDO can occur through several inflammatory signaling pathways including signal transducer and activator of transcription 1 (STAT1), interferon regulatory factor (IRF)-1, p38 mitogen-activated protein kinase (MAPK), and nuclear factor-kappaB (NF-kB) (Fujigaki et al. 2006). Once activated, IDO transforms L-tryptophan into L-kynurenine (KYN), thereby reducing the availability of tryptophan, which not only plays an important role in the regulation of T cells (Mellor et al. 2003) but also is a precursor for serotonin, a monoamine that is believed to play a prominent role in the neurobiology of mood disorders (Owens and Nemeroff 1994; Flores et al. 2004). Increase in the levels of proinflammatory cytokines also induces the production of neurotoxic end products of the tryptophan–kynurenine pathway, such as

3-hydroxykynurenine and quinolinic acid. 3-Hydroxykynurenine is neurotoxic because of its ability to generate oxidative radicals and quinolinic acid that acts as an agonist for NMDA receptor. It is also reported that quinolinic acid inhibits glutamate uptake by astrocytes leading to an accumulation of glutamate in the microenvironment inducing slow excitotoxicity (Tavares et al. 2002).

To date, no studies on humans have been addressed on this interesting topic. However, in animal models two important components of MetS (hyperinsulinemia and hypertension) have been related to the activation of glutamatergic receptors in the rostral ventrolateral medulla, increasing the sympathoexcitatory response activity (Bardgett et al. 2010). Furthermore, an altered balance of γ-aminobutyric acidergic (GABAergic) and glutamatergic afferent inputs in rostral ventrolateral medulla, projecting neurons in the PVN of the hypothalamus, has also been observed (Biancardi et al. 2010). However, further investigations are needed to clarify this relationship and for the better understanding if metabolic changes could be predictive of an increased risk to develop mood disorders and vice versa.

## 9.9 Conclusion

Depression is a multisystem, multifactorial, widespread, and debilitating disorder characterized by variety of symptoms (behavioral, affective, cognitive, and somatic) and a high risk of relapse/recurrence. Neurochemically, depression is characterized by the reduction in plasma levels of BDNF, elevation in blood levels of IL-6 and TNF-α, and alterations in neurotransmitters and neuropeptide levels. Interactions between genes and environmental factors may also contribute to the pathogenesis of major depression. The MetS is a multifactorial condition characterized by insulin resistance, central obesity, atherogenic dyslipidemia, hypertension, and proinflammatory and prothrombotic states, with or without glucose intolerance. The major risk factors leading to MetS are physical inactivity and an atherogenic diet and cornerstone clinical feature is abdominal obesity or adiposity. Each of these characteristics is a significant risk factor for development of vascular dysfunction and cardiovascular and cerebrovascular diseases. Multiple biological links potentially mediate the adverse effect of MetS on depression and related cardiovascular and cerebrovascular diseases. These include increase in proinflammatory cytokines (TNF-α and IL-1β), chronic hyperactivation of HPA, and changes in homeostasis between the sympathetic and parasympathetic nervous systems under prolonged stress. Both HPA axis and sympathetic nervous system are activated by stress. The increase in cortisol levels is associated with HPA activity and elevations in catecholamine and cytokine levels. If depression is chronic, cortisol levels may stay consistently high. High levels of cortisol have been shown to impair insulin sensitivity and contribute to fat distribution in the waistline area, which secretes adipocytokines and inflammatory cytokines, which have been implicated in insulin resistance, an important component of the metabolic syndrome.

# References


aan het Rot M, Mathew SJ, Charney DS (2009) Neurobiological mechanisms in major depressive disorder. CMAJ 180:305–313

Akbaraly TN, Ancelin ML, Jaussent I, Ritchie C, Barberger-Gateau P, Dufouil C, Kivimaki M, Berr C, Ritchie K (2011) Metabolic syndrome and onset of depressive symptoms in the elderly: findings from the three-city study. Diabetes Care 34:904–909

Akil H, Haskett RF, Young EA, Grunhaus L, Kotun J, Weinberg V, Greden J, Watson SJ (1993) Multiple HPA profiles in endogenous depression: effect of age and sex on cortisol and beta-endorphin. Biol Psychiatry 33:73–85

Alexopoulos GS, Meyers BS, Young RC, Campbell S, Silbersweig D, Charlson M (1997) "Vascular depression" hypothesis. Arch Gen Psychiatry 54:915–922

Ali S, Stone MA, Peters JL, Davies MJ, Khunti K (2006) The prevalence of co-morbid depression in adults with type 2 diabetes: a systematic review and meta-analysis. Diabet Med 23:1165–1173

Almeida RD, Manadas BJ, Melo CV, Gomes JR, Mendes CS, Graos MM, Carvalho RF, Carvalho AP, Duarte CB (2005) Neuroprotection by BDNF against glutamate-induced apoptotic cell death is mediated by ERK and PI3-kinase pathways. Cell Death Differ 12:1329–1343

Alonso G (2000) Prolonged corticosterone treatment of adult rats inhibits the proliferation of oligodendrocyte progenitors present throughout white and gray matter regions of the brain. Glia 31:219–231

Anda RF, Whitfield CL, Felitti VJ, Chapman D, Edwards VJ, Dube SR (2002) Adverse childhood experiences, alcoholic parents, and later risk of alcoholism and depression. Psychiatr Serv 53:1001–1009

Anisman H (2009) Cascading effects of stressors and inflammatory immune system activation: implications for major depressive disorder. J Psychiatry Neurosci 34:4–20

Anisman H, Matheson K (2005) Stress, depression, and anhedonia: caveats concerning animal models. Neurosci Biobehav Rev 29:525–546

Arai K, Matsuki N, Ikegaya Y, Nishiyama N (2001) Deterioration of spatial learning performances in lipopolysaccharide-treated mice. J Pharmacol 87:195–201

Astorg P, Couthouis A, Bertrais S, Arnault N, Meneton P, Guesnet P, Alessandri JM, Galan P, Hercberg S (2008) Association of fish and long-chain n-3 polyunsaturated fatty acid intakes with the occurrence of depressive episodes in middle-aged French men and women. Prostaglandins Leukot Essent Fatty Acids 78:171–182

Atmaca M, Kuloglu M, Tezcan E, Ustundag B, Gecici O, Firidin B (2002) Serum leptin and cholesterol values in suicide attempters. Neuropsychobiology 45:124–127

Bachis A, Cruz MI, Nosheny RL, Mocchetti I (2008) Chronic unpredictable stress promotes neuronal apoptosis in the cerebral cortex. Neurosci Lett 442:104–108

Balschun D, Wetzel W, Del Rey A, Pitossi F, Schneider H, Zuschratter W, Besedovsky HO (2004) Interleukin-6: a cytokine to forget. FASEB J 18:1788–1790

Banasr M, Valentine GW, Li XY, Gourley S, Taylor J, Duman RS (2007) Chronic stress decreases cell proliferation in adult cerebral cortex of rat: reversal by antidepressant treatment. Biol Psych 62:496–504

Banks WA, Kastin AJ (1991) Blood to brain transport of interleukin links the immune and central nervous systems. Life Sci 48:PL117–PL121

Bardgett ME, McCarthy JJ, Stocker SD (2010) Glutamatergic receptor activation in the rostral ventrolateral medulla mediates the sympathoexcitatory response to hyperinsulinemia. Hypertension 55:284–290

Bayatti N, Hermann H, Lutz B, Behl C (2005) CRH-mediated induction of intracellular signaling pathways and BDNF expression is inhibited by the activation of the endocannabinoid system. Endocrinology 146:1205–1213

Bazan NG (2005) Lipid signaling in neural plasticity, brain repair, and neuroprotection. Mol Neurobiol 32:89–103

Behl C, Skutella T, Lezoualc'h F, Post A, Widmann M, Newton CJ, Holsboer F (1997) Neuroprotection against oxidative stress by estrogens: structure-activity relationship. Mol Pharmacol 51:535–541

Bellinger FP, Madamba SG, Campbell IL, Siggins GR (1995) Reduced long-term potentiation in the dentate gyrus of transgenic mice with cerebral overexpression of interleukin-6. Neurosci Lett 198:95–98

Biancardi VC, Campos RR, Stern JE (2010) Altered balance of gamma-aminobutyric acidergic and glutamatergic afferent inputs in rostral ventrolateral medulla-projecting neurons in the paraventricular nucleus of the hypothalamus of renovascular hypertensive rats. J Comp Neurol 518:567–585

Bjorntorp P (1993) Visceral obesity: a "civilization syndrome". Obes Res 1:206–222

Black SA (1999) Increased health burden associated with comorbid depression in older diabetic Mexican Americans. Results from the hispanic established population for the epidemiologic study of the elderly survey. Diabetes Care 22:56–64

Blitzer RD, Iyengar R, Landau EM (2005) Postsynaptic signaling networks: cellular cogwheels underlying long-term plasticity. Biol Psychiat 57:113–119

Bornstein SR, Schuppenies A, Wong ML, Licinio J (2006) Approaching the shared biology of obesity and depression: the stress axis as the locus of gene-environment interactions. Mol Psychiat 11:892–902

Braida D, Sacerdote P, Panerai AE, Bianchi M, Aloisi AM, Losue S, Sala M (2004) Cognitive function in young and adult IL (interleukin)-6 deficient mice. Behav Brain Res 153:423–429

Bremner JD, Vythilingam M, Vermetten E, Nazeer A, Adil J, Khan S, Staib LH, Charney DS (2002) Reduced volume of orbitofrontal cortex in major depression. Biol Psychiatry 51:273–279

Brown ES, Varghese FP, Mc Ewen B (2004) Association of depression with medical illness: does cortisol play a role? Biol Psychiat 55:1–9

Brummett BH, Boyle SH, Siegler IC, Kuhn CM, Ashley-Koch A, Jonassaint CR, Züchner S, Collins A, Williams RB (2008) Effects of environmental stress and gender on associations among symptoms of depression and the serotonin transporter gene linked polymorphic region (5-HTTLPR). Behav Genet 38:34–43

Buettner C, Pocai A, Muse ED, Etgen AM, Myers MG Jr, Rossetti L (2006) Critical role of STAT3 in leptin's metabolic actions. Cell Metab 4:49–60

Buettner C, Muse ED, Cheng A, Chen L, Scherer T, Pocai A, Su K, Cheng B, Li X, Harvey-White J, Schwartz GJ, Kunos G, Rossetti L, Buettner C (2008) Leptin controls adipose tissue lipogenesis via central, STAT3-independent mechanisms. Nat Med 14:667–675

Campbell S, Macqueen G (2004) The role of the hippocampus in the pathophysiology of major depression. J Psychiatry Neurosci 29:417–426

Carney RM, Freedland KE, Miller GE, Jaffe AS (2002) Depression as a risk factor for cardiac mortality and morbidity: a review of potential mechanisms. J Psycho Res 53:897–902

Carter F, Bulik C, Joyce P (1994) Direction of weight change in depression. J Affect Disord 30:57–60

Cassidy K, Kotynia-English R, Acres J, Flicker L, Lautenshlager N, Almeida O (2004) Association between lifestyle factors and mental health measures among community-dwelling older women. Aust NZ J Psychiatry 38:940–947

Cenquizca LA, Swanson LW (2006) Analysis of direct hippocampal cortical field CA1 axonal projections to diencephalon in the rat. J Comp Neurol 497:101–114

Ceriello A (2006) Oxidative stress and diabetes-associated complications. Endocr Pract 12(Suppl 1):60–62

Chesik D, Wilczak N, De Keyser J (2008) IGF-1 regulates cAMP levels in astrocytes through a beta2-adrenergic receptor-dependant mechanism. Int J Med Sci 5:240–243

Chrapko WE, Jurasz P, Radomski MW, Lara N, Archer SL, Le Mellédo JM (2004) Decreased platelet nitric oxide synthase activity and plasma nitric oxide metabolites in major depressive disorder. Biol Psychiatry 56:129–134

Considine RV, Sinha MK, Heiman ML, Kriauciunas A, Stephens TW, Nyce MR, Ohannesian JP, Marco CC, McKee LJ, Bauer TL (1996) Serum immunoreactive-leptin concentrations in normal-weight and obese humans. N Engl J Med 334:292–295

Cotter D, Mackay D, Landau S, Kerwin R, Everall I (2001) Reduced glial cell density and neuronal size in the anterior cingulated cortex in major depressive disorder. Arch Gen Psych 58:545–553

Cotter D, Mackay D, Chana G, Beasley C, Landau S, Everall IP (2002) Reduced neuronal size and glial cell density in area 9 of the dorsolateral prefrontal cortex in subjects with major depressive disorder. Cereb Cortex 12:386–394

Cowell A-M, Buckingham JL (2001) Glucocorticoids and the HPA axis in glucocorticoids. In: Goulding NJ, Flower RJ (eds) Glucocorticoids: Milestones in Drug Therapy. Birkhauser, Basel, pp 129–145

Cryan JF, Slattery DA (2010) GABA B receptors and depression: current status. Adv Pharmacol 58:427–451

Dantzer R, O'Connor JC, Freund GG, Johnson RW, Kelley KW (2008) From inflammation to sickness and depression: when the immune system subjugates the brain. Nat Rev Neurosci 9:46–56

Davidson RJ (2003) Affective neuroscience and psychophysiology: toward a synthesis. Psychophysiology 40:655–665

Davidson RJ, Lewis DA, Alloy LB, Amaral DG, Bush G, Cohen JD, Drevets WC, Farah MJ, Kagan J, McClelland JL, Nolen-Hoeksema S, Peterson BS (2002a) Neural and behavioral substrates of mood and mood regulation. Biol Psychiatry 52:478–502

Davidson RJ, Pizzagalli D, Nitschike JB, Putnam K (2002b) Depression: perspectives from affective neuroscience. Annu Rev Psychol 53:545–574

Davidson TL, Kanoski SE, Schier LA, Clegg DJ, Benoit SC (2007) A potential role for the hippocampus in energy intake and body weight regulation. Curr Opin Pharmacol 7:613–616

De Kloet ER, Vreugdenhil E, Oitzl MS, Joels M (1998) Brain corticosteroid receptor balance in health and disease. Endocr Rev 19:269–301

de Kloet ER, Joels M, Holsboer F (2005) Stress and the brain: from adaptation to disease. Nat Rev Neurosci 6:463–475

de Pablos RM, Villaran RF, Arguelles S, Herrera AJ, Venero JL, Ayala A, Cano J, Machado A (2006) Stress increases vulnerability to inflammation in the rat prefrontal cortex. J Neurosci 26:5709–5719

De Vriese SR, Christophe AB, Maes M (2004) In humans, the seasonal variation in polyunsaturated fatty acids is related to the seasonal variation in violent suicide and serotonergic markers of violent suicide. Prostaglandins Leukot Essent Fatty Acids 71:13–18

Delgado PL, Price LH, Miller HL, Salomon RM, Aghajanian GK, Heninger GR, Charney DS (1994) Serotonin and the neurobiology of depression. Effects of tryptophan depletion in drug-free depressed patients. Arch Gen Psychiatry 51:865–874

Delgado PL, Miller HL, Salomon RM, Licinio J, Krystal JH, Moreno FA, Heninger GR, Charney DS (1999) Tryptophan-depletion challenge in depressed patients treated with desipramine or fluoxetine: implications for the role of serotonin in the mechanism of antidepressant action. Biol Psychiatry 46:212–220

Dowlati Y, Herrmann N, Swardfager W, Liu H, Sham L, Reim EK, Lanctôt KL (2010) A meta-analysis of cytokines in major depression. Biol Psychiatry 67:446–457

Drevets WC (1998) Functional neuroimaging studies of depression: the anatomy of melancholia. Annu Rev Med 49:341–361

Duman RS (2004) Depression: a case of neuronal life and death? Biol Psychiatry 56:140–145

Duman RS (2005) Neurotrophic factors and regulation of mood: role of exercise, diet and metabolism. Neurobiol Aging 26(Suppl 1):88–93

Ehlert U, Gaab J, Heinrichs M (2001) Psychoneuroendocrinological contributions to the etiology of depression, posttraumatic stress disorder, and stress-related bodily disorders: the role of the hypothalamus-pituitary-adrenal axis. Biol Psychol 57:141–152

Elovainio M, Keltikangas-Järvinen L, Pulkki-Råback L, Kivimäki M, Puttonen S, Viikari L (2006) Depressive symptoms and C-reactive protein: the Cardiovascular Risk in Young Finns Study. Psychol Med 36:797–805
Esch T, Stefano GB, Fricchione GL, Benson H (2002) Stress in cardiovascular diseases. Med Sci Monit 8:RA93–RA101
Farooqui AA (2009) Beneficial effects of fish oil on human brain. Springer, New York
Farooqui AA (2011) Lipid mediators and their metabolism in the brain. Springer, New York
Farooqui AA (2012) n-3 fatty acid-derived lipid mediators in the brain: new weapons against oxidative stress and inflammation. Curr Med Chem 19:532–543
Farooqui AA, Horrocks LA (2007) Glycerophospholipids in brain: phospholipases $A_2$ in neurological disorders. Springer, New York
Farooqui AA, Horrocks LA, Farooqui T (2007) Modulation of inflammation in brain: a matter of fat. J Neurochem 101:577–599
Farooqui AA, Farooqui T, Panza F, Frisardi V (2012) Metabolic syndrome as a risk factor for neurological disorders. Cell Mol Life Sci 69:741–762
Fava M, Kendler KS (2000) Major depressive disorder. Neuron 28:335–341
Finger BC, Dinan TG, Cryan JF (2011) High-fat diet selectively protects against the effects of chronic social stress in the mouse. Neuroscience 192:351–360
Flores BH, Musselman DL, DeBattista C, Garlow SJ, Schatzberg AF, Nemeroff CB (2004) Biology of Mood Disorders in Textbook of Psychopharmacology. In: Schatzberg AF, Nemeroff CB (eds), Third edition (pp. 717–763), America Psychiatric Publishing, Inc.; Washington DC
Frank MG, Baratta MV, Sprunger DB, Watkins LR, Maier SF (2007) Microglia serve as a neuroimmune substrate for stress-induced potentiation of CNS pro-inflammatory cytokine responses. Brain Behav Immun 21:47–59
Franko DL, Striegel-Moore RH, Bean J, Tamer R, Kraemer HC, Dohm FA, Crawford PB, Schreiber G, Daniels SR (2005) Psychosocial and health consequences of adolescent depression in black and white young adult women. Health Psychol 24:586–593
French S, Jeffery R, Sherwood NN (1999) Prevalence and correlates of binge eating in a nonclinical sample of women enrolled in a weight gain prevention program. Int J Obes Relat Metab Disord 23:576–585
Frisardi V, Panza F, Farooqui AA (2011) Late-life depression and Alzheimer's disease: the glutamatergic system inside of this mirror relationship. Brain Res Rev 67:344–355
Fu X, Zunich SM, O'Connor JC, Kavelaars A, Dantzer R, Kelley KW (2010) Central administration of lipopolysaccharide induces depressive-like behavior in vivo and activates brain indoleamine 2,3 dioxygenase in murine organotypic hippocampal slice cultures. J Neuroinflammation 7:43
Fujigaki H, Saito K, Fujigaki S, Takemura M, Sudo K, Ishiguro H, Sesihima M (2006) The signal transducer and activator of transcription 1alpha and interferon regulatory factor 1 are not essential for the induction of indoleamine 2,3-dioxygenase by lipopolysaccharide: involvement of p38 mitogen-activated protein kinase and nuclear factor-kappaB pathways, and synergistic effect of several proinflammatory cytokines. J Biochem 139:655–662
Gardner A, Salmaso D, Nardo D, Micucci F, Nobili F, Sanchez-Crespo A, Jacobsson H, Larsson SA, Pagani M (2008) Mitochondrial function is related to alterations at brain SPECT in depressed patients. CNS Spectrums 13:805–814
Godbout JP, Berg BM, Krzyszton C, Johnson RW (2005) Alpha-tocopherol attenuates NFkappaB activation and pro-inflammatory cytokine production in brain and improves recovery from lipopolysaccharide-induced sickness behavior. J Neuroimmunol 169:97–105
Gold PW, Chrousos GP (1999) The endocrinology of melancholic and atypical depression: relation to neurocircuitry and somatic consequences. Proc Assoc Am Physicians 111:22–34
Goldbacher EM, Matthews KA, Salomon K (2005) Central adiposity is associated with cardiovascular reactivity to stress in adolescents. Health Psychol 24:375–384
Golden SH (2007) A review of the evidence for a neuroendocrine link between stress, depression, and diabetes mellitus. Curr Diabetes Rev 34:252–259
Goldstein DS, Kopin IJ (2007) Evolution of concepts of stress. Stress 10:109–120

Gracia-Garcia P, Rao V, Haughey NJ, Ratnam Banduru VV, Smith G, Rosenberg PB, Lobo A, Lyketsos CG, Mielke MM (2011) Elevated plasma ceramides in depression. J Neuropsychiatry Clin Neurosci 23:215–218

Grundy SM, Brewer HB Jr, Cleeman JI, Smith SC Jr, Lenfant C (2004) Definition of metabolic syndrome: Report of the National Heart, Lung, and Blood Institute/American Heart Association conference on scientific issues related to definition. Circulation 109:433–438

Haffner SM (2006) The metabolic syndrome: inflammation, diabetes mellitus, and cardiovascular disease. Am J Cardiol 97(Suppl):3A–11A

Hallikainen M, Kolehmainen M, Schwab U, Laaksonen DE, Niskanen L, Rauramaa R et al (2007) Serum adipokines are associated with cholesterol metabolism in the metabolic syndrome. Clin Chim Acta 383:126–132

Hauger RL, Risbrough V, Oakley RH, Olivares-Reyes JA, Dautzenberg FM (2009) Role of CRF receptor signaling in stress vulnerability, anxiety, and depression. Ann N Y Acad Sci 1179:120–143

Heine VM, Maslam S, Zareno J, Joëls M, Lucassen PJ (2004) Suppressed proliferation and apoptotic changes in the rat dentate gyrus after acute and chronic stress are reversible. Eur J Neurosci 19:131–144

Heiskanen TH, Niskanen LK, Hintikka JJ et al (2006) Metabolic syndrome and depression: a cross-sectional analysis. J Clin Psychiatry 67:1422–1427

Heyman E, Gamelin FX, Goekint M, Piscitelli F, Roelands B, Leclair E, Di Marzo V, Meeusen R (2012) Intense exercise increases circulating endocannabinoid and BDNF levels in humans-Possible implications for reward and depression. Psychoneuroendocrinology 37:844–851

Hibbeln JR (1998) Fish consumption and major depression. Lancet 351:1213

Hu F, Pace TW, Miller AH (2009) Interferon-alpha inhibits glucocorticoid receptor-mediated gene transcription via STAT5 activation in mouse HT22 cells. Brain Behav Immun 23:455–463

Huang EJ, Reichardt LF (2001) Neurotrophins: roles in neuronal development and function. Annu Rev Neurosci 24:677–736

Ibanez CF (1995) Neurotrophic factors: from structure–function studies to designing effective therapeutics. Trends Biotechnol 13:217–227

Ikenouchi-Sugita A, Yoshimura R, Hori H, Umene-Nakano W, Ueda N, Nakamura J (2009) Effects of antidepressants on plasma metabolites of nitric oxide in major depressive disorder: comparison between milnacipran and paroxetine. Prog Neuropsychopharmacol. Biol Psychiatry 33:1451–1453

Jaracz J (2008) The anatomy of depression in light of evidence from neuroimaging studies. Psychiatr Pol 42:875–888

John S, Sabo PJ, Johnson TA, Sung MH, Biddie SC, Lightman SL, Voss TC, Davis SR, Meltzer PS, Stamatoyannopoulos JA, Hager GL (2008) Interaction of the glucocorticoid receptor with the chromatin landscape. Mol Cell 29:611–624

Johnson JD, O'Connor KA, Hansen MK, Watkins LR, Maier SF (2003) Effects of prior stress on LPS-induced cytokine and sickness responses. Am J Physiol Regul Integr Comp Physiol 284:R422–R432

Jow GM, Yang TT, Chen CL (2006) Leptin and cholesterol levels are low in major depressive disorder, but high in schizophrenia. J Affect Disord 90:21–27

Karatsoreos IN, Bhagat SM, Bowles NP, Weil ZM, Pfaff DW, McEwen BS (2010) Endocrine and physiological changes in response to chronic corticosterone: a potential model of the metabolic syndrome in mouse. Endocrinology 151:2117–2127

Karmiris K, Koutroubakis IE, Xidakis C, Polychronaki M, Voudouri T, Kouroumalis EA (2006) Circulating levels of leptin, adiponectin, resistin, and ghrelin in inflammatory bowel disease. Inflamm Bowel Dis 12:100–105

Kasahara T, Kubota M, Miyauchi T, Noda Y, Mouri A, Nabeshima T, Kato T (2006) Mice with neuron-specific accumulation of mitochondrial DNA mutations show mood disorder-like phenotypes. Mol Psychiatry 11:577–593

Kasckow JW, Baker D, Geracioti TD Jr (2001) Corticotropin-releasing hormone in depression and post-traumatic stress disorder. Peptides 22:845–851

Kempermann G, Kronenberg G (2003) Depressed new neurons adult hippocampal neurogenesis and a cellular plasticity hypothesis of major depression. Biol Psychiatry 54:499–503

Kimonides VG, Spillantini MG, Sofroniew MV, Fawcett JW, Herbert J (1999) Dehydroepiandrosterone antagonizes the neurotoxic effects of corticosterone and translocation of stress-activated protein kinase 3 in hippocampal primary cultures. Neuroscience 89:429–436

Kinder LS, Carnethon MR, Palaniappan LP, King AC, Fortmann SP (2004) Depression and the metabolic syndrome in young adults: findings from the Third National Health and Nutrition Examination Survey. Psychosom Med 66:316–322

Koenigs M, Grafman J (2009) The functional neuroanatomy of depression: distinct roles for ventromedial and dorsolateral prefrontal cortex. Behav Brain Res 201:239–243

Kopf D, Westphal S, Luley CW, Ritter S, Gilles M, Weber-Hamann B, Lederbogen F, Lehnert H, Henn FA, Heuser I, Deuschle M (2004) Lipid metabolism and insulin resistance in depressed patients: significance of weight, hypercortisolism, and antidepressant treatment. J Clin Psychopharmacol 24:527–531

Krabbe KS, Nielsen AR, Krogh-Madsen R, Plomgard P, Rasmussen P, Erikstrup C, Fischer CP, Lindegaard B, Petersen AMW, Taudorf S, Secher NH, Pilegaard H, Bruunsgaard H, Pedersen BK (2007) Brain-derived neurotrophic factor (BDNF) and type 2 diabetes. Diabetologia 50:431–438

Krysinska K, Lester D (2010) Post-traumatic stress disorder and suicide risk: a systematic review. Arch Suicide Res 14:1–23

Krystal JH, Sanacora G, Blumberg H, Anand A, Charney DS, Marek G, Epperson CN, Goddard A, Mason GF (2002) Glutamate and GABA systems as targets for novel antidepressant and mood-stabilizing treatments. Mol Psychiatry 7:S71–S80

Kyrou I, Chrousos GP, Tsigos C (2006) Stress, visceral obesity, and metabolic complications. Ann N Y Acad Sci 1083:77–110

Lawson EA, Miller KK, Blum JI, Meenaghan E, Misra M, Eddy KT, Herzog DB, Kibanski A (2012) Leptin levels are associated with decreased depressive symptoms in women across the weight spectrum, independent of body fat. Clin Endocrinol (Oxf) 76:520–525

Lee HC, Wei YH (2007) Oxidative stress, mitochondrial DNA mutation, and apoptosis in aging. Exp Biol Med (Maywood) 23:2592–2606

Leonard BE, Myint A (2009) The psychoneuroimmunology of depression. Hum Psychopharmacol 24:165–175

Leuner B, Gould E (2010) Structural plasticity and hippocampal function. Annu Rev Psychol 61:111–140

Leuner B, Gould E, Shors TJ (2006) Is there a link between adult neurogenesis and learning? Hippocampus 16:216–224

Levinson DF (2006) The genetics of depression: a review. Biol Psychiatry 60:84–92

Lightman SL, Wiles CC, Atkinson HC, Henley DE, Russell GM, Leendertz JA, McKenna MA, Spiga F, Wood SA, Conway-Campbell BL (2008) The significance of glucocorticoid pulsatility. Eur J Pharmacol 583:255–262

Lin MT, Simon DK, Ahn CH, Kim LM, Beal MF (2002) High aggregate burden of somatic mtDNA point mutations in aging and Alzheimer's disease brain. Hum Mol Genet 11:133–145

Linde J, Jeffery R, Levy R, Sherwood NE, Utler J, Pronk NP, Boyle RG (2004) Binge eating disorder, weight control self-efficacy, and depression in overweight men and women. Int J Obes Relat Metab Disord 28:418–425

Lu XY (2007) The leptin hypothesis of depression: a potential link between mood disorders and obesity? Curr Opin Pharmacol 7:648–652

Lu T, Pan Y, Kao SY, Li C, Kohane I, Chan J, Yankner BA (2004) Gene regulation and DNA damage in the ageing human brain. Nature 429:883–891

Lu XY, Kim CS, Frazer A, Zhang W (2006) Leptin: a potential novel antidepressant. Proc Natl Acad Sci USA 103:1593–1598

Lucassen PJ, Vollmann-Honsdorf GK, Gleisberg M, Czeh B, De Kloet ER, Fuchs E (2001) Chronic psychosocial stress differentially affects apoptosis in hippocampal subregions and cortex of the adult tree shrew. Eur J Neurosci 14:161–166
Lucassen PJ, Meerlo P, Naylor AS, van Dam AM, Dayer AG, Fuchs E, Oomen CA, Czeh B (2010a) Regulation of neurogenesis by stress, sleep, exercise and inflammation; implications for depression and antidepressant action. Eur Neuropsychopharmacol 20:1–17
Lucassen PJ, Stumpel MW, Wang Q, Aronica E (2010b) Decreased numbers of progenitor cells but no response to antidepressant drugs in the hippocampus of elderly depressed patients. Neuropharmacology 58:940–949
Maes M, Yirmyia R, Noraberg J, Brene S, Hibbeln J, Perini G, Kubera M, Bob P, Lerer B, Maj M (2009) The inflammatory & neurodegenerative (I&ND) hypothesis of depression: leads for future research and new drug developments in depression. Metab Brain Dis 24:27–53
Maes M, Mihaylova I, Kubera M, Uytterhoeven M, Vrydags N, Bosmans E (2010) Increased plasma peroxides and serum oxidized low density lipoprotein antibodies in major depression: markers that further explain the higher incidence of neurodegeneration and coronary artery disease. J Affect Disord 125:287–294
Maier SF, Watkins LR (1998) Cytokines for psychologists: implications of bidirectional immune-to-brain communication for understanding behavior, mood, and cognition. Psychol Rev 105:83–107
Makar TK, Hungund BL, Cook GA, Kashfi K, Cooper AJ (1995) Lipid metabolism and membrane composition are altered in the brains of type II diabetic mice. J Neurochem 64:2159–2168
Maletic V, Robinson M, Oakes T, Iyengar S, Ball SG, Russell J (2007) Neurobiology of depression: an integrated view of key findings. Int J Clin Pract 61:2030–2040
McEwen BS (1992) Re-examination of the glucocorticoid hypothesis of stress and aging. Prog Brain Res 93:365–381
McEwen BS, Sapolsky RM (1995) Stress and cognitive function. Curr Opin Neurobiol 5:205–216
McIntosh LJ, Sapolsky RM (1996) Glucocorticoids increase the accumulation of reactive oxygen species and enhance adriamycin-induced toxicity in neuronal culture. Exp Neurol 141:201–206
McIntyre RS, Soczynska JK, Konarski JZ, Woldeyohannes HO, Law CWY, Miranda A, Fulgosis D, Kennedy SH (2007) Should depressive syndromes be reclassified as "Metabolic Syndrome Type II"? Ann Clin Psychiatry 19:257–264
McKay LI, Cidlowski JA (1999) Molecular control of immune/inflammatory responses: interactions between nuclear factor-kappa B and steroid receptor-signaling pathways. Endocr Rev 20:435–459
Meek TH, Wisse BE, Thaler JP, Guyenet SJ, Matsen ME, Fischer JD, Taborsky GJ Jr, Schwartz MW, Morton GJ (2013) BDNF Action in the Brain Attenuates Diabetic Hyperglycemia via Insulin-Independent Inhibition of Hepatic Glucose Production. Diabetes 62:1512–1518
Mellor AL, Munn D, Chandler P, Keskin D, Johnson T, Marshall B, Jhaver K, Baban B (2003) Tryptophan catabolism and T cell responses. Adv Exp Med Biol 527:27–35
Miller AH, Raison CL (2008) Immune system contributions to the pathophysiology of depression. Focus 6:36–45
Miller AH, Maletic V, Raison CL (2009) Inflammation and its discontents: the role of cytokines in the pathophysiology of major depression. Biol Psychiatry 65:732–741
Monje ML, Toda H, Palmer TD (2003) Inflammatory blockade restores adult hippocampal neurogenesis. Science 302:1760–1765
Morand EF, Leech M (2001) Hypothalamic-pituitary-adrenal axis regulation of inflammation in rheumatoid arthritis. Immunol Cell Biol 79:395–399
Morton GJ, Cummings DE, Baskin DG, Barsh GS, Schwartz MW (2006) Central nervous system control of food intake and body weight. Nature 443:289–295
Müller N (2010) COX-2 inhibitors as antidepressants and antipsychotics: clinical evidence. Curr Opin Investig Drugs 11:31–42

Muller N, Schwarz MJ (2007a) The immune-mediated alteration of serotonin and glutamate: towards an integrated view of depression. Mol Psychiatry 12:988–1000

Muller N, Schwarz MJ (2007b) The immunological basis of glutamatergic disturbance in schizophrenia: towards an integrated view. J Neural Transm Suppl 72:269–280

Muller D, Nitsch RM, Wurtman RJ, Hoyer S (1998) Streptozotocin increases free fatty acids and decreases phospholipids in rat brain. J Neural Transm 105:1271–1281

Munhoz CD, Lepsch LB, Kawamoto EM, Malta MB, Lima LS, Avellar MC, Sapolsky RM, Scavone C (2006) Chronic unpredictable stress exacerbates lipopolysaccharide-induced activation of nuclear factor-kappaB in the frontal cortex and hippocampus via glucocorticoid secretion. J Neurosci 26:3813–3820

Musselman D, Betan E, Larsen H, Phillips LS (2003) Relationship of depression to diabetes types 1 and 2: epidemiology, biology, and treatment. Biol Psychiatry 54:317–329

Nadjar A, Bluthe RM, May MJ, Dantzer R, Parnet P (2005) Inactivation of the cerebral NFkappaB pathway inhibits interleukin-1beta-induced sickness behavior and c-Fos expression in various brain nuclei. Neuropsychopharmacology 30:1492–1499

Nair A, Bonneau RH (2006) Stress-induced elevation of glucocorticoids increases microglia proliferation through NMDA receptor activation. J Neuroimmunol 171:72–85

Niculescu MD, Lupu DS (2009) High fat diet-induced maternal obesity alters fetal hippocampal development. Int J Develop Neurosci 27:627–633

Norata GD, Ongari M, Garlaschelli K, Raselli S, Grigore L, Catapano AL (2007) Plasma resistin levels correlate with the determinants of metabolic syndrome. Eur J Endocrinol 156:279–284

Owens MJ, Nemeroff CB (1994) Role of serotonin in the pathophysiology of depression: focus on the serotonin transporter. Clin Chem 40:288–295

Pace TW, Miller AH (2009) Cytokines and glucocorticoid receptor signaling. Relevance to major depression. Ann N Y Acad Sci 1179:86–105

Pace TW, Hu F, Miller AH (2007) Cytokine-effects on glucocorticoid receptor function: relevance to glucocorticoid resistance and the pathophysiology and treatment of major depression. Brain Behav Immun 21:9–19

Palazidou E (2012) The neurobiology of depression. Br Med Bull 101:127–145

Pariante CM, Lightman SL (2008) The HPA axis in major depression: classical theories and new developments. Trends Neurosci 31:464–468

Paskitti ME, McCreary BJ, Herman JP (2000) Stress regulation of adreno-corticosteroid receptor gene transcription and mRNA expression in rat hippocampus: time-course analysis. Brain Res Mol Brain Res 80:142–152

Patterson CM, Leshan RL, Jones JC, Myers MG Jr (2011) Molecular mapping of mouse brain regions innervated by leptin receptor-expressing cells. Brain Res 1378:18–28

Peeke PM, Chrousos GP (1995) Hypercortisolism and obesity. Ann N Y Acad Sci 771:665–676

Pêgo JM, Sousa JC, Almeida OF, Sousa N (2010) Stress and the neuroendocrinology of anxiety disorders. Curr Top Behav Neurosci 2:97–117

Pini M, Gove ME, Senello JA, van Baal JW, Chan L, Fantuzzi G (2008) Role and regulation of adipokines during zymosan-induced peritoneal inflammation in mice. Endocrinology 149:4080–4085

Pittenger C, Duman RS (2008) Stress, depression, and neuroplasticity: a convergence of mechanisms. Neuropsychopharmacol 33:88–109

Poluektova L, Meyer V, Walters L, Paez X, Gendelman HE (2005) Macrophage-induced inflammation affects hippocampal plasticity and neuronal development in a murine model of HIV-1 encephalitis. Glia 52:344–353

Pomara N, Bruno D, Sarreal AS, Hernando RT, Nierenberg J, Petkova E, Sidtis JJ, Wisniewski TM, Mehta PD, Pratico D, Zetterberg H, Blennow K (2012) Lower CSF amyloid beta peptides and higher F2-isoprostanes in cognitively intact elderly individuals with major depressive disorder. Am J Psychiatry 169(5):523–530

Raison CL, Capuron L, Miller AH (2006) Cytokines sing the blues: inflammation and the pathogenesis of major depression. Trend Immunol 27:24–31

Rezin GT, Cardoso MR, Gonçalves CL, Scaini G, Fraga DB, Riegel RE, Commim CM, Quevedo J, Streck EL (2008) Inhibition of mitochondrial respiratory chain in brain of rats subjected to an experimental model of depression. Neurochem Int 53:395–400
Rezin GT, Amboni G, Zugno AI, Quevedo J, Streck EL (2009) Mitochondrial dysfunction and psychiatric disorders. Neurochem Res 34:1021–1029
Robertson SD, Matthies HJ, Owens WA, Sathananthan V, Christianson NS, Kennedy JP, Lindsley CW, Daws LC, Galli A (2010) Insulin reveals Akt signaling as a novel regulator of norepinephrine transporter trafficking and norepinephrine homeostasis. J Neurosci 30:11305–11316
Robinson CM, Hale PT, Carlin JM (2006) NF-kappa B activation contributes to indoleamine dioxygenase transcriptional synergy induced by IFN-gamma and tumor necrosis factor-alpha. Cytokine 35:53–61
Rogatsky I, Ivashkiv LB (2006) Glucocorticoid modulation of cytokine signaling. Tissue Antigens 68:1–12
Rosmond R, Bjorntorp P (2000) The hypothalamic-pituitary-adrenal axis activity as a predictor of cardiovascular disease, type 2 diabetes and stroke. J Intern Med 247:188–197
Rubin RT, Rhodes ME, Czambel RK (2002) Sexual diergism of baseline plasma leptin and leptin suppression by arginine vasopressin in major depressives and matched controls. Psychiatry Res 113:255–268
Rustad JK, Musselman DL, Nemeroff CB (2011) The relationship of depression and diabetes: pathophysiological and treatment implications. Psychoneuroendocrinol 36:1276–1286
Sandoval D, Cota D, Seeley RJ (2008) The integrative role of CNS fuel-sensing mechanisms in energy balance and glucose regulation. Annu Rev Physiol 70:513–535
Sapolsky RM (2000) The possibility of neurotoxicity in the hippocampus in major depression: a primer on neuron death. Biol Psychiatry 48:755–765
Schule C, Zill P, Baghai TC, Eser D, Zwanzger P, Wenig N, Rupprecht R, Bondy B (2006) Brain-derived neurotrophic factor Val66Met polymorphism and dexamethasone/CRH test results in depressed patients. Psychoneuroendocrinology 31:1019–1025
Schulz R, Beach SR, Ives DG, Martire LM, Ariyo AA, Kop WJ (2000) Association between depression and mortality in older adults: the Cardiovascular Health Study. Arch Intern Med 160:1761–1768
Scott MM, Lachey JL, Sternson SM, Lee CE, Elias CF, Friedman JM, Elmquist JK (2009) Leptin targets in the mouse brain. J Comp Neurol 514:518–532
Selek S, Savas HA, Gergerlioglu HS, Bulbul F, Uz E, Yumru M (2008) The course of nitric oxide and superoxide dismutase during treatment of bipolar depressive episode. J Affect Disord 107:89–94
Sen S, Duman R, Sanacora G (2008) Serum brain-derived neurotrophic factor, depression, and antidepressant medications: meta-analyses and implications. Biol Psychiatry 64:527–532
Serhan CN (2005) Novel ω-3-derived local mediators in anti-inflammation and resolution. Pharmacol Ther 105:7–21
Shelton RC, Miller AH (2010) Eating ourselves to death (and despair): the contribution of adiposity and inflammation to depression. Prog Neurobiol 91:275–299
Singh RB, Gupta S, Dherange P, De Meester F, Wilczynska A, Alam SE, Pella D, Wilson DW (2012) Metabolic syndrome: a brain disease. Can J Physiol Pharmacol 90:1171–1183
Skilton MR, Moulin P, Terra JL, Bonnet F (2007) Associations between anxiety, depression, and the metabolic syndrome. Biol Psychiat 62:1251–1257
Smith SM, Vale WW (2006) The role of the hypothalamic-pituitary-adrenal axis in neuroendocrine responses to stress. Dialogues Clin Neurosci 8:383–395
Sparkman NL, Buchanan JB, Heyen JR, Chen J, Beverly JL, Johnson RW (2006) Interleukin-6 facilitates lipopolysaccharide-induced disruption in working memory and expression of other proinflammatory cytokines in hippocampal neuronal cell layers. J Neurosci 26:10709–10716
Sullivan PF, Neale MC, Kendler KS (2000) Genetic epidemiology of major depression: review and meta-analysis. Am J Psychiatry 157:1552–1562
Swaab DF, Bao AM, Lucassen PJ (2005) The stress system in the human brain in depression and neurodegeneration. Ageing Res Rev 4:141–194

Taha AY, Gao F, Ramadan E, Cheon Y, Rapoport SI, Kim H-W (2012) BMC Neurosci 13:131
Takeuchi T, Nakao M, Nomura K, Inoue M, Tsurugano S, Shinozaki Y, Yano E (2009) Association of the metabolic syndrome with depression and anxiety in Japanese men: a 1-year cohort study. Diabetes Metab Res Rev 25:762–767
Talbot F, Nouwen A (2000) A review of the relationship between depression and diabetes in adults: is there a link? Diabetes Care 23:1556–1562
Tancredi V, D'Antuono M, Cafe C, Giovedi S, Bue MC, D'Arcangelo G, Onofri F, Benfenati F (2000) The inhibitory effects of interleukin-6 on synaptic plasticity in the rat hippocampus are associated with an inhibition of mitogen-activated protein kinase ERK. J Neurochem 75:634–643
Tavares RG, Tasca CI, Santos CE et al (2002) Quinolinic acid stimulates synaptosomal glutamate release and inhibits glutamate uptake into astrocytes. Neurochem Int 40:621–627
Tilg H, Moschen AR (2008) Inflammatory mechanisms in the regulation of insulin resistance. Mol Med 14:222–231
Tozuka Y, Wada E, Wada K (2009) Diet-induced obesity in female mice leads to peroxidized lipid accumulations and impairment of hippocampal neurogenesis during the early life of their offspring. FASEB J 23:1920–1934
Tozuka Y, Kumon M, Wada E, Onodera M, Mochizuki H, Wada K (2010) Maternal obesity impairs hippocampal BDNF production and spatial learning performance in young mouse offspring. Neurochem Int 57:235–247
Tracey KJ (2002) The inflammatory reflex. Nature 420:853–859
Tremblay A (2004) Dietary fat and body weight set point. Nutr Rev 62:S75–S77
Trifunovic A, Wredenberg A, Falkenberg M, Spelbrink JN, Rovio AT, Bruder CE, Bohlooly YM, Gidiofs S, Oldfors A, Wibom R, Tornell J, Jacob HT, Larsson NG (2004) Premature ageing in mice expressing defective mitochondrial DNA polymerase. Nature 429:417–423
Tyrovolas S, Lionis C, Zeimbekis A, Bountziouka V, Micheli M, Katsarou A, Papairakleous N, Metallinos G, Makri K, Polychronopoulos E, Panagiotakos DB (2009) Increased body mass and depressive symptomatology are associated with hypercholesterolemia, among elderly individuals; results from the MEDIS study. Lipids Health Dis 8:10
Unger TJ, Calderon GA, Bradley LC, Sena-Esteves M, Rios M (2007) Selective deletion of BDNF in the ventromedial and dorsomedial hypothalamus of adult mice results in hyperphagic behavior and obesity. J Neurosci 27:14265–14274
Uranova NA, Vostrikov VM, Orlovskaya DD, Rachmanova VI (2004) Oligodendroglial density in the prefrontal cortex in schizophrenia and mood disorders: a study from the Stanley Neuropathology Consortium. Schizophr Res 67:269–275
Ustun TB, Ayuso-Mateos JL, Chatterji S, Mathers C, Murray CJ (2004) Global burden of depressive disorders in the year 2000. Br J Psychiatry 184:386–392
Varghese FP, Brown ES (2001) The hypothalamic-pituitary-adrenal axis in major depressive disorder: a brief primer for primary care physicians. Prim Care Campanion J Clin Psychiat 3:151–155
Vilcek J (2003) The cytokines: an overview. In: Thompson MT (ed) The cytokines handbook, vol 1, 4th edn. Elsevier, Amsterdam, pp 3–9
Vogelzangs N, Suthers K, Ferrucci L, Simonsick EM, Ble A, Schrager M, Bandinelli S, Lauretani F, Giannelli SV, Penninx BW (2007) Hypercortisolemic depression is associated with the metabolic syndrome in late-life. Psychoneuroendocrinology 32:151–159
Wang G-J, Yang J, Volkow ND, Telang F, Ma Y, Zhu W, Wong CT, Tomasi D, Thanos PK, Fowler JS (2006) Gastric stimulation in obese subjects activates the hippocampus and other regions involved in brain reward circuitry. Proc Natl Acad Sci U S A 103:15641–15645
Wang S, Sun Z, Guo Y, Yuan Y, Yang B (2009) Diabetes impairs hippocampal function via advanced glycation end product mediated new neuron generation in animals with diabetes-related depression. Toxicol Sci 111:72–79
Weber-Hamann B, Hentschel F, Kniest A, Deuschle M, Colla M, Lederbogen F, Heuser I (2002) Hypercortisolemic depression is associated with increased intra-abdominal fat. Psychosom Med 64:274–277

Winokur A, Maislin G, Phillips JL, Amsterdam JD (1988) Insulin resistance after oral glucose tolerance testing in patients with major depression. Am J Psychiatry 145:325–330

Wise LA, Zierler S, Krieger N, Harlow BL (2001) Adult onset of major depressive disorder in relation to early life violent victimisation: a case-control study. Lancet 358:881–887

Wu F, Tyml K, Wilson JX (2008) iNOS expression requires NADPH oxidase-dependent redox signaling in microvascular endothelial cells. J Cell Physiol 217:207–214

Wurtman RJ (2005) Genes, stress, and depression. Metabolism 54(Suppl 1):16–19

Xu LL, Warren MK, Rose WL, Gong W, Wang JM (1996) Human recombinant monocyte chemotactic protein and other C-C chemokines bind and induce directional migration of dendritic cells in vitro. J Leukoc Biol 60:365–371

Yager S, Forlenza MJ, Miller GE (2010) Depression and oxidative damage to lipids. Psychoneuroendocrinology 35:1356–1362

Yamada N, Katsuura G, Ochi Y, Ebihara K, Kusakabe T, Hosoda K, Nakao K (2011) Impaired CNS leptin action is implicated in depression associated with obesity. Endocrinology 152:2634–2643

Yamagishi S, Nakamura K, Matsui T (2008) Agents that block advanced glycation end product (AGE)-RAGE (receptor for AGEs)-oxidative stress system: a novel therapeutic strategy for diabetic vascular complications. Expert Opin Investig Drugs 17:983–996

Zheng H, Liu Y, Li W, Yang B, Chen D, Wang X, Jiang Z, Wang H, Wang Z, Cornelisson G, Halberg F (2006) Beneficial effects of exercise and its molecular mechanisms on depression in rats. Behav Brain Res 168:47–55

# Chapter 10
# Perspective and Direction for Future Research

## 10.1 Introduction

The metabolic syndrome (MetS) is a condition characterized by clustering of interrelated risk factors of metabolic origin (insulin resistance, hyperinsulinemia, hypertension, dyslipidemia, impaired glucose disposal, type II diabetes, abnormal blood fat levels, fatty liver disease, and abdominal obesity) (Cornier et al. 2008). In the United States, the estimated rate of the MetS in adults aged 60–70 years is 43.5 % (Ford et al. 2002). Insulin resistance, hyperinsulinemia, and impaired glucose disposal increase the risk of developing cardiovascular, cerebrovascular, and neurological disorders diseases. Approximately one-fourth of European, American, and Canadian adults suffer from MetS (Grundy 2008). Clustering of the above-mentioned components reflects overnutrition, sedentary lifestyles, and resultant excess adiposity. According to worldwide estimates, 1.7 billion people are classified as either overweight or obese, more than 1 billion have hypertension, and more than 500 million have either diabetes or the prediabetes with impaired glucose tolerance (Hossain et al. 2007).

The development of insulin resistance involves not only aging, obesity, body fat distribution, and dietary factors but also genetic factors, cardiorespiratory and cerebrovascular fitness, and physical inactivity (Assah et al. 2008). The clustering of these factors is not considered as a random event: rather, these factors probably have common molecular mechanisms, which are interconnected and regulated by several common parameters. Most molecular details on these clustering factors have been obtained from studies with experimental models of MetS, particularly from rodents (Clee and Attie 2007; Wang et al. 2008). The complexity and heterogeneity of MetS have made mechanistic studies in humans very difficult not only due to ethical reasons but also because of dietary restrictions and lack of genetic manipulations.

The pathophysiology of MetS involves insulin resistance caused not only by excessive flux of fatty acids and increase in levels of ceramide (Chew et al. 2006; Sesti 2006) but also by the induction of a proinflammatory state resulting from the production of adipokines and cytokines from adipocytes and macrophages along with

A.A. Farooqui, *Metabolic Syndrome: An Important Risk Factor for Stroke, Alzheimer Disease, and Depression*, DOI 10.1007/978-1-4614-7318-3_10,

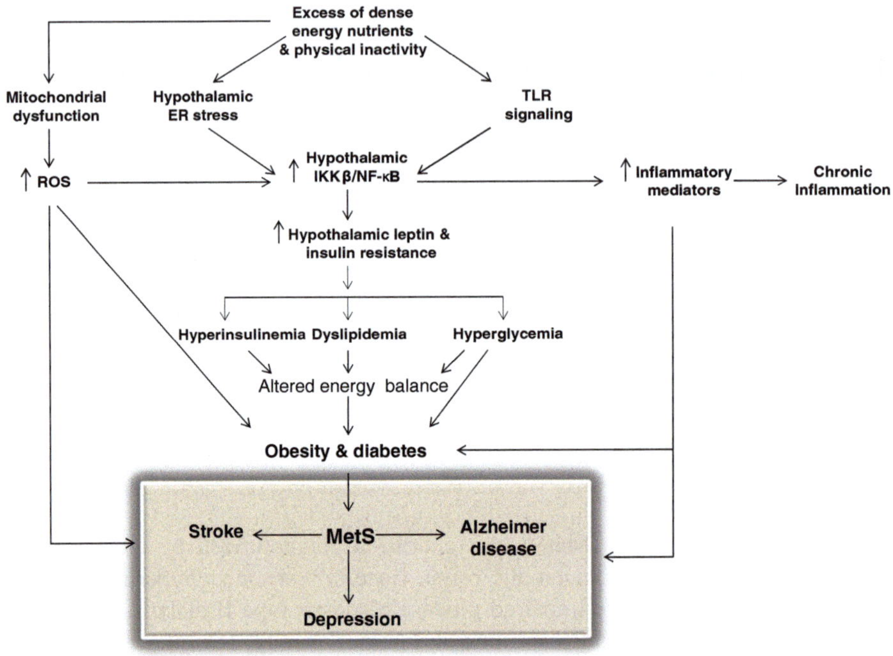


**Fig. 10.1** Effects of energy-dense food and physical inactivity on metabolic syndrome and neurological disorders

induction of chronic oxidative stress and low-grade inflammation (Chew et al. 2006; Fulop et al. 2006). Thus, elevation in inflammatory mediators causes increase in generation of reactive oxygen species (ROS) and advanced glycation products (AGEs) that are closely associated with detrimental effects of cardiovascular and cerebrovascular alterations in MetS (Hugel et al. 2005; Martinez et al. 2005). Chronic and low-grade inflammation and oxidative stress are common denominators linking obesity, insulin resistance, and dyslipidemia with excessive glucose metabolism in MetS (Fig. 10.1). In addition, continuously elevated insulin concentrations impair normal vascular function (Muniyappa et al. 2007). Among the above-mentioned characteristics, family history and insulin resistance are involved in the pathogenesis of type II diabetes, which is closely linked with the pathogenesis of MetS. MetS is a pronounced risk factor not only for the development of cardiovascular diseases but also for the pathogenesis of neurodegenerative diseases (Alzheimer disease, AD, and Parkinson disease, PD) and neuropsychiatric diseases (depression and schizophrenia) and several types of cancers (Aggarwal 2010; Shi et al. 2005; Farooqui et al. 2012).

Diet influences brain function. Diet high in saturated fat not only increases insulin resistance (Tschop and Thomas 2006) but also negatively effects cognitive processing and increases the risk of neurological disorders (Luchsinger et al. 2002; Greenwood and Winocur 2005; Convit et al. 2003). Consumption of n-3 fatty acids modulates neural membrane fluidity and permeability and improves spatial learning by regulating synaptic and cognitive function (Farooqui 2009; Gomez-Pinilla 2008). High intake of fruits, vegetables, fish, and whole grains (Mediterranean-type diet) along with phytochemicals (resveratrol and other polyphenols) produces beneficial effects on the human brain (Engelhart et al. 2002; Scarmeas et al. 2009; Farooqui 2012). Although primary determinants of total caloric intake and the largest component of any diet consist of macronutrients (carbohydrates, fat, and protein), the role of macronutrient intake relative to total caloric intake on cognitive function in older population has received little attention, and more studies are needed on this topic in large human population.

Hyperglycemia, a major pathological characteristic of type II diabetes and MetS, may have toxic effects on neurons in the brain through osmotic insults and oxidative stress. The maintenance of chronic high glucose also leads to the enhanced formation of advanced glycation end products (AGEs), which have potentially toxic effects on neurons. As stated in Chap. 8, insulin accelerates AD-related pathology through its effects on the amyloid-β (Aβ) metabolism and tau phosphorylation (Li and Hölscher 2007). Insulin raises Aβ concentrations in the plasma in AD subjects, and these effects may contribute to the risk of developing AD in patients with type II diabetes. The desensitization of insulin receptors reduces the synthesis of several proteins, including insulin-degrading enzyme (IDE), the enzyme that degrades Aβ as well as insulin. The reduction in IDE activity may result in greater amyloid deposition. Reduction in insulin signaling may also produce increase in glycogen synthase kinase-3β activity leading to enhancement in phosphorylation of tau protein and the formation of neurofibrillary tangles (NFTs). Thus, among the different lifestyle factors related to the pathogenesis of diabetes and AD, insulin resistance is closely associated with reduction in cognition function among older people (Schubert et al. 2004). Impaired insulin signal transduction is assumed (Hoyer 2002) to cause improper brain glucose metabolism since an upregulation of insulin receptor density has been observed in AD brains and type II diabetes (Hoyer 1998). These changes not only lead to reduction in the quality of life given their protracted nature but also result in premature death (Umegaki 2010).

Like diabetes and MetS, obesity is also a fast growing problem that is reaching epidemic proportions worldwide and is associated with an increased risk of premature death (Kopelman 2000). According to the National Institute of Diabetes and Digestive and Kidney Diseases, approximately two-thirds of US adults—nearly 167 million—are overweight, and nearly one-third (31.4 %) are obese (http://www2.niddk.nih.gov/AboutNIDDK/OfficeofObesityResearch). Typical eating patterns among obese populations in the United States or most Western countries include foods that are high in saturated fat, flours, sugars, and cholesterol, along with high-fructose corn syrup (HFCS)-containing soft drinks, which are very popular these days among young consumers. Cholesterol and saturated fatty acids have been linked to activation of proinflammatory signaling cascades, alterations in insulin signaling, changes in leptin signaling in neural and nonneural cells (Lee et al. 2003), and HFCS. These processes and components contribute to the generation of AGEs and ROS resulting in hypertension, obesity, diabetes, and MetS (Farooqui et al. 2012; Spreadbury 2012). Obesity is a condition of energy imbalance between intake and expenditure. When an organism encounters food excess, it conserves the nutrients either as glycogen for short-term storage or as lipids for longer storage duration, which may result in a consequent state of obesity. Obesity is positively associated with chronic systemic low-grade inflammation, oxidative stress, and risks of other metabolic diseases, such as MetS and cardiovascular and cerebrovascular diseases (Shoelson et al. 2007; Styskal et al. 2012).

AGEs are a group of heterogeneous compounds formed by the nonenzymic reactions between aldehydic group of reducing sugars with proteins, lipids, or nucleic acids. Generation and accumulation of AGEs are related to the aging process and accelerated in MetS. AGEs not only contribute to vascular diabetic complications (Puddu and Viviani 2011) but also promote generation of ROS and subsequently cause inflammatory and thrombogenic reactions in various types of cells through interactions with AGE receptors (RAGE). Since ROS generation and inflammation are closely associated with insulin resistance as well, it is conceivable that the AGEs–RAGE system may be involved in the pathogenesis of insulin resistance and subsequently the development of diabetes and MetS. In addition, mitochondrial superoxide generation has been shown to play an important role in the formation and accumulation of AGEs under diabetic conditions (Yagihashi and Sima 1986). It should be noted that high-calorie, high-fat, and low-fiber diet usually promotes insulin resistance, obesity, and diabetes whereas caloric restriction, exercise, and wholesome foods reverse these pathological conditions (Fig. 10.1) (Furukawa et al. 2004; Reeds 2009; Chaput et al. 2011; Varela-Moreiras 2006; Bassett et al. 2008). Other factors that contribute to obesity and MetS include obesogenic lifestyle such as physical inactivity, sleep deprivation, television watching, and alcohol consumption (Monasta et al. 2010).

## 10.2 Links Between Metabolic Syndrome and Neurological Disorders

It is well known that glucose metabolism and homeostasis contribute to energy maintenance, neurogenesis, neuronal survival, and synaptic plasticity, which are required for learning and memory. During insulin resistance, one develops reduced sensitivity to insulin, resulting in hyperinsulinemia, and this impairment in insulin signaling is closely associated with the pathogenesis of AD, a neurological disorder, which is accompanied by neuroinflammation, oxidative stress, accumulation of Aβ, and neurodegeneration. It is obvious from Chaps. 1 to 9 that at the molecular level, MetS is accompanied not only by insulin resistance caused by increase in lipid mediators (fatty acids, ceramide, acylcarnitines) but also by alterations in levels of adipokines (leptin and adiponectin). These changes modulate immune response and inflammation that lead to alterations in the hypothalamic "body weight/appetite/satiety set point" resulting in the initiation and development of the MetS, a complex multifactorial condition, which is a risk factor for neurological disorders such as stroke, depression, and AD (Farooqui et al. 2012). Family history, age, environmental factors, and lifestyle factors (diet and physical inactivity) are closely

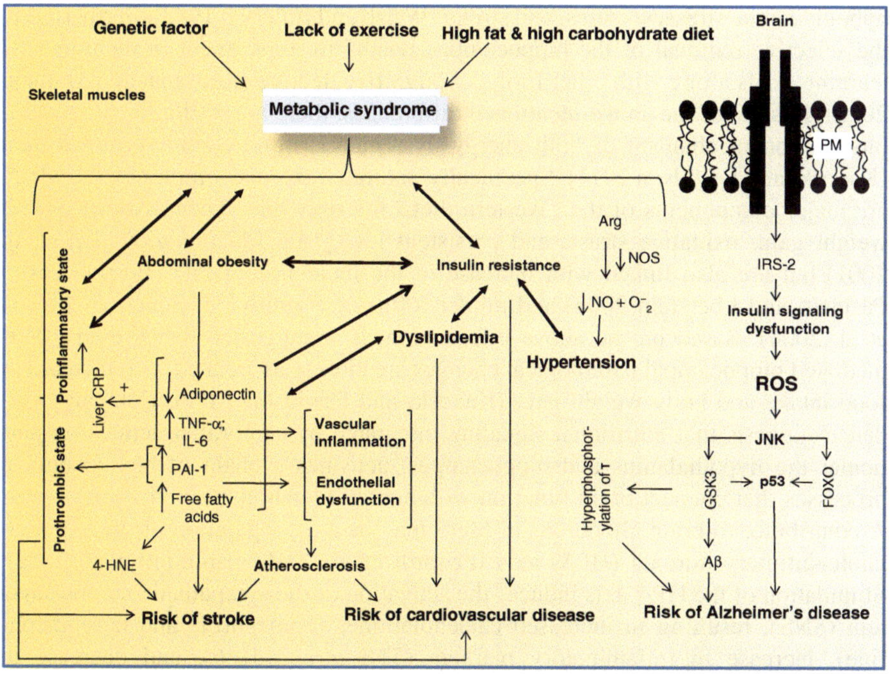


**Fig. 10.2** Metabolic syndrome is an important risk factor for stroke, heart disease, and Alzheimer disease. Genetic factors, sedentary lifestyle, and high unhealthy diet induce central obesity and insulin resistance and may cause dyslipidemia, hypertension, and inflammation. These are risk factors for stroke, heart disease, and Alzheimer disease. C-reactive protein (CRP); tumor necrosis factor-α (TNF-α); plasminogen activator inhibitor-1 (PAI-1); 4-hydroxynonenal (4-HNE); insulin receptor substrate-2 (IRS-2); reactive oxygen species (ROS); c-Jun N-terminal kinase (JNK); protein 53 or tumor protein 53 (p53); forkhead transcription factor (FOXO); glycogen synthase kinase 3 (GSK3); and β-amyloid (Aβ). Upward arrows indicate increase and downward arrows indicate decrease

associated with predisposition for the development of MetS as well as the above-mentioned neurological disorders (Fig. 10.2). Weight loss and its maintenance is an important strategy to reduce the incidence of obesity-related comorbidities. Lifestyle modification focused on diet and physical activity represents the essential component of any kind of weight management.

Metabolic diseases and neurological disorders are controlled largely by separate brain structures and circuits. For example, it is well known that surgical, genetic, and hormonal alterations in the hypothalamus not only produce profound effects on eating behavior and body weight gain in experimental animals but also produce increase in energy intake and body weight in humans due to marked changes in hypothalamic neurohormonal signaling pathways (Benoit et al. 2009; Kanoski and Davidson 2011). In contrast, the hippocampus is the site of structural abnormalities associated with early stages of AD and other cognitive dementias (Mungas et al. 2001; Price et al. 2001). In fact, the hippocampus is preferentially susceptible compared to other brain regions to a variety of insults including hypoxia, hypoperfusion, hypoglycemia, stress, seizures, and stroke (Walsh and Emerich 1988). Furthermore, the selective removal of the hippocampus results in impairment in memory and learning skills along with other forms of cognitive decline (Rothman and Mattson 2010). In spite of the above-mentioned differences, there are similarities in the etiologies and mechanism of both energy dysregulation and cognitive impairment (Kanoski and Davidson 2011). Specifically, saturated fats and simple sugars, which are major components of the "Western diet," not only have been associated with weight gain, oxidative stress, and persistent low-grade inflammation (Hu et al. 2001) but are also linked with increase in the incidence of AD (Berrino 2002; Pasinetti and Eberstein 2008) and milder forms of cognitive dysfunction (Morris et al. 2004). Based on the above information, it is suggested that Western diet-mediated hippocampal pathological changes are closely associated with increase in food intake and body weight gain (Kanoski and Davidson 2011). Collective evidence suggests that nutritional signaling integrated by cerebral structures, in particular the hypothalamus and hippocampus, activates a chain of neurochemical processes that connect brain function with energy metabolism in visceral organs. A contributor to brain alterations in MetS may be a dysregulation of the hypothalamic–pituitary–adrenal (HPA) axis (Frisardi et al. 2010; Farooqui et al. 2012). Stimulation of the HPA axis induces the activation of the sympathetic nervous system (SNS), resulting in increased catecholamines (epinephrine and norepinephrine), increase in inflammatory markers (TNF-α and IL-6), and decrease in neurovegetative functions (eating and sleeping), along with other changes in the brain (Tsigos and Chrousos 2002). In addition in humans, stimulation of the SNS, induced by chronic stress, elevates pulse rate and cardiac minute output and also activates the renin–angiotensin–aldosterone system, which is another important pressor mechanism (Soucek and Kara 2001). Increased activity of the SNS also plays a part in the development of impaired glucose (Seematter et al. 2000) and lipid metabolism (Morimoto et al. 2008; Bachen et al. 2002).

Both AD and type II diabetes have several common risk factors, such as higher cholesterol, deposition of Aβ, changes in glycogen synthesis kinase 3 activity, phosphorylation of τ protein, increase in oxidative stress, along with induction of insulin resistance, elevation in inflammatory cytokines, and induction of apoptotic cell death (Fig. 10.2) (Doble and Woodgett 2003; Ristow 2004; Haan 2006). A common observation in both of these pathological conditions is the desensitization of insulin receptors in the brain. As stated above, insulin acts as a growth factor in the brain and its neuroprotective action activates dendritic sprouting, regeneration, and stem cell proliferation. The impairment of insulin signaling is known to promote the onset and development of AD. Patients with type II diabetes have a two- to threefold increased risk for developing dementia and AD (Zemva and Schubert 2011). To explain these observations, a novel hypothesis has been proposed. In an in vitro model of AD-type amyloidogenesis, the elevation of amyloidogenic Aβ peptides coincides with hyperglycemic culture conditions in a dose-dependent manner (Pasinetti and Eberstein 2008), supporting the causal role for hyperglycemic environment along with abnormal oxidative phosphorylation associated with MetS as risk factors in AD (Craft and Watson 2004; Craft 2007). Although evidence exists that hyperglycemia in type II diabetes produces up to a fourfold increase in neuronal glucose levels (Tomlinson and Gardiner 2008), the molecular mechanisms through which impaired glucose metabolism/mitochondrial oxidative phosphorylation contributes to AD amyloid neuropathology are not known (Lin and Beal 2006), especially in respect to the potential role of dietary lifestyle factors in the prevention of AD dementia. In addition, the brains from AD and type II diabetes patients show significantly lower expression of the insulin receptor (IR), the IGF-1 receptor (IGF-1R), and the insulin receptor substrate (IRS) proteins supporting the view that decrease in insulin/IGF-1 signaling (IIS) may be associated with the pathogenesis of both AD and type II diabetes. Thus, neuronal resistance for insulin/insulin-like growth factor 1 (IGF-1) may represent a molecular link between type II diabetes and AD, characterizing AD as "brain-type diabetes" or diabetes type III (de la Monte and Wands 2005; Zemva and Schubert 2011). Furthermore, alterations in leptin signaling are also associated with obesity, type II diabetes, MetS, and AD. In AD, transport of leptin across the blood–brain barrier (BBB) is also significantly impaired through the downregulation of megalin, a protein to which leptin must bind in order to enter the brain (Dietrich et al. 2008). Leptin has also been reported to inhibit the activity of β-secretase in neuronal cells, possibly by altering the lipid composition of membrane rafts and thereby affecting Aβ generation. In fact, chronic administration of leptin actually reduces Aβ load in the brains of AD-transgenic mice, supporting the view that leptin can be used for the treatment of AD (Fewlass et al. 2004). Recent neuroimaging studies indicate that diabetic patients show diffuse cortical atrophy with notable vulnerability in the temporal lobe (Hirano et al. 2012), and it is proposed that diabetes may facilitate the onset of AD via vascular lesion, poor vascular responsiveness, and aberrant amyloid metabolisms, probably through atherosclerosis (Hirano et al. 2012).

Hyperinsulinemia and hypoglycemia increase cerebral blood flow and produces diffuse cortical atrophy, while hyperglycemia decreases cerebral blood flow and causes poor vascular responsiveness. To meet anabolic and catabolic demands efficiently, the blood flow to the brain needs to be maintained to approximately 750 ml per minute or about 50 ml/100 ml of brain tissue per minute (Walters 1998; Ito et al. 2005). Studies on patients with diabetes and MetS have indicated that these patients have reduced global rates of cerebral blood flow as measured using xenon, and the magnitude of reduction correlates with the duration of the disease (Rodriguez et al. 1993; Kodl and Seaquist 2008). This reduced rate of cerebral blood flow in type II diabetes and MetS patients is similar to that found in AD patients with dementia (Biessels et al. 1994; Kodl and Seaquist 2008). Reduction in glucose uptake due to reduction in cerebral blood flow may have a direct effect on the effect of insulin on regulation of hippocampal function in these patients. Accumulating evidence in animal models of AD and type II diabetes indicates that (a) type II diabetes enhances cognitive dysfunction with cerebrovascular changes, such as vascular inflammation and cerebral amyloid angiopathy, (b) neuropathological changes are associated with impairment of brain insulin signaling, and (c) amyloid pathology may adversely affect diabetic phenotypes. In addition, chronic peripheral hyperinsulinemia in type II diabetes results in a reduction of insulin transport across the BBB and reduction in insulin signaling in the brain. Peripheral administration of insulin at optimal doses produces an increase in brain insulin levels resulting in a cognition-enhancing effect in patients with AD. It is suggested that insulin dysregulation in type II diabetes and AD may act in a variety of ways including decrease in cortical glucose utilization, induction of oxidative stress along with the formation of AGE-linked proteins, elevation in τ phosphorylation, increase in NFT formation, and increase in β-amyloid aggregation through inhibition of IDE (de la Monte and Wands 2005; Farooqui et al. 2012).

Obesity has also been linked mechanistically to AD pathogenesis. This suggestion is based on abnormal metabolism of the obesity-related protein leptin (Fewlass et al. 2004). Leptin attenuates β-secretase processing of APP in neuronal cells, possibly through mechanisms involving altered lipid composition of membrane lipid rafts. Chronic administration of leptin to AD-transgenic animals reduces the brain Aβ load supporting the view that leptin may have some therapeutic potential in AD (Fewlass et al. 2004). Since circulating leptin is transported into the brain by binding to the lipoprotein receptor megalin at the choroid plexus epithelium, it is proposed that attenuation of megalin expression in physiological and pathological conditions, such as during aging or in AD dementia, may be correlated with poor entry of leptin into the brain (Dietrich et al. 2008).

Interestingly, obesity which suppresses circulating levels of adiponectin, an adipocyte-derived hormone with antidiabetic and anti-inflammatory actions in peripheral cells, has now been identified as a risk factor for AD (Luchsinger 2008). It is becoming increasingly evident that adiponectin provides neuroprotection in several pathways. Firstly, adiponectin increases insulin sensitivity and inhibits gonadotropin-releasing hormone by modulating AMPK (Tomas et al. 2002; Wen et al. 2008), which is a sensor of cellular energy status in almost all eukaryotic cells. The activation of AMPK not only protects neurons against stroke (McCullough et al. 2005) and decreases glutamate toxicity in the hippocampus but also increases hippocampal neurogenesis (Dagon et al. 2005). Secondly, adiponectin has been reported to improve endothelial cell function of the brain vasculature which can be beneficial for brain function (Jeon et al. 2009). Furthermore, dietary restriction increases adiponectin levels in mammals, supporting the view that this adipose-derived hormone may provide neuroprotection in stroke and AD (Shinmura et al. 2007; Zhu et al. 2007).

Metabolic syndrome (MetS) increases the risk of vascular disease, including stroke and ischemic heart disease (Boden-Albala et al. 2008). The prevalence of MetS in patients with a history of stroke is 44 % compared to 23 % without stroke (Boden-Albala et al. 2008; Arenillas et al. 2007). MetS increases the risk of vascular disease and stroke through the involvement of hypertension, insulin insensitivity, and dyslipidemia (Abarquez 2003). These processes lead to atherogenesis and the prothrombotic state, which increases the risk of stroke (Uchino et al. 2010). Thus, hypertension is a known risk factor for stroke, but very little is known about the link between hypertension and AD and vascular dementia. Cholinergic projections from the nucleus basalis of Meynert (NBM) extend into the cortex and are a major mediator for vasodilation of cortical arterioles and capillaries. Stimulation of the NBM is known to activate muscarinic and nicotinic acetylcholine receptors in perivascular neurons and astrocytes to release the vasodilator nitric oxide (Roman and Kalaria 2006; van Beek and Claassen 2011). Vasoconstriction is mediated in part by catecholaminergic and serotonergic innervation emanating from the locus ceruleus and dorsal raphe nucleus, respectively (Kalaria et al. 1989; Tejani-Butt et al. 1995). The mechanisms linking hypertension to AD remain to be elucidated. Based on white matter lesions seen on cerebral magnetic resonance imaging, it is becoming increasingly evident that chronic hypertension damages brain endothelial cells and smooth muscle cells that control cerebral blood flow leading to changes in the cerebral microvasculature (Pickering 2001; O'Rourke and Safar 2005). It is not yet clearly established whether lowering blood pressure reduces the risk of white matter lesions and dementia (Tzourio 2007), so large trials dealing with this question are eagerly awaited.

The molecular mechanism associated with increased risk of stroke in MetS is not fully understood. However, it is suggested that leptin not only stimulates brain metabolic rate (Villanueva and Myers 2008) but also mediates the activation of the pituitary adrenal sympathetic system and the rennin–angiotensin system (Shek et al. 1998; Esler et al. 2001). Increased plasma norepinephrine and angiotensin II produce vasoconstriction through the involvement of NADPH oxidase (Kazama et al. 2004) which also increases oxygen demand through activation of uncoupling of mitochondrial oxidative phosphorylation. This sequence of events may ultimately lead to stroke (Uchino et al. 2010).

Although studies on the association between depression and the MetS have been mostly cross-sectional or limited to restricted groups, such as middle-aged or clinical populations, it is becoming increasingly evident that depression is an important risk factor for cardiovascular and cerebrovascular diseases (Kinder et al. 2004; Muhtz et al. 2009; Toker et al. 2008; Vaccarino et al. 2008; Zeugmann et al. 2010). The molecular mechanisms associated with the link among depression, MetS, and cardiovascular and cerebrovascular diseases remain unclear (Nicholson et al. 2006). However, multiple mechanisms have been proposed to explain the association between depression and MetS. One mechanism involves the overstimulation of the hypothalamic–pituitary–adrenal (HPA) axis which leads to excess cortisol, a hormone which has been closely associated with the development of depression and the MetS components (Rosmond and Björntorp 2000). Lifestyle is another important factor that may contribute to the link among depression, MetS, and cardiovascular and cerebrovascular diseases. For example, depression and the above-mentioned conditions are associated with cigarette smoking, excessive use of alcohol, and a sedentary lifestyle (Cohen et al. 2010; Vaccarino et al. 2008). Finally, changes in the functioning of the immune system may be another factor that may be linked to the association between depression and MetS. C-reactive protein is a marker of systemic inflammation, which is closely associated with the development of depression (Elovainio et al. 2006) and the MetS (Haffner 2006). It is proposed that depression initiates, facilitates, and maintains inflammatory responses by downregulating glucocorticoid receptor expression, resulting in the immune system's inability to benefit from cortisol's anti-inflammatory actions (Carney et al. 2002). Another mechanism that explains the association between depression and MetS is sedentary lifestyle of subjects with depression and/or enduring dysregulation of the adrenocortical and autonomic nervous system, which may lead to increased visceral adiposity and insulin resistance (Kinder et al. 2004; Rosmond and Björntorp 2000;

Vaccarino et al. 2008). Despite the above-mentioned plausible mechanisms, it is not known whether depression is a cause, a consequence, or simply a marker of the metabolic syndrome. For example, physical limitations or social stigma caused by obesity may predispose subjects to depression. Emotional problems in MetS subjects may also be secondary to excess adipocytokine production in adipose tissue and cytokines in the brain, which may contribute to the etiology of depression (Raison et al. 2006). Nonetheless, mounting evidence supports the view that there is a causal relationship between MetS and depression.

## 10.3 Effect of Diet on Metabolic Syndrome and Neurological Disorders

It is well known that the brain plays fundamental roles in the regulation of dietary energy (calorie) intake, and long-term reductions in energy intake can enhance cognitive performance in rats and mice (Fontán-Lozano et al. 2007; Adams et al. 2008; Mattson 2010). Conversely, overeating (consumption of elevated levels of simple sugars and saturated fats) not only has a negative impact on cognitive performance in animal models but also promotes increased risks for a number of diseases including cardiovascular disease, insulin resistance/diabetes, neurological disorders, and many types of cancers (Wu et al. 2004; Stranahan et al. 2008; Mattson 2010). Based on animal model studies, it is also shown that (a) dietary energy restriction can enhance neural plasticity and cognitive function and may protect against stroke, AD, and depression; (b) excessive energy intake can impair, whereas energy restriction can enhance, adaptive cellular stress response pathways that protect neurons against aging, injury, and disease; and (c) among the adaptive stress response pathways, those involving neurotrophic factors, protein chaperones, and antioxidant enzymes may be particularly important (Fig. 10.3) (Mattson 2010). It can also be speculated that postponing or delaying the onset of neurological disorders through diet may lead to improvement in health status and quality of life in older age. Thus, over-consumption of sugars and grains results in alterations in insulin and leptin signaling causing insulin resistance in both the visceral tissues and brain. High glucose levels, a major pathological characteristic of type II diabetes, may produce toxic effects on neurons in the brain through osmotic insults and oxidative stress (Brownlee 2001). Chronically high glucose also leads to enhanced production of AGEs, which have potentially toxic effects on neurons (Valente et al. 2010). These processes along with decrease in insulin production in the brain may not only contribute to neurodegeneration but also impair cognition including problems with memory, language, thinking, and judgment (de la Monte 2012). Similarly, consumption of saturated fatty acid- and cholesterol-enriched diet may produce decline in cognition and mild cognitive impairment (MCI). In contrast, consumption of Mediterranean diet and diet enriched in n-3 fatty acids reduces the risk of cognitive decline and dementia in rodents and human subjects. Consumption of phytochemicals along with light-to-moderate red wine drinking may also lower the risk of incident stroke, dementia, and AD (Farooqui 2012). Similarly, fruit and vegetable

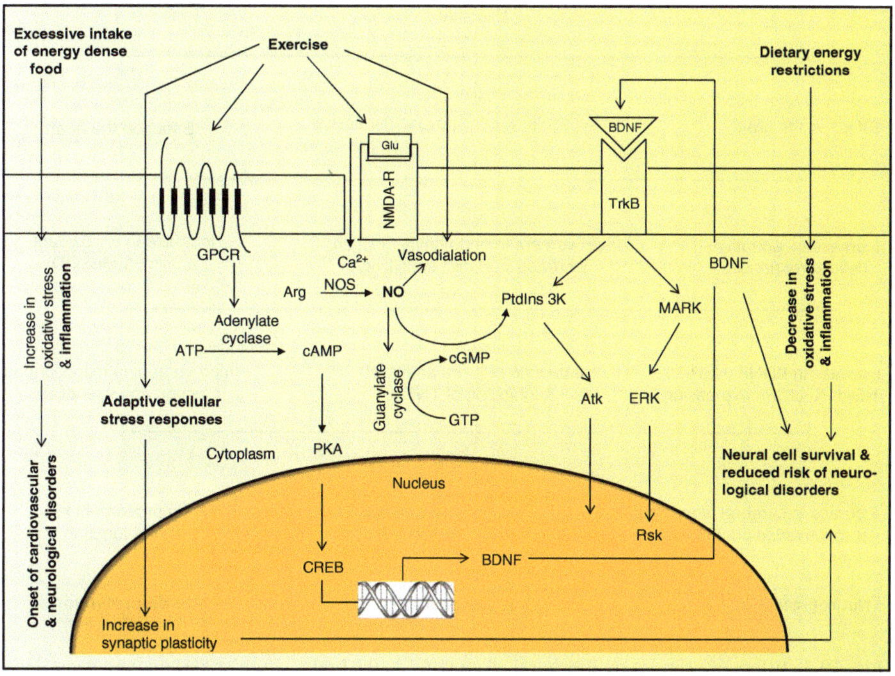


**Fig. 10.3** Hypothetical diagram showing the effect of exercise, dietary caloric restrictions, and excessive intake of energy-dense food on synaptic plasticity and neural cell survival. *N*-Methyl-D-ASPARTATE receptor (NMDA-R); brain-derived growth factor (BDNF); trak B receptor (trkB); phosphatidylinositol 3-kinase (PtdIns 3 K); protein kinase B (Akt); mitogen-activated protein kinase (MARK); extracellular signal-regulated kinase (ERK); cAMP regulatory element binder (CREB); adenosine triphosphate (ATP); adenylyl cyclase (AC); cAMP-activated protein kinase (PKA); protein kinase C (PKC); guanosine 5′-triphosphate (GTP); arginine (Arg); nitric oxide synthase (NOS); and nitric oxide (NO)

consumption may also reduce the risk of the above-mentioned neurological disorders (Farooqui 2012; Farooqui et al. 2012).

## 10.4 Can Exercise Delay the Onset of Metabolic Syndrome and Neurological Disorders?

Exercise is known to reduce obesity, improve glucose tolerance, and decrease the risk of diabetes, MetS, cardiovascular diseases, and neurological disorders including stroke, AD, and depression (Fig. 10.4). Regular exercise produces anti-inflammatory actions through the induction of adaptive cellular stress responses (Mattson 2010). During exercise induction of interleukin-6 (IL-6) in muscle fibers not only results in production of anti-inflammatory cytokines, such as IL-Ira (interleukin-1 receptor antagonist) and interleukin-10 (IL-10), but also blocks the production of the proinflammatory cytokine TNF-α (Pedersen 2006). In addition, IL-6

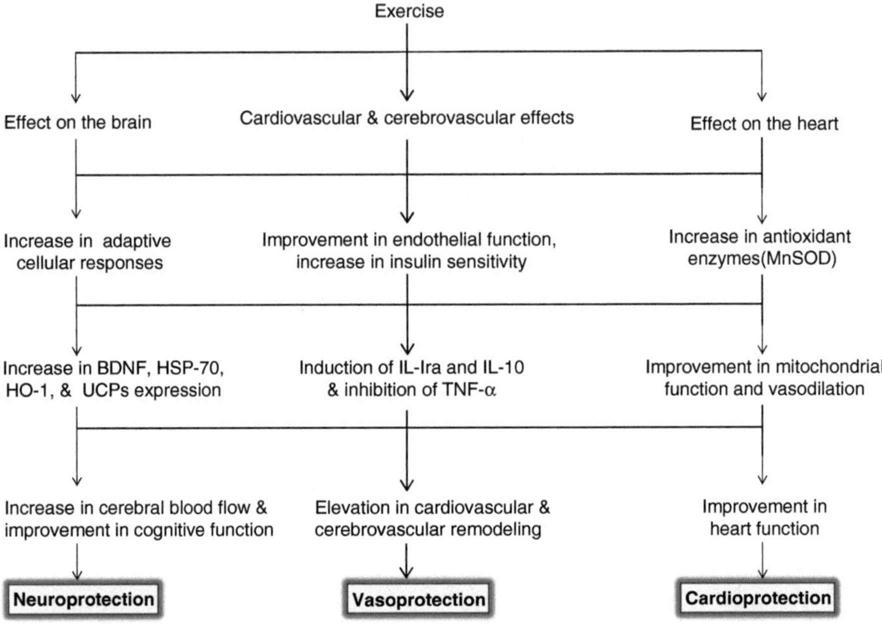


**Fig. 10.4** Effect of exercise on biochemical changes in the brain, vascular system, and heart. Heat shock protein-70 (HSP-70); brain-derived neurotrophic factor (BDNF); interleukin-1 receptor antagonist (IL-Ira ); interleukin-6 (IL-6); interleukin-10 (IL-10); uncoupling proteins (UCPs); and tumor necrosis factor-α (TNF-α)

enhances lipid turnover, stimulating lipolysis as well as fat oxidation. It is suggested that regular exercise causes suppression of TNF-α and thereby offering protection against TNF-α-mediated insulin resistance (Pedersen 2006). Strong evidence supports the view that exercise promotes neuroprotective and cardioprotective effects due to the activation of nitric oxide synthase and generation of NO, which results in vasodilation (Webb et al. 2004). Exercise elevates endoplasmic reticulum stress proteins, increases nitric oxide production, increases autophagy, and improves function of sarcolemmal and/or mitochondrial ATP-sensitive potassium (sarco $K_{ATP}$ and/or mito $K_{ATP}$) channels (Golbidi and Laher 2011). Exercise also facilitates the production of BDNF, which promotes the survival of neurons in the brain by binding to tyrosine kinase B (TrkB) receptors (Ibanez 1995) and activating MAP kinase pathway. BDNF also promotes and regulates neuronal survival via the PtdIns 3K/Akt pathway, which phosphorylates (Fig. 10.3) and regulates a number of cell survival-related proteins, including IkB, the forkhead transcription factor FKHRL1, glycogen synthase kinase 3β (GSK3β), and Bad, a proapoptotic member of the bcl-2 family (Huang and Reichardt 2001). In addition, exercise also elevates energy consumption, improves insulin sensitivity, and improves lifespan. Another possibility is that regular exercise reduces oxidative stress by upregulating endogenous antioxidant defense systems (Chung et al. 2009). At the molecular level, the beneficial effects of exercise are mediated by the transcriptional coactivator peroxisome proliferator-activated receptor γ coactivator-1α (PGC-1α) (Puigserver et al. 1998).

PGC-1α is transiently induced by a single bout of exercise and chronically elevated in endurance-trained muscle (Pilegaard et al. 2003). Activated PGC-1α then controls the expression of genes encoding proteins involved in mitochondrial biogenesis, oxidative phosphorylation, and other features of oxidative muscle fibers (Puigserver et al. 1999). Mechanistically, PGC-1α has been reported to lower the phosphorylation of NF-κB family member p65, which limits its transcriptional activation potential. This proposal is further supported by studies on muscle-specific PGC-1α transgenic animals that also exhibit reduction in p65 phosphorylation (Wenz et al. 2009).

Exercise also increases vagal tone (Routledge et al. 2010), which according to the cholinergic anti-inflammatory reflex leads to reductions in systemic inflammation (Tracey 2009). Furthermore, exercise not only induces antidepressive qualities but also increases performance on cognitive tasks in older individuals (Colcombe and Kramer 2003). Regular exercise also delays the onset of dementia and AD (Larson et al. 2006; Rovio et al. 2005) and reduces risk of developing AD (Lindsay et al. 2002). Collective evidence suggests that exercise also enhances cognitive processes in demented and nondemented older adults (Middleton and Yaffe 2009; Colcombe and Kramer 2003). Some information is available on molecular mechanisms underlying these beneficial effects of exercise on MetS, stroke, AD, and depression, but detailed information on the effect of exercise-induced changes on various physiological parameters, such as hormone production, cardiovascular and cerebrovascular blood flow, and expression/activity of related genes along with adaptive cellular stress responses in human brain, remains unknown (Mattson 2010; Texel and Mattson 2011). Thus, more mechanistic studies are required on large human population to judge the beneficial effects of exercise.

## 10.5 Directions for Future Research

Although some progress has been made on the relationship between MetS and neurological disorders, but this information is still in the infancy (Farooqui et al. 2012). More studies are needed on molecular mechanisms that MetS with neurological disorders not only in animal models of these diseases, but also in large human population. Effect of diet and lifestyle on insulin resistance, lipid mediators, oxidative stress (generation of ROS and AGEs), and inflammation in MetS and neurological disorders has not been studied in large human populations. This information is necessary to control the epidemic of MetS and neurological disorders in humans. Induction of insulin resistance and increase in oxidative stress and low-grade inflammation are key features of MetS that may modulate the onset of neurological disorders (Farooqui 2010). Markers of insulin resistance, oxidative damage, and low-grade inflammation precede pathological changes in experimental models of MetS and neurological disorders and are observed very early in the clinical course of these disorders. Studies on risk profiles or risk-related patterns of biomarkers in MetS and neurological disorders from midlife to old age have not been performed in large human population. This information may provide strategies for early diagnosis of MetS and neurological disorders. Discovering the treatment of neurological disorders is daunting but a necessary task. Some antidiabetic, insulin-sensitizing, antihypertensive, anti-inflammatory drugs reduce the risk of neurological disorders by reducing oxidative stress, inflammation, and improving cerebral blood flow. Considerable interest has also arisen regarding the effects of lifestyle interventions such as exercise and dietary/nutriceutical manipulations.

## 10.6 Conclusion

The present-day rise in the prevalence of type II diabetes and MetS is caused by increase in sedentary lifestyles (lack of physical activity) combined with availability of energy-rich food sources and consumption of fructose corn syrup-containing soft drinks in large human population. Healthy humans respond to positive energy balance primarily by storing excess energy as triglyceride in the adipose tissue. While this response enables humans to cope efficiently with fluctuating energy supplies, it predisposes persistently over-nourished individuals to weight gain, induces insulin resistance, and ultimately develops MetS, which is a risk factor for stroke, cognitive deterioration and dementia, AD, and depression. AD patients often have hyperinsulinemia and hyperglycemia as compared with healthy controls. There is evidence of both peripheral and central insulin resistance in AD. There are several striking similarities among MetS, stroke, AD, and depression including impaired glucose and energy metabolism, elevated AGEs and ROS, sustained expression of proinflammatory cytokines, cellular injury (i.e., pancreatic β cells vs. neurons), amyloidogenesis (Aβ vs. islet amyloid polypeptide deposits), impairment in long-term contextual and spatial memory, and impaired expression and activity of insulin-degrading enzyme (IDE). Since IDE can degrade Aβ, this activity can potentially be compromised by excessive insulin, a competing substrate, resulting in increased Aβ. However, information on the relationship among MetS, stroke, AD, and depression is still in infancy, and more population-based studies are needed in the elderly in order to accurately diagnose and estimate the risk of developing MetS-mediated stroke, AD, and depression.

## References

Abarquez RF (2003) Microvascular disease relevance in the hypertension syndrome. Clin Hemosheol Microcirc 29:295–3000

Adams MM, Shi L, Linville MC, Forbes ME, Long AB, Bennett C, Newton IG, Carter CS, Sonntag WE, Riddle DR, Brunso-Bechtold JK (2008) Caloric restriction and age affect synaptic proteins in hippocampal CA3 and spatial learning ability. Exp Neurol 211:141–149

Aggarwal BB (2010) Targeting inflammation-induced obesity and metabolic diseases by curcumin and other nutraceuticals. Ann Rev Nutr 30:173–199

Arenillas JF, Moro MA, Dávalos A (2007) The metabolic syndrome and stroke: potential treatment approaches. Stroke 38:2196–2203

Assah FK, Brage S, Ekelund U, Wareham NJ (2008) The association of intensity and overall level of physical activity energy expenditure with a marker of insulin resistance. Diabetologia 51:1399–1407

Bachen EA, Muldoon MF, Matthews KA, Manuck SB (2002) Effects of hemoconcentration and sympathetic activation on serum lipid responses to brief mental stress. Psychosom Med 64:587–594

Bassett DR Jr, Pucher J, Buehler R, Thompson DL, Crouter SE (2008) Walking, cycling, and obesity rates in Europe, North America, and Australia. J Phys Act Health 5:795–814

Benoit SC, Kemp CJ, Elias CF, Abplanalp W, Herman JP, Migrenne S, Lefevre AL, Cruciani-Guglielmacci C, Magnan C, Yu F, Niswender K, Irani BG, Holland WL, Clegg DJ (2009) Palmitic acid mediates hypothalamic insulin resistance by altering PKC-theta subcellular localization in rodents. J Clin Invest 119:2577–2589

Berrino F (2002) Western diet and Alzheimer's disease. Epidemiol Prev 26:107–115

Biessels GJ, Kappelle AC, Bravenboer B, Erkelens DW, Gispen WH (1994) Cerebral function in diabetes mellitus. Diabetologia 37:643–650

Boden-Albala B, Sacco RL, Lee HS, Grahame-Clarke C, Rundek T, Elkind MV, Wright C, Giardina EG, DiTullio MR, Homma S, Paik MC (2008) Metabolic syndrome and ischemic stroke risk: Northern Manhattan Study. Stroke 39:30–35

Brownlee M (2001) Biochemistry and molecular cell biology of diabetic complications. Nature 414:813–820

Carney RM, Freedland KE, Miller GE, Jaffe AS (2002) Depression as a risk factor for cardiac mortality and morbidity: a review of potential mechanisms. J Psychosom Res 53:897–902

Chaput JP, Klingenberg L, Astrup A, Sjödin AM (2011) Modern sedentary activities promote overconsumption of food in our current obesogenic environment. Obes Rev 12:e12–e20

Chew GT, Gan SK, Watts GF (2006) Revisiting the metabolic syndrome. Med J Aust 185:445–449

Chung HY, Cesari M, Anton S, Marzetti E, Giovannini S, Seo AY, Carter C, Yu BP, Leeuwenburgh C (2009) Molecular inflammation: underpinnings of aging and age-related diseases. Ageing Res Rev 8:18–30

Clee SM, Attie AD (2007) The genetic landscape of type 2 diabetes in mice. Endocr Rev 28:48–83

Cohen BE, Panguluri P, Na B, Whooley MA (2010) Psychological risk factors and the metabolic syndrome in patients with coronary heart disease: findings from the Heart and Soul Study. Psychiat Res 175:133–137

Colcombe S, Kramer AF (2003) Fitness effects on the cognitive function of older adults: a meta-analytic study. Psychol Sci 14:125–130

Convit A, Wolf OT, Tarshish C, de Leon MJ (2003) Reduced glucose tolerance is associated with poor memory performance and hippocampal atrophy among normal elderly. Proc Natl Acad Sci USA 100:2019–2022

Cornier MA, Dabelea D, Hernandez TL, Lindstrom RC, Steig AJ, Stob NR, Van Pelt RE, Wang H, Eckel RH (2008) The metabolic syndrome. Endocr Rev 29:777–822

Craft S (2007) Insulin resistance and Alzheimer's disease pathogenesis: potential mechanisms and implications for treatment. Curr Alzheimer Res 4:147–152

Craft S, Watson GS (2004) Insulin and neurodegenerative disease: shared and specific mechanisms. Lancet Neurol 3:169–178

Dagon Y, Avraham Y, Magen I, Gertler A, Ben-Hur T, Berry EM (2005) Nutritional status, cognition, and survival: a new role for leptin and AMP kinase. J Biol Chem 280:42142–42148

de la Monte SM (2012) Metabolic derangements mediate cognitive impairment and Alzheimer's disease: role of peripheral insulin-resistance diseases. Panminerva Med 54:171–178

de la Monte SM, Wands JR (2005) Review of insulin and insulin-like growth factor expression, signaling, and malfunction in the central nervous system: relevance to Alzheimer's disease. J Alzheimers Dis 7:45–61

Dietrich MO, Spuch C, Antequera D, Rodal I, de Yébenes JG, Molina JA, Bermejo F, Carro E (2008) Megalin mediates the transport of leptin across the blood-CSF barrier. Neurobiol Aging 29:902–912

Doble BW, Woodgett JR (2003) GSK-3: tricks of the trade for a multi-tasking kinase. J Cell Sci 116:1175–1186
Elovainio M, Keltikangas-Järvinen L, Pulkki-Råback L, Kivimäki M, Puttonen S, Viikari L, Rasanen L, Mansikkaniemi K, Vilkari J, Raitakari OT (2006) Depressive symptoms and C-reactive protein: the Cardiovascular Risk in Young Finns Study. Psychol Med 36:797–805
Engelhart MJ, Geerlings MI, Ruitenberg A, van Swieten JC, Hofman A, Witteman JC, Breteler MM (2002) Dietary intake of antioxidants and risk of Alzheimer disease. JAMA 287:3223–3229
Esler M, Rumantir M, Wiesner G, Kaye D, Hastings J, Lambert G (2001) Sympathetic nervous system and insulin resistance: from obesity to diabetes. Am J Hypertens 14:304S–309S
Farooqui AA (2009) Beneficial effects of fish oil on human brain. Springer, New York
Farooqui AA (2010) Neurochemical aspects of neurotraumatic and neurodegenerative diseases. Springer, New York
Farooqui AA (2012) Phytochemicals, signal transduction, and neurological disorders. Springer, New York
Farooqui AA, Farooqui T, Panza F, Frisardi V (2012) Metabolic syndrome as a risk factor for neurological disorders. Cell Mol Life Sci 69:741–762
Fewlass DC, Noboa K, Pi-Sunyer FX, Johnston JM, Yan SD, Tezapsidis N (2004) Obesity-related leptin regulates Alzheimer's Abeta. FASEB J 18:1870–1878
Fontán-Lozano A, Sáez-Cassanelli JL, Inda MC, de los Santos-Arteaga M, Sierra-Domínguez SA, López-Lluch G, Delgado-García JM, Carrión AM (2007) Caloric restriction increases learning consolidation and facilitates synaptic plasticity through mechanisms dependent on NR2B subunits of the NMDA receptor. J Neurosci 27:10185–10195
Ford ES, Giles WH, Dietz WH (2002) Prevalence of the metabolic syndrome among US adults: findings from the third National Health and Nutrition Examination Survey. JAMA 287:356–359
Frisardi V, Solfrizzi V, Seripa D, Capurso C, Santamato A, Sancarlo D, Vendemiale G, Pilotto A, Panza F (2010) Metabolic-cognitive syndrome: a cross-talk between metabolic syndrome and Alzheimer's disease. Ageing Res Rev 9:399–417
Fulop T, Tessier D, Carpentier A (2006) The metabolic syndrome. Pathol Biol (Paris) 54:375–386
Furukawa S, Fujita T, Shimabukuro M, Iwaki M, Yamada Y, Nakajima Y, Nakayama O, Makishima M, Matsuda M, Shimomura I (2004) Increased oxidative stress in obesity and its impact on metabolic syndrome. J Clin Invest 114:1752–1761
Golbidi S, Laher I (2011) Molecular mechanisms in exercise-induced cardioprotection. Cardiol Res Pract 2011:972807
Gomez-Pinilla F (2008) Brain foods: the effects of nutrients on brain function. Nat Rev Neurosci 9:568–578
Greenwood CE, Winocur G (2005) High-fat diets, insulin resistance and declining cognitive function. Neurobiol Aging 26(Suppl 1):42–45
Grundy SM (2008) Metabolic syndrome pandemic. Arterioscler Thromb Vasc Biol 28:629–636
Haan MN (2006) Therapy insight: type 2 diabetes mellitus and the risk of late-onset alzheimer's disease. Nat Clin Pract Neurol 2:159–166
Haffner SM (2006) The metabolic syndrome: inflammation, diabetes mellitus, and cardiovascular disease. Am J Cardiol 97(Suppl):3A–11A
Hirano S, Shimada H, Yoshiyama Y (2012) Neuroimaging studies of diabetes and risk of Alzheimer's disease. Brain Nerve 64:1411–1419
Hossain P, Kawar B, El NM (2007) Obesity and diabetes in the developing world—a growing challenge. N Engl J Med 356:213–215
Hoyer S (1998) Is sporadic Alzheimer disease the brain type of non-insulin dependent diabetes mellitus? A challenging hypothesis. J Neural Transm 105:415–422
Hoyer S (2002) The brain insulin signal transduction system and sporadic (type ii) Alzheimer disease: an update. J Neural Transm 109:341–360

Hu F, van Dam R, Liu S (2001) Diet and risk of type II diabetes: the role of types of fat and carbohydrate. Diabetologia 44:805–817
Huang EJ, Reichardt LF (2001) Neurotrophins: roles in neuronal development and function. Annu Rev Neurosci 24:677–736
Hugel B, Martinez MC, Kunzelmann C, Freyssinet JM (2005) Membrane microparticles: two sides of the coin. Physiology (Bethesda) 20:22–27
Ibanez CF (1995) Neurotrophic factors: from structure–function studies to designing effective therapeutics. Trends Biotechnol 13:217–227
Ito H, Kanno I, Fukuda H (2005) Human cerebral circulation: positron emission tomography studies. Ann Nucl Med 19:65–74
Jeon BT, Shin HJ, Kim JB, Kim YK, Lee DH, Kim KH, Kim HJ, Kang SS, Cho GJ, Choi WS, Roh GJ (2009) Adiponectin protects hippocampal neurons against kainic acid-induced excitotoxicity. Brain Res Rev 61:81–88
Kalaria RN, Stockmeier CA, Harik SI (1989) Brain microvessels are innervated by locus ceruleus noradrenergic neurons. Neurosci Lett 97:203–208
Kanoski SE, Davidson TL (2011) Western diet consumption and cognitive impairment: links to hippocampal dysfunction and obesity. Physiol Behav 103:59–68
Kazama K, Anrather J, Zhou P, Girouard H, Frys K, Milner TA, Iadecola C (2004) Angiotensin II impairs neurovascular coupling in neocortex through NADPH-oxidase-derived radicals. Circ Res 95:1019–1026
Kinder LS, Carnethon MR, Palaniappan LP, King AC, Fortmann S (2004) Depression and the metabolic syndrome in young adults: findings from the Third National Health and Nutrition Examination Survey. Psychosom Med 66:316–322
Kodl CT, Seaquist ER (2008) Cognitive dysfunction and diabetes mellitus. Endocr Rev 29:494–511
Kopelman PG (2000) Obesity as a medical problem. Nature 404:635–643
Larson EB, Wang L, Bowen JD, McCormick WC, Teri L, Crane P, Kukull W (2006) Exercise is associated with reduced risk for incident dementia among persons 65 years of age and older. Ann Intern Med 144:73–81
Lee JY, Ye J, Gao Z, Youn HS, Lee WH, Zhao L, Sizemore N, Hwang DH (2003) Reciprocal modulation of Toll-like receptor-4 signaling pathways involving MyD88 and phosphatidylinositol 3-kinase/AKT by saturated and polyunsaturated fatty acids. J Biol Chem 278:37041–37051
Li L, Hölscher C (2007) Common pathological processes in Alzheimer disease and type 2 diabetes: a review. Brain Res Rev 56:384–402
Lin MT, Beal MF (2006) Mitochondrial dysfunction and oxidative stress in neurodegenerative diseases. Nature 443:787–795
Lindsay J, Laurin D, Verreault R, Hébert R, Helliwell B, Hill GB, McDowell I (2002) Risk factors for Alzheimer's disease: a prospective analysis from the Canadian study of health and aging. Am J Epidemiol 156:445–453
Luchsinger JA (2008) Adiposity, hyperinsulinemia, diabetes and Alzheimer's disease: an epidemiological perspective. Eur J Pharmacol 585:119–129
Luchsinger JA, Tang MX, Shea S, Mayeux R (2002) Caloric intake and the risk of Alzheimer disease. Arch Neurol 59:1258–1263
Martinez MC, Tesse A, Zobairi F, Andriantsitohaina R (2005) Shed membrane microparticles from circulating and vascular cells in regulating vascular function. Am J Physiol Heart Circ Physiol 288:H1004–H1009
Mattson MP (2010) The impact of dietary energy intake on cognitive aging. Front Aging Neurosci 2:5
McCullough LD, Zeng Z, Li H, Landree LE, McFadden J, Ronnett GV (2005) Pharmacological inhibition of AMP-activated protein kinase provides neuroprotection in stroke. J Biol Chem 280:20493–20502
Middleton LE, Yaffe K (2009) Promising strategies for the prevention of dementia. Arch Neurol 66:1210–1215

Monasta L, Batty GD, Cattaneo A, Lutje V, Ronfani L, Van Lenthe FJ, Brug J (2010) Early-life determinants of overweight and obesity: a review of systematic reviews. Obes Rev 11:695–708
Morimoto K, Morikawa M, Kimura H, Ishii N, Takamata A, Hara Y, Uji M, Yoshida K (2008) Mental stress induces sustained elevation of blood pressure and lipid peroxidation in postmenopausal women. Life Sci 82:99–107
Morris MC, Evans DA, Bienias JL, Tangney CC, Wilson RS (2004) Dietary fat intake and 6-year cognitive change in an older biracial community population. Neurology 62:1573–1579
Muhtz C, Zyriax BC, Klähn T, Windler E, Otte C (2009) Depressive symptoms and metabolic risk: effects of cortisol and gender. Psychoneuroendocrinology 34:1004–1011
Mungas D, Jagust WJ, Reed BR, Kramer JH, Weiner MW, Schuff N, Norman D, Mack WJ, Willis L, Chui HC (2001) MRI predictors of cognition in subcortical ischemic vascular disease and Alzheimer's disease. Neurology 57:2229–2235
Muniyappa R, Montagnani M, Koh KK, Quon MJ (2007) Cardiovascular actions of insulin. Endocr Rev 28:463–491
Nicholson A, Kuper H, Hemingway H (2006) Depression as an aetiologic and prognostic factor in coronary heart disease: a meta-analysis of 6362 events among 146 538 participants in 54 observational countries. Eur Heart J 27:2763–2774
O'Rourke MF, Safar ME (2005) Relationship between aortic stiffening and microvascular disease in brain and kidney: cause and logic of therapy. Hypertension 46:200–204
Pasinetti GM, Eberstein JA (2008) Metabolic syndrome and the role of dietary lifestyles in Alzheimer's disease. J Neurochem 106:1503–1514
Pedersen BK (2006) The anti-inflammatory effect of exercise: its role in diabetes and cardiovascular disease control. Essays Biochem 42:105–117
Pickering TG (2001) Mental stress as a causal factor in the development of hypertension and cardiovascular disease. Curr Hypertens Rep 3:249–254
Pilegaard H, Saltin B, Neufer PD (2003) Exercise induces transient transcriptional activation of the PGC-1alpha gene in human skeletal muscle. J Physiol 546:851–858
Price JL, Ko AI, Wade MJ, Tsou SK, McKeel DW, Morris JC (2001) Neuron number in the entorhinal cortex and CA1 in preclinical Alzheimer disease. Arch Neurol 58:1395–1402
Puddu A, Viviani GL (2011) Advanced glycation endproducts and diabetes. Beyond vascular complications. Endocr Metab Immune Disord Drug Targets 11:132–140
Puigserver P, Wu Z, Park CW, Graves R, Wright M, Spiegelman BM (1998) A cold-inducible coactivator of nuclear receptors linked to adaptive thermogenesis. Cell 92:829–839
Puigserver P, Adelmant G, Wu Z, Fan M, Xu J, O'Malley B, Spiegelman BM (1999) Activation of PPARgamma coactivator-1 through transcription factor docking. Science (New York, NY) 286:1368–1371
Raison CL, Capuron L, Miller AH (2006) Cytokines sing the blues: inflammation and the pathogenesis of depression. Trends Immunol 27:24–31
Reeds DN (2009) Nutrition support in the obese, diabetic patient: the role of hypocaloric feeding. Curr Opin Gastroenterol 25:151–154
Ristow M (2004) Neurodegenerative disorders associated with diabetes mellitus. J Mol Med 82:510–529
Rodriguez G, Nobili F, Celestino MA, Francione S, Gulli G, Hassan K, Marenco S, Rosadini G, Cordera R (1993) Regional cerebral blood flow and cerebrovascular reactivity in IDDM. Diabetes Care 16:462–468
Roman GC, Kalaria RN (2006) Vascular determinants of cholinergic deficits in Alzheimer disease and vascular dementia. Neurobiol Aging 27:1769–1785
Rosmond R, Björntorp P (2000) The hypothalamic-pituitary-adrenal axis activity as a predictor of cardiovascular disease, type 2 diabetes and stroke. J Internal Med 247:188–197
Rothman SM, Mattson MP (2010) Adverse stress, hippocampal networks, and Alzheimer's disease. Neuromolecular Med 12:56–70
Routledge FS, Campbell TS, McFetridge-Durdle JA, Bacon SL (2010) Improvements in heart rate variability with exercise therapy. Can J Cardiol 26:303–312

Rovio S, Kareholt I, Helkala EL, Viitanen M, Winblad B, Tuomilehto J, Soininen H, Nissinen A, Kivipelto M (2005) Leisure-time physical activity at midlife and the risk of dementia and Alzheimer's disease. Lancet Neurol 4:705–711

Scarmeas N, Stern Y, Mayeux R, Manly JJ, Schupf N, Luchsinger JA (2009) Mediterranean diet and mild cognitive impairment. Arch Neurol 66:216–225

Schubert M, Gautam D, Surjo D, Ueki K, Baudler S, Schubert D, Kondo T, Alber J, Galldiks N, Kustermann E, Arndt S, Jacobs AH, Krone W, Kahn CR, Brüning JC (2004) Role for neuronal insulin resistance in neurodegenerative diseases. Proc Natl Acad Sci USA 101:3100–3105

Seematter G, Guenat E, Schneiter P, Cayeux C, Jequier E, Tappy L (2000) Effects of mental stress on insulin-mediated glucose metabolism and energy expenditure in lean and obese women. Am J Physiol Endocrinol Metab 279:E799–E805

Sesti G (2006) Pathophysiology of insulin resistance. Best Pract Res Clin Endocrinol Metab 20:665–679

Shek EW, Brands MW, Hall JE (1998) Chronic leptin infusion increases arterial pressure. Hypertension 31:409–414

Shi Q, Vandeberg JF, Jett C, Rice K, Leland MM, Talley L, Kushwaha RS, Rainwater DL, Vandeberg JL, Wang XL (2005) Arterial endothelial dysfunction in baboons fed a high-cholesterol, high-fat diet. Am J Clin Nutr 82:751–759

Shinmura K, Tamaki K, Saito K, Nakano Y, Tobe T, Bolli R (2007) Cardioprotective effects of short-term caloric restriction are mediated by adiponectin via activation of AMP-activated protein kinase. Circulation 116:2809–2817

Shoelson SE, Herrero L, Naaz A (2007) Obesity, inflammation, and insulin resistance. Gastroenterology 132:2169–2180

Soucek M, Kara T (2001) Stress-induced hypertension and diabetes mellitus. Vnitr Lek 47:315–319

Spreadbury I (2012) Comparison with ancestral diets suggests dense acellular carbohydrates promote an inflammatory microbiota, and may be the primary dietary cause of leptin resistance and obesity. Diabetes Metab Synd Obes 5:175–189

Stranahan AM, Lee K, Pistell PJ, Nelson CM, Readal N, Miller MG, Spangler EL, Ingram DK, Mattson MP (2008) Accelerated cognitive aging in diabetic rats is prevented by lowering corticosterone levels. Neurobiol Learn Mem 90:479–483

Styskal J, Remmen HV, Richardson A, Salmon AB (2012) Oxidative stress and diabetes: what can we learn about insulin resistance from antioxidant mutant mouse models? Free Radic Biol Med 52:46–58

Tejani-Butt SM, Yang J, Pawlyk AC (1995) Altered serotonin transporter sites in Alzheimer's disease raphe and hippocampus. Neuroreport 6:1207–1210

Texel SJ, Mattson MP (2011) Impaired adaptive cellular responses to oxidative stress and the pathogenesis of Alzheimer's disease. Antioxid Redox Signal 14:1519–1534

Toker S, Shirom A, Melamed S (2008) Depression and the metabolic syndrome: gender-dependent associations. Depress Anxiety 25:661–669

Tomas E, Tsao TS, Saha AK, Murrey HE, Zhang CC, Itani SI, Lodish HF, Ruderman NB (2002) Enhanced muscle fat oxidation and glucose transport by ACRP30 globular domain: acetyl-CoA carboxylase inhibition and AMP-activated protein kinase activation. Proc Natl Acad Sci USA 99:16309–16313

Tomlinson DR, Gardiner NJ (2008) Glucose neurotoxicity. Nat Rev Neurosci 9:36–45

Tracey KJ (2009) Reflex control of immunity. Nat Rev Immunol 9:418–428

Tschop M, Thomas G (2006) Fat fuels insulin resistance through Toll-like receptors. Nat Med 12:1359–1361

Tsigos C, Chrousos GP (2002) Hypothalamic-pituitary-adrenal axis, neuroendocrine factors and stress. J Psychosom Res 53:865–871

Tzourio C (2007) Hypertension, cognitive decline, and dementia: an epidemiological perspective. Dialogues Clin Neurosci 9:61–70

Uchino K, Lin R, Zaidi SF, Kuwabara H, Sashin D, Bircher N, Chang YF, Hammer MD, Reddy V, Jovin TG, Vora N, Jumaa M, Massaro L, Billigen J, Boada F, Yonas H, Nemoto EM (2010)

Increased cerebral oxygen metabolism and ischemic stress in subjects with metabolic syndrome-associated risk factors: preliminary observations. Transl Stroke Res 1:178–183
Umegaki H (2010) Pathophysiology of cognitive dysfunction in older people with type 2 diabetes: vascular changes or neurodegeneration? Age Ageing 39:8–10
Vaccarino V, McClure C, Johnson BD, Sheps D, Bittner V, Rutledge T, Shaw LJ, Sopko G, Olson MB, Krantz DS, Parashar S, Marroquin OC, Merz CN (2008) Depression, the metabolic syndrome and cardiovascular risk. Psychosom Med 70:40–48
Valente T, Gella A, Fernàndez-Busquets X, Unzeta M, Durany N (2010) Immunohistochemical analysis of human brain suggests pathological synergism of Alzheimer's disease and diabetes mellitus. Neurobiol Dis 37:67–76
van Beek AH, Claassen JA (2011) The cerebrovascular role of the cholinergic neural system in Alzheimer's disease. Behav Brain Res 221:537–542
Varela-Moreiras G (2006) Controlling obesity: what should be changed? Int J Vitam Nutr Res 76:262–268
Villanueva EC, Myers MG Jr (2008) Leptin receptor signaling and the regulation of mammalian physiology. Int J Obes (Lond) 32(Suppl 7):S8–S12
Walsh TJ, Emerich DF (1988) The hippocampus as a common target of neurotoxic agents. Toxicology 49:137–140
Walters FJM (1998) Intracranial pressure and cerebral blood flow. Physiology 8:1–4
Wang MY, Grayburn P, Chen S, Ravazzola M, Orci L, Unger RH (2008) Adipogenic capacity and the susceptibility to type 2 diabetes and metabolic syndrome. Proc Natl Acad Sci USA 105:6139–6144
Webb A, Bond R, McLean P, Uppal R, Benjamin N, Ahluwalia A (2004) Reduction of nitrite to nitric oxide during ischemia protects against myocardial ischemia-reperfusion damage. Proc Natl Acad Sci USA 101:13683–13688
Wen J-P, Lv W-S, Yang J, Nie A-F, Cheng X-B, Yang Y, Ge Y, Li X-Y, Ning G (2008) Globular adiponectin inhibits GnRH secretion from GT1-7 hypothalamic GnRH neurons by induction of hyperpolarization of membrane potential. Biochem Biophys Res Commun 373:756–761
Wenz T, Rossi SG, Rotundo RL, Spiegelman BM, Moraes CT (2009) Increased muscle PGC-1alpha expression protects from sarcopenia and metabolic disease during aging. Proc Natl Acad Sci USA 106:20405–20410
Wu A, Ying Z, Gomez-Pinilla F (2004) The interplay between oxidative stress and brain-derived neurotrophic factor modulates the outcome of a saturated fat diet on synaptic plasticity and cognition. Eur J Neurosci 19:1699–1707
Yagihashi S, Sima AA (1986) Diabetic autonomic neuropathy in BB rat. Ultrastructural and morphometric changes in parasympathetic nerves. Diabetes 35:733–743
Zemva J, Schubert M (2011) Central insulin and insulin-like growth factor-1 signaling: implications for diabetes associated dementia. Curr Diabetes Rev 7:356–366
Zeugmann S, Quante A, Heuser I, Schwarzer R, Anghelescu I (2010) Inflammatory biomarkers in 70 depressed in patients with and without the metabolic syndrome. J Clin Psychiatry 71:1007–1016
Zhu M, Lee GD, Ding L, Hu J, Qiu G, Cabo R, Bernier M, Ingram DK, Zou S (2007) Adipogenic signaling in rat white adipose tissue: modulation by aging and calorie restriction. Exp Gerontol 42:733–744

# About the Author

**Dr. Akhlaq A. Farooqui** is a leader in the field of signal transduction, brain phospholipases $A_2$, bioactive ether lipid metabolism, polyunsaturated fatty acid metabolism, glycerophospholipid-, sphingolipid-, and cholesterol-derived lipid mediators, glutamate-induced neurotoxicity, and modulation of signal transduction by phytochemicals. Dr. Farooqui has discovered the stimulation of plasmalogen-selective phospholipase $A_2$ (PlsEtn-$PLA_2$) and diacyl- and monoacylglycerol lipases in brains from patients with Alzheimer's disease. Stimulation of PlsEtn-$PLA_2$ produces plasmalogen deficiency and increases levels of eicosanoids that may be related to the loss of synapses in brains of patients with Alzheimer's disease. Dr. Farooqui has published cutting-edge research on the generation and identification of glycerophospholipid-, sphingolipid-, and cholesterol-derived lipid mediators in kainic acid-mediated neurotoxicity by lipidomics. Dr. Farooqui has authored eight monographs: *Glycerophospholipids in Brain: Phospholipase $A_2$ in Neurological Disorders* (2007); *Neurochemical Aspects of Excitotoxicity* (2008); *Metabolism and Functions of Bioactive Ether Lipids in Brain* (2008); and *Hot Topics in Neural Membrane Lipidology* (2009); *Beneficial Effects of Fish Oil in Human Brain* (2009); *Neurochemical Aspects of Neurotraumatic and Neurodegenerative Diseases* (2010); *Lipid Mediators and Their Metabolism in the Brain* (2011); and *Phytochemicals, Signal Transduction, and Neurological Disorders*. All monographs are published by Springer, New York. In addition, Dr. Akhlaq A. Farooqui has edited six books (*Biogenic Amines: Pharmacological, Neurochemical and Molecular Aspects in the CNS* (2010) Nova Science Publisher, Hauppauge, New York; *Molecular Aspects of Neurodegeneration and Neuroprotection*, Bentham Science Publishers Ltd (2011); *Phytochemicals and Human Health: Molecular and Pharmacological Aspects* (2011), Nova Science Publisher, Hauppauge, New York; *Molecular Aspects of Oxidative Stress on Cell Signaling in Vertebrates and Invertebrates* (2012), Wiley Blackwell Publishing Company, New York; *Beneficial Effects of Propolis on Human Health in Chronic Diseases* (2012) Vol. 1, Nova Science Publishers, Hauppaauge, New York; and *Beneficial Effects of Propolis on Human Health in Chronic Diseases* (2012) Vol. 2, Nova Science Publishers, Hauppaauge, New York.

A.A. Farooqui, *Metabolic Syndrome: An Important Risk Factor for Stroke, Alzheimer Disease, and Depression*, DOI 10.1007/978-1-4614-7318-3,

# Index



A.A. Farooqui, *Metabolic Syndrome: An Important Risk Factor for Stroke, Alzheimer Disease, and Depression,* DOI 10.1007/978-1-4614-7318-3,

## H

## O

MIX
Papier aus verantwortungsvollen Quellen
Paper from responsible sources
FSC® C105338
www.fsc.org

If you have any concerns about our products,
you can contact us on
**ProductSafety@springernature.com**

In case Publisher is established outside the EU,
the EU authorized representative is:
**Springer Nature Customer Service Center GmbH**
**Europaplatz 3, 69115 Heidelberg, Germany**

Printed by Libri Plureos GmbH
in Hamburg, Germany